ʏʏʏʏʏʏʏʏʏʏʏʏ

THE BUSINESS
OF ENLIGHTENMENT

THE BUSINESS
OF ENLIGHTENMENT

ᵞᵞᵞᵞᵞᵞᵞᵞᵞᵞᵞᵞ

A Publishing History
of the *Encyclopédie*
1775-1800

ROBERT DARNTON

The Belknap Press of
Harvard University Press

Cambridge, Massachusetts, and London, England

1979

Library of Congress Cataloging in Publication Data

Darnton, Robert.
    The business of enlightenment.

    Bibliography: p.
    Includes index.
    1. Encyclopédie; ou Dictionnaire raisonné des sciences.
2. Encyclopedias and dictionaries, French. I. Title.
AE25.A6D37   034′.1   78-23826
ISBN 0-674-08785-2

To Susan

ACKNOWLEDGMENTS

I would like to express my gratitude to two research centers that provided support and fellowship during the labor on this book—the Center for Advanced Study in the Behavioral Sciences at Stanford, California, where I began writing the book in 1973, and the Netherlands Institute for Advanced Study at Wassenaar, The Netherlands, where I finished in 1977, having laid it aside for other work in the interim. The research goes back to 1965, when I first began to explore the archives of the Société typographique de Neuchâtel and the Chambre syndicale and Anisson-Duperron collections of the Bibliothèque nationale. Thanks to generous help from the Society of Fellows of Harvard University in 1965–1968 and the Guggenheim Foundation in 1970–1971, these explorations have continued over many years and have led to other publications as well as this one. By spending a semester as a directeur d'études associé in the VIe Section of the Ecole pratique des hautes études (now the Ecole des hautes études en sciences sociales) in 1971, I learned a great deal from the French masters of *histoire du livre.* And I learned most of all from Jacques Rychner, now director of the Bibliothèque de la ville de Neuchâtel, who explained the mysteries of analytical bibliography to me over countless cups of coffee in the shops around the library in Neuchâtel. He permitted me to pursue the study of the *Encyclopédie* into the printing shop of the STN, territory that belongs to him and that will become known in all its richness when he publishes his doctoral thesis. My work in Neuchâtel also benefited greatly from the encouragement of the late Charly Guyot, and I hope that this book may help to perpetuate the memory of his kindness. Another Neuchâtelois also helped me enormously, although I never met him. He was Jean Jeanprêtre, a retired chemist, who devoted the last years of his life to cataloguing the papers of the STN. His labor and the cheerful cooperation of the staff of the Bibliothèque de la ville made it a joy to work in the STN archives.

Finally, I would like to acknowledge the help of several persons who aided in the preparation of this book. Caroline Hannaway provided information on five obscure contributors to the *Encyclopédie méthodique.* Marie-Claude Lapeyre of the Laboratoire cartographique in the Ecole des hautes études en sciences sociales prepared the maps, and Charlotte Carlson drew the graphs. Giles Barber read Chapter V and Raymond Birn read the entire manuscript with great care. Elizabeth Suttell edited it. Susan Darnton prepared the index. Marianne Perlak designed the book. American–Stratford Graphic Services, Inc., did the composition. And The Maple-Vail Book Manufacturing Group did the printing and binding.

ᵛᵛᵛᵛᵛᵛᵛᵛᵛᵛᵛᵛ

CONTENTS

Contents

V. Bookmaking   177

VI. Diffusion   246

VII. Settling Accounts   324

VIII. The Ultimate *Encyclopédie*   395

IX. Encyclopedism, Capitalism, and Revolution   460

Contents

X. Conclusion   520

The Production and Diffusion of Enlightenment / 520   Enlightenment Publishing and the Spirit of Capitalism / 531   The
Encyclopédie and the State / 535   The Cultural Revolution / 539

Appendices   549

A. Contracts of the *Encyclopédie* Publishers, 1776–1780 / 549
B. Subscriptions to the Quarto *Encyclopédie* / 586
C. Incidence of Subscriptions in Major French Cities / 594
D. Contributors to the *Encyclopédie Méthodique* / 597

Bibliographical Note   611

Index   619

Figures

xi

Contents

Illustrations from the *Encyclopédie*

Table

THE BUSINESS
OF ENLIGHTENMENT

A Note on Terminology and Spelling

In the eighteenth century, the French did not have an equivalent of "publisher" in English or *éditeur* in modern French. They normally spoke of *libraires, libraires-imprimeurs,* or simply *entrepreneurs.* Of course many *libraires* sold books without becoming involved in their production, so "publisher" and "publishing" have been used in their modern, English sense throughout this book. "Edition" is also an ambiguous term. Modern bibliographers distinguish between "editions," "printings," "states," and other units in the production and reproduction of texts. But eighteenth-century *libraires* and *imprimeurs* talked loosely of *éditions,* which were partial reruns of incomplete printings and sometimes did not exist at all, as will be seen in the discussion of the "missing" second editions of the quarto and octavo *Encyclopédies.* To avoid confusion, and at the cost of some bibliographical impurity, the term "edition" has been used in the casual, eighteenth-century manner. In this way, it will be possible to follow the publishers' discussions of their work without becoming entangled in anachronistic terms or distracted by the excessive use of quotation marks. As this book is based almost entirely on manuscript material, which has a rich, original flavor, quotations have been given in French. Spelling and punctuation have been modernized, except in a few cases, where the original is so primitive that it indicates a significantly poor mastery of the written word. Place names such as Lyons and Marseilles have been spelled according to English usage, except where they occur in French passages.

I

vvvvvvvvvvvv

INTRODUCTION:
THE BIOGRAPHY OF A BOOK

By recounting the life story of the *Encyclopédie,* this book is meant to dispel some of the obscurity surrounding the history of books in general. A book about a book: the subject seems arcane, and it could contract into the infinitely small, like a mirror reflected in a mirror. If done properly, however, it should enlarge the understanding of many aspects of early modern history, for *l'histoire du livre,* as it is known in France, opens onto the broadest questions of historical research. How did great intellectual movements like the Enlightenment spread through society? How far did they reach, how deeply did they penetrate? What form did the thought of the philosophes acquire when it materialized into books, and what does this process reveal about the transmission of ideas? Did the material basis of literature and the technology of its production have much bearing on its substance and its diffusion? How did the literary market place function, and what were the roles of publishers, book dealers, traveling salesmen, and other intermediaries in cultural communication? How did publishing function as a business, and how did it fit into the political as well as the economic systems of pre-revolutionary Europe? The questions could be multiplied endlessly because books touched on such a vast range of human activity—everything from picking rags to transmitting the word of God. They were products of artisanal labor, objects of economic exchange, vehicles of ideas, and elements in political and religious conflict.

Yet this inviting subject, located at the crossroads of so

many avenues of research, hardly exists in the United States today. We do not have a word for it. *Histoire du livre* sounds awkward as "history of the book," and the awkwardness betokens unfamiliarity with what has emerged as a distinct historical genre, with its own methods, its special journals, and its allotted place among sister disciplines, on the other side of the Atlantic. In the United States, book history has been relegated to library schools and rare book collections. Step into any rare book room and you will find aficionados savoring bindings, epigones contemplating watermarks, *érudits* preparing editions of Jane Austen; but you will not run across any ordinary, meat-and-potatoes historian attempting to understand the book as a force in history.

It is a pity, for the generalist could learn a great deal from the specialists in the treasure houses of books. They could teach him to sift through their riches and to tap the vein of information that runs through their periodicals: *The Library, Studies in Bibliography, Papers of the Bibliographical Society of America, Revue Française d'histoire du livre, Den gulden passer,* the *Gutenberg Jahrbuch,* and many others. Admittedly, these publications seem to be written by bibliographers for bibliographers, and it can be difficult to see issues of substance beneath the esoteric language and the antiquarianism. But bibliography need not be confined to problems such as how consistently compositor B misspelled the text of *The Merchant of Venice* or whether the patterns of skeleton formes reveal regularity in compositorial practices. Bibliography leads directly into the hurly-burly of working-class history: it provides one of the few means of analyzing the work habits of skilled artisans before the Industrial Revolution.

Curiously, however, it has not attracted much attention among the French, who have done the most to bring the history of books out of the realm of mere erudition and into the broad paths of *histoire totale.* French research tends to be statistical and sociological. It usually takes the form of macroscopic surveys of book production or microscopic analyses of individual libraries, but it neglects the processes by which books were produced and distributed. Those processes have been studied best in Britain, where researchers have pursued their quarry into the account books of publishers and the ledgers of booksellers, not merely into state and notarial archives, as in France. By mixing British empiricism with the French concern for broad-gauged social history it might be possible to develop an original blend of the history of books in America.[1]

Of course it is easier to pronounce on how history ought to be written than to write it; and once the historian of books has equipped himself with prolegomena and methodologies and has ventured into the field, he is likely to stumble on the greatest difficulty of all: inadequate sources. He may work in a library overflowing with ancient volumes, but he cannot know where they circulated before they reached him and whether they really represent the reading habits of the past. State archives show how books appeared to the authorities in charge of controlling them. Auction catalogues and *inventaires après décès* give glimpses of private libraries. But the official sources do not reveal much about the lived experience of literature among ordinary readers. In fact the catalogues as well as the books had to pass the censorship in eighteenth-century France, so it does not seem surprising that the Enlightenment fails to appear in research based on catalogues and requests for *privilèges,* a kind of royal copyright. The Enlightenment existed elsewhere, first in the speculations of philosophes, then in the speculations of publishers, who invested in the market place of ideas beyond the boundaries of French law.

How these speculations came together in books and how the books acquired readers has remained a mystery because the papers of the publishers have almost entirely disappeared. But the records of the Société typographique de Neuchâtel, one of the most important publishers of French books in the eighteenth century, have survived in the Swiss city of Neuchâtel, and they contain information about every aspect of book history They show how authors were treated, paper manufactured, copy processed, type set, sheets printed, crates shipped, authorities courted, police circumvented, booksellers provisioned, and readers satisfied everywhere in Europe between 1769 and 1789. The information is vast enough to overwhelm the researcher. A few letters from a bookseller can reveal more than a whole monograph about the book trade, yet the papers in Neuchâtel contain 50,000 letters by all kinds

1. For examples of the different areas of research in this field and for further reading see the Bibliographical Note.

of persons who lived by the book trade in all kinds of ways. It would be impossible to do justice to the material and to reconstruct the world of eighteenth-century books in a single volume. Therefore, after some reconnoitering in 1963, I decided to go through the entire collection in Neuchâtel, to supplement it with research in other archives, and to write a series of studies about intellectuals, books, and public opinion in the age of the Enlightenment.

The present volume constitutes the first installment. It is intended to explore the ways of Enlightenment publishing by tracing the life cycle of a single book—not just any book, to be sure, but the supreme work of the Enlightenment, Diderot's *Encyclopédie*. Given the richness of the sources and the complexity of the subject, it seemed better to attempt an *histoire totale* of one publication than to treat the totality of publishing. By following a single theme wherever it leads, one can branch out in many directions and cut into unmapped territory. This approach has the advantage of specificity: better, at a preliminary stage of groping in the unknown, to find out precisely how publishers drew up contracts, editors handled copy, printers recruited workers, and booksellers pitched sales talk while making and marketing one book than to withdraw into hazy statements about books in general. There is also the appeal of novelty: it has never before been possible to trace the production and diffusion of an eighteenth-century book. And finally, the publishing history of the *Encyclopédie* deserves to be told because it is a good story.

The story can be pieced together from the letters of the publishers—not very businesslike letters, most of them. They abound in denunciations of conspiracies and epithets like "pirate," "corsaire," and "brigand," which suggest the flavor of the book trade in the Old Regime. Driven by an unlimited appetite for lucre, uninhibited by compunctions about stabbing partners in the back and tossing competitors to the sharks, the publishers of the *Encyclopédie* epitomized the phase of economic history known as "booty capitalism." Perhaps they had more in common with the merchant adventurers of the Renaissance than with modern executives, but then how much is known about the inside history of business in any period? What other enterprise can be studied as closely as the *Encyclopédie,* not only from its commercial correspondence but also from account books, the secret memoranda of the managers, the diaries of traveling salesmen, the complaints of customers, and the reports of industrial spies—a whole series of industrial spies the publishers used against allies and enemies alike? The *Encyclopédie* gave rise to so many alliances and alignments that its contracts and codicils —*traités,* the publishers called them—need to be studied in the same way as diplomatic documents. And its publishers wrote so many letters that one can investigate their way of thinking as well as their behavior. To see how they reached decisions, how they calculated strategy, and what they cared about is to enter into the mental world of early entrepreneurs. The story of the *Encyclopédie* suggests the possibility of an intellectual history of businessmen as well as a diplomatic history of business. But it is difficult to tell a story and to analyze behavior patterns at the same time. This book will switch from the narrative to the analytical mode when it seems appropriate, and the reader who prefers one to the other can jump around in the text, using chapter subheadings as signposts.

The story begins around the time that Diderot ended his connection with the *Encyclopédie*—that is, in 1772, when the last volume of plates came out. It may seem strange to embark on a history of the *Encyclopédie* just after Diderot had steered it safely into port, but this procedure can be justified by two considerations. First, a huge literature on Diderot and the original *Encyclopédie* already exists. The text of the book has been analyzed and anthologized dozens of times: to recapitulate all the studies of its intellectual content would be redundant, even if it were important for the purposes of publishing history.[2] Secondly, very little can be learned about the production and diffusion of the first edition. A few fragments from the account books of the original publishers have been found, and some of the publishers' commercial activities can be deduced from material assembled by Luneau de Boisjermain, a cranky subscriber who unsuccessfully sued them for swindle. Although several scholars have combed through these documents with great care, they have failed to find out

2. This statement should not be construed to imply that publishing history can ignore the contents of books. On the contrary, this study is meant to show the importance of understanding not only texts but also the meaning of texts for their audience at specific points in the past. For references to the literature on the *Encyclopédie,* especially studies of the early editions, see the Bibliographical Note.

how the first edition was manufactured, where it was sold, and who bought it. The history of the second edition remains almost equally obscure, despite some revealing material that George B. Watts and John Lough have excavated from archives in Geneva. And although Italian scholars have uncovered some of the politics surrounding the editions of Lucca and Leghorn, they have not found out how much the Italian reprints cost and how many copies they contained.

As far as the diffusion of the *Encyclopédie* is concerned, however, the first four editions were relatively unimportant. They were luxurious folio publications ordinary readers could not afford and, when taken together, accounted for only about 40 percent of the *Encyclopédies* in existence before 1789. The great mass of the *Encyclopédies* in prerevolutionary Europe came from the cut-rate quarto and octavo editions printed between 1777 and 1782. Between 50 and 65 percent of the copies in France were quartos, and all of them can be traced, thanks to the papers of the Société typographique de Neuchâtel (STN). The archives in Neuchâtel also make it possible to explain the history of the octavo edition and the origins of the *Encyclopédie méthodique,* the ultimate encyclopedia of the Enlightenment, whose fate can be followed through the Revolution from other sources. Furthermore, the Neuchâtel papers reveal the connecting links between all the *Encyclopédie* speculations, including some that never materialized, from 1750 to 1800. They show how the book changed in shape as the publishers adapted it to an ever-widening audience and how publishing consortia succeeded one another as the speculators scrambled to exploit the biggest best seller of the century. From the viewpoint of book history, therefore, the story of the *Encyclopédie* took its most important turn in the 1770s. Only then did it move into a phase that represented the diffusion of Enlightenment on a massive scale. If the documentation will not permit much study of the book's previous incarnations, it is rich enough to show how Diderot's work reached the vast majority of his readers after he had finished with it.

Before attempting to follow the later transmigrations of the text, it is important to take account of a basic fact that became apparent to the authorities in France as soon as the first volume of the first edition reached the subscribers: the book was dangerous. It did not merely provide information about everything from A to Z; it recorded knowledge according to philosophic principles expounded by d'Alembert in the Preliminary Discourse. Although he formally acknowledged the authority of the church, d'Alembert made it clear that knowledge came from the senses and not from Rome or Revelation. The great ordering agent was reason, which combined sense data, working with the sister faculties of memory and imagination. Thus everything man knew derived from the world around him and the operations of his own mind. The *Encyclopédie* made the point graphically, with an engraving of a tree of knowledge showing how all the arts and sciences grew out of the three mental faculties. Philosophy formed the trunk of the tree, while theology occupied a remote branch, next to black magic. Diderot and d'Alembert had dethroned the ancient queen of the sciences. They had rearranged the cognitive universe and reoriented man within it, while elbowing God outside.

They knew that tampering with world views was a dangerous business, so they hid behind subterfuge, irony, and false protestations of orthodoxy. But they did not hide the epistemological basis of their attack on the old cosmology. On the contrary, the Preliminary Discourse made it explicit in a brief history of philosophy that established the intellectual pedigree of the philosophes and struck down orthodox Thomism on one side and neo-orthodox Cartesianism on the other, leaving only Locke and Newton standing. Thus Diderot and d'Alembert presented their work as both a compilation of information and a manifesto of *philosophie*. They meant to merge those two aspects of the book, to make them seem like two sides of the same coin: Encyclopedism. This strategy served as a way of legitimizing the Enlightenment because the Encyclopedists identified their philosophy with knowledge itself—that is, with valid knowledge, the kind derived from the senses and the faculties of the mind as opposed to the kind dispensed by church and state. Traditional learning, they implied, amounted to nothing but prejudice and superstition. So beneath the bulk of the *Encyclopédie*'s twenty-eight folio volumes and the enormous variety of its 71,818 articles and 2,885 plates lay an epistemological shift that transformed the topography of everything known to man.

It was this break with the established notions of knowledge

and intellectual authority that made the *Encyclopédie* so heretical. Having made the break and having learned to look at the world of knowledge from the viewpoint of the Preliminary Discourse, readers could see smaller heresies scattered throughout the book. Finding them became a game. It would not do to look in obvious places, where the Encyclopedists had to be most careful about the censorship, although they even smuggled some impiety into the article CHRISTIANISME. Better to search through out-of-the-way articles with absurd headings like ASCHARIOUNS and EPIDÉLIUS for remarks about the absurdities of Christianity. Of course the remarks had to be veiled. The Encyclopedists draped the pope in Japanese robes before mocking him in SIAKO; they diguised the Eucharist as an extravagant pagan ritual in YPAINI; they dressed up the Holy Spirit as a ridiculous bird in AIGLE; and they made the Incarnation look as silly as a superstition about a magic plant in AGNUS SCYTHICUS. At the same time, they produced a parade of high-minded, law-abiding Hindus, Confucians, Hottentots, Stoics, Socinians, deists, and atheists, who usually seemed to get the better of the orthodox in arguments, although orthodoxy always triumphed in the end, thanks to non sequiturs or the intervention of ecclesiastical authorities, as in UNITAIRES. In this way, the Encyclopedists stimulated their readers to seek for meaning between the lines and to listen for double-entendre.

Once a reader learned to exercise his reason in this manner, he would discover unreason in all spheres of life, including the social and political. The *Encyclopédie* treated the state with more respect than the church, and it did not contest the supremacy of the privileged orders. But mixed among its conventional and sometimes contradictory articles, the attentive reader could find a good deal of irreverence for the masters of the secular world. Not only did Diderot seem to reduce the authority of the king to the consent of the people in AUTORITÉ POLITIQUE but also d'Holbach advocated a bourgeois-type constitutional monarchy in REPRÉSENTANTS; Rousseau anticipated the radical arguments of his *Contrat social* in ÉCONOMIE (*Morale et Politique*); and Jaucourt popularized natural law theory in dozens of articles that implicitly challenged the ideology of Bourbon absolutism. Several articles mocked the pomp and pretensions of the aristocracy. Although the tax exemptions of the privileged orders were defended in some places (EXEMPTIONS and PRIVILÈGE), they were attacked in others (VINGTIÈME and IMPÔT). And the dignity of ordinary persons was affirmed at many points, not only in articles about bourgeois (NÉGOCE) but also in impassioned descriptions of the hard life of laborers (PEUPLE).

It would be wrong to construe such remarks as a call for revolution. The *Encyclopédie* was a product of its time, of mid-century France, when writers could not discuss social and political questions openly, in contrast to the prerevolutionary era, when a tottering government permitted a good deal of frank discussion. The *Encyclopédie* did not even favor an advanced form of capitalism. Despite its emphasis on technology and physiocracy, it discouraged the concentration of men and machines in factories, and it presented an archaic picture of manufacturing rather than a preview of the industrial revolution in articles like INDUSTRIE and MANUFACTURES. The radical element in the *Encyclopédie* did not come from any prophetic vision of the far-off French and industrial revolutions but from its attempt to map the world of knowledge according to new boundaries, determined by reason and reason alone. As its title page proclaimed, it pretended to be a "dictionnaire *raisonné* des sciences, des arts et des métiers" —that is, to measure all human activity by rational standards and so to provide a basis for rethinking the world.

Contemporaries had no difficulty in detecting the purpose of the book, which its authors acknowledged openly in key articles like Diderot's ENCYCLOPÉDIE and d'Alembert's Avertissement to volume 3. From the appearance of the first volume in 1751 until the great crisis of 1759, the *Encyclopédie* was denounced by defenders of the old orthodoxies and the Old Regime, by Jesuits, Jansenists, the General Assembly of the Clergy, the Parlement of Paris, the king's council, and the pope. The denunciations flew so thick and fast, in articles, pamphlets, and books as well as official edicts, that the *Encyclopédie* seemed doomed. But the publishers had invested a fortune in it, and they had powerful protectors, notably Chrétien-Guillaume de Lamoignon de Malesherbes, the liberal Directeur de la librairie, who superintended the book trade during the crucial years between 1750 and 1763.

Malesherbes saved the *Encyclopédie* several times, first in 1752, when it became implicated in the de Prades affair. One of Diderot's collaborators, the abbé Jean-Martin de Prades, had submitted a thesis for a licentiate in theology at the Sorbonne that seemed to come straight out of the Preliminary Discourse, if not hell itself, as de Prades's bishop observed. In the course of the subsequent scandal, de Prades fled to Berlin, where Frederick II made him a reader; the *Encyclopédie* was denounced to the king as evidence of creeping atheism; Diderot, who had spent four painful months in Vincennes only two years earlier for his *Lettre sur les aveugles,* seemed likely to be imprisoned once more; and rumors had it that the Jesuits would take over the *Encyclopédie* as a reward for their diligence in exposing the conspiracy to destroy religion. Thanks to Malesherbes, this crisis resulted only in an arrêt du Conseil, which condemned the first two volumes for "plusieurs maximes tendantes à détruire l'autorité royale, à établir l'esprit d'indépendance et de révolte et, sous des termes obscurs et équivoques, à élever les fondements de l'erreur, de la corruption des moeurs, de l'irréligion et de l'incrédulité."[3] That sounded terrible enough, but it had little effect because the volumes had already been distributed to the subscribers, and the government permitted the work to continue, without revoking its privilege.

The scandal continued to sizzle and spread for the next seven years, as volumes 3 through 7 appeared and as skillful polemicists like Charles Palissot and Jacob-Nicolas Moreau fanned the flames on the side of the priests. On the other side, Voltaire loaned his pen and his prestige to the cause; and Diderot and d'Alembert found the ranks of their collaborators swelling with other illustrious writers, including most of the men who were beginning to be identified as philosophes: Duclos, Toussaint, Rousseau, Turgot, Saint-Lambert, d'Holbach, Daubenton, Marmontel, Boulanger, Morellet, Quesnay, Damilaville, Naigeon, Jaucourt, and Grimm. They also claimed Montesquieu and Buffon, whose works they cited constantly, although it seems that neither wrote anything expressly for the *Encyclopédie.* (Montesquieu died in 1755, leaving a fragment which was published posthumously in the article GOÛT, and Buffon kept his distance from the Encyclopedists, perhaps because he had enough difficulty defending the unortho-

---

3. Arrêt du Conseil of Feb. 7, 1752, quoted in John Lough, *The "Encyclopédie"* (New York, 1971), p. 21. Lough's book provides a good survey of the early history of the *Encyclopédie.* It can be supplemented by the works of Watts, Proust, and Wilson, cited in the Bibliographical Note.

dox passages in his *Histoire naturelle,* which began to appear in 1749.)

Nothing could have been better for business than the continued controversy and the volunteer corps of authors. The publishers, André-François Le Breton and his associates, Antoine-Claude Briasson, Michel-Antoine David, and Laurent Durand, had envisaged an edition of 1,625 copies, but the subscriptions poured in so fast that they increased it three times, until it reached 4,255 copies in 1754. In their prospectus of 1751, they had promised to provide eight folio volumes of text and two of plates, at a total cost of 280 livres, by the end of 1754. The prospectus did envisage the possibility of an additional volume, which would be sold at a 29 percent reduction, but it reassured the subscribers that the text and plates had been completed, even though Diderot was more than twenty years away from the end of his labors and he would produce almost three times as many volumes as the prospectus had promised. This stroke of false advertising set a standard that the *Encyclopédie* publishers were to maintain without flagging for the next fifty years. Indeed, if the public had known that the book would grow to seventeen volumes of text and eleven of plates, that its price would inflate to 980 livres, and that its last volume would not appear until 1772, the enterprise would never have got off the ground. Although Luneau de Boisjermain did try, unsuccessfully, to bring it down by suing the publishers for swindle, the real threat came once again from the French authorities during a second crisis, from 1757 to 1759.

That was a dark period in French history. It began with Damiens's attempt to assassinate Louis XV. The country, already bleeding from the Seven Years' War, filled with rumors about atheists and regicides; and the crown stirred up fears of conspiracies by a Declaration of April 16, 1757, which threatened to put to death anyone who wrote or printed anything against church or state—indeed, anything even tending to "émouvoir les esprits." At this point, the anti-Encyclopedists opened fire with their heaviest barrage of propaganda, not only denouncing the heresies in volumes 4 and 7 of the *Encyclopédie* but also associating them with bold-faced atheism, which, they charged, had broken out shamelessly in public, and with a censor's approval, when Helvétius published *De l'Esprit* in July 1758. This book caused an even greater scandal than the thesis of the abbé de Prades; and although Helvétius had not contributed to the *Encyclopédie,* most of the indignation he aroused fell on it. In January 1759, the procureur général of the Parlement of Paris warned that behind *De l'Esprit* lurked the *Encyclopédie* and behind the *Encyclopédie* hovered a conspiracy to destroy religion and undermine the state. The parlement promptly banned the sale of the *Encyclopédie* and appointed a commission to investigate it. But though it had hunted witches for centuries, the parlement had never gained control over the printed word in France.

That authority belonged to the king, who exercised it through his chancellor, who delegated it to the Directeur de la librairie, who in this case happened to be Malesherbes. On March 8, 1759, the Conseil d'Etat reaffirmed the king's authority by taking the destruction of the *Encyclopédie* into its own hands. It revoked the book's privilege and forbade the publishers to continue it, noting, by way of explanation, the strategy that its authors had pursued: ''Ladite *Encyclopédie,* étant devenue un dictionnaire complet et un traité général de toutes les sciences, serait bien plus recherchée du public et bien plus souvent consultée, et que par là on répandrait encore davantage et on accréditerait en quelque sorte les pernicieuses maximes dont les volumes déjà distribués sont remplis.''[4] The *Encyclopédie* went onto the Index on March 5, 1759, accompanied by *De l'Esprit,* and on September 3 Pope Clement XII warned all Catholics who owned it to have it burned by a priest or to face excommunication. It was hardly possible for a book to be condemned more completely. The *Encyclopédie* had run afoul of the most important authorities of the Old Regime, yet it survived. Its survival marked a turning point in the Enlightenment and in the history of books in general.

Sometime in the course of this crisis, Diderot, who had been writing away behind locked doors, learned from Malesherbes that his papers were about to be seized by the police —and that they could be saved by being deposited with Malesherbes himself, who had just issued the order for their confiscation. Malesherbes also seemed to be behind the compromise that finally saved the entire enterprise. On July 21,

4. Arrêt du Conseil of March 8, 1759, quoted ibid., p. 26.

1759, an arrêt du Conseil required the publishers to refund 72 livres to each subscriber, ostensibly as a way of closing their accounts. In fact, however, the government permitted them to apply the money to a *Recueil de mille planches . . . sur les Sciences, les Arts libéraux et les Arts mécaniques,* which was nothing but the plates of the *Encyclopédie* under a new title. Having regained a legal hold on their speculation by a new privilege, issued for the *Recueil de planches* on September 8, 1759, the publishers proceeded to print the last ten volumes of text. In order to minimize scandal, the volumes appeared all at once in 1765, under the false imprint "A NEUFCHASTEL,/ CHEZ SAMUEL FAULCHE & Compagnie, Libraires & Imprimeurs." And to make doubly sure, Le Breton purged the text in page proofs, while Diderot's back was turned. Although he never forgave the publisher for this *atrocité,* Diderot continued to labor on the plates, and the last two volumes appeared in 1772. But the joy had gone out of the work. Deserted by d'Alembert, Voltaire, and most of the other writers who had rallied around him in the early 1750s, Diderot threw the last volumes together haphazardly, leaning more and more on the faithful Jaucourt, who copied and compiled tirelessly and saw the book through to the end. Diderot ended it in a state of disappointment and disillusion. Looking back at the result of twenty-five years of labor, he described the *Encyclopédie* as a monstrosity, which needed to be rewritten from beginning to end.[5] His verdict touched off a series of projects to remodel the book that culminated in the even more monstrous *Encyclopédie méthodique,* for Le Breton's successors, and booksellers everywhere in Europe, considered Diderot's work too faulty to be left untouched and too lucrative to be left alone. But whatever its faults, its completion counts as one of the great victories for the human spirit and the printed word.

In permitting Diderot's text to appear in print, despite its formal illegality, the state gave the philosophes an opportunity to try their wares on the market place of ideas. But what resulted from this breakthrough in the traditional re-

5. Diderot produced his criticism for Charles Joseph Panckoucke, a publisher who was soliciting permission to produce a completely revised edition of the *Encyclopédie* in 1768. The original text of Diderot's memoir is missing, but part of it was published during the lawsuit of Luneau de Boisjermain and reprinted in Diderot's *Oeuvres complètes,* ed. J. Assézat and M. Tourneux (Paris, 1875–77), XX, 129–33.

straints on the press in France? By concentrating on the duel between the Encyclopedists and the powers of the Old Regime, historians have told only half the story. The other half concerns some basic questions in the history of eighteenth-century books. First, is it possible to situate the work in a social context? Where did the Encyclopedists come from, and where did the *Encyclopédies* go? Second, how did the later editions emerge from the first, and what do they reveal about the operations of the publishing industry?

Research on the social background of the Encyclopedists has turned on the question of whether they can be considered bourgeois who shaped the consciousness of their class and helped to establish industrial capitalism in the eighteenth century. To an older generation of Marxist scholars, the answer to that question was an unqualified—and an undocumented—yes.[6] But a younger generation of social historians has found all sorts of complexities and contradictions within the eighteenth-century bourgeoisie, while economic historians have failed to turn up much evidence of industrialization in France before the second half of the nineteenth century. Faced with so much ambiguity in the sister disciplines and with a general change in the intellectual climate, literary scholars have been challenged to adopt a *révolution conceptuelle* in the study of the *Encyclopédie*. The call has come from Jacques Proust, the leading authority on the *Encyclopédie* in France, who argues that the Encyclopedists must be understood as a peculiar group, a *société encyclopédique* with an underlying *formation structurée,* although they also can be identified with the bourgeoisie.[7] This analytical approach has led to some important research, but after wading into the

6. Albert Soboul, *Encyclopédie ou Dictionnaire raisonné des Sciences, des Arts et des Métiers* (Paris, 1952), pp. 7–24. Soboul goes so far as to treat Diderot's esthetics as a prophetic version of socialist realism in painting (p. 179), although he concedes that the philosophes failed to attain Stalin's concept of the nation (p. 149) and that Encyclopedism had to wait for Stalinism to reach full perfection: ''L'esprit encyclopédique se réalise librement et pleinement dans la seule société affranchie du capitalisme et de l'exploitation de l'homme par l'homme, la société sans classes dont l'*Encyclopédie soviétique* est le reflet'' (p. 23).

7. Jacques Proust, ''Questions sur l'*Encyclopédie*,'' *Revue d'histoire littéraire de la France*, LXXII (Jan.–Feb. 1972), 45. Proust's ''revolution'' apparently would proceed from a social analysis of the Encyclopedists to a structuralist analysis of their texts. For research on the Encyclopedists as a group see the Bibliographical Note.

data, the researchers have generally found that "structures" and "bourgeois" disappear in the welter of information about individuals; and even so, their information is incomplete. The authors of almost two-fifths of the articles cannot be identified; and almost one-third of the identifiable authors wrote only one article, while workhorses like Diderot, the abbé Mallet, and Boucher d'Argis produced the bulk of the book. The chevalier de Jaucourt, a nobleman who could trace his lineage well back into the Middle Ages, wrote about one-fourth of the entire text, but no one would argue that the *Encyclopédie* was a quarter aristocratic, especially as many of Jaucourt's contributions contain only a few lines and look trivial in comparison with a treatise like VINGTIÈME by Damilaville, who wrote only three articles.

Given the unrepresentativeness of the articles whose authors can be identified and the unevenness of the contributions of those authors, how can one find a meaningful standard of measurement in order to study the Encyclopedists sociologically? Even if one lumps them all together and sorts them into socio-occupational categories, they do not look very bourgeois, at least not in the modern, capitalist sense of the term. Only 4 percent were merchants or manufacturers. The same proportion came from the titled nobility, and both groups seem small in comparison with doctors and surgeons (15 percent), administrative officials (12 percent), and even clerics (8 percent). What identified the Encyclopedists as a group was not their social position but their commitment to a cause. To be sure, many of them retreated when the cause was most in danger, but they left their mark on the book, and the book came to epitomize the Enlightenment. Through scandal, persecution, and sheer survival, the *Encyclopédie* became recognized, by friends and enemies alike, as the summa of a great intellectual movement, and the men behind it became known not merely as collaborators but as *Encyclopédistes*. Their work signaled the emergence of an "ism."[8]

How Encyclopedism fared on the market place immediately after the Encyclopedists had finished their work is difficult to say because the papers of Le Breton and his as-

8. The above percentages have been calculated from the information on contributors in Jacques Proust, *Diderot et l'"Encyclopédie"* (Paris, 1967), chap. 1 and Annexe 1 and in John Lough, *The Contributors to the "Encyclopédie"* (London, 1973). For further details see Chapters VIII and IX.

sociates have almost completely disappeared. Some evidence in the rather unreliable material produced during the lawsuit of Luneau de Boisjermain indicates that the first folio edition did not sell widely in France: only one-half or perhaps even one-quarter of the copies remained within the kingdom.[9] But the publishers made a fortune from it. On an initial investment of about 70,000 livres, their profit may have reached as much as 2,500,000 livres. Net income came to approximately 4,000,000 livres and net costs to something in the range of 1,500,000 to 2,200,000 livres, of which about 80,000 went to Diderot.[10] Those were spectacular sums for the eighteenth century, and the publishers were only able to deal in them by tapping capital from the subscribers. Thanks to this flow of cash, the *Encyclopédie* financed itself by 1751, although the paper and printing for the last ten volumes of text, which were issued simultaneously, must have required a heavy outlay of cash.

The business seems to have been run like many speculations in publishing. On October 18, 1745, Le Breton and his three associates signed a *traité de société*, establishing a capital fund of 20,000 livres and dividing shares among themselves according to the proportions of their contributions: Le Breton

9. The publishers claimed that three-quarters of the edition went to foreign subscribers, but they probably exaggerated the importance of foreign sales in order to indicate that in opposing Luneau they were contributing to the welfare of the entire nation by promoting a favorable balance of trade. See John Lough, ''Luneau de Boisjermain v. the Publishers of the *Encyclopédie*,'' *Studies on Voltaire and the Eighteenth Century,* ed. Theodore Besterman, XXIII (1963), 132–133.

10. These estimates are based on fragments of the publishers' accounts and other material connected with the Luneau case, later published by Louis-Philippe May: ''Histoire et sources de l'*Encyclopédie* d'après le registre de délibérations et de comptes des éditeurs et un mémoire inédit,'' *Revue de synthèse*, XV (1938), 7–110. In ''The *Encyclopédie* as a Business Venture,'' *From the Ancien Régime to the Popular Front: Essays in the History of Modern France in Honor of Shepard B. Clough,* ed. Charles K. Warner (New York and London, 1969), pp. 19–20, Ralph H. Bowen argues that these documents confirm Diderot's contention that receipts totaled 4,000,000 livres, expenses 1,500,000 livres, and profits 2,500,000 livres. But Luneau manipulated the evidence to suggest that the publishers gouged the subscribers, and a closer reading of it by Lough (''Luneau de Boisjermain v. the Publishers of the *Encyclopédie*,'' p. 167) shows that the expenses probably came to at least 2,205,839 livres. In fact, the Luneau material is too contentious to support firm conclusions, especially as difficulties concerning bill collecting cut badly into the profits of eighteenth-century publishers, who allowed for them in their financial statements under rubrics like *recouvrement* and *mauvais débiteurs.*

acquired an interest of three-sixths and the others one-sixth apiece. Supplementary articles allotted Le Breton a fixed sum per sheet for printing expenses, so the associates delegated the responsibility of production to him, and he did as well as he could within the terms fixed by the contract.[11] Precisely how he managed this enormous task cannot be known, nor is it possible to learn much about how he supplied the customers and who they were. The Luneau material contains the names of about seventy-five subscribers. Most of them were noblemen, including several eminent courtiers—the Duc de Noailles, the Maréchal de Mouchy, the Duc de la Vallière— and several magistrates of the parlements and bailliages. The rest came mainly from the law, the clergy, and the upper echelons of the royal administration. Only two were merchants.[12] Of course those few names, bandied about in the polemics of a lawsuit, hardly constitute a representative sample of all 4,000 subscribers. About all one can conclude from the publishing history of the first edition is that its text came from a disparate group of writers who were united by a common commitment to the task; that its luxurious folio volumes went to wealthy and well-born readers scattered across Europe; and that it was extremely lucrative.

One of the first persons to draw that last conclusion was an aggressive publisher from Lille named Charles Joseph Panckoucke, who had set up business in Paris in 1762 after a brief apprenticeship with Le Breton. Panckoucke cultivated philosophes, especially Buffon, Voltaire, and Rousseau; he also courted protectors in the government. By 1768 he had become the official bookseller of the Imprimerie Royale and the Académie royale des Sciences, and he was well on his way to becoming the dominant figure in the French press, thanks to an interlocking set of government-granted monopolies covering periodical literature. On December 16, 1768—four years before the final volumes of plates were published—Panckoucke and two associates, a bookseller named Jean Dessaint and a papermaker called Chauchat, bought the rights to fu-

11. For the texts of the contracts and additions see May, ''Histoire et sources de l'*Encyclopédie*,'' pp. 15–17, 25.

12. Lough, ''Luneau de Boisjermain v. the Publishers of the *Encyclopédie*,'' pp. 133–140.

ture editions of the *Encyclopédie* and the copper plates for the illustrations from Le Breton and his partners.[13]

While the original publishers completed their printing of the plates, the new consortium lobbied for permission to produce a *refonte* or totally revised edition. Panckoucke recruited Diderot to help in this effort, and Diderot complied with an eloquent memoir that argued the case for a new *Encyclopédie* by recounting the faults of the old one. The chancellor, Maupeou, refused this request, although the Duc de Choiseul, a more liberal figure who was about to be ousted from the government, authorized a reprint of the original text. These difficulties scared off Dessaint and Chauchat, but Panckoucke bought back their shares, transformed them into shares in a speculation on the reprint, and on June 26, 1770, sold them again to a new set of partners. These partners eventually included Voltaire's publisher, Gabriel Cramer, and Samuel de Tournes in Geneva; Pierre Rousseau, the director of the Société typographique de Bouillon; and two Parisians, a notary called Lambot and a bookseller called Brunet. Nine months later, on April 12, 1771, Panckoucke formed another separate association, this time for a set of *Suppléments,* which would correct the erors and fill the gaps of the original text. This Société was made up of the speculators on the reprint, except for Lambot, who probably had sold out to Panckoucke early in 1771, and the two Genevans, who had meant to join but finally dropped out; it also included Marc-Michel Rey, Rousseau's publisher in Amsterdam, and Jean-Baptiste Robinet, a man of letters who was to edit the *Supplément.*[14] Thus what had begun as a modest partnership

13. Durand died in 1763 and the other partners divided his 1/6 share, so in 1768 Le Breton owned 10/18 of the speculation and David and Briasson each owned 4/18. Panckoucke and his two partners each acquired 1/3 of the new speculation. The following account of the publishing history of the *Encyclopédie* and the *Supplément* from 1768 to 1776 is derived mainly from the work of Watts, Lough, Clément, and Birn cited in the Bibliographical Note.

14. In their correspondence, the publishers usually referred to the *Supplément* in the singular, as it appeared on its title page, but they sometimes talked of *Suppléments* in the plural. According to the original "Acte de Bouillon" of April 12, 1771, in dossier Marc-Michel Rey, Bibliotheek van de Vereeniging ter Bevordering van de Belangen des Boekhandels of Amsterdam, the shares in the association for the *Supplément* were to be divided as follows: 6/24 to Cramer and de Tournes, 6/24 to Rousseau, 3/24 to Rey, 3/24 to Robinet, 4/24 to Panckoucke, and 2/24 to Brunet. After the Genevans withdrew from the enterprise, their shares were divided between Panckoucke and Brunet. It is impossible to follow

among three Parisian booksellers grew into two international consortia, built on a system of overlapping alliances among the most powerful publishers of the Enlightenment.

The subsequent history of the *Encyclopédie* has a good deal in common with eighteenth-century diplomacy: baroque intrigue and sudden reversals mixed with warfare. The first publishers had been attacked by some pirates from England in 1751; and although they apparently put an end to the threat of an English *Encyclopédie* by paying ransom, they could not prevent two folio editions from being produced in Italy. The first began to appear in 1758 in the republic of Lucca, and the second followed suit from Leghorn, beginning in 1770. Although both became mired in delays and difficulties, they conquered some of the *Encyclopédie* market outside France, especially south of the Alps. The northern market then fell in large part to a renegade Italian monk called, in the French version of his name, Fortuné-Barthélemy de Félice. After setting up shop in the Swiss city of Yverdon near Neuchâtel, Félice announced that he would produce the much-desired *refonte* of the *Encyclopédie*—that is, a completely rewritten version in quarto format, which would draw on contributions from savants all over Europe in order to correct errors, fill gaps, and, as it developed, substitute some sober Protestantism for the impieties of the original. *Encyclopédie* buyers therefore faced a choice: they could take Diderot's text with or without Robinet's *Supplément,* or they could order the purged and perfected version from Félice.

As millions of livres hung on those decisions, the publishers soon became embroiled in a trade war. Against Panckoucke's dual alliance, which covered Geneva, Bouillon, and Amsterdam, Félice mobilized two allies of his own: the Société typographique de Berne, which had helped him to found his business (Yverdon was located in Bernois territory), and a powerful bookdealer in The Hague named Pierre Gosse, who traded extensively throughout northern Europe. Gosse and the Bernois bought up Félice's entire edition, leaving him to

---

the selling and reselling of shares in the separate speculation on the reprint, but Panckoucke retained only a fraction of a fraction of his original interest in it. By Oct. 26, 1770, Cramer and de Tournes owned 2/6 of it, Lambot apparently owned 2/6, Rousseau owned 1/6, and Brunet owned part of the remaining 1/6 with Panckoucke.

do the editing and printing while they handled the marketing.[15] In prospectuses, circular letters, and journal advertisements, they hammered away about the deficiencies of Diderot's work and the excellence of Félice's. Now that Diderot's reputation has obliterated the memory of his rival, it is difficult to appreciate the effectiveness of this propaganda. But the *Encyclopédie d'Yverdon* had a good reception in the eighteenth century, and not only in pietistic corners of Germany and Holland. Voltaire, whose *Questions sur l'Encyclopédie* resulted from a broken promise to contribute to the *Supplément,* said he would take Félice's text over Diderot's if he were shopping for *Encyclopédies.*[16] And Félice's backers promoted that attitude through journals like the *Gazette de Berne* and the *Gazette de Leyde,* where they could manipulate literary notices. In 1771, for example, Gosse rebuked the STN for printing an unfavorable review of Félice's first volume in its own periodical, the *Journal helvétique,* and the STN immediately changed its tack, for the simple reason that Gosse was its biggest customer in the Low Countries.[17]

The Panckoucke group replied in kind through its journals,

15. In a letter to the STN of July 16, 1779, Félice said he was printing 1,600 copies. On Jan. 18, 1771, Gosse informed the STN that he had bought 3/4 of the edition and the Société typographique de Berne had bought 1/4. And in a letter of July 30, 1771, he noted that he had taken over the entire edition. His son, Pierre Gosse Junior, who succeeded him in 1774, told the STN in a letter of July 16, 1779, that he was still receiving all 1,600 copies, as Félice neared the end of his labor. These and all subsequent references to the STN come from the papers of the Société typographique de Neuchâtel, Bibliothèque de la ville de Neuchâtel, unless specified otherwise.

16. At first Voltaire showed nothing but scorn for Félice and his *Encyclopédie.* Voltaire to d'Alembert, June 4, 1769, *Voltaire's Correspondance,* ed. Theodore Besterman (Geneva, 1962), LXXII, 60. But by 1771 he had decided that Félice had got the better of Panckoucke: ''Ils [Félice's contributors] ont l'avantage de corriger dans leur édition beaucoup de fautes grossières, qui fourmillent dans l'*Encyclopédie* de Paris et que Panckoucke et Dessaint ont eu l'imprudence de réimprimer. Cette faute capitale les force à donner un supplément, qui renchérit le livre, et on aura l'édition d'Yverdon à une fois meilleur marché. Pour moi, je sais bien que j'acheterai l'édition d'Yverdon et non l'autre.'' Voltaire to Gabriel Cramer, [Dec. 1770], ibid., LXXVII, 163. In 1777 Voltaire proposed that his *Questions sur l'Encyclopédie,* which he had originally undertaken for Panckoucke's *Supplément,* be incorporated into Panckoucke's quarto edition, but this project never came to anything. See Voltaire to Henri Rieu, Jan. 13, 1777, ibid., XCVI, 27.

17. Gosse to STN, Jan. 18, July 1, and July 30, 1771. For further information on Félice and his conflicts with the rival *Encyclopédie* publishers see E. Maccabez, *F. B. de Félice (1723–1789) et son Encyclopédie (Yverdon, 1770–1780)* (Basel, 1903) and J. P. Perret, *Les Imprimeries d'Yverdon au XVIIe et au XVIIIe siècle* (Lausanne, 1945).

mainly Panckoucke's *Journal des savants* and Rousseau's *Journal encyclopédique*. Cramer even sent Rousseau detailed instructions about how to ridicule Félice: instead of seeming to take the Protestant *Encyclopédie* seriously, the *Journal encyclopédique* should stress the absurdity of an obscure Italian, who could not even write decent French, attempting to correct a text produced by the finest philosophes in all of France.[18] Félice answered that he merely expunged the absurdities of Diderot's text and assembled articles supplied by authorities like Albrecht von Haller and Charles Bonnet, who made Diderot's contributors look outmoded. He went on to offer his subscribers a supplement of their own that would incorporate anything worthwhile from Robinet's *Supplément* in a more up-to-date survey of the current arts and sciences. And in 1775 he went further: he announced that he would produce his supplement in a folio as well as a quarto edition and that he would fill the folio with the most important original material from the main text of the *Encyclopédie d'Yverdon*. This move struck at the heart of the rival publication because Robinet had aimed his *Supplément* at the owners of all the folio editions—the *Encyclopédies* of Lucca, Leghorn, and Paris as well as the Panckoucke-Cramer reprint. By ordering their supplements from Félice, the owners of the folios could combine the standard version of the *Encyclopédie* with the modern revisions of it, and the bottom would drop out of Robinet's market.

Panckoucke then attempted to snare Félice's subscribers by announcing a quarto edition of Robinet's *Supplément*. This counterattack never got anywhere because an open supplement war was certain to hurt Robinet far more than Félice, as the owners of folio *Encyclopédies* outnumbered the subscribers of Félice's quarto by a factor of at least six to one. In the end, therefore, Panckoucke sued for peace. He agreed to withdraw his quarto if Félice would withdraw the folio, and both sides promised to exchange their printed sheets so that they could crib from one another with maximum efficiency.

Meanwhile, Panckoucke ran into greater difficulties with his more important enterprise, the reprint itself. In February

---

18. Cramer to Rousseau, July 23, 1771, quoted in John Lough, *Essays on the "Encyclopédie" of Diderot and d'Alembert* (London, 1968), 88.

1770, following a denunciation by the General Assembly of the French Clergy, the Parisian police seized 6,000 copies of the first three volumes and walled them up under a vault in the Bastille, where they remained for six years, despite everything Panckoucke could do to get them released by pulling strings and greasing palms. After this catastrophe, the publishers of the reprint decided to move it from Paris to the printing shops of Cramer and de Tournes in Geneva. But no sooner had Cramer and de Tournes begun setting type than the Genevan Venerable Company of Pastors tried to force them to stop by denouncing them to the civil authorities. While Cramer argued his case before the city's Magnificent Council, Panckoucke secretly maneuvered to cut him out of the speculation and to transfer it to Bouillon and Amsterdam, where it could be reconstructed by Robinet, Rousseau, and Rey as a *refonte* once again. But Rey refused to go along with such a spectacular and costly reversal of policy, and Cramer eventually won over the city fathers of Geneva, who appreciated the importance of his operation for the local economy. Cramer placated the pastors with an offer to tone down d'Alembert's controversial article GENÈVE, which made them look like deists, and to let them purge anything that wounded their Calvinism in the text of the *Supplément*. This arrangement did not settle all the problems of the reprint because the French authorities continued to keep the first three volumes in the Bastille and Panckoucke continued to flirt with other printers. But these difficulties did not lead to anything more than some sharp remarks in the correspondence between Geneva and Paris. In the end, the Genevans not only kept the lucrative printing job in their own hands but also redid volumes 1 through 3 and made an attempt to take over the *Supplément*.

When he put together the complementary speculation on the *Supplément* in April 1771, Panckoucke had offered Cramer and de Tournes a 6/24 share in it. They were also to get the printing commission, but before accepting, they demanded that they be given control over the subscription and finances and that Robinet move his editorial operation to Geneva. Hoping to keep Robinet in Bouillon and to get the printing transferred there, Rousseau vetoed this proposal, and Rey supported him. The Genevans responded in November 1771 by withdrawing from the *Supplément* altogether. Then, for almost a year, the remaining partners squabbled about how to divide the outstanding shares of 6/24 and where to locate the printing. Panckoucke and Brunet finally bought the shares and agreed to advance the capital for the printing operation. In return they forced Rousseau and Rey to let them negotiate with the French authorities for a Parisian printing or, failing that, to abandon the printing to Cramer, for they insisted that Geneva would serve as a better base for smuggling than Bouillon. Having buried this bone of contention, Panckoucke and Rousseau then became entangled in a dispute over their journals. Panckoucke wanted to reserve the French market for his newest acquisition, the *Journal historique et politique de Genève,* while Rousseau fought to keep France open for the *Journal politique* and the *Journal encyclopédique,* which he published in Bouillon. Thanks to the protection of the foreign minister, Panckoucke finally forced Rousseau to pay 5,000 livres a year for the right to distribute the Bouillon journals in France, and at the same time he won over Robinet, who dabbled in the intrigues against Rousseau's journals while putting together the copy for the *Supplément* in Bouillon. By February 1776, the plots and subplots had become more than Rousseau could bear. He sold his 6/24 share in the *Supplément* to the Parisian printer Jean-Georges-Antoine Stoupe, who proceeded to print it in Paris, while Rey produced an edition in Amsterdam. The two editions, each containing four volumes of text and one of plates, were completed in 1777.

By this time Cramer had finished the reprint. Although he had filled his letters to Panckoucke with lamentations about the difficulties of the enterprise, it probably succeeded well enough because at several points he and de Tournes offered to buy out all the other partners. No one would part with his shares, however, and by June 13, 1775, Panckoucke and the Genevans felt ready to reach a settlement, even though the printing would continue for another year and Panckoucke would have to make a later settlement with the partners to whom he had sold portions of his original interest. In the Genevan agreement of June 13, 1775, Panckoucke terminated the partnership by paying the Genevans 200,000 livres against their one-third share in the profits, while they promised to administer the final stage of production and sales in his interest. At that point, the profits came to only 71,039 livres, but about 670 of the 2,000 copies remained to be sold. If they could be marketed at the subscription price of 840 livres apiece, they would fetch 562,800 livres. Of course much of that sum would be eaten away by delays, booksellers' discounts, defaults in the payments, and the loss of the 6,000 copies of volumes 1 to 3, which Panckoucke valued at 45,000 livres. But even if Panckoucke and his hidden partners cleared only 400,000 livres (the equivalent, considering their two-thirds share, of the Genevans' 200,000 livres), they would have had a good return on their investment.[19]

As the Geneva folio had a relatively low pressrun and a high price, it did not represent much of an expansion of the *Encyclopédie* market. Nor did the *Suppléments,* which filled some of the gaps in Diderot's text but without Diderot's verve. In the long run Félice's work probably did not have a great impact on the audience of the *Encyclopédie*. It never penetrated France because the authorities successfully prohibited it, and it even floundered elsewhere in Europe because Félice kept expanding its size, raising its price, and delaying its completion. By 1780, when he issued the last of his fifty-eight volumes, ten years after the first, he had lost a great many subscribers, and the publishers of the cheaper quarto and octavo editions of Diderot's *Encyclopédie* had cut into his market. It was through those editions that the original text, and also the *Supplément,* reached ordinary readers everywhere in Europe. Having put together and taken apart several international consortia, having done battle against partners and competitors alike, and having learned to operate with the backing of the government rather than in defiance of

19. The full text of the complex contract signed by Panckoucke, Cramer, and de Tournes in Geneva on June 13, 1775, is printed in Lough, *Essays,* pp. 102–108. It is difficult to say why Panckoucke bought out the Genevans instead of settling the speculation by apportioning the profits according to shares after the distribution of the final volumes. He probably wanted to wind up the Genevan enterprise quickly and cleanly so that he could move on to other speculations. His payments actually come to 130,000 livres, spread out over three years because he deducted the current profits, evaluated at 70,000 livres, from the 200,000 he agreed to pay for the Genevan shares. He also took over uncollected credits with a paper value of 152,020 livres. The exact number of unsold *Encyclopédies* that he acquired cannot be known because the original contract for the edition had set the pressrun at 2,000, with a surplus of 150 to cover spoiled sheets, and there is no way to find out the actual spoilage.

it, Panckoucke was ready to speculate on Encyclopedism for the *grand public.*

Before taking up the story of how the *Encyclopédie* reached the general reading public, it is worth looking back over the early history of the book to see whether any connecting themes run through its twists and turns. From 1749, when Le Breton and his associates petitioned the government to release Diderot from the prison in Vincennes, until 1776, when Panckoucke persuaded it to free the 6,000 volumes from the Bastille, two objectives stand out in the maneuvers of the publishers: they wanted to appease the state, and they wanted to make money.[20] But the *Encyclopédie* sold for the same reason that prompted the government to confiscate it: it challenged the traditional values and established authorities of the Old Regime. The publishers sought a way out of this dilemma by toning down the text. Not only did Le Breton emasculate the last ten volumes, but Panckoucke planned to restrain the *philosophie* of his *refonte* as well, when he lobbied for permission to print it in 1768—or so the backers of the *Encyclopédie d'Yverdon* claimed during the early battles of their commercial war. In a printed circular, Gosse warned the booksellers of Europe to beware of bowdlerization:

C'est sur des avis reçus de très bonne part de Paris qu'il a été fait mention dans nos avis que Messieurs les libraires de Paris, en demandant un nouveau privilège, s'étaient engagés de retrancher dans cette nouvelle édition tous les articles qui ont pu choquer le gouvernement dans la première édition, tout comme nous tenons des avis de très bonne part que ce nouveau privilège leur est refusé et que Monseigneur le Chancelier et le Parlement s'opposent à la réimpression de l'*Encyclopédie* en France. Tous ceux qui sont instruits des persécutions que les auteurs et les premiers éditeurs ont essuyées en France comprendront facilement qu'un pays de liberté convient seul pour la perfection de cet ouvrage.[21]

20. On the day of Diderot's arrest, July 24, 1749, the publishers appealed to the Comte d'Argenson, the minister who had ordered it, by emphasizing its economic consequences: "Cet ouvrage, qui nous coûtera au moins deux cent cinquante mille livres, était sur le point d'être annoncé au public. La détention de M. Diderot, le seul homme de lettres que nous connaissons capable d'une aussi vaste entreprise et qui possède seul la clef de toute cette opération, peut entraîner notre ruine." Letter quoted in John Lough, *The ''Encyclopédie''* (New York, 1971), p. 18.

21. Circular from Pierre Gosse and Daniel Pinet of The Hague, dated Aug. 2, 1769, and sent to the STN. In a letter to Marc-Michel Rey of Oct. 26, 1770, Panckoucke indicated that Gosse's version of his activities was not far from the

Of course *perfection* for Félice also meant cutting *philosophie* and, furthermore, substituting Protestantism for Catholicism in the articles that had won the blessings of the French censors—a tactic that was designed to please the authorities in Bern but not those in Versailles. When he reviewed the idea of a *refonte* in his unsuccessful proposal to transfer the printing operation from Geneva to Amsterdam and Bouillon in 1770, Panckoucke made it clear that he put commercial considerations above everything else: "Il ne faudra point se permettre aucune hardiesse impie qui puisse effrayer les magistrats. Au contraire il faudra que tout l'ouvrage soit écrit avec beaucoup de sagesse, de modération, qu'il puisse même mériter des encouragements de votre gouvernement . . . C'est ici une affaire d'argent, de finance, où tout le monde peut s'intéresser."[22] Business was business, even if it involved Enlightenment. Similarly, the *Supplément* turned into a cautious venture, ideologically if not commercially. The agreement of April 12, 1771, envisaged a cast of savants rather than philosophes, and Robinet promised to direct them toward the natural sciences rather than philosophy. The contract bound him to "écrire les *Suppléments* avec sagesse et à n'y rien admettre contre la religion, les bonnes moeurs et le gouvernement, les *Suppléments* ayant pour principal objet la perfection des sciences naturelles."[23] Given this emphasis, it hardly seems surprising that Panckoucke succeeded

---

truth because he explained that he (Panckoucke), Dessaint, and Chauchat had lobbied for permission to do the *refonte* for six months, hoping "que le gouvernement permettrait la refonte de l'ouvrage en supprimant les articles qui avaient pu déplaire." Letter quoted in Fernand Clément, "Pierre Rousseau et l'édition des *Suppléments* de l'*Encyclopédie*," *Revue des sciences humaines*, LXXXVI (April–June 1957), 140.

22. Panckoucke to Rey, Oct. 26, 1770, quoted ibid., p. 141.

23. Ibid., p. 136. The prospective contributors named in the agreement included d'Alembert for physics, Albrecht von Haller for anatomy, J.-J. de Lalande and Jean Bernouilli the younger for astronomy, Antoine Louis for surgery, Antoine Petit for medicine, L.-F.-G. de Keralio for tactics, Philibert Gueneau de Montbéliard for artillery, Nicolas de Beauzée for grammar, and J.-F. de La Harpe for literature. Nearly all these men were recruited by Panckoucke several years later to write for the *Encyclopédie méthodique*, in some respects an extension of the *Supplément*. Robinet failed to recruit several of the writers mentioned in the agreement, and he recruited many more who were not mentioned—about fifty in all, including Condorcet and Marmontel as well as hacks like J.-L. Carra and J.-L. Castilhon, who each wrote about 400 articles. See Lough, *The Contributors to the "Encyclopédie,"* pp. 54–69.

not only in transferring the printing to Paris but also in getting a privilege for it.[24]

While Panckoucke steered the *Encyclopédie* toward official orthodoxy, the officials moved closer to Encyclopedism. During the last years of Louis XV's reign, the government had actually increased the severity of its policy toward books, but the reign of Louis XVI began under the influence of an Encyclopedist, Turgot. Panckoucke's confiscated *Encyclopédies* were released from the Bastille, and his later speculations thrived under a series of reforming ministers who not only relaxed the state's control of the book trade but also consulted him about how to do it. The Malesherbes tradition, which had lapsed after Malesherbes left the Direction de la librairie in 1763, revived in time to stimulate an *Encyclopédie* boom, which began in 1776 and continued until the Revolution.

The legalization of the *Encyclopédie* also helps explain the connecting links in the series of speculations on the book between 1745 and 1789. Legality in publishing derived from a privilege, the exclusive right to reproduce a text, granted by the grace of the king, administered through the Direction de la librairie, and registered with the Communauté des libraires et des imprimeurs of Paris. Although they had something in common with modern copyrights, book privileges, like privileges in general under the Old Regime, involved ancient notions and institutions—the authority of the king, a baroque bureaucracy, and a monopolistic guild. By granting a privilege, the king did not merely allow a book to come into being: he put his stamp of approval on it; he recommended it to his subjects, speaking through one or more censors who expatiated on its importance and even its style in long-winded *permissions* and *approbations* that were usually printed in the book along with a formal *lettre de privilège* from the king. Privileges were also properties, which could be bought and sold, divided into shares, and willed from husband to wife and father to son. But they extended only as far as the king's authority. Outside the kingdom, other publishers could reprint a French text as often as they pleased, unless their own

24. The privilege, conveying the exclusive right to print and reprint the work for twelve years, was entered in the Registre des privilèges of the Communauté des libraires et des imprimeurs de Paris (referred to hereafter as the booksellers' guild) on Feb. 10, 1776. Bibliothèque nationale, ms. Fr. 21967, p. 94.

governments objected. The privileged publisher in France might cry out about piracy, but he could only ask the Directeur de la librairie, the customs officials, the guild inspectors, and the police to close the borders to the rival edition and to confiscate any copies that might reach the domestic market.

The whole system stimulated the production of French books outside France because the spread of the French language had created a demand for cheap, pirated editions everywhere in Europe and because only books of unalloyed orthodoxy could be published legally within the kingdom. By its very nature, the organization of publishing in France forced the Enlightenment underground and into exile—into the printing shops of Amsterdam, Bouillon, Geneva, and Neuchâtel; for how could the king sanction the printing of texts that challenged the basic values of the regime? The rigidity of privilege kept a multimillion-livre industry beyond the pale of the law. Faced with this dilemma, administrators like Malesherbes encouraged the development of a grey area of quasi-legality in publishing. They granted *permissions tacites, permissions simples, tolérances,* and *permissions de police*—that is, authorizations for books to appear without the royal imprimatur, though also without formal and exclusive property rights attached to them. If the clergy or parlements protested against an unorthodox book, the government would not seem to have sponsored it and could promise to have it confiscated, taking care, on some occasions, to warn its publishers in time for them to save their stock.[25]

The struggle to print and reprint the *Encyclopédie* took place at the center and all around the edges of this complex and contradictory system. The original publishers actually took out three privileges for the text, one in April 1745, one in January 1746, and one in April 1748. Each corresponded to a stage in the expansion of the original plan to publish a four-volume translation of Ephraim Chambers's *Cyclopaedia,*

25. For a general discussion of the various degrees of legality in eighteenth-century publishing see Robert Darnton, ''Reading, Writing, and Publishing in Eighteenth-Century France: A Case Study in the Sociology of Literature,'' *Daedalus* (winter, 1971), pp. 214–256. A great deal can be learned about the institutional aspects of publishing from the *Almanach de l'auteur et du libraire* (Paris, 1777) and the *Almanach de la librairie* (Paris, 1781) as well as from the royal decrees on the book trade in A. J. L. Jourdan, O. O. Decrusy, and F. A. Isambert, eds., *Recueil général des anciennes lois françaises* (Paris, 1822–33), XVI, 217–251; XXV, 108–128.

or Universal Dictionary of the Arts and Sciences, which had
first appeared in England in 1728. On March 8, 1759, however,
the government destroyed the rights to the final, full-blown
Encyclopédie by revoking its privilege. True, the publishers
continued production, but only under the cover of "une
tolérance tacite, inspirée par l'intérêt national," as Diderot
put it.[26]

How, then, was it possible for Panckoucke to claim he had
bought the exclusive "droits" to the book from the Le Breton
association? This claim served as the basis for most of the
Encyclopédie speculations between 1768 and 1800, and Panckoucke asserted it in the most absolute manner, in all his letters
and contracts. In writing to Marc-Michel Rey, for example,
he stated, "Vous n'ignorez pas que j'ai acquis il y a environ
18 mois avec M. Dessaint et un papetier de Paris nommé
M. Chauchat tous les droits et cuivres de l'*Encyclopédie*."[27]
In his contract with Cramer and de Tournes for the Genevan
folio edition, he described himself as "propriétaire des droits
et cuivres de l'ouvrage intitulé *Dictionnaire encyclopédique*."[28]
Far from questioning those proprietary rights, other publishers acknowledged them. Thus the Société typographique
de Neuchâtel observed in 1779 that Panckoucke could market
the *Encyclopédie* everywhere in France, owing to his "privilège exclusif pour cet ouvrage."[29] Eighteenth-century publishers did not use such language lightly. They knew that
droits derived from *privilèges,* yet they recognized Panckoucke's right to a book whose privilege had been destroyed.

The explanation of this paradox stands out in a contract
that Panckoucke signed with the Société typographique de
Neuchâtel on July 3, 1776 (see Appendix A.I). In it, Panckoucke identified himself, in his usual manner, as "propriétaire des droits et cuivres du *Dictionnaire encyclopédique*,"
and, as usual, he traced his ownership of the rights and the

26. Diderot, *Au public et aux magistrats* as quoted in Lough, "Luneau de
Boisjermain v. the Publishers of the *Encyclopédie*," p. 132. Strictly speaking,
the arrêt du Conseil of March 8, 1759, revoked the second of the three privileges,
and in his lawsuit, Luneau de Boisjermain argued that the contractual obligations
of the publishers, which were based on the final privilege, therefore remained in-
tact. But his argument turned on a technicality or an oversight by the Conseil
d'Etat, and the court did not uphold it.
27. Panckoucke to Rey, Oct. 26, 1770, in Clément, "Pierre Rousseau et l'édition des *Suppléments* de l'*Encyclopédie*," p. 140.
28. See the text of the contract printed in Lough, *Essays*, p. 67.
29. STN to Maréchal of Metz, Aug. 22, 1779.

plates to his contract of December 16, 1768, with Le Breton, David, and Briasson. He then noted that he had bought out his own partners, Dessaint and Chauchat, in 1769 and 1770 and that his exclusive rights to the book had been confirmed by a royal privilege dated May 20, 1776, "sous le titre de *Recueil de Planches sur les Sciences, Arts et Métiers*." The register of the Parisian booksellers' guild for 1776 contains a privilege under Panckoucke's name for a work with precisely this title, and a similar privilege appears in the first edition of the *Encyclopédie*—not in volumes 1–7 of the text, which carry the privilege that was revoked in 1759, but in volume 6 of the plates, which appeared in 1768, when Panckoucke bought the rights to the book from Le Breton and his associates.[30] The privilege in the plates states that it had been registered in the booksellers' guild on September 8, 1759—that is, just at the time when the government saved the *Encyclopédie*, after ostensibly destroying it, by letting Le Breton apply the subscribers' money to the volumes of plates.[31] Thus the rescue operation of 1759 was not merely an attempt to preserve the publishers' capital while permitting them to continue the printing in a semiclandestine manner. It restored their claim to the "rights" of the book, property rights, which had enormous commercial value in the book trade. Consequently, when Panckoucke's group bought out Le Breton's group on December 16, 1768, they paid 200,000 livres for "la totalité des droits dans les réimpressions futures et dans la totalité des planches en cuivre." This huge sum covered far more than the value of the copper plates, as the

30. Bibliothèque nationale, ms. Fr. 21967, p. 122, entry for March 29, 1776: "Notre aimé le Sr. Panckoucke, libraire, Nous a fait exposer qu'il désirerait faire imprimer et donner au public les ouvrages intitulés *Recueil des planches sur les sciences, arts et métiers* in-folio, *Histoire générale des voyages* par M. l'abbé Prevôt [that is, Prévost], s'il Nous plaisait lui accorder nos lettres de privilège pour ce nécessaires. A ces causes, voulant favorablement traiter l'exposant, Nous lui avons permis et permettons par ces présentes de faire imprimer lesdits ouvrages autant de fois que bon lui semblera et de les vendre et débiter partout Notre royaume pendant le temps de douze années consécutives." No entry for any such work exists under the date of May 20, 1776, but Panckoucke's reference to that date might have concerned his final acquisition of the privilege in the Chancellerie rather than its registry with the guild.

31. As their title pages proclaimed, the plates appeared "avec approbation et privilège du roi"; but their title, *Recueil de planches sur les sciences, les arts libéraux, et les arts mécaniques, avec leur explication,* did not indicate that they had any connection with the *Encyclopédie,* which had been banned three years before the appearance of their first volume.

contract made clear, though it resorted to tortuous phrasing when it described the nature of the "droits."[32]

In Panckoucke's next speculation, the partnership of June 26, 1770, which resulted in the Genevan folio reprint, he let his partners make use of his "droits" for one edition only, but he insisted that the rights remain his.[33] Similarly, when he formed the association for the *Supplément* on April 12, 1771, he required that the rights to the *Supplément* revert to him after the completion of one edition.[34] Then, in the spring of 1776, he confirmed his rights to the *Recueil de planches* by taking out a new privilege, thereby substantiating his claim to be "seul propriétaire" of the entire *Encyclopédie,* texts, plates, and supplement. From this point on, he spoke of his "privilège" as well as his "droits." And finally, when he formed his partnership with the Société typographique de Neuchâtel, he was able to sell a half share in the "cuivres, droits, et privilège" for 108,000 livres. Thus, after being outlawed, the *Encyclopédie* gradually regained a degree of legality that had cash value in the eyes of publishers, even if it did not protect Panckoucke's 6,000 volumes from confiscation in 1770; and it served as the basis for a series of speculations that stretched into the 1770s and beyond, as one consortium succeeded another and the publishers passed on the book's pedigree for ever-increasing sums.

32. According to the text printed in Lough, *Essays*, p. 59, Le Breton and his partners stated that "nous vendons pour toujours aux sieurs Dessaint, Panckoucke et Chauchat tous nos droits dans les réimpressions à faire à l'avenir dudit ouvrage de l'*Encyclopédie*, nos dits droits tels qu'ils se poursuivent et qu'ils se comportent, que lesdits sieurs acquéreurs ont dit bien connaître et dont ils sont contents; en conséquence de quoi ledit objet est par nous vendu sans aucune garantie." It might be thought that ownership of the copper plates meant de facto control over future editions, as the text would be worthless without the illustrations. Cramer once developed this argument in a letter to Panckoucke (ibid., pp. 94–95), but it did not carry much weight. The quarto and octavo editions included only three of the eleven volumes of plates, and the publishers in Leghorn offered to sell the text without the plates if their customers preferred. What Panckoucke wanted was a legal claim to the book so that he could sell shares in it and use it to fight off competitors. His behavior seems puzzling only because modern concepts of legality and property do not suit eighteenth-century practices.

33. The contract of June 26, 1770, printed in Lough, *Essays*, pp. 67–73, specified that "Messieurs Cramer et de Tournes n'entendent s'intéresser que dans l'édition actuelle de deux mille exemplaires, et ne prétendent aucun droit de propriété perpétuelle sur les droits et cuivres dudit ouvrage."

34. Article 22 of the contract in dossier Marc-Michel Rey, Bibliotheek van de Vereeniging ter Bevordering van de Belangen des Boekhandels.

Of course the pedigree remained ambiguous, and the succession of contracts and partnerships seems bizarre and confusing today: a legalized illegal book? A privilege for a text replaced by a privilege for some plates, even though the plates appeared under a different title and the title did not include such key words as *Encyclopédie* and *Dictionnaire?* "Rights" to this baroque hybrid, half illegitimate, half fictitious being divided into tiny fractions and hawked around publishing circles, not only in France, where privileges had some meaning, but also in neighboring states, where publishers existed by infringing them? It took an eighteenth-century mind to devise such expedients, but they made sense in an eighteenth-century context. The publishers needed to protect their investment, not merely to get on with their printing. They wanted to buy and sell the rights to books as well as the books themselves, to divide the rights into shares, and to deal the shares out in partnerships, which could be taken apart and put together again according to changes in circumstances. That was how the publishing game was played—by endless *combinaisons,* as Panckoucke put it.[35]

To speculate on *combinaisons* for such high stakes required something more than money: it called for *protections,* to use another of Panckoucke's favorite expressions. Publishers needed protectors to make their rights stick, and rights without protectors often proved to be worthless. The history of the *Encyclopédie* therefore involved a great deal of lobbying and influence peddling—successful in 1752 and 1759, when the government saved the first edition; unsuccessful in 1770, when it sacrificed the second edition to the clergy; and successful again in 1776, when Panckoucke installed the *Supplément* in Paris with a privilege. From then until the end of the century, Panckoucke and his allies fought to defend their "rights" by currying favor with the government. Their defense—and the attacks on it from Yverdon, Lyons, Lausanne, Bern, and Liège—constitute a central theme in the following pages. It is worth noting that the *Encyclopédie* depended on *combinaisons* of money and power from the very beginning; that political and economic interests interwined throughout the earliest stages of its history; and that it worked its way into the social fabric of France because its backers knew how

35. On Panckoucke's notion of *combinaisons* see Chapter IX.

to weave around the contradictions that characterized the culture of the Old Regime.

To help the reader keep his bearings, it might be useful to list the editions of the *Encyclopédie* with some of the basic facts about them.

(1) The Paris folio (1751–1772) : it consisted of seventeen volumes of text issued from 1751 to 1765 (the last ten appeared simultaneously under the false imprint of Neuchâtel in 1765) and eleven volumes of plates issued from 1762 to 1772. The publishers—an association formed on October 18, 1745, by Le Breton, David, Briasson, and Durand—set the pressrun at 4,225 copies; but the number of complete sets must have been smaller, owing to spoilage and attrition among the subscribers, who did not always claim the later volumes. Although the subscription price was originally set at 280 livres, it eventually came to 980. In later years, the market price increased to as much as 1,400 livres, but that figure, quoted by the publishers of the cheaper editions, may have included the *Supplément* and *Table* and even the binding.[36]

The *Supplément,* in four folio volumes of text and one of plates, was published in Paris and Amsterdam in 1776 and 1777, followed by a two-volume *Table analytique* in 1780. The pressrun of the *Supplément* apparently came to 5,250 copies, its price to 160 livres. It had no formal connection with the original *Encyclopédie* and involved a new group of contributors and publishers.[37]

36. The *Gazette de Leyde* of Jan. 3, 1777, carried an advertisement by the publishers of the quarto edition saying the Paris folio was then selling for 1,400 livres. The same figure occurs often in the correspondence of the STN. On June 8, 1777, for example, the STN told Considérant, a bookseller in Salins, that the first edition had become extremely rare and commonly sold for 1,100 to 1,500 livres. The prospectus for the Genevan reprint, dated Feb. 1771, said that the first edition "coûte aujourd'hui jusqu'à soixante louis [that is, 1,440 livres], quand on peut la trouver, car les premiers volumes entr'autres sont d'une rareté extrême." Lough, *Essays*, p. 76. That price probably included binding but not the *Supplément* and *Table*, which had not yet appeared.

37. The pressrun of the *Supplément* was set by the original contract of April 12, 1771, but it might have been modified later. See Clément, "Pierre Rousseau et l'édition des *Suppléments* de l'*Encyclopédie*," p. 136 and Raymond F. Birn, *Pierre Rousseau and the philosophes of Bouillon* in *Studies on Voltaire and the Eighteenth Century*, ed. Theodore Besterman, XXIX (1964), 122. On the price and background of the *Table* see George B. Watts, "The *Supplément* and the *Table analytique et raisonnée* of the *Encyclopédie*," *French Review*, XXVIII (Oct. 1954), 4–19.

(2) The Geneva folio (1771–1776): it was a reprint of the first edition, at a pressrun of 2,150 copies, including the *chaperon* or extra sheets to cover spoilage. The subscription price was 840 livres, but by 1777 competition from the quarto edition had driven the market price down to 700 livres and even less.[38]

(3) The Lucca folio (1758–1776): because it followed the original edition from an early date, this reprint became bogged down in delays. From what little can be learned about its history, it seems to have had a pressrun of 3,000 copies, at least during the printing of the first volumes, and a price of about 737 livres. Although no international copyright law existed in the eighteenth century, the French publishers probably considered it a pirated work and tried to keep it out of the kingdom. In the tiny republic of Lucca, however, it was an important and legitimate enterprise, directed by an adventurous patrician named Ottaviano Diodati, with the financial backing of some wealthy notables and the political protection of Lucca's senate, to which it was dedicated.[39]

(4) The Leghorn folio (1770–1778): this was the last of the folio reprints, followed by an edition of the *Supplément* (1778–1779). It included 1,500 copies and may have cost only 574 livres, without the *Supplément*. Its publisher was Giu-

---

38. On June 8, 1777, the STN noted in a letter to Droz of Besançon that the current price of the Geneva folio had fallen to 700 livres. It got this information from letters from booksellers like Pavie of La Rochelle, who offered to sell a Geneva folio for 700 livres on Feb. 8, 1777. By this time the subscription for the cheaper quarto *Encyclopédie* had opened, and Panckoucke, who had taken a half interest in the quarto, had sold 200 of his last Geneva folios to a Parisian speculator in the book trade named Batilliot for 100,000 livres. Batilliot then offered them to retailers at the rock-bottom price of 600 livres each with three months credit for payment. See Batilliot to STN, Feb. 6, 1777, and Batilliot's printed circular of Dec. 1, 1776, in the STN papers. The price was 620 livres, including the *Supplément* and *Table*, in 1786 when Thomas Jefferson went shopping for *Encyclopédies* in Paris (see Chapter VI).

39. The only well-documented account of the Lucca *Encyclopédie* is Salvatore Bongi, "L'*Enciclopedia* in Lucca," *Archivio storico italiano*, 3d ser., XVIII (1873), 64–90, which has little to say about the commercial aspects of the enterprise. Bongi does note, however, that in launching the subscription in Nov. 1756, Diodati set the price at 2 zecchini for each volume of text and 3 zecchini for each volume of plates. As the full set contained seventeen volumes of text and eleven of plates, its price probably came to 67 zecchini. According to the conversion tables in Samuel Ricard, *Traité général de commerce* (The Hague and Amsterdam, 1781), II, 289, 293, the zecchino was worth 11 livres tournois at this time; so a set would have cost 737 livres, unbound, at the subscription price in Lucca. That figure seems low, but transport costs would have increased it appreciably north of the Alps.

seppe Aubert, a specialist in Enlightenment literature, who persuaded three wealthy bourgeois to put up the capital. More important, the enlightened archduke of Tuscany, Peter Leopold, accepted the dedication of the work, shielded it against the pope, and even provided loans and a building for the presses.[40]

(5) The Geneva and Neuchâtel quartos (1777–1779) : these were really two editions with the *Supplément* blended into the original text. Each set contained thirty-six volumes of text and three plates and cost 384 livres at the subscription price. Owing to competition from the octavo edition, the last sets sold on the open market for as little as 240 livres by 1781. The quartos were printed at a total pressrun of 8,525 copies, including *chaperon*. Because of extensive spoilage and mishaps, however, only 8,011 complete sets could be assembled and sold, according to Joseph Duplain, a Lyonnais bookseller who managed the enterprise for a consortium made up of Duplain, Panckoucke, the Société typographique de Neuchâtel, Clément Plomteux of Liège, Gabriel Regnault of Lyons, and some minor partners.[41]

(6) The Lausanne and Bern octavos (1778–1782) : although these were advertised as two editions, they really were one expanded edition based on two subscription campaigns. Their

40. On Aubert and his relations with the archduke see Ettore Levi-Malvano, ''Les éditions toscanes de l'*Encyclopédie*,'' *Revue de littérature comparée*, III (April–June 1923), 213–256 and Adriana Lay, *Un editore illuminista: Giuseppe Aubert nel carteggio con Beccaria e Verri* (Turin, 1973). Neither work, however, provides information about the price and pressrun. According to a prospectus of 1769, Aubert promised to supply the subscribers with his edition for 36 zecchini, 10 zecchini less than the price of the Lucca edition. But at that time the original publishers had issued only six of the eleven volumes of plates, so Aubert's price must have been much higher ten years later when he finished the printing. As he originally set his price at 78 percent of the price charged in Lucca, 574 livres represents a fair estimate, perhaps a little on the low side, of its level in livres tournois. The information on the pressrun comes from a circular letter by Aubert, sent to the STN in a letter from Gentil and Orr, shipping agents in Leghorn, on March 6, 1775.

41. The drop in the price of the quarto affected only some leftover sets, which the publishers divided among themselves toward the end of the enterprise. On Nov. 19, 1780, the STN informed Batilliot of Paris that it had just sixty copies left and was selling them at 240 livres each for cash or 294 livres with a year or so of credit. Three months later it sold thirty of them to Batilliot at a special price of 200 livres each, but otherwise it maintained its price at 240 livres until it cleared its stock. Panckoucke, however, sold off his surplus at a lower price, so that by March 1780 the quarto could be bought in Paris for 200 livres, according to a report that two of the partners of the STN sent to Neuchâtel from Paris on March 31, 1780.

combined pressrun came to 5,500 or 6,000 copies; they cost 225 livres at the subscription price; and they contained thirty-six volumes of text and three of plates. The allied sociétés typographiques of Lausanne and Bern produced the octavos jointly by reprinting the text of the quarto edition in reduced format. They were therefore treated as pirates by Panckoucke and his partners, who owned the rights to the text and the *Supplément.*[42]

This enumeration of facts and figures suggests a surprising conclusion: there were far more *Encyclopédies* in pre-revolutionary Europe than anyone—except eighteenth-century publishers—has ever suspected. And in addition to the six versions of Diderot's basic text, there were two quite different works that used it as a point of departure: Félice's *Encyclopédie d'Yverdon,* printed between 1770 and 1780 at 1,600 copies, and Panckoucke's *Encyclopédie méthodique,* begun in 1782 at a pressrun of approximately 5,000 copies. Some publishers probably also put together small scrap editions from the leftover sheets of the *chaperons.* So the total number of *Encyclopédies,* excluding the *Encyclopédie d'Yverdon* and the *Encyclopédie méthodique,* can be estimated as follows:[43]

42. Although the octavo publishers originally announced that their subscription would cost 195 livres, it eventually came to 225. See *Gazette de Berne,* April 8, 1780. During negotiations for a marketing agreement with the quarto publishers, they consistently said they would double their pressrun from 3,000 to 6,000. Société typographique de Lausanne to STN, Oct. 16 and Nov. 11, 1779, and Bérenger of Lausanne to STN, Nov. 23, 1779. But after the agreement was finally concluded early in 1780, one of the STN's partners reported that the increase had amounted to only 2,500 copies. Ostervald of the STN to Bosset of the STN, June 4, 1780: ''Je sais de science certaine que les gens de Lausanne et de Berne, qui ne la tiraient d'abord qu'à 3,000, la tirent présentement à 5,500, depuis l'entrée en France obtenue.''

43. These estimates involve much guesswork, especially when it comes to calculating *chaperons* and the size of French versus non-French sales, as the question marks indicate on the table. But the guesses can be supported by a great deal of quantitative and qualitative evidence from the papers of the STN, which provide the key for calculating *Encyclopédie* diffusion in general (see Chapter VI). The size of the *chaperons* varied, although printers' manuals said it was customary to include one *main de passe* (25 sheets) for every two reams or every ream (1,000 or 500 sheets). S. Boulard, *Le Manuel de l'imprimeur* (Paris, 1791), p. 72 and A.-F. Momoro, *Traité élémentaire de l'imprimerie, ou le manuel de l'imprimeur* (Paris, 1793), p. 91. Their total number seems impressive: 625 folios, 514 quartos, and 300 octavos, at a conservative estimate. (The references to the size of the Lucca, Leghorn, and Lausanne-Bern editions are in round figures, so the estimated *chaperons* have been placed within parentheses.) Most of these sheets were spoiled, but many of the unspoiled ones could have been combined to form complete sets, especially if a few of the gaps were filled by reprinting. As some publishers almost certainly put together scrap editions in this manner, and as the octavos may

	Total	*In France*	*Outside France*	*Chaperon*
Paris folio	4,225	2,000(?)	2,050	175
Geneva folio	2,150	1,000(?)	1,000	150
Lucca folio	3,000 (plus 200)	250(?)	2,750	(200)
Leghorn	1,500 (plus 100)	0(?)	1,500	(100)
Geneva-Neuchâtel quarto	8,525	7,257	754	514
Lausanne-Bern octavo	5,500 (plus 300)	1,000(?)	4,500	(300)
	24,900 (plus 600)	11,507	12,554	839 (plus 600)

All the presses of all the publishers turned out about 25,000 copies of the *Encyclopédie* before 1789. At least 11,500 of them reached readers in France, and 7,257 of the French copies were quartos. Thus the *Encyclopédie* became a best seller in the country where it originated and where it suffered most from persecution. Fortunately, the majority of the *Encyclopédies* in prerevolutionary France (about 60 percent) came from the only editions whose sales can be traced in detail. Therefore, by studying the production and diffusion of the quartos, one should be able to understand how the *Encyclopédie* penetrated the Old Regime.

---

have been printed at 6,000 instead of 5,500 copies, 25,000 represents a conservative estimate of the total number of *Encyclopédies* in existence before the French Revolution. On the question of *chaperons* and scrap editions see Robert Darnton, "True and False Editions of the *Encyclopédie*, a Bibliographical Imbroglio," forthcoming in the proceedings of the Colloque international sur l'histoire de l'imprimerie et du livre à Genève.

II

ᵞᵞᵞᵞᵞᵞᵞᵞᵞᵞᵞ

THE GENESIS OF A
SPECULATION IN PUBLISHING

The directors of the Société typographique de Neuchâtel planned to produce an *Encyclopédie* as soon as they set up business. On July 25, 1769, before they had printed a single book and when they possessed only three secondhand presses and a few dilapidated fonts of type, they sent a memorandum to the most powerful publisher in France, Charles-Joseph Panckoucke:

L'Encyclopédie, traversée en France dans son origine, encore aujourd'hui arrêtée par les mêmes obstacles, ne pourra peut-être jamais être publiée dans le royaume avec la liberté nécessaire. Le public, avide de connaître les sentiments des divers savants de l'Europe, attend avec impatience que cet ouvrage destiné à instruire les hommes soit imprimé sans aucune gêne . . . Il reste un moyen infaillible d'éviter les oppositions que l'on a lieu de craindre dans le royaume et de procurer à l'ouvrage toute la supériorité qu'il peut avoir. La Société Typographique nouvellement établie à Neuchâtel en Suisse et dirigée par un certain nombre de gens de lettres, offre de se charger de l'impression pour le compte de Messieurs les libraires de Paris. Contente pour cette fois d'un profit très modique pour l'impression, la Société, qui désire de donner quelque célébrité à son début, s'engagera à abandonner toute l'édition et à n'en faire aucune offre, ni en Angleterre, ni en Hollande, ni en Allemagne, ni en Italie—en un mot à n'en tirer que le nombre d'exemplaires convenu. On sait que le Comté de Neuchâtel est un des pays les plus libres de la Suisse, en sorte qu'il n'y aurait aucun obstacle à redouter de la part du gouvernement et du magistrat.[1]

1. STN to Panckoucke, July 25, 1769, included as a memoir in a letter from the STN to Jean-Frédéric Perregaux of July 25, 1769.

Genesis of a Speculation

The Neuchâtel Reprint Plan

The Neuchâtelois may have been obscure and inexperienced, but they had a case. Their town offered an ideal setting for the production of books that could not be printed safely in France. Though as Swiss in character as Lausanne or Geneva to the south, Neuchâtel had been a Prussian principality since 1707. Its printers therefore owed allegiance to a philosopher king, Frederick II, who left them to the lax supervision of their own local authorities and shielded them against the giant across the Jura mountains. France was capable of raiding print shops beyond its borders,[2] but the Neuchâtelois saw France more as a market than a menace. Swiss *porteurs* had backpacked forbidden books over the Juras to French readers since the sixteenth century. By 1769, censorship, the monopolistic practices of the Parisian booksellers' guild, and the state apparatus for controlling the book trade had forced the philosophes to publish their works in the dozens of *sociétés typographiques* that sprang up like mushrooms in a ring around the French borders. Having watched the publishing industry flourish throughout the Rhineland and Switzerland, the Neuchâtelois decided to found a publishing house of their own. As they announced in circular letters to booksellers everywhere in Europe, the Société typographique de Neuchâtel (STN) would produce "good" books of all kinds; and their first ventures showed a willingness to speculate on works by Voltaire, Rousseau, and even d'Holbach.

To some extent, this willingness might have resulted from the tastes of the STN's three founders, Frédéric-Samuel Ostervald, Jean-Elie Bertrand, and Samuel Fauche.[3] Ostervald was a civic leader—*banneret* or head of the local militia and a member of the governing Conseil de ville—and a man of letters, having published two learned works on geography. His son-in-law, Bertrand, was professor of belles-lettres at

2. On Dec. 11, 1764, a Parisian police inspector and a company of French troops raided three printing shops in the theoretically independent duchy of Bouillon. Raymond F. Birn, *Pierre Rousseau and the ''Philosophes'' of Bouillon* in *Studies on Voltaire and the Eighteenth Century,* ed. Theodore Besterman, XXIX (Geneva, 1964), 93.

3. A fourth founder, Jonas-Pierre Berthoud, withdrew within a year. For background on the firm and its founders see John Jeanprêtre, ''Histoire de la Société typographique de Neuchâtel 1769–1798,'' *Musée neuchâtelois* (1949), pp. 1–22 and Jacques Rychner, ''Les archives de la Société typographique de Neuchâtel,'' *Musée neuchâtelois* (1969), pp. 1–24.

the Collège de Neuchâtel and a pastor. Bertrand abandoned his ecclesiastical functions in 1769 in order to devote himself fully to the STN, where his encyclopedic knowledge proved especially useful in an encyclopedic project: an expanded, pirated edition of the multivolume *Description des arts et métiers,* which was being produced in Paris under the sponsorship of the Academy of Sciences. Fauche represented the commercial and technical aspects of the business. He had been publishing and selling books in Neuchâtel for several years before joining forces with Ostervald and Bertrand, and he had developed a specialty in prohibited books—a branch of the trade where profits and risks were greatest. In 1772 Fauche attempted to market an obscene, anticourt libel behind his partners' backs, and they retaliated by forcing him out of the company. But they had gone along with his efforts to produce an edition of d'Holbach's atheistic *Système de la nature* in 1771—a venture that proved both profitable and humiliating because it produced such a scandal that Ostervald and Bertrand were temporarily forced out of their positions in the Conseil de ville and the Compagnie des Pasteurs.

Whether or not its directors felt partial to the ideas in the books they published, the STN never specialized in the literature of the Enlightenment. It printed and traded in all kinds of books—books about travel, romance, medicine, history, and law, books like the *Voyage autour du monde* by Bougainville and *Lettres de Sophie* by Madame Riccoboni, which appealed to an educated but not especially highbrow readership. Essentially, the directors of the STN wanted to make money rather than to spread *lumières.* But they knew that there were profits in Enlightenment. Pierre Rousseau, a third-rate actor and playwright had made a fortune by popularizing the work of the philosophes—and especially the *Encyclopédie* —from the Société typographique de Bouillon. And just at the other end of the Lake of Neuchâtel, Barthélémy de Félice had put together a publishing business that was doing very well by producing his expurgated, Protestant version of the *Encyclopédie.* Fauche himself had collected 834 livres merely for lending his name to the false imprint under which volumes 8–17 of the original *Encyclopédie* appeared: "A Neufchastel, chez Samuel Faulche & Compagnie, libraires & imprimeurs." Ostervald and Bertrand were men of substance and influence who considered themselves several cuts above Rousseau, Félice, and Fauche. They were eager to speculate in *Encyclopédisme,* and their eagerness came from enlightened self-interest as well as interest in the Enlightenment when they offered to print the *Encyclopédie* for Panckoucke.

Instead of approaching Panckoucke directly, they attempted to negotiate through Jean-Frédéric Perregaux, a Neuchâtelois who was to be a founder of the Banque de France in 1800 and was beginning his career as a financier in Paris in 1769. They sent their memorandum to Perregaux with a covering letter, explaining that "nous savons que l'interdiction lancée contre la première édition de l'Encyclopédie en France n'a pu être levée par les libraires qui viennent d'en annoncer une seconde. Nous leur offrons nos presses dans le mémoire que vous trouverez ici et que nous vous prions de vouloir bien communiquer à M. Panckoucke." Not only could they do the printing safely, they added; they would also improve the quality of the text, for they were men of letters, not merely printers, and they could call upon other erudite Swiss to help them.[4]

After several weeks of soundings and *pourparlers,* Perregaux finally learned that the STN had made its bid too late. "Voici le secret de l'affaire que je n'ai découvert que hier et avec toute la peine imaginable. N'ayant pu obtenir la permission pour Paris, ils se sont accommodés de l'édition d'Hollande qu'on a commencé à y imprimer, et tous les beaux arrangements faits ici ne serviront que pour les suppléments à celle de Paris . . . Malgré la permission que les libraires intéressés ont actuellement de faire venir les volumes d'Hollande, jugez de toutes les révolutions auxquelles cet ouvrage est encore sujet d'ici à deux ans, époque pour laquelle il doit être prêt."[5] Perregaux judged correctly: the STN could be thankful that it never became involved with the folio reprint, which was produced in Geneva, not Amsterdam, in 1771–1776. As explained above, this edition had a stormy history. It provoked quarrels between Panckoucke and his partners; 6,000 copies of its first three volumes were confiscated by the French government and had to be reprinted; and the subscription campaign floundered. In June 1775, when its backers

4. STN to Perregaux, July 25, 1769.
5. Perregaux to STN, Sept. 13, 1769.

met for a preliminary settling of accounts in Geneva, a third of its 2,000 sets had not been sold; and profits looked thin, although they seemed likely to grow in the next few years.

Just when Panckoucke was liquidating the second folio enterprise in Geneva, Ostervald arrived in Paris with a proposal to join him in a new speculation on the *Encyclopédie*. Upon learning that his man was in Switzerland, Ostervald wrote home suggesting that the STN lure Panckoucke to Neuchâtel, where it could negotiate from a position of strength: "S'il se rend à votre invitation, faites-lui boire du meilleur, c'est à dire de deux niches tout à fait à gauche du fond de ma cave."[6] Panckoucke was too entangled in Geneva to make the trip, but he wrote encouragingly that he would be "charmé d'être instruit de l'affaire que vous avez en vue."[7] He had repulsed an effort by the STN to do business with him in April 1770 and this new note of affability suggested new respect for the STN, which had grown into a major publishing firm in the six years since it had first attempted to collaborate on his *Encyclopédie* speculations. Not only did the STN print a great many books of all kinds by 1775, it also did an enormous wholesale trade with booksellers everywhere in Europe, from Moscow to Naples and Dublin to Pest. And it had increased its capital, while expanding, by taking on Abram Bosset-DeLuze, one of Neuchâtel's wealthiest businessmen, as a third partner. Panckoucke's position also improved in the mid–1770s. His 6,000 *Encyclopédies*, which had been confiscated in 1770, were returned in February 1776—an indication of the new atmosphere in Versailles and of his influence within it. The accession of Louis XVI on May 10, 1774, brought a new brand of reformist ministers into power. They favored a freer trade in books as well as in wheat, and they showered favors on Panckoucke, who helped them liberalize the publishing industry. With support from Versailles, he elbowed competitors aside, pushed his way to the center of journalism as well as the book trade, subsidized entire stables of authors, and made and unmade publishing consortia on a gigantic scale. But he piled speculation on speculation so precariously that he strained his resources; and he needed a

6. Ostervald to STN, June 2, 1775.
7. Panckoucke to STN, from Geneva, June 12, 1775.

fresh infusion of capital in the summer of 1776 when he came to Neuchâtel to talk *Encyclopédies* with the STN.

This time, on its third attempt to speculate on the *Encyclopédie,* circumstances favored the STN. Not only was the government developing a more liberal policy toward the book trade, it was also being pushed in this direction by Panckoucke, who owned the rights to the *Encyclopédie.* Having liquidated the second-folio association a year ago, Panckoucke thought the market was ripe for a new edition and needed backers to finance it. The STN could negotiate the deal from its own territory, where it could demonstate the size and solidity of its business and where Ostervald could at last make use of the secret corners of his wine cellar. Thus, on the day before the American colonies declared their independence, these entrepreneurs of Enlightenment formed a partnership to produce a new *Encyclopédie.*

The contract of July 3, 1776 (see Appendix A.I) created the first in a series of alliances and alignments that shaped the history of the *Encyclopédie* for the last twenty-five years of the eighteenth century. It made Panckoucke and the STN equal partners in a reprint of a reprint. They planned to incorporate the 2,000 copies of Volumes 1–3, which Panckoucke had recently recovered from the Bastille, in a new version of the Geneva folio, so the terms of their agreement resembled those of Panckoucke's contract with the Genevans of 1770. The STN was to print volumes 4–17 at the same pressrun (2,000 and 150 sheets as *chaperon* to cover spoilage) for a fixed price of 34 livres per sheet; it would use the same quality of paper (*grand bâtard fin,* at 10 livres per ream) and of type (all from the foundry of Fournier le jeune in Paris); and Panckoucke would have the illustrations printed from the same plates. The new edition would be cheaper (720 livres for subscribers instead of 840 livres), and it would be an STN affair: the Neuchâtelois would handle accounts, publicity, and sales. Their partnership cost them 108,000 livres, which they promised to pay to Panckoucke in sixteen notes of 6,750 livres each, payable at specified intervals over four years, beginning on April 1, 1777. In this way, Panckoucke acquired some badly needed capital, and the STN became co-owner of the most important book of the Enlightenment.[8]

8. Cramer and de Tournes had paid 76,451 livres for only a third interest in the second folio edition, while the STN paid 108,000 livres for a half interest in

The association created in Neuchâtel differed from Panckoucke's earlier partnership in one crucial way: it gave the STN a permanent half interest in the *Encyclopédie* itself, not merely in one edition of it. The contract for the second folio edition, which Panckoucke had signed with Cramer and de Tournes in Geneva on June 26, 1770, specifically exempted the Genevans from anything more than a one-third interest in the edition that they were to print. The Neuchâtelois, however, contracted not merely to print a third folio edition but to acquire half of Panckoucke's holdings "dans la totalité des cuivres, droits, et privilèges du Dictionnaire encyclopédique, tant pour le présent que pour l'avenir." The agreement stipulated that the STN might reprint the *Supplément,* which Panckoucke was beginning to publish with another set of associates, and it held out the possibility that "dans quelques années" Panckoucke and the STN would have equal interests in a speculation on "une nouvelle édition corrigée dudit Dictionnaire encyclopédique dans laquelle on fondrait tous les suppléments." Thus instead of merely becoming Panckoucke's printer, as it had attempted to do in 1769, the STN became his ally. This alliance had momentous consequences, for although the plan for the third folio edition soon dissolved, Panckoucke and the STN remained united in a long-term effort to wring a profit from their common property: the text and plates that had been put together so painfully by Diderot and his collaborators.

From the Reprint to the Revised Edition

Soon after he returned to Paris in mid-July 1776, Panckoucke decided to scrap the Neuchâtel plan for a new edition

---

the proposed third edition and in the *droits et cuivres* as well. The STN probably got a better price because a great many sets of the earlier edition remained to be sold. Panckoucke, who owned them all, protected his future sales in the contract with the STN by emphasizing that the new edition had to remain *le plus profond secret* until Jan. 1, 1777, thereby giving himself enough time to dispose of the Genevan copies. The contract also committed the STN to pay Panckoucke 35,400 livres for half the value of the first three volumes of text and the first volume of plates, as well as the frontispiece and engraved portraits of d'Alembert and Diderot, which Panckoucke had had printed in Paris. The Genevans had also paid for those volumes at virtually the same price, and Panckoucke had reimbursed them in the settling of accounts of June 13, 1775. He therefore did not swindle the STN by exacting payment for the same thing twice, as it might appear, but he did well to cover the cost of the confiscated volumes by incorporating them in a new edition.

of the old text and to create instead an *Encyclopédie* that would be so completely revised as to be virtually a new book. How and why he reached this decision cannot be known with absolute certainty because many of the documents from this period are missing, but enough of them remain for one to follow the general lines of his rapidly changing course of action.

After leaving Neuchâtel in early July 1776, Panckoucke stopped in Geneva to see Samuel de Tournes, his former partner in the Geneva folio edition, who had agreed to administer the sales of the 670 sets that were left over when the partnership was disbanded in June 1775. De Tournes reported that about 300 sets were still sitting unsold in his warehouse. The slowness in the liquidation of the old reprint did not bode well for the sales of the new one; and as the new *Encyclopédie* would be 120 livres cheaper than its predecessor, Panckoucke seemed certain to ruin one speculation by rushing into another.[9] While his enthusiasm for the Neuchâtel agreement cooled, Panckoucke began to favor a grander plan, which he tried out, after his arrival in Paris, on some of his philosophe friends, notably his brother-in-law Jean-Baptiste-Antoine Suard, a prominent member of the Académie française. Having ingratiated himself with influential philosophes, won admission to salon society, and put together a handsome income from pensions and sinecures, Suard represented the Enlightenment at its most mature and most mundane—the kind of Enlightenment that was championed by d'Alembert and that found its spiritual home in the Academy.[10] Panckoucke apparently suggested that Suard enlist a group of philosophes to rework Diderot's text for a revised edition. Suard seized eagerly on the proposal, and he persuaded France's two most important academicians, d'Alembert and Condorcet, to direct the enterprise with him. The three philosophes outlined their plan in a memorandum dated July 27, 1776, which Panckoucke sent to the STN. Although this document has disappeared, its main points seem clear enough from Panckoucke's subsequent exchanges with Neuchatel:

9. Panckoucke's discussions in Geneva can only be known from some notes Bosset wrote under the title ''Observations de M. Bosset sur la refonte,'' STN papers, ms. 1233.

10. On Suard and the integration of the High Enlightenment in the upper echelons of the Old Regime see Robert Darnton, ''The High Enlightenment and the Low-Life of Literature in Prerevolutionary France,'' *Past and Present*, no. 51 (1971), 81–115.

1. The new editors would blend into the text the five-volume *Supplément,* which was then being published by Panckoucke; 2. they would correct errors and omissions and would improve the poor coordination between the text and the plates; 3. they would include a great many new articles; and 4. they would incorporate a "Dictionnaire de la langue française," which Suard was then preparing for publication. In short, Suard, d'Alembert, and Condorcet proposed to overhaul the original *Encyclopédie* from top to bottom. They planned to put together a whole team of philosophes to do the work. And they expected to be paid liberally by Panckoucke and his partners.

The origins of this proposal went back far beyond the formation of the Suard group to the beginning of Panckoucke's plans to speculate on the *Encyclopédie.* Shortly before or after his purchase of the rights to the work on December 16, 1768, Panckoucke sought Diderot's help in persuading the French authorities to permit the publication of a completely revised edition. Diderot complied by writing an extraordinary memoir about the imperfections of the work on which he had labored so hard for the last twenty years The book had been marred, he explained, by the mediocrity of its contributors—and he named them, along with the vast sections of the *Encyclopédie* that they had spoiled. Some of the contributors were incompetent. Others subcontracted their assignments to hack writers, who produced hack work. And these flawed articles made the good ones look incongruous. There was no consistency in the quality of the writing and little coordination in the allotment of the work. Thus important subjects were omitted because some contributors thought they were being treated by others, the cross-references were neglected, and the text was not carefully related to the plates. Diderot put it bluntly; the *Encyclopédie* was a mess: "*L'Encyclopédie fut un gouffre, où ces espèces de chiffoniers jèterent pêle-mêle une infinité de choses mal digérées, bonnes, mauvaises, détestables, vraies, fausses, incertaines, et toujours incohérentes et disparates.*"[11] The new publishers could pro-

---

11. Diderot, *Oeuvres complètes,* ed. J. Assézat and M. Tourneux (Paris, 1875–77), XX, 130. For the context and reception of Diderot's memoir see L. P. de Bachaumont and others, *Mémoires secrets pour servir à l'histoire de la République des Lettres en France de 1762 jusqu'à nos jours* (London, 1777–89), entry for June 29, 1772; cited hereafter as Bachaumont.

duce a much better *Encyclopédie,* Diderot continued, if they confided it to a director who would plan the rewriting with great care, holding the contributors to a strict schedule, paying a copyist to produce legible copy, coordinating the plates and the text, and choosing only the best authors, who would be well paid. Writing with the frustrations of his own directorship vividly in mind, Diderot showed how the *Encyclopédie* could be transformed into a new and vastly superior work.

The Diderot memorandum reveals the thinking that shaped Panckoucke's *Encyclopédie* enterprises from the very beginning, namely, a conviction that the original book was badly flawed and needed to be reworked into a revised edition—a *refonte* as Panckoucke called it in his correspondence. Unlike modern literary scholars, Panckoucke did not approach the *Encyclopédie* as if it were a sacred text or an untouchable classic. From the very beginning he meant to remold it into something better. Circumstances prevented him from realizing his original intention, but he held fast to his plan until the very end, when he was putting out the *Encyclopédie méthodique,* a work that was not completed until 1832, after it had run to 202 volumes and Panckoucke had been dead for thirty-three years.

If Diderot's memorandum belongs to a vision that haunted Panckoucke throughout his career, it also had an immediate and self-avowed purpose in 1768: it was intended to convince the French authorities that the original *Encyclopédie* was so riddled with faults that they should grant Panckoucke permission to publish a revised edition. Panckoucke's request was refused and Diderot's memorandum forgotten—until it was published for an entirely different purpose in 1772 and again in 1776 by Luneau de Boisjermain. Luneau was a cantankerous man of letters who had embroiled the original publishers of the *Encyclopédie* in a celebrated lawsuit. He wanted to convict them of defrauding the subscribers of the work because, he claimed, they had supplied a shoddy book at a much higher price than had been set by the subscription. Having somehow got his hands on Diderot's memorandum, he used it as evidence to support his case. Luneau lost his suit, but Panckoucke never gave up in his determination to produce a new *Encyclopédie.* When he revived his pet project in July 1776, he dredged up Diderot's memorandum once again, this time to convince the STN to accept the change in plans. "Je vous envoie le mémoire de Diderot, qui n'aurait jamais dû être publié. C'est un abus de confiance qui y a donné lieu. Luneau a supprimé tout ce qui est à l'avantage de l'*Encyclopédie,* comme de raison, mais la lecture de ce mémoire vous convaincra de la nécessité de la refonte. Nous y avions pensé il y a 8 ans [that is, in 1768], mais Diderot est aussi une mauvaise tête qui nous demandait cent mille écus et qui nous aurait désespéré."[12]

This tantalizingly brief note—one of only two letters from Panckoucke that survive from this period—shows that the Diderot memorandum was continuously used as a weapon in the process of lobbying, quarreling, and intriguing that made publishing such a rough business in the eighteenth century. The original version of it has disappeared and Panckoucke noted that Luneau cut passages from it in order to damn the *Encyclopédie* more effectively. So Diderot did not have quite so critical an attitude toward his book as Luneau claimed, but he did criticize it—and very trenchantly, too—because he entertained thoughts about editing the revised edition that Panckoucke originally wanted to publish. Panckoucke's letter indicates that he offered Diderot the editorship in 1768 and that Diderot demanded 300,000 livres for the job. Perhaps Diderot took the offer seriously enough to write the memorandum, which provided the principal argument in Panckoucke's campaign to get the government's permission for the new work. Thus, while ending his association with the original publishers, Diderot apparently began planning a new *Encyclopédie,* one which would redeem all the mistakes that made him feel so bitter about his twenty-five years of labor for Le Breton. The labor for Panckoucke would be more rewarding, though Diderot may not have seriously expected to receive as much as 300,000 livres. More important, perhaps, Panckoucke would not mutilate the copy: he was a friend of the philosophes and would leave Diderot free to realize the *Encyclopédie* of his dreams.[13]

12. Panckoucke to STN, Aug. 4, 1776. The purpose of the memoir stands out clearly in an introduction to it written by or for Panckoucke. Diderot, *Oeuvres complètes,* XX, 129–130.

13. The version of the memoir in the Assézat-Tourneux edition of Diderot's *Oeuvres complètes* was taken from a published factum or judicial brief by Luneau, which has several ellipsis dots where passages and person's names were cut. Unfortunately, the copy sent by Panckoucke has not remained in the STN's papers.

Genesis of a Speculation

Panckoucke's project gives one an intriguing glimpse of the great Encyclopedist in his old age, preparing to redo the work that had consumed his middle years; but it came to nothing because the authoritarian Maupeou ministry refused to permit such an ambitious undertaking, which might well have resulted in a more outspoken *Encyclopédie* than Le Breton's. A year later, Panckoucke came back with a plan to reprint the original text and to correct its errors and omissions by producing some supplementary volumes—the plan that eventually led to the second or Genevan folio edition and the *Supplément*. He asked Diderot to direct the *Supplément* and received the famous reply: "Allez vous faire f . . . , vous et votre ouvrage, je n'y veux plus travailler. Vous me donneriez 20,000 louis et je pourrais expédier votre besoin en un clin d'oeil, que je n'en ferais rien. Ayez pour agréable de sortir d'ici et de me laisser en repos."[14] No wonder that Panckoucke described Diderot as a *mauvaise tête* and that he did not turn to him seven years later when he revived the *refonte* project. The fact remains, however, that Diderot had helped to shape that project in the first place and should be considered as the father or grandfather of the Suard plan.

The plan took final form in a contract signed by Panckoucke and Suard on August 14, 1776 (see Appendix A. II). According to this agreement, d'Alembert and Condorcet would "preside" over the new folio *Encyclopédie* but Suard would be held responsible for its preparation. He would put together a team of distinguished writers to produce the text. The contract listed Saint Lambert, Thomas, Morellet, d'Arnaud, Marmontel, La Harpe, Petit, and Louis as likely candidates—men whose names have lost their luster today but who

---

But in his covering letter, Panckoucke did not challenge the accuracy of the passages that Luneau printed, so the version in the *Oeuvres complètes* is probably accurate, as far as it goes. Diderot scholars have correctly pointed out the polemical background of the memoir, but they have not seen its implications for Diderot's biography. Panckoucke's letter suggests that Diderot seriously considered assuming the editorship of an entirely new version of the *Encyclopédie*, not just the *Supplément*, despite the disclaimers in the memoir itself, p. 131.

14. Diderot to Sophie Volland, Aug. 31 (?) in Diderot, *Correspondance*, ed. G. Roth (Paris, 1955——), IX, 123–124. Diderot's dislike of Panckoucke is also suggested by a remark that Ostervald reported to Bosset in a letter of June 4, 1780: "Harlé [Ostervald's son-in-law, a merchant in Saint-Quentin] vous en aura peut-être parlé [that is, about Panckoucke] et vous aura dit comme à moi que Diderot l'avait assuré que c'était un homme de mauvaise foi, offrant d'en fournir la preuve."

commanded the most prestigious positions in the Republic of Letters during the 1770s. They included so many academicians that the *refonte* would have appeared as a product of the Académie française, which d'Alembert and Voltaire had packed with philosophes of their own stripe.[15]

Suard and his colleagues were to rewrite the text, incorporating new material from the *Supplément,* from certain articles of Félice's *Encyclopédie d'Yverdon,* and from other sources such as Suard's proposed Dictionnaire de la langue française. They would take special care to correct the poor coordination of the plates and the text and the cross-references, as Diderot had recommended in his memoir. And as Diderot had also suggested, they would be held to a rigid schedule, a copyist would produce a neat version of all their work, and they would be well paid. By giving Suard complete control over the rewriting, Panckoucke probably meant to correct the unevenness and incongruities that Diderot had found so objectionable. But Panckoucke required Suard to produce a steady stream of copy—at least three volumes a year—from May 1, 1777, when the first two volumes were to be in the printer's hands, until the end of 1781, when presumably the last volume would be finished. Suard would have to pay 500 livres for every week that the printshop remained idle owing to a lack of copy. By maintaining this strict production schedule, Suard would receive 5,000 livres for each volume and 20,000 livres when the work had been completed. The contract did not specify how many volumes were to make up a set in the revised edition, but Panckoucke evidently planned on about twenty volumes in text. If so, Suard would receive 120,000 livres, of which he was obligated to pay at least 40,000 to the writers working under him.

To set a whole stable of philosophes to work for four and a half years, at a cost of 120,000 livres, was a major enterprise, and Panckoucke knew that he needed intellectual and political as well as financial backing for it. So he probably attached great importance to d'Alembert's support of the project. As the ruler of the Académie française and as one of

15. All of the men named in the contract belonged to the Académie française except the two who would be least well known today: Antoine Petit was a famous doctor and member of the Académie des sciences, and Antoine Louis was the distinguished secrétaire perpétuel of the Académie de chirurgie. Both had contributed articles on medical science to the original *Encyclopédie.*

France's most prestigious philosophes, d'Alembert would attract the best talent and would make the new *Encyclopédie* appear as the legitimate successor of the old one, which he had originally edited with Diderot. Also, d'Alembert's patronage could attract that of still greater figures. On December 8, 1776, d'Alembert wrote the following letter to the STN:

Messieurs,

Quoique ma santé d'une part et de l'autre des occupations indispensables ne me permettent pas d'avoir la même part qu'autrefois à l'ouvrage important dont vous me parlez, vous pouvez être persuadés de tout l'intérêt que j'y prends et du désir que j'ai d'y concourir autant qu'il sera en moi, tant à cause de l'utilité de l'ouvrage que par les liens d'estime et d'amitié qui m'unissent depuis longtemps à M. Suard, mon digne confrère, qui conduira sûrement cette entreprise à votre satisfaction et à celle du public. Je compte aller à Berlin au mois de mai prochain, et je ferai pour vous auprès du roi de Prusse tout ce qui dépendra de mon faible crédit et des bontés dont ce prince m'honore. Vous pouvez faire et vous ferez sans doute de cet ouvrage, grâce à la liberté honnête dont vous jouissez, un des plus beaux monuments de la littérature ancienne et moderne, et je n'ai d'autre regret que de ne pouvoir pas mettre à ce bel édifice autant de pierres que je désirerais. Mais je porterai du moins un peu de mortier aux architectes, et je voudrais seulement qu'il fût meilleur et plus abondant.

J'ai l'honneur d'être avec respect,

> Messieurs,
> Votre très humble et très
> obéissant serviteur
> d'Alembert[16]

D'Alembert never made this trip to Berlin, but he promised to lobby for Frederick II's support of the new *Encyclopédie*. This point mattered a great deal to the STN because Frederick was sovereign of Neuchâtel and could protect them against interference by local or French authorities. In their original contract with Panckoucke, the Swiss printers had stipulated that Frederick's protection was to be sought. And after d'Alembert canceled his trip, they sent him a memorandum on their need for a formal statement (*rescrit*) from Frederick that they could use to ward off any attempt to interrupt the printing. It also stressed their hopes that Freder-

16. D'Alembert was answering a letter from the STN, which is missing, as is most of its correspondence concerning the *Encyclopédie* during this period.

ick would accept the dedication of the work. Ostervald and Bosset discussed these plans with d'Alembert in Paris in the spring of 1777. Their letters at that time and later remarks in their correspondence show that d'Alembert had made a serious commitment to promote the project. He evidently shared some of Diderot's feelings about the need to improve the original work, and for his part he promised to help by writing a "Histoire de l'*Encyclopédie*" for the revised edition.[17]

For Panckoucke, therefore, the revised *Encyclopédie* did not represent a casual side bet but a serious speculation on the kind of work that he hoped to produce in the first place, the kind that Diderot had recommended to him in 1768 and that looked more feasible in 1776, when the new reign of Louis XVI seemed to promise a more tolerant attitude toward publishing, when Frederick II might extend protection from abroad, and when d'Alembert, Condorcet, and Suard could be counted upon to recruit the most distinguished writers in Paris. It was for such an *Encyclopédie* that he opted in July 1776. Next he had to persuade the STN to go along with him.

Panckoucke could expect his Swiss partners to resist such a drastic change in plans. The contract of July 3, 1776, did envisage an eventual joint speculation on a revised edition, but it committed the associates to start work right away on the folio reprint. Soon after Panckoucke's departure, the

---

17. "Mémoire envoyé à Paris le 1e. juin 1777" in STN papers, dossier *Encyclopédie:* "En rendant nos très humbles actions de grâce à Monsieur d'Alembert du soin qu'il daigne prendre pour nos intérêts auprès de S. M. le Roi de Prusse, nous le supplions de nous favoriser de sa puissante recommandation, dans la vue d'obtenir de S.M. qu'il lui plaise adresser un rescrit au Conseil d'Etat de sa Principauté de Neuchâtel et Valangin, portant qu'informée qu'il s'est établi dans la capitale une imprimerie considérable sous le nom de la Société typographique, Elle la prend sous sa haute protection, pour qu'elle puisse travailler avec tout le succès possible, lui accordant non seulement la permission d'imprimer librement la nouvelle édition de l'*Encyclopédie* à laquelle on travaille, mais agréant de plus que ce grand ouvrage lui soit dédié.

Il sera convenable que ce rescrit nous soit adressé et envoyé directement, afin que nous puissions en faire usage au besoin et le produire seulement dans le cas où l'on voudrait nous gêner pour ce travail."

Ostervald and Bosset knew d'Alembert and discussed their *Encyclopédie* projects with him during a trip to Paris in the spring of 1777. No record of those discussions survives, but the STN alluded to them in a letter to Panckoucke of Feb. 8, 1778: "Rappelez-vous aussi que M. d'Alembert nous avait fait espérer au printemps dernier de nous fournir une histoire de l'*Encyclopédie*, morceau neuf et qui produit par une telle plume donnerait un merveilleux relief à notre affaire?"

STN had bought a house adjoining its workshop in order to have room to execute the enormous printing job. It began searching for new workers, presses, type, and paper supplies, for it expected at least to double its printing capacity in a few months. The prospectus accompanying its agreement with Panckoucke committed it to a tight production schedule, which it needed to maintain, in any case, in order to bring in enough capital to pay the first of its sixteen notes to Panckoucke as they became due. The papers of the STN do not reveal how Panckoucke presented his proposal, but they indicate that he sent five items to support his case: Diderot's memorandum of 1768; some critical ''Réflexions'' on the Panckoucke–STN prospectus for the proposed third folio reprint; the draft of a contract between Panckoucke and Suard for the preparation of the revised edition; a proposed amendment to the Panckoucke–STN contract, which would commit the STN to the Suard plan; and a memorandum of July 27, 1776, by d'Alembert, Condorcet, and Suard, which argued the need for rewriting instead of reprinting the original text. None of these survives, but the Neuchâtel papers contain a more revealing document: a memorandum Bosset sent to the other directors of the STN on the eve of a conference on Panckoucke's proposal, which shows how eighteenth-century publishers confronted crucial decisions.[18] Should the STN accept the Suard plan? Hundreds of thousands of livres and many years of labor would be determined by that decision, which the directors were to make the next day at two o'clock in the afternoon. Bosset considered the issue so important that he wrote down his thoughts as they occurred to him and sent his notes to Ostervald and Bertrand. Writing memos, scheduling conferences, going over the pros and cons of complex questions of finance and marketing—the directors of the STN operated like modern businessmen, although their business was Enlightenment.

First, Bosset argued, the STN should face the fact that Panckoucke was acting out of self-interest: he needed to postpone the reprint in order to have more time to market his 300 unsold sets (a matter of 210,000 livres to him). But Panc-

18. Bosset's memorandum, ''Observations de M. Bosset sur la refonte,'' goes over the material sent by Panckoucke closely enough for one to have a good idea of its contents, especially as Panckoucke's draft contracts served as the basis for the contracts that do survive in the STN's papers.

koucke's motive was irrelevant to the real issue facing the STN. Would the increased cost of the Suard project result in substantially larger profits? Bosset was inclined to think so because he found that the "Réflexions" sent by Panckoucke exposed a dangerous weakness of the reprint strategy: the market for the original edition might well be sated. It would be safe to assume that sufficient demand existed for a substantially new *Encyclopédie*—provided the price were right. But here Bosset detected a flaw in the Suard plan. It would price the revised edition out of the range of all but the wealthiest book buyers. Bosset believed that the greatest profit was to be made by tapping the demand for the *Encyclopédie* among ordinary readers: "Ce ne peut donc être que par le bas prix auquel on établira cette nouvelle édition qu'on pourra en faciliter l'écoulement en le mettant plus à la portée de chacun." The subsequent history of the *Encyclopédie* would prove that Bosset had perceived a profound truth about the literary market, but the low-price policy appealed to him for another reason. In satisfying the STN's interest as a shareholder, it would do even more for the STN's interest as a printer. The Neuchâtelois expected to print the entire revised edition and to be paid from the receipts according to their output. So they would make far more from a cheap edition of three or four thousand sets than from an expensive edition of two thousand. A large, inexpensive edition also would diminish the danger of pirating. And it even might be more advantageous to Panckoucke, Bosset believed. He argued that the eleven volumes of plates, which Panckoucke proposed to sell at 36 livres each, could be compressed into six somewhat larger volumes to be sold at 40 livres each for a total cost of 240 livres instead of 396 livres. He thought that the text could be kept to twenty volumes (the seventeen original volumes and three volumes of text from the *Supplément*) because the addition of new material would balance evenly with the deletion of errors and repetitions from the old. Each volume of text could sell for 24 livres, making 480 livres for the text, plus 240 livres for the plates or 720 livres for the entire set, the same price that Panckoucke and the STN originally had planned to charge for their folio reprint. At that price, they could sell twice as many sets as Panckoucke had projected. And three or four thousand sets of 720 livres would fetch more in profits than

2,000 at 864 livres, the price Panckoucke had proposed for
the revised edition. They could do best by aiming their edition
at the *grand public.*

Of course the profit margin would not increase if costs went
up beyond a certain point. The STN would have to force Panckoucke to reduce the 100,000 to 120,000 livres that he proposed to pay the philosophes. Bosset argued that the revising
required merely "de l'ordre et du goût," not genius, and he
particularly objected to the "prétentions excessives" of
Suard. He said nothing about Suard's being Panckoucke's
brother-in-law, but he proposed that Suard be paid for every
sheet of new prose instead of for every volume—a policy
that would prevent him from receiving payment for any material that he took over without modification from Diderot's
text. Bosset also suggested that at the end of the rewriting
Suard receive 12,000 livres and twenty free sets instead of
the 40,000 livres that Panckoucke had proposed. And finally,
he recommended that the STN demand three modifications
in Panckoucke's suggested amendments to the contract of
July 3, 1776. First, Panckoucke had wanted to broaden the
financial base of the enterprise by selling an interest of one-
third in it to other book dealers. He therefore had proposed
that he and the STN each sell one-third of their half interest
for 25,000 livres. Bosset considered the idea good and the
price bad, because a third of the price that the STN had paid
to Panckoucke was 36,000 livres: Panckoucke therefore was
asking the Swiss to take a 25 percent loss without compensation and should be opposed. Secondly, the STN had paid
Panckoucke 35,400 livres to cover the production cost of the
three volumes from the Genevan edition that he had recovered from the Bastille and that he meant to use for the proposed third folio reprint (that sum also covered one volume
of plates, the frontispiece, and the portraits of Diderot and
d'Alembert). If he abandoned the plan for the reprint, the STN
should not be expected to pay for half the loss of those volumes. Thirdly, Panckoucke had required that the revised edition be kept secret until July 1, 1777, when it would be announced publicly and its first two volumes would appear.
That would give him six additional months in which to sell
his 300 surplus sets of the Genevan reprint because his contract with the STN committed him to publish the prospectus
for the Neuchâtel reprint on January 1, 1777. Bosset consid-

ered speed extremely important. The STN had committed too much capital to sacrifice six months without receiving any return on its money. It had bought an entire house, at an inflated price, in order to expand its plant immediately, and Bosset felt it should fight the demand for the delay. Apparently sitting at home toward the end of the day, Bosset brought his argument to a close: "Voilà, Messieurs, en gros mes réflexions sur cette affaire que je soumets absolument à votre décision et à vos lumières. J'aurai l'honneur de vous voir, Messieurs, demain contre les deux heures pour en conférer ensemble . . . J'ai l'honneur, Messieurs, de vous souhaiter le bon soir."

There is no record of what happened at the two o'clock conference, but the next piece of the puzzle shows that the Neuchâtelois accepted the revision proposal. On August 31, 1776, they signed an agreement to adapt their earlier contract with Panckoucke to the Suard plan (see Appendix A. III). In this "Addition au traité avec M. Panckoucke," they consented to delay the announcement of the new edition until July 1, 1777. At that time the first two volumes of text and the first volume of plates would appear, and the rest of the work would be published at the rate specified in the contract for the reprint. The STN had to accept the loss of its share in Panckoucke's old copies of volumes 1–3 of the Geneva edition, which were to be sold as scrap paper, except for some salvageable tables and art work. It apparently also gave in to Panckoucke's position on pricing because the agreement made no provision for reducing the number of volumes in the set, and it priced each volume at the level favored by Panckoucke, that is, 24 livres for each volume of text and 36 livres for each volume of plates. (These were retail prices; booksellers were to buy the text at 20 livres per volume and the plates at 30 livres per volume.) The STN agreed to let Panckoucke sell one-sixth instead of one-third of its half interest, at a price specified in some missing letters. Panckoucke was to sell half of his own interest. The shares would be divided into twelfths, so that the ownership of the enterprise would be be apportioned as follows: STN, 5/12; Panckoucke, 3/12; other book dealers, 4/12. This arrangement would ease the strain on Panckoucke's finances and would lessen the danger of pirating by recruiting powerful backers like Marc Michel Rey, whom Panckoucke planned to see in Amsterdam in the autumn. Instead of entering into a contractual relationship with Suard, the STN merely authorized Panckoucke to negotiate with him according to guidelines that it specified in a memorandum, which is missing from the Neuchâtel papers. Since Panckoucke had already fixed the terms of Suard's operation by the contract of August 14, the STN failed in this attempt to trim the budget. Thus, as explained above, Suard was authorized to put his philosophes to work rewriting Diderot's text at 5,000 livres per volume. After the completion of the work, he was to receive an additional 20,000 livres, instead of 40,000, as Panckoucke had originally proposed.[19] Bertrand of the STN was also to receive 20,000 livres for copyreading and proofreading. And the STN was to print the entire work but only at a pressrun of 2,000. So Panckoucke conceded enough to mollify the Neuchâtelois; but he gave very little ground, and he forced his reluctant partners to accept the plan for an ambitious and expensive reworking of Diderot's text, which he had originally formulated with Diderot himself. Despite the cogency of Bosset's arguments, the STN had lost the second round of its bargaining with the most powerful publisher in France.

Joseph Duplain and His Quarto *Encyclopédie*

In the autumn of 1776 Panckoucke made a business trip to Holland and England. Upon his return he reported to the STN that he had sold 200 sets of the Geneva edition, "mais il m'en reste encore et vous sentez que je ne puis pas m'occuper sérieusement de notre affaire que ces exemplaires ne soient placés. Mais cela doit être fait incessament." He also had sold shares in the revised edition, and he named Marc Michel Rey of Amsterdam, C. Plomteux of Liège, and Gabriel Regnault of Lyons as the "autres associés." Each man evidently bought a single share of 1/12, because Panckoucke had had four 1/12 shares to sell, and he wrote that Rey had bought only 1/12 instead of 2/12 as he had hoped. Regnault

19. Although Suard had accepted the contract for the revised edition on Aug. 14, 1776, he did not accept the modifications of it imposed by the Panckoucke–STN agreement until much later. On Nov. 4, 1776, Panckoucke wrote to the STN: "Je joins ici l'acte que M. Suard a enfin signé. Il exigeait des changements. Je lui ai représenté qu'ils entraîneraient des longueurs et que nous n'avions déjà perdu que trop de temps. Je lui ai fait sentir la nécessité de l'article du chômage, la perte immense qu'une suspension, ne fût-elle que d'un mois, entraînerait etc."

had actually purchased his share in July 1776, when Panckoucke had stopped in Lyons after concluding the original contract with the STN in Neuchâtel and after seeing de Tournes in Geneva. And Panckoucke returned to the STN the remaining 1/12 share, "que vous m'avez rétrocédé avec tant de peine." Thus the enterprise had backers from important dealers in Lyons, Liège, and Amsterdam as well as in Neuchâtel and Paris.[20]

It was proceeding smoothly, though at a slower pace than anticipated, Panckoucke reported. Suard, who had tried to haggle over details in his contract, could not complete the first two volumes before August 1777, but he promised to produce the rest on schedule. Panckoucke would try to hurry Fournier, who was to supply the STN's font of type but was overburdened with orders. The first three volumes of plates should be finished by the end of 1777: it was slow work because they had to be reduced in scale and re-engraved in order to fit the more compact format that had been planned. None of these delays bothered Panckoucke, who still had at least 100 of his old *Encyclopédies* to sell, but he presented them as a blessing in disguise: "M. Rey est d' avis de ne rien annoncer d'ici à un an. Trop de précipitation peut gâter la plus excellente affaire. Le public pourrait prendre une médiocre confiance dans une entreprise où l'on met tant de diligence . . . On verra que c'est une entreprise de librairie mal conçue. On nous accusera d'avidité. En ne nous pressant pas trop au contraire, nous aurons l'agrément du public, des connaisseurs, et nous ne pourrons manquer de faire une belle et utile entreprise." Meanwhile, Panckoucke continued to feel the strain on his finances. The capital from the new associates had helped somewhat, but he had 80,000 livres outstanding in his payments for 1777, so he had to hold the STN to a strict schedule in the payments of the 108,000 livres it owed him. He had already begun to pay wages to Suard. Later letters revealed that Suard had received 1,000 livres a month from

---

20. Panckoucke to STN, Nov. 4, 1776. Regnault was one of the most important bookdealers in Lyons. He apparently bought a 1/12 share in the reprint and agreed to convert it to the *refonte*, just as the STN did. See Regnault to STN, Aug. 27, 1776. Rey sold his 1/12 back to Panckoucke some time later. During its brief period as his partner, the STN tried to develop close ties with him, but he did not want to become aligned with a rival firm that sometimes pirated his own books. See STN to Rey, Jan. 25, 1777, dossier Marc-Michel Rey, Bibliotheek van de Vereeniging ter Bevordering van de Belangen des Boekhandels.

September 1776, had rented an apartment at 300 livres a year to serve as a "bureau de travail," had hired a "commis intelligent" at 1,200 livres a year and a copyist at 800 livres, and had set to work diligently combing other encyclopedias and reference works, correcting errors in the original *Encyclopédie,* and collecting material for new articles. The enterprise was therefore well off the ground when a bookseller from Lyons called Joseph Duplain threatened to bring it crashing down.[21]

Duplain was one of the scrappiest book dealers in one of the toughest towns of the book trade. Lyons served as the main conduit for the *mauvais livres* and *livres philosophiques* produced in Geneva and Lausanne and smuggled into France to satisfy the demand for illegal literature. Lyonnais booksellers thought nothing of ordering wagonloads of works like *La Naissance du dauphin dévoilée* and *Le Système de la nature* and shepherding them through their guild hall, where their syndics were supposed to confiscate them and turn them over to the public hangman for laceration and burning. To be sure, some booksellers in Lyons—the houses of Bruysset and Périsse, for example—kept their hands clean and vouched for the purity of the trade in their city in long-winded memorandums to Versailles, where governmental agents, who knew the Lyonnais very well, refuted them point by point.[22] But more forbidden books probably passed through Lyon than through any other provincial city in France. The town had a penchant for the underground trade because not only was it a natural outlet for the Swiss and Avignonese printers, it also had led the French provincial publishers in a long, losing battle against the guild in Paris. The state had given the Parisians a stranglehold on the publishing industry in the late seventeenth century, and they never relaxed their grip in the eighteenth. Because the Parisians monopolized legal books—books with privileges—the provincials retaliated by trading in pirated books, which were cheaper in any case, thanks to the cut-rate, cutthroat capitalism of the pirates who operated in havens like Neuchâtel. These operators found

21. Panckoucke described his financial situation in a letter to the STN of Nov. 4, 1776, and Suard discussed his own operations in letters of April 18, 1777, and Jan. 11, 1779.

22. See, for example, the collection of memorandums in the Archives de la Chambre syndicale des libraires et imprimeurs de Paris, Bibliothèque nationale, ms. Fr. 21833.

plenty of allies in Lyons, where book dealers sometimes commissioned pirated editions or wholesaled large portions of them or helped with the smuggling. The Lyonnais could be adversaries, too, because they sometimes printed illegal books secretly in their own shops.

They also were tough customers. Traveling salesmen testified that it took courage and caution to beard a Lyonnais bookseller in his shop. Before confronting his clients, Jean-Elie Bertrand of the STN filled a notebook with sketches of their characters and points to be covered in the negotiations. He warned himself about "J. M. Bruysset, homme froid et habile," for example, noting that it would be wisest to steer their discussion toward three themes, which he outlined in detail and probably rehearsed before taking the plunge into Bruysset's back room. He seemed less intimidated by "les frères Périsse, gens d'esprit, se piquant de littérature," with whom he proposed to discuss six, carefully planned subjects. And he placed "Jacquenod père et fils" near the bottom of the hierarchy of booksellers in Lyons. They warranted only a quick chat: "une simple visite, traiter légèrement avec eux; le fils vaut mieux."[23]

The Lyonnais left similar impressions on Emeric David, a printer from Aix-en-Provence, who recorded his reactions in a diary during a business trip in 1787: "Vu le célèbre De Los Rios: triste mine . . . n'est guère qu'un bouquiniste . . . est, dit-on, charlatan, menteur." "Cizeron: homme agé et indolent." "Vu M. Regnault, maître homme: air assuré, volonté ferme; paraît avoir le coup d'oeil juste et les idées nettes." "Dîné au Château Périsse en table de 25 couverts. Politesse excessive et qui ne se relâche jamais. Ton cérémonieux, même entre proche parents . . . Périsse Duluc passe avec raison pour homme d'esprit." Despite such occasional sumptuosity, David concluded that a spirit of crassness and duplicity reigned in the book trade of Lyons: "Douze imprimeries—les trois quarts ne s'occupent qu'aux contrefaçons . . . Point d'imprimeur qui cherche à bien faire . . . amour de l'argent . . . brigandage."[24]

Other inside observers drew the same conclusions. One clandestine bookdealer, who wrote a full account of the under-

23. "Carnet de voyage, 1773, J. E. Bertrand," STN papers, ms. 1058.
24. Emeric David, "Mon voyage de 1787," a diary in the Bibliothèque de l'Arsénal, Paris, ms. 5947.

ground trade for the police during a spell in the Bastille, characterized the Lyonnais as specialists in the "noble métier de fripons"—that is, pirating: "Les Réguilliat, Regnault . . . de Lyon sont les pestes de la librairie de Paris, d'autant plus dangereux qu'ils sont protégés."[25] A bill collector for the STN found the Lyonnais booksellers so shifty that he could rarely get them to pay without threatening to drag them into court: "Nous avons presque usé une paire de souliers après Cellier, lequel est un vrai étourdi, barbouillon et menteur."[26] And Panckoucke not only fulminated against individual Lyonnais dealers like Jean-Marie Barret—"un homme d'une insigne mauvaise foi"—but he also pronounced an anathema against them as a group: "Si j'avais à faire choix d'un malhonnête homme, il faudrait le chercher dans la librairie de Lyon. Il n'y a ni foi ni pudeur."[27]

Joseph Duplain grew up and flourished in this milieu. His father, Benoît, and his cousin, Pierre-Joseph, were booksellers;[28] and when he took over the family business, he was known among his friends as a particularly sharp operator. One of them, a smuggler, tried to recommend himself to the STN by stressing how much his character differed from Duplain's: "Nous ne ressemblons point aux Duplain et aux Le Roy, avec [sic] lesquels, quoiqu'amis intimes depuis l'enfance, pour nous être livrés à eux de bonne foi et nous être fiés à leur parole, voudraient nous escroquer un objet de 4000 livres et plus qui nous sont dûs."[29] By this time, the STN had got to know Duplain very well from its own dealings with him, which illustrate the symbiosis between provincial booksellers and foreign publishers.

In the spring of 1773, Duplain and the STN agreed to exchange two illegal works. Duplain promised to send eighty-four copies of a twelve-volume duodecimo edition of Rous-

25. "Mémoire sur la librairie de France fait par le sieur Guy pendant qu'il était à la Bastille," Feb. 8, 1767, Bibliothèque nationale, ms. Fr. 22123.

26. Jean Schaub to STN, Jan. 10, 1775.

27. Panckoucke to STN, Nov. 6, 1779.

28. Pierre-Joseph Duplain dealt heavily in the illegal trade until a colleague denounced him in 1773 and a *lettre de cachet* forced him to flee to Switzerland. In 1777 he turned up as a *commissionnaire* and clandestine dealer in Paris, where he fought off bankruptcy by handling the most lucrative and dangerous kinds of forbidden books and manuscripts. See the P.-J. Duplain dossier in the STN papers.

29. Revol to STN, June 24, 1780. Amable and Thomas Le Roy were also Lyonnais booksellers, who worked with Duplain on the quarto *Encyclopédie*.

seau's works, which he had had printed, as a trade for the equivalent value in the STN's edition of Voltaire's *Questions sur l'Encyclopédie*. Exchanges in kind were common among wholesalers in the book trade, and the STN sent its books off punctually. But it had to wait three months before receiving Duplain's books. The Neuchâtelois interpreted the delay as an attempt by Duplain to keep them from competing with him in the market for Rousseau, while he competed with them in selling Voltaire. After they finally received the Rousseau, they asked Duplain to do them a favor, which would prove his good will and compensate them for his bad behavior. They needed to get a shipment of the prohibited *Encyclopédie d'Yverdon* to the fair at Beaucaire and requested Duplain to clear it through the guild in Lyons. He complied, thereby acquiring a debt of his own to collect at a future date from the STN. In the autumn of 1773, the Neuchâtelois learned that Duplain was producing a pirated edition of the *Dictionnaire des arts et métiers* (five volumes, octavo), which they also had begun to print. They suppressed their counterfeit edition in favor of his—and then came back at him with another request. They needed help in getting the release of three crates of forbidden books that had been seized in the Lyons guild. Duplain did so and forwarded them on to the STN's customer, Gaude of Nîmes. A few months later he agreed to clear another shipment of the *Encyclopédie d'Yverdon* through the guild: "L'*Encyclopédie* ne passe plus ici. Notre Chambre syndicale a reçu à cet égard des ordres très précis, mais comme je n'ai point oublié le service que vous m'avez rendu, adressez-moi ceux que vous voulez faire passer et ils passeront."[30]

Next it was Duplain's turn to request a service. He had printed a pirated edition of *Les lois ecclésiastiques*. The widow Dessaint, a powerful Parisian bookseller who owned the privilege for the book, had managed to get a shipment of Duplain's edition seized and was prosecuting him for piracy. In order to save himself from a heavy fine and perhaps disbarment, Duplain asked the STN to send a fake petition to the lieutenant general of police in Paris. It should state that the STN had bought the shipment from a publisher outside

30. Duplain to STN, Nov. 3, 1775. The other information in this paragraph has been culled from the eighty-four letters in Duplain's dossier in the STN papers.

France and had sent it to Duplain, who had discovered that it lacked some sheets. The STN should explain that it had persuaded Duplain—after a great deal of pleading—to have the sheets printed locally, thereby saving the value of the book without becoming entangled in negotiations with the real pirate publisher. The widow Dessaint had learned of this small repair job and had accused Duplain of publishing the entire work—a calumny that could cause a disastrous miscarriage of justice. Therefore (the STN should say), the French authorities ought to have the shipment sent back to Neuchâtel and ought to clear Duplain "d'une accusation et d'un procès où il ne doit pas entrer." The STN had no desire to make a false confession to the Paris police, but it appreciated the value of an ally in the guild of Lyons, so it sent the petition to Duplain: "Vous trouverez sous ce pli la requête que vous désirez. Nous souhaitons qu'elle fasse l'effet que vous en attendez, le tout sans notre préjudice, et serons charmés d'avoir souvent occasion de vous prouver notre dévouement."[31] Such was the way relations evolved between Lyonnais dealers and Swiss publishers—a matter of accumulating obligations while driving hard bargains, of steering between extremes of competition and cooperation, and of holding mutual mistrust well enough in check to inflict damage on common enemies in Paris.

In December 1776 this man, who epitomized the Lyonnais style of book dealing at its toughest, issued a prospectus for a cheap reprint of the *Encyclopédie* in quarto format. Duplain had no right to do such a thing; the "rights" to the book were owned by Panckoucke and his associates, and even they did not dare to print it on French soil. So Duplain was taking a gamble. He was announcing the publication of an enormous, illegal work before he had any assurance that he could bring it into being or get it into France. But prospectuses were cheap: Duplain merely announced his terms in a handbill, which he mailed to his clients and contacts. He evidently wanted to see what the response would be before spending money on type and paper. Since he proposed the work by subscription, he could apply the subscribers' down payments to this expensive initial investment. And while sounding the market, he could keep himself hidden. For he

31. Duplain to STN, Nov. 3, 1775, and STN to Duplain, Nov. 9, 1775.

issued the prospectus under the name of Jean Léonard Pellet, a Genevan publisher, who agreed to serve as straw man for 3,000 livres.

This process of sounding—or "taking the pulse of the public," as it was known among eighteenth-century publishers—also involved printing *annonces* and *avis* or advertising notices in certain journals. On January 3, 1777, the *Gazette de Leyde* ran an *avis* about Duplain's speculation that is the richest source of information about its original character. The notice showed that Duplain meant to follow the same strategy as Bosset had recommended, that is, to tap a wide market by producing a relatively cheap *Encyclopédie*. It lamented the fact that the supreme work of the century—a book that was a library in itself—had been priced beyond the range of the persons who could most profit from it. The *nouveaux éditeurs*—whom it did not mention by name—therefore were offering it at a spectacular reduction: for 344 livres instead of 1,400 livres, its current selling price. They could slash the price so drastically, they explained, by producing only three volumes of plates—no great loss, because most of the plates in the original eleven folio volumes had little utility. The new edition would contain re-engraved versions of the truly important plates, and any reader who wanted to collect illustrations of trades could buy the inexpensive *Cahiers des arts et métiers* sponsored by the Académie des sciences. The text of the new edition, however, would be far superior to the old. Printed (appropriately) in a type called *philosophie* and on handsome paper, it would incorporate the *Supplément,* would correct the numerous errors of the folio editions, and would contain some new material, which the notice described vaguely as *"quelques morceaux que leur rareté ou leur utilité rendent précieux."* Subscribers should make a down payment of 12 livres and should send 10 livres after receiving each volume of text and 18 livres after each volume of plates (the last volume of plates would cost only 6 livres). They would receive twenty-nine volumes of text and three volumes of plates at a rate of six to eight volumes a year. The publishers would limit their printing to the subscriptions they received; so it would be impossible to take advantage of this bargain after the subscription closed. Anyone interested should rush his down payment to a book dealer in Geneva called Téron, who apparently was serving as a marketing agent for Duplain.

To announce a cheap *Encyclopédie* while Panckoucke was producing an expensive one was like holding a pistol to his head, but Panckoucke was not the sort who would wait for the enemy to fire first. He counterattacked with another *Encyclopédie* project, which he and the STN formulated in a contract dated January 3, 1777 (see Appendix A. IV–V). First, they acknowledged that the announcement of Duplain's *prétendue nouvelle édition* made it necessary for them to reduce the printing of their revised edition from 2,000 to 1,000 sets. But they would retaliate with a counterquarto, which would force Duplain's off the market because it would be a cheap version of the revised text printed at a large pressrun of 3,150. It would have only three or four volumes of plates and thirty-six to forty volumes of text, which would cost 12 livres apiece. It would therefore be somewhat larger and more expensive than Duplain's edition, but not so expensive as to be in another price range. Potential buyers could be expected to shy away from Duplain's edition if they knew that a distinguished group of philosophes were preparing a superior work. And Panckoucke would make sure that the buyers stayed away from the rival quarto by enforcing his *droits et privilèges*.

Privilege protection was a crucial element in Panckoucke's strategy. He owned the exclusive right to reproduce the book—a right so valid, according to the standards of the day, that he could divide it into portions and sell it at a very high price throughout western Europe. It would be vain to translate a legal fiction into a palpable asset, if legality could not be enforced. Therefore the state had to be made to hunt down Duplain's quartos as rigorously as if they were contraband salt. A few exemplary confiscations, and even the publication of a fierce interdiction, would make many of Duplain's subscribers desert to the Panckoucke group. So Panckoucke got Le Camus de Néville, the Directeur de la librairie, to send a circular letter to the various booksellers' guilds and the inspecteurs de la librairie, warning that Duplain's quarto was an illegal, pirated edition and that all copies of it would be confiscated by the authorities. Thus Panckoucke and his associates struck back in two ways: they tried to woo away Duplain's subscribers and potential subscribers by offering their own, superior quarto *Encyclopédie,* and they tried to crush Duplain's quarto with the power of the French state. But this counteroffensive raises a question important enough to merit a digression: how was it that the French government, which had nearly destroyed the first two editions of this book, could serve as the main line of defense in an effort to save the third?

Publishing, Politics, and Panckoucke

This paradox seems less puzzling if one considers the differences between the government that locked Panckoucke's 6,000 *Enyclopédies* in the Bastille in 1770 and the government that released them in 1776. The political situation had become more and more oppressive during the last years of Louis XV's reign. From the costly and humiliating experience of the Seven Years' War and the dissolution of the Jesuits in 1764 to the Brittany Affair and the parliamentary crisis of 1771–1774, the government had aroused increasing opposition, which it put down with increasing authoritarianism. It was especially severe in policing the book trade, as Panckoucke learned at considerable expense. But the accession of Louis XVI in May 1774 brought an end to the tough "triumvirate" ministry of Maupeou, Terray, and d'Aiguillon. Turgot, a contributor to the *Encyclopédie* and a friend of the philosophes, set the tone of the new reign. Even after his fall in June 1776, the government remained intermittently reformist, and it was especially liberal in its policy on publishing. On August 30, 1777, it issued several edicts that were intended to tighten measures against pirating and to loosen the monopoly on privileges held by the members of the Parisian booksellers' guild (Communauté des libraires et des imprimeurs de Paris).

The guild looked like a vestige of Louisquatorzean statecraft in 1777. The state had used it in the second half of the seventeenth century to gain control of the printed word. Colbert had eliminated a great many provincial presses, had concentrated publishing in Paris under the authority of the guild, and had enlisted the guild's help in suppressing all nonprivileged books. With the censorship, the royal bureaucracy, and the police reinforcing their economic monopoly, the guild members had taken over most book privileges, forcing their provincial rivals into the arms of publishers like the STN, which specialized in pirated and prohibited works. By the accession of Louis XVI, this policy had proved to be counterproductive. It had produced a boom in the illegal book trade, while giving the Parisian patriciate a monopoly of orthodox literature. The reformers of Louis XVI wanted to liberalize the orthodoxy and to create a limited free trade in books. Their legislation provided that instead of conferring an unlimited and perpetual right to a work, a privilege should normally expire after ten years or the death of its author. Authors themselves and their heirs could hold privileges indefinitely (the power of the guild had made it almost impossible for them to possess the privileges for their own works), and provincial printers could produce any book that fell into the public domain—that is, the great bulk of the literature that had been reserved for the Parisians. The August edicts conceded that the Parisian monopoly had forced the provincials into piracy, and so it allowed them to liquidate their current stock of counterfeit books but set up a system of severe penalties and policing to prevent any further trade in pirated as well as prohibited books.[32]

Inspired as they were by a desire to instill a modern, entrepreneurial spirit to an industry that had languished under an archaic, Colbertist organization, these edicts produced consternation in the Parisian oligarchy. The guild responded with petitions, protests, pamphlets, lawsuits, and a kind of informal strike, which created chaos in the publishing industry until the Revolution resolved the dispute by destroying privilege and corporatism altogether. Throughout the protesting, the most powerful member of the Parisian guild was conspicuous by his absence. In December 1777, the Paris agent of the STN reported: "Les libraires d'ici font feu et flamme contre les nouveaux règlements. 100 d'entr'eux s'assembleront ici à quelques jours à la chambre syndicale et don-

---

32. See the text of the edicts of Aug. 30, 1777, in *Recueil des anciennes lois françaises*, ed. F. A. Isambert, Decrusy, and A. H. Taillandier (Paris, 1822–33), XXV, 108–128 and, for a general discussion of guild versus entrepreneurial publishing, Robert Darnton, "Reading, Writing, and Publishing in Eighteenth-Century France: A Case Study in the Sociology of Literature," *Daedalus* (winter, 1971), pp. 214–256.

neront une requête au Garde des Sceaux. Si elle n'a pas le succès désiré, ils s'adresseront au Roi. Il s'agit principalement de la conservation des privilèges . . . Panckoucke . . . n'était point à l'assemblée des libraires, qui l'accusent d'être l'auteur de tous ces règlements.''[33] Panckoucke did not discuss his role in the reforms openly in his own letters, but he did not hide his bad relations with the other members of the guild and his support of the new legislation. "On parle beaucoup d'un nouveau règlement, mais j'ignore encore quand il paraîtra,'' he wrote on July 4, 1777. "La librairie a besoin d'une réforme. Les abus ont produit les excès, qui à son tour ont fait tout le mal dont nous sommes témoins.'' On November 19, 1777, he wrote, "Les arrêts font ici beaucoup de sensation. Il y a des représentations de toutes parts. Les gens de lettres et les libraires paraissent avoir mis leur raison sous leurs pieds. Il est impossible de plus mal voir et de plus mal raisonner.''[34] In 1791, he tried to prove his *civisme* by emphasizing his opposition to "les vautours de la librairie, les despotes des chambres syndicales'' before 1789. He claimed that he had fought against the guild's esprit de corps by campaigning for the reforms of 1777.[35] He must have campaigned discreetly, however, because all he could produce by way of evidence for his prerevolutionary progressivism was a memoir, apparently written for the Direction de la librairie, which argued the government's case against a memoir submitted by the guild. But Panckoucke could not be expected to break openly with the guild in 1777; and he may well have been a key figure in the reform of the book trade, as contemporaries believed, even if he remained behind the scenes advising and lobbying in the style that he preferred.[36]

One reason for Panckoucke's alienation from the other

33. Perregaux to STN, Dec. 17, 1777.

34. Indicating his alignment with the provincial booksellers against the Parisians, he added, "Les Rouennais ont fait des remerciements par députations. Il serait à désirer que Lyon et les autres grandes villes en fissent d'autant. Quoiqu'il en arrive, ces arrêts ne font rien à notre affaire [that is, the *Encyclopédie*].''

35. *Lettre de M. Panckoucke à Messieurs le président et électeurs de 1791* (Paris, Sept. 9, 1791), pp. 23, 14.

36. Panckoucke described his role as a reformer and his memoir in some "Observations de M. Panckoucke,'' which he published in the *Mercure* of Nov. 21, 1789. He printed the memoir itself, or at least part of it—a low-keyed, reasonable argument for limiting book privileges—in the *Encyclopédie méthodique, Jurisprudence,* VI, 813–817.

members of the guild was that he did not do business as they did. Except for a few adventurers, they tended to be conservative, to milk privileges for safe books—classics, legal treatises, religious works, and the like—which brought them a relatively secure, regular, and restricted income. He speculated extravagantly on new books and enormous compilations: the thirty-volume *Grand vocabulaire français,* the twenty-three volume *Abrégé de l'histoire générale des voyages,* and the eighty-six-volume *Répertoire universel et raisonné de jurisprudence.* In 1763 when he set himself up in business in Paris by acquiring the stock in trade of the bookseller Michel Lambert, Panckoucke assumed responsibility for selling the massive works being turned out by the Imprimerie royale: Buffon's *Histoire naturelle,* which eventually ran to thirty-six volumes in quarto, the forty-one-volume *Mémoires de l'Académie des inscriptions et belles-lettres,* and the *Mémoires de l'Académie royale des sciences,* which had been printed since 1699 and reached volume 188 in 1793.

Panckoucke did not finance these elephantine enterprises single-handed. He set up consortia, sold shares, and spun together credits and debits in such complex combinations that it is impossible to form a clear idea of the extent of his wealth. He clearly made enough from his books to pay Anisson Duperron, the Directeur de l'Imprimerie royale, 70,000 to 80,000 livres a year, and he speculated still more heavily in journalism. At various stages of his career, he invested in sixteen periodicals. He merged nine journals in the *Mercure,* and swallowed up others at such a rate that he can be considered the first press baron in French history. In the 1770s he followed a general policy of shifting from the book trade to journalism, as he explained to the STN: "Je viens d'avoir le brevet du *Mercure* pour 25 ans avec des advantages que n'avait pas mon prédécesseur. J'y ai remis le *Journal de politique* et les souscriptions de cinq journaux que je supprime. Cette opération me porte à réaliser le plan que j'ai toujours eu de vendre mon fonds, hors *L'Histoire naturelle.*"[37] In 1776 he even considered establishing a residence in Neuchâtel, so that he could speculate on books from a safe

37. Panckoucke to STN, July 7, 1778. In a letter to the STN of Nov. 4, 1776, Panckoucke explained, "Je suis chaque année avec M. Duperron, Directeur de l'Imprimerie royale, pour 70 à 80 mille livres . . . Je vous prie de faire attention que je serais en avance de plus de 80 mille livres à la fin de 1777."

base outside the kingdom during the summer months and manage his journalistic empire from Paris during the winter.[38]

Contemporaries saw this empire building as an attempt to monopolize the entire book trade. After Panckoucke bought the nation's oldest journal, *La Gazette de France,* in 1786, the *Mémoires secrets* commented, "L'avidité du sieur Panckoucke est insatiable: à lui seul, s'il pouvait, il envahirait toute la librairie."[39] Ten years earlier the STN asked him if it were true as rumored that he had offered to pay the crown 8 million livres a year in order to completely take over the printing industry. Panckoucke attributed the rumor to resentment over his role in the reform of the book trade: "L'offre de 8 millions pour être seul imprimeur n'a pas le sens commun. Le bruit en a aussi couru [ici], et mille autres

38. Panckoucke outlined this plan, which would have involved d'Alembert's intervention to get the blessing of Frederick II, in a draft of a letter to the STN dated Dec. 25, 1776, in the Bibliothèque publique et universitaire de Genève, ms. suppl. 148: "Mandez-moi, Messieurs, si un français catholique romain peut acheter chez vous, dans le Comté de Neuchâtel et de Valangin, des terres, des biens-fonds. Faites-moi aussi savoir si le roi de Prusse peut lui donner des places, quelles sont celles qui sont à sa nomination, quelles sont celles qui s'achetent, celles qui se donnent, s'il ne faut point être protestant pour les occuper, s'il y en a actuellement quelques unes de vacantes, celles qui exigent résidence, s'il ne suffirait pas d'y être quelques mois de l'année, etc. Je ne vous cache pas, Messieurs, que je serais assez d'avis d'aller m'établir avec ma femme et une fille six mois de l'année auprès de vous, c'est à dire la belle saison, et de revenir passer l'hiver à Paris. Comme M. d'Alembert compte faire un voyage en Prusse au mois de mai, nous profiterons de ce moment pour notre dédicace [that is, of the *Encyclopédie*], et j'aurais envie d'en profiter moi-même pour me donner dans votre ville un état qui pourrait donner plus de consistance à nos opérations. J'ai d'ailleurs ici le projet de vendre tout mon fonds hors *L'Histoire naturelle* et mon journal. On fait même actuellement mon inventaire, et cette vente peut être faite dans un mois, de sorte que si elle avait lieu, je n'hésiterais point à acheter la petite maison de M. Bosset De Luze, avec quelques dépendances. Mais comme nous dépendons dans le monde de l'opinion et qu'en vendant ici les trois quarts de mon fonds, je ne voudrais point avoir l'air d'aller établir auprès de vous une librairie nouvelle, je voudrais une place honnête, qui motivât cet arrangement. Nous avons eu des libraires dans ce pays-ci qui ont formé des établissements chez plusieurs souverains, mais tous ont eu des places, qui ont autorisé leurs démarches, et je voudrais être dans le même cas." Panckoucke excluded everything except the first two sentences of this part of the letter in the final draft sent to the STN on the next day, but later allusions in letters from the STN to him show that he had discussed the possibility of taking residence in Neuchâtel with Ostervald and Bosset.

39. Bachaumont, entry for Dec. 2, 1786. And in its entry for July 6, 1778, the *Mémoires secrets* remarked, "Le sieur Panckoucke, en vertu du brevet qui lui accorde l'entreprise du *Mercure*, élève les plus grandes prétentions. Il ne se contente pas d'avoir déjà englobé le *Journal français,* celui *Des dames,* celui *De politique et de littérature;* il voudrait que les autres devinssent encore au moins tributaires du sien à cause de sa primatie."

calomnies. Comme on croit que j'ai bonne part à ces arrêts, les libraires sont irrités contre moi. Cela se calmera."[40]

Panckoucke's peculiar position as the "Atlas de la librairie"[41] made him a natural ally of the government against the guild. Whether or not he collaborated on the edicts of 1777, he represented the entrepreneurial spirit that they attempted to instill in the book trade. Of course Duplain was also an entrepreneur, but he operated as a pirate, outside the law. Panckoucke, the bookseller of the Imprimerie royale, speculated from the center of the system—a system whose enlightened reformism and liberal trade policy harmonized perfectly with his own interests and attitudes. But Panckoucke's speculation on the *Encyclopédie* seemed to contradict his general principles and policies. He favored reforms to restrict privileges and to open up the market while using his privilege for the *Encyclopédie* to close the market to Duplain. There was no contradiction in Panckoucke's own interests, however, because the ten-year limit on privileges decreed by the reforms of 1777 did not threaten his stake in the *Encyclopédie*. The slack sales of the Geneva edition indicated that there would be little future demand for the book in its original form, especially if the Neuchâtel reprint were to be executed. But Panckoucke expected the revised edition to do very well, and the edicts of 1777 stipulated that any book whose text had been increased by at least one quarter should be exempted from the expiration of its privilege. Moreover, the new legislation provided stronger protection against pirating, which was the greatest danger facing the revised edition. So the revised edition would fare better under the new laws than under the old. Panckoucke's general plan to shift his investments from books to periodicals would protect him from the expiration of his other privileges, in case any of them were endangered, because the edicts of 1777 did not affect privileges for journals. And above all, his support of the government in its confrontation with the guild put him in a position to defend all of his interests by pulling strings in Versailles.

This last consideration was probably the most important because publishing in the Old Regime had none of the gen-

---

40. STN to Panckoucke, Dec. 18, 1777, and Panckoucke to STN, Dec. 22, 1777.
41. Bachaumont, entry for Dec. 5, 1781.

tlemanly veneer it later developed and Old Regime politics took the form of court intrigue, unrestrained by popular participation. Administration involved the exploitation of office unencumbered by a civil service tradition. And officeholders expected a yield on their investment without any modern compunctions about graft and bribery. Conflicts of interest therefore were resolved by influence peddling or "protection" as it was known in the eighteenth century.[42]

Panckoucke's protectors included the most powerful men in Versailles. His letters often alluded to his influence in the highest quarters, especially among men like Jean-Charles-Pierre Lenoir, the lieutenant-général de police of Paris, who could confiscate counterfeit copies of Panckoucke's books; Le Camus de Néville, the Directeur de la librairie, who could look after Panckoucke's interests in the bureaucracy in charge of the book trade; and the comte de Vergennes, the foreign minister, who could open France's borders to books Panckoucke wanted to import and close them to his rivals. Panckoucke had such influence in the government that Linguet accused him of tyranny and, to prove the point, published a letter to him from Vergennes, which was "écrite avec tant de cordialité, d'affection, de politesse et de considération, qu'elle sort absolument du protocole ordinaire," according to the *Mémoires secrets*.[43] So deeply did Panckoucke ingratiate himself in Versailles that contemporaries considered him as a kind of ex officio minister of culture: "Sa voiture le portait chez les ministres à Versailles, où il était reçu comme un fonctionnaire ayant portefeuille."[44]

Panckoucke naturally used his protections to defend his interests. When the most outspoken edition of Raynal's *Histoire philosophique et politique des établissements et du commerce des Européens dans les deux Indes* (Geneva, 1780) was prohibited in France, Panckoucke called on Vergennes and Maurepas: soon afterwards his agents were selling it in the

42. A study of politics and influence peddling in the court of Louis XV and Louis XVI remains to be written, but a great deal about this strangely neglected subject may be learned from Michel Antoine, *Le conseil du roi sous le règne de Louis XV* (Geneva, 1970) and J. F. Bosher, *French Finances 1770–1795: From Business to Bureaucracy* (Cambridge, Eng., 1970).

43. Bachaumont, entry for Sept. 17, 1776. The *Mémoires secrets* found the letter so familiar as to be of dubious authenticity, but noted, "On explique cela en disant que M. de Vergennes l'a écrite lui-même, d'abondance de coeur."

44. D.-J. Garat, *Mémoires historiques sur la vie de M. Suard, sur ses écrits, et sur le XVIIIe siècle* (Paris, 1820), I, 274.

Palais royal while the police looked the other way. He alone succeeded in getting permission to market the quarto Genevan edition of Voltaire's works in 1776. The ministers of Louis XVI not only returned his confiscated volumes of the Genevan *Encyclopédie* but also gave him permission to import huge shipments of it directly to his warehouses in Paris, by-passing the customs and the inspectors of the guild.[45] His influence in Versailles was so notorious that booksellers trembled before him. J. M. Barret, one of the canniest dealers in Lyons, warned the STN not to attempt to smuggle a pirated edition of Buffon's *Histoire naturelle* into France because Panckoucke owned the privilege for it: "Vous n'ignorez pas que M. Panckoucke, furieusement jaloux de cet article, obtiendra facilement des ministres, avec qui il est bien, les ordres les plus sévères pour en arrêter le cours; et le libraire de France qui serait surpris, serait écrasé."[46]

Panckoucke deployed his protections to greatest effect in defending his journalistic empire. The STN published a small literary journal and tried for years, using all manner of machinations and bribes, to get it permitted in France. Nothing worked: Panckoucke would not allow the slightest incursion into the market of his *Mercure*.[47] In 1779, Panckoucke claimed that the *Journal de littérature, des sciences et des*

45. In his *Lettre* of Sept. 9, 1791, Panckoucke boasted about his success in circulating the works of Voltaire, Rousseau, and Raynal: "Je sus si bien manier les ministres du roi que je les ai fait librement circuler dans le royaume" (p. 9; see also p. 16 on the sales of Raynal's *Histoire philosophique* and the similar remarks in Bachaumont, entry for Feb. 16, 1776). Of course Panckoucke did not pull strings merely to spread Enlightenment. A document in the Bodleian Library, Oxford, ms. Fr. c.31, a contract between Panckoucke and Stoupe dated May 7, 1781, shows that he bought a controlling interest in a Genevan edition of the *Histoire philosophique* for about 250,000 livres. Panckoucke mentioned his special permission to import the Geneva *Encyclopédie* in a letter to the STN of Aug. 5, 1777.

46. Barret to STN, Oct. 24, 1779. Four years later the STN made the same proposition to Amable Le Roy of Lyons and received the same reply (Le Roy to STN, Dec. 17, 1783): "Je n'hésiterais pas de m'y intéresser pour mon industrie, si elles [vos spéculations] n'étaient pas dirigées contre M. Panckoucke, qui est le favori de tous les ministres. Il a un privilège authentique sur cet ouvrage, et je crois qu'il écraserait de son crédit un libraire national qui tremperait dans votre projet."

47. The STN's campaign occurred for the most part after it had finished its speculation on the *Encyclopédie* with Panckoucke. Several of its agents reported that he blocked every attempt to negotiate a permission. See Thiriot to STN, May 5, 1781 ("Panckoucke jette feu et flammes, rien n'avance chez le Garde des Sceaux") and similar remarks in Le Senne to STN, undated letter, evidently from May 1780.

arts had trespassed on his territory when it published some political news disguised as letters to the editor. The government ruled that the journal had violated his privilege and would have to pay him a prohibitively expensive indemnity if it continued to discuss subjects that were reserved for the *Mercure*. The *Journal de Paris* became involved in a similar quarrel with the *Mercure* in 1786 and was likely to lose, the *Mémoires secrets* commented, because Panckoucke "a distribué environ mille louis dans les bureaux des Affaires étrangères, du Ministre de Paris et de la Police."[48] Panckoucke did force the *Journal de Paris* to discontinue printing for a while in 1777 and also got the *Journal encyclopédique* suspended in 1773 for printing some remark that displeased a minister. The real reason for this severity, according to the *Mémoires secrets,* was that the *Journal encyclopédique* controlled a market that Panckoucke wanted to take over with his *Journal historique et politique; the Journal encyclopédique* had no legal status in France, because it was published in Bouillon. It only saved itself by paying a ransom of 51,500 livres to Panckoucke. Panckoucke himself had to make regular payments to various ministries in order to maintain his monopolies. In January 1777 he found that he could not produce 22,000 livres that were due to the foreign ministry, and three months later he was 340,000 livres in the red. This time he used his protections to save himself from bankruptcy. Amelot, minister for the département de Paris and the maison du roi, permitted him to suspend payments temporarily, and soon afterwards Panckoucke reestablished his finances well enough to reassert his special relations with Versailles.

Lobbying was therefore essential to publishing as Panc-

48. Bachaumont, entry for Aug. 31, 1786. On the other incidents mentioned here see Bachaumont, entries for Nov. 5, 1786; Nov. 13, 1786; July 2, 1773; and the additional remarks printed as an appendix in Bachaumont, vol. 27, pp. 278–279 without an entry date. Bachaumont is more reliable for information about public opinion than events, but most historians of eighteenth-century journalism have had to rely on it for lack of a better source, except the journals themselves. For general background on the subject see Eugène Hatin, *Histoire politique et littéraire de la presse en France,* 8 vols. (Paris, 1859–61) and Hatin, *Bibliographie historique et critique de la presse périodique française* (Paris, 1866), which have not been superseded by the more recent work of Claude Bellanger, Jacques Godechot, Pierre Guiral, and Fernand Terrou, *Histoire générale de la presse française* (Paris, 1969), vol. I. On Panckoucke's financial crisis of 1777 see his *Lettre* of Sept. 9, 1791, pp. 11, 29.

koucke practiced it, and he was a notoriously tough practitioner. He turned the apparatus of the state against his competitors. But he did not call on the government lightly. He refused the STN's constant requests for help in its own, comparatively insignificant attempts at lobbying by explaining that he hoarded his influence in order to use it in moments of supreme importance: "Je ne puis pas encore vous rendre service auprès de M. de Néville, qui protégera notre grande affaire. Je ne dois pas l'importuner de petites demandes . . . Je vous servirai mieux dans les choses importantes quand je conserverai auprès des magistrats une bonne réputation que les criailleries de mes confrères ne pourront entamer."[49] "Notre grande affaire" meant the *Encyclopédie*. Panckoucke planned to defend his privilege for the book by invoking his protections.

The confrontation between the revised and the quarto *Encyclopédie* was therefore as complex as any conflict of interest in the Old Regime. It can not be interpreted simply as a contest between privilege and enterprise because Panckoucke was a privileged entrepreneur who fought off rivals by enlisting the government on his side but sided with the underprivileged in the government's attempt to open up the publishing industry. That the state should defend a book it had prohibited eighteen years earlier may seem paradoxical, but no less paradoxical than the fact that it based its defense on a principle—privilege—which it called into question by its reforms and which the *Encyclopédie* itself undermined. The Old Regime was shot full of such contradictions, especially during its last years, when reformers attempted to remodel elements of the system without changing its structure. But one consistent motive ran through all the twists and turns of Panckoucke's *Encyclopédie* policy: self-interest. Whatever his personal values and his friendships with the philosophes, he kept to an old-fashioned strategy of greasing palms and twisting arms. He was even ready to embrace his enemy if it would increase his profit margin.

49. Panckoucke to STN, May 5, 1777. The STN had asked Panckoucke's help in getting the release of some confiscated copies of its pirated edition of the *Description des arts et métiers*. Panckoucke refused on the grounds that "ce serait me compromettre. On a permis l'entrée directe dans mes magasins de plusieurs balles *Encyclopédie*. Mes injustes confrères, sachant les liaisons que j'ai avec votre maison, ont soupçonné qu'elles pouvaient contenir de vos *Arts*."

From the Revised Edition to the Quarto

It was therefore perfectly natural for Panckoucke to get the Directeur de la librairie to strike down Duplain's quarto *Encyclopédie* by decree. But soon after setting this counterattack in motion, Panckoucke began to consider a greater temptation: was there not more to be gained by joining Duplain than by beating him? This issue arose because of another peculiarity of eighteenth-century publishing: industrial espionage.

On December 26, 1776, Panckoucke sent to the STN a secret report on Duplain's enterprise, which he had received from Gabriel Regnault, their associate in Lyons: "C'est par son canal que je suis instruit de toutes les démarches de Duplain, mais il ne faut pas qu'il soit compromis." Regnault, who was one of the craftiest book dealers in Lyons, had received confidential information about the success of the quarto subscription, presumably from contacts inside Duplain's shop. The information was so convincing and the success so spectacular that Panckoucke suddenly decided to reverse his plans. Two-and-a-half weeks later he and Duplain met in Dijon and agreed to exploit the quarto in common instead of waging war on one another.[50]

Just how Panckoucke came to make this fourth drastic change in his *Encyclopédie* policy cannot be determined because the documentation is too sparse. But his contract with Duplain of January 14, 1777, which they usually referred to later as the Traité de Dijon, provides an unusually rich account of how a publishing enterprise was organized in the eighteenth century. Since it determined the character of the *Encyclopédie* that finally emerged from all of Panckoucke's machinations, it deserves to be studied in detail (see Appendix A. VI). The contract created a *société* or association between Panckoucke and Duplain, allotting to each a half interest in the quarto edition. Panckoucke received half the income from the subscriptions that had already arrived and that would continue to accumulate in response to the pro-

50. Panckoucke to STN, Dec. 26, 1776. Regnault's report and the other letters to and from Panckoucke at this time are missing from the STN's papers, but the Dijon agreement and the subsequent correspondence of the STN make it clear that Panckoucke considered the quarto such a commercial success that he preferred to cash in on it rather than to destroy it.

spectuses already spread through "toutes les provinces" of France. In return, he bestowed upon the enterprise "tous les droits qu'il peut transmettre"—that is a legal status deriving from his ownership of the "droits [et] cuivres du *Dictionnaire encyclopédique* et du privilège du recueil de planches sur les sciences, arts et métiers." He was to oversee the production of the three volumes of plates and to handle shipments for the Paris market—a delicate business in which his protections could be crucial. Duplain was to manage the production and distribution of the twenty-nine volumes of text. Each partner would gather subscriptions and would report on their progress every month. After a semiannual tallying of accounts, profits would be divided equally. The subscriptions had poured in so quickly that Duplain expected to accumulate enough capital from the down payments to cover the initial costs of production; but if those costs temporarily outstripped revenues, each partner would advance half the capital necessary to continue with the printing.[51] To simplify the accounting, Duplain would receive a fixed amount for every sheet printed at the pressrun of 4,250, and Panckoucke would be compensated in the same way for the retouching and re-engraving of the plates. Since the production of the text would be the most demanding phase of the operation, Duplain was to receive 2,000 livres a year for expenses. The new association also would pay 600 livres a year to a *rédacteur*, who would blend the four-volume *Supplément* into the text, "sans addition ni correction"—that is, he was not to tamper with the text but to act more as a copyist than a copy editor. This was the only respect in which the contract deviated from the provisions of Duplain's original prospectus, which promised to include some new material in the new edition. Panckoucke probably insisted on maintaining the original text in order to avoid spoiling the market for the revised edition.

The contract specified the character of the type (*philosophie*) and the paper (it was to weigh between eighteen and twenty pounds, *poids de Lyon,* and to cost 9 livres per ream) and set up a tough production schedule. Duplain was to put

---

51. Actually, Duplain would make the necessary payments and would be compensated for the use of his capital by receiving interest on it at 5 percent from Panckoucke, who also would provide him with a promissory note of 20,000 livres as security.

out four volumes every six months, beginning on July 1, 1777. The text would be distributed from a central warehouse in Geneva and would be sold under Pellet's name, but Duplain would have it printed as he pleased, contracting the work to printers in Geneva and other Swiss towns and perhaps even in Lyons. The retail price was set at the same level as in Duplain's prospectuses: 10 livres for each of the twenty-nine volumes of text and 18 livres for each of the three volumes of plates, making 344 livres in all. But booksellers could subscribe at a wholesale rate of 7 livres 10 sous per volume of text and 15 livres 10 sous per volume of plates or 264 livres in all, and they would receive a free set for every twelve that they bought.

In short, Panckoucke bartered his monopoly on legality for a half interest in a sure success. In his later letters to the STN, he explained that the subscription rate proved that Duplain might well sell twice as many *Encyclopédies* as the 4,000 specified in the contract, and that it would be wiser to cash in on this coup than to try to destroy it. One can reconstruct his calculations: the total revenue of the enterprise would come to about a million livres, the total costs to about a half million, leaving a half million in profits—profits to be collected almost effortlessly over four years, with no risk and little outlay of capital.[52]

But wouldn't the quarto ruin the sales of the remaining sets of Panckoucke's Geneva folio *Encyclopédies?* He had

---

52. This calculation is based on the terms of the contract and represents costs and income as projected in 1777 rather than the final figures, which turned out to be much greater. The only information that is missing concerns the cost of the plates, and this can be estimated from a ''Résumé des frais des planches'' produced by Panckoucke in 1780 (STN papers, ms. 1233). If the 4,000 sets all sold at the wholesale price, they would fetch 968,088 livres, taking account of the maximum number of free thirteenth copies. If they all sold at the retail price, they would bring in 1,376,000 livres. Therefore Panckoucke could safely estimate that the gross revenue would be over a million livres. His estimate of the costs would have been more complicated, but skipping some of the details and mathematics, it would have been roughly as follows:

Engraving and retouching the plates	34,916 livres
Printing the plates	16,414
Paper for the plates	17,737
Composition and printing of the text	180,090
Paper for the text	237,600
Duplain's expenses for four years	8,000
Salary of the *rédacteur*, four years	2,400
Total cost	497,157 livres

forced the STN to delay the announcement of the revised edition until July 1, 1777, in order to protect the market for the Genevan edition and had reiterated his determination to maintain that policy after returning from his sales trip in November 1776. But on December 1, a banker called Batilliot, who specialized in discounting notes of bookdealers, published a circular announcing that he had bought up Panckoucke's folio *Encyclopédies* and would sell them to book dealers at 600 livres apiece. That was 240 livres less than the subscription price; so Batilliot could expect to find buyers—and also to make a killing, because it later turned out that Panckoucke had sold him 200 sets for 100,000 livres or 500 livres apiece. That bargain made it possible for him to clear 20,000 livres from the transaction; yet it also served Panckoucke's interests because Panckoucke needed capital badly, and he knew from his sales trip that he could no longer market the leftover folios at the subscription price. The Batilliot deal seemed to be a rare case of profit sharing instead of profiteering in the book trade. But six weeks after concluding it, Panckoucke joined forces with Duplain, knowing full well that the quarto could ruin the market for Batilliot's folios. Had Batilliot learned about the Traité de Dijon, he would certainly have cried swindle. But Panckoucke kept his partnership with Duplain secret. A year later he saved Batilliot from bankruptcy, and Batilliot eventually did sell all his *Encyclopédies*. So it seems unlikely that Panckoucke meant to defraud his friend. Events moved fast. Panckoucke changed strategy rapidly in order to keep up with them—rapidly and ruthlessly, but not dishonestly. Thus by January 1777 he had freed himself from the folio and was ready to capitalize on the quarto.[53]

He needed the capital desperately at that time. In April 1777 he found himself 340,000 livres behind in his payments and secured a royal decree authorizing him to legally suspend payments. In June he offered to let the STN buy the *Table analytique* of the *Encyclopédie*, the gigantic index-cum-digest to Diderot's text, which Pierre Mouchon, a pastor of Basel, just finished after five years of labor. Panckoucke's offer

53. This account is based on Batilliot's rich dossier of 101 letters in the STN papers. On March 13, 1778, Batilliot informed the STN that he had sold all but one of the Geneva *Encyclopédies*, although his profit had been eroded by the need to clear their way into Paris by bribery.

illustrates the strain on his finances—and also the eighteenth-century version of what is known as hard sell today. The *Table* was a sure money maker, he stressed. In fact many persons would buy it who did not own the *Encyclopédie,* and he was so sure of its success that he would refund the STN's money if it failed. He described the manuscript in detail and expatiated on the best way to produce and market it. "Soyez sûrs que tout se placera et que vous ferez une bonne affaire et sûre," he concluded. "Mais il ne faut pas perdre le temps à tergiverser." He had bought the manuscript for 30,000 livres and would sell it for 60,000—really a bargain price, considering its market value. But he needed the money fast and was only making this offer because of the pressure on his finances: "Vous savez, Messieurs, les malheurs que j'ai éprouvés depuis un an. Je me suis trouvé pour près de 300,000 livres de faillites. Je perdrai 100 mille livres avec [*illegible name*]. Boisserand de Roanne vient de manquer et je m'y trouve pour une somme considérable. Cependant je n'ai point suspendu mes paiements, mais j'ai été obligé de modérer mes entreprises, et c'est cette position qui m'oblige à vous faire l'offre de cette table . . . Je puis me vanter qu'ayant fait les entreprises de librairie les plus grandes et les plus hardies, aucune n'a manqué, et que toutes les personnes qui ont travaillé avec moi ont beaucoup gagné."[54]

It is easy enough to understand Panckoucke's eagerness to cut in on Duplain's profits. But where did the new quarto association leave the old plan for producing a revised *Encyclopédie?* The Traité de Dijon mentioned the revised edition only once, in a clause that bound Panckoucke to delay the publication of its prospectus for two years so that it would not spoil the market for the quarto. The contract also gave Duplain the option of buying a three-twelfths interest in it but said nothing about the possibility of Panckoucke's associates buying shares in the quarto. Panckoucke could hardly exclude them from Duplain's enterprise because in January 1777 he owned only three-twelfths of the *droits et privilège,*

---

54. Panckoucke to STN, June 16, 1777. In setting a price of 60,000 livres on the *Table,* Panckoucke was indulging in some typical fast talk: "J'en ai acheté la copie aux associés de l'*Encyclopédie* 30 mille livres . . . Je vous propose de vous vendre cette table, mais je veux doubler mon argent." He actually bought it from de Tournes for 22,000 livres. See the text of the contract in Lough, *Essays,* p. 104.

which he exchanged for a half interest in it. So Regnault, Rey, Plomteux, and the STN could expect cuts in that half interest which would be proportionate to their holdings in the project for the revised edition. On February 3, Panckoucke wrote a *rétrocession* into the Traité de Dijon which specified that the STN's half interest in the revised edition entitled it to a quarter interest in the quarto (see Appendix A. VI). Presumably he sent similar subcontracts to the other three associates. But would that concession satisfy them?

Looked at from the STN's point of view, the Traité de Dijon was a disaster. It hurt them the greatest in the most important aspect of their business, their printing. In dropping the original reprint plan for the revised edition, Panckoucke had refused most of their demands, but he had mollified them with the prospect of a gigantic printing job. And that job had suddenly doubled in size on January 3, 1777, when Panckoucke agreed to meet Duplain's threat by producing a quarto as well as a folio edition of the revised *Encyclopédie*. The Traité de Dijon canceled that arrangement and postponed the revised edition for two years. What was the STN to do meanwhile with its vastly expanded plant? Panckoucke himself, in his dealings with Suard, had emphasized the importance of keeping the STN's workshop occupied, and he later stressed this consideration in his offer to sell the *Table analytique*.[55] The STN refused his proposition because it wanted to get a return on its investment, not to be drawn into further speculation. So it must have been appalled at article 4 of the Traité de Dijon, which specified that all the printing of the quarto would be done "à la convenance de M. Duplain." Duplain had already hired two Genevans to begin the job and was planning to contract some of it to printers in Lyons. He had no reason to hire the STN; and even if he did, he could do so at a lower price than the 54 livres per sheet allotted him by the Traité. The previous agreement of January 3 had given the STN a similar set price for printing every volume in 1,000 folio sets and 3,150 quarto sets of the revised edition, no matter what its actual costs. The Traité de Dijon seemed to cut it out of the printing operation altogether and even to deny it any role in the management of the enterprise, for the contract only concerned Panckoucke and

55. Panckoucke to STN, Nov. 4, 1776, and June 16, 1777.

Duplain. As Duplain later made it clear, it created no obliga-tion between him and the STN. This sudden reversal of policy therefore threatened to damage the Neuchâtelois as much as it would benefit Panckoucke, and they had reason to fear that the two fast-moving Frenchmen had outmaneuvered them while their backs were turned.

The Paris Conference of 1777

Ostervald and Bosset considered the situation so serious that they traveled to Paris in mid-February to conduct their own investigation. They planned to arrive secretly and to gather as much information about Panckoucke as possible before confronting him—that is, they proposed to investigate him as he had investigated Duplain, with the help of spies. They explained their plan in a letter to Perregaux: "Faites-nous l'amitié de vouloir prendre quelques informations par-ticulières avant notre arrivée de M. Panckoucke, libraire, Hôtel de Thou, rue des Poitevins, mais qui soient des gens qui puissent connaître non seulement sa fortune mais encore ce qui peut regarder sa bonne foi, probité etc. Nous com-prenons que cela n'est pas absolument aisé; mais comme ces informations nous importent essentiellement à notre arrivée à Paris, nous vous demandons instamment la grâce de ne rien négliger pour cela, et surtout qu'il ne soit point informé, di-rectement ni indirectement de ces informations, ne souhaitant point qu'il sache notre arrivée à Paris." A week later, Per-regaux replied, "J'ai deux personnes aux informations pour l'homme dont vous désirez connaître les facultés, le coeur, etc." And soon after the arrival of the two Swiss, he re-ported to the home office in Neuchâtel that he had accom-plished his mission: "J'ai communiqué à vos associés les informations que vous désiriez qu'ils prissent." Ostervald and Bosset never gave a full account of what they learned, but they wrote that it was favorable as to Panckoucke's wealth and connections, if not his "heart": "D'abord nous vous dirons que les informations les plus exactes prises sur la solvabilité de l'homme avec qui nous avons à traiter se sont réunies en sa faveur. Nous ne pouvons pas douter sur nos propres observations qu'il ne soit très entendu, très actif, bien vu de ses supérieurs et jouissant de beaucoup de crédit.''[56]

In order to travel from Neuchâtel to Paris, Ostervald and Bosset had to cover an enormous cultural distance. They were sophisticated Swiss, who had already made several trips to the French capital. Bosset had business contacts throughout France and the Low Countries, and Ostervald, who was sixty-four in 1777, corresponded regularly with booksellers in every major European city; their ,mental horizon must have been vast. But their daily routine kept them confined within a small town where people had an Alpine air and spoke a slow, Germanic French. Neuchâtel had nothing approaching a café society, though its inhabitants had learned to drink coffee—much to the regret of visitors from Paris, who came in search of rustic simplicity and Rousseauistic pastoralism.[57] The principal cultural nourishment of the Neuchâtelois still came from Sunday sermons, delivered in the old Calvinist style from the pulpit of Guillaume Farel. From Farel's Romanesque hilltop church they could easily see their entire town, enclosed within medieval walls between the Alps to the east and the south and the Juras to the west and the north.

Ostervald and Bosset left this tiny world on Monday, February 17, and after two days of rough riding arrived at Besançon, the main outpost of French culture on the rugged western slopes of the Juras. As the crow flies, the journey from Besançon to Paris was five times as long as the Neuchâtel-Besançon journey. But as coaches traveled, it required only twice the time, owing to a transformation of the facilities for travel in France, completed just a year earlier and already helping to change the kingdom from a heterogeneous mosaic of provinces into a unified nation. The vehicle of this ''revolution'' was the *diligence,* a comfortable, light coach fitted with springs and carried at a gallop over a superb new road system by horses that were changed at regular *relais.* Ostervald

56. Quotations from STN to Perregaux, Feb. 11, 1777; Perregaux to STN, Feb. 19 and 28, 1777; Ostervald and Bosset to STN, March 7, 1777. The letters from France did not often mention persons by name because the French government was notorious for opening mail.
57. Charly Guyot, *De Rousseau à Mirabeau. Pèlerins de Môtiers et prophètes de 89* (Neuchâtel, 1936), p. 103. Rousseau had lived in the area in the 1760s and had written eloquent descriptions of it.

and Bosset stepped into their diligence at Besançon on February 20. Four days later, having dashed through Dole, Dijon, Châtillon, Troyes, and Provins, they stepped out in Paris.[58]

In racing across the country at unprecedented speed to renegotiate their speculation on the *Encyclopédie,* Ostervald and Bosset seemed to be agents of modernity, of the forces epitomized by the diligence and the book. But they also traveled in the style of gentlemen under the Old Regime—not aristocrats, that is, but men whose manners derived from an international code of gentility. In some respects, therefore, they had more in common with Panckoucke than with the peasants of their own estates. Before leaving Neuchâtel, they requested Perregaux to provide them with a necessary prop for gentlemanly life in Paris, "un bon domestique, intelligent, actif, et sûr," and also asked him to reserve two adjoining rooms for them "du prix d'environ 30 sous, petites mais propres et chez gens sûrs."[59] Once settled in, they made the rounds of the capital and court. They went to cafés and theaters. They dined with worldly abbés and beautiful ladies. They had audiences with potentates in Versailles and learned whose secretary to cultivate, whose favorite to flatter, and whose valet to bribe. That was how one did business at the nerve center of the publishing industry. But for all their experience and sophistication, Ostervald and Bosset felt like aliens in Paris—and indeed they were. Swiss by nationality, French provincial by culture, they sounded somewhat bewildered in their letters home: "Nous irons aujourd'hui à l'audience de M. de Néville et à celle de M. Boucherot et vous quittons pour nous habiller . . . C'est une vie bien étrange que celle que nous menons." Ostervald cut one letter off short by explaining that he had just dined "chez M. l'abbé Fouchet et avec un autre abbé qui l'ont fait trop boire et que par ainsi il n'a rien de mieux à faire qu'à aller se coucher."[60]

The wining and dining was incidental to the main business

58. Arrangements for the journey are clear from the STN's correspondence with Pellier and Pochet, *commissionnaires* of Besançon, in Feb. 1777. On the "revolution" in travel see Guy Arbellot, "La grande mutation des routes de France au milieu du XVIIIe siècle," *Annales. E.S.C.,* XXVIII (May–June 1973), 765–791.

59. STN to Perregaux, Feb. 11, 1777.

60. Ostervald and Bosset to STN, March 12 and March 20, 1777. See also the similar remarks in their letter of March 23.

of discovering whether Panckoucke had duped them and of attempting to get better terms from him for the *Encyclopédie* speculations. After they had received the favorable report on Panckoucke, Ostervald and Bosset began to negotiate with him. They wrote home about their sessions in such detail that it is easy to imagine the three men squirming in their seats as the arguments flew around the table : "Notre homme [Panckoucke] prend diverses formes, prétend avoir fait un coup de maître pour lui et nous à Dijon. Nous avons exigé qu'il écrivît de la manière la plus pressante à Duplain pour que nous imprimions la moitié de son affaire. La crainte que nous n'allions lâcher une annonce est un épouvantail pour lui. Nous le lui présenterons au besoin et le ménageons cependant, parce que cela est indispensable . . . Panckoucke a pris [ses sûretés] vis-à-vis de Duplain en se réservant d'expédier les planches gravées d'ici. Mais nous devons en prendre contre l'un et l'autre, crainte de devenir leurs dupes . . . Notre homme est un vrai protée. On a meilleure opinion de sa fortune—que du reste il faut le manier avec délicatesse et tenir souvent sa patience à deux mains. Nos conseils sont le fils aîné du voisin et l'abbé G."[61]

Ostervald and Bosset sounded so suspicious because they assumed that Panckoucke was attempting to cut them out of a promising market by colluding with Duplain. They knew how roughly "their man" had treated the unsuspecting Batilliot, and a report that they received from their home office upon their arrival in Paris made Duplain's enterprise seem even more suspect. Charmet, a veteran bookdealer of Besançon and an old ally of the STN, had stopped by Neuchâtel while making a business trip around the circuit of Swiss publishers. He told Bertrand (the third partner of the STN, who had remained behind to mind the business) that Duplain never had any serious intention of producing the *Encyclopédie* but had only published his prospectus "pour tenter le goût du public." Moreover, Charmet believed that the public had failed to respond. Contrary to the claims of Duplain and Panckoucke, he asserted that "Capel, libraire à Dijon, n'est pas plus en état d'y faire 150 souscriptions pour cet ouvrage que l'on ne pourrait en faire dans le plus petit hameau." Bertrand concluded, "Il résulterait de ce fait sup-

---

61. Ostervald and Bosset to STN, Feb. 28, 1777. The "abbé G." was probably Grosier, a minor litterateur with whom the STN was in contact.

posé vrai que M. P. a été trompé ou qu'il a voulu vous tromper, qu'il a en effet le dessin de vendre à deux acheteurs la même chose."[62]

The Neuchâtelois found their suspicions confirmed and their position reinforced from an unexpected quarter: Suard and the philosophes. Despite their earlier opposition to his plans for the revision, Suard greeted Ostervald and Bosset warmly and offered them a "dîner académique."[63] The academician-philosophes stood to lose almost as much as the STN by the Traité de Dijon because it postponed their work for two years. They therefore appealed over Panckoucke's head to the government; and they succeeded well enough to frighten Duplain into sending his associate, Thomas Le Roy, on an emergency mission to Paris. Le Roy and Panckoucke decided to pacify the opposion with bribes. On January 23, they signed an Addition to the Traité de Dijon, which authorized Panckoucke to distribute 240 livres before the appearance of each volume in order to smooth its path into France.[64] Whether Suard became seriously disaffected with

62. Bertrand to Ostervald and Bosset, Feb. 23, 1777. Bertrand's letter illustrates a factor that complicated negotiations among early-modern publishers: mistrust compounded by misperception. By 1777 Charmet was an old man who had lost his grip on his business. He was wrong about Capel, who eventually collected 152 subscriptions in Dijon, and he grossly underestimated the demand for the *Encyclopédie* in his own territory, where a younger bookseller called Lépagnez eventually sold 338 subscriptions. At the same time, however, Charmet's report seemed to be confirmed by other reports that reached Ostervald and Bosset in Paris. On March 10, 1777, Panckoucke wrote to Duplain that "M. Boucher, libraire de Rouen qui est actuellement ici, ne croit point au succès de votre enterprise. Il l'a dit ici à ces Messieurs, et ces rapports leur font croire que j'ai trop légèrement cru à vos souscriptions. Ils se persuadent toujours que s'ils annonçaient leur édition, tous vos souscripteurs déserteraient." Bibliothèque publique et universitaire de Genève, ms. suppl. 148.

63. Ostervald and Bosset to STN, Feb. 28, 1777.

64. The Addition expressed this arrangement, somewhat elliptically, as follows: "Attendu les difficultés qu'éprouve l'exécution dudit acte, le ministère le regardant comme contraire aux intérêts des gens de lettres, M. Thomas Le Roy, associé aux Srs. Duplain et Cie., étant de retour à Paris pour lever les difficultés élevées au sujet dudit acte, a chargé le Sr. Panckoucke de faire toutes les démarches convenables pour surmonter les obstacles qui se rencontrent en cette occasion: et à cet effet il l'autorise à offrir à qui il appartiendra une somme de cent pistoles par chaque volume de discours, à l'effet d'obtenir les facilités nécessaires pour l'entrée de cette édition en France." A margin note in Panckoucke's handwriting next to article 17 of the Traité de Dijon (the article mentioning the need to postpone the announcement of the revised edition for two years) said, "C'est l'ordre exprès que m'a donné le magistrat [that is, Néville]. Il a même désiré qu'on allât plus vite." Panckoucke evidently meant that the Directeur de la librairie had applied pressure on him to speed up the quarto so as to minimize the

his brother-in-law is doubtful, but he and his philosophes tended to support Ostervald and Bosset in their debates with Panckoucke: "Nous avons vu deux fois M. Suard et verrons aujourd'hui MM. d'Alembert et de Condorcet," the Neuchâtelois reported after a week of negotiations. "M. Suard est assez dans nos idées, mais pense toujours que l'annonce de la refonte et rédaction sera en tout temps favorablement accueillie du public. Il a mauvaise opinion de l'entreprise de Duplain. Notre homme [Panckoucke] soutient toujours que les souscriptions actuelles sont en très grand nombre et que le bon marché fera écouler tout de suite cette édition-là."[65]

The sparring continued for almost four weeks. Panckoucke insisted on the importance of cashing in on a best seller while keeping the revised edition in reserve. Ostervald and Bosset objected that the Traité de Dijon deprived them of a lucrative printing job, for which they had already sacrificed a great deal of capital in the expansion of their plant. And Suard argued against the dispersal of his editorial team. The debate put Panckoucke in an awkward position because he seemed to have sold the same half interest twice—once to the STN for the *refonte* and once to Duplain for the quarto— and he could not reconcile the contradictory obligations of his contracts unless he persuaded the Neuchâtelois to accept a secondary partnership in Duplain's enterprise. They could hold him to his original commitment and undercut the quarto by publishing the prospectus for the *refonte*. And if he dumped them for Duplain, he expected them to produce a pirated quarto of their own.

The only way to prevent the crossed speculations from exploding was to persuade Duplain to concede enough of the printing to mollify the STN. On February 28, Panckoucke explained the gravity of the situation in an urgent letter to Duplain. But Duplain failed to reply because while the atmosphere thickened in Paris, he was getting married in Ly-

---

harm it would cause to Suard's stable of writers. Contemporaries believed that Panckoucke and Duplain, like other publishers, resorted to bribery. An anonymous pamphlet, *Lettre d'un libraire de Lyon à un libraire de Paris* (March 1, 1779), reported as current gossip (p. 1), "Je vous ai mandé dans le temps, et toute la librairie de Lyons en est informée, que Duplain a donnée 40,000 livres pour avoir la permission d'imprimer l'*Encyclopédie*." See also the similar remarks on Panckoucke, p. 8.

65. Ostervald and Bosset to STN, March 7, 1777. The original contains only the first letter in each of the proper names.

ons. Unable to withstand the pressure from Ostervald and Bosset much longer, Panckoucke wrote again on March 10. After some quick congratulations and a perfunctory tribute to matrimony—"le véritable état de bonheur quand on sait bien s'y gouverner"—he sketched the terms of a "lettre ostensible," which Duplain was to write to him so that he could show it to the Neuchâtelois. Duplain should offer the STN as much of the printing job as possible; he should present the quarto as a get-rich-quick speculation, which would not cause much delay in the *refonte;* and he should provide plenty of convincing information about the abundance of the subscriptions. "Il ne faut point les effaroucher. Donnez-leur à imprimer, et tout ira selon vos désirs . . . Ne mettez pas un mot dans cette lettre qui puisse m'empêcher de la leur montrer. Ne regardez point encore une fois cette réponse comme indifférente.'"[66]

Meanwhile the Neuchâtelois tried to soften up Panckoucke by working on the philosophes. On the same day as Panckoucke's final appeal to Duplain, they reported to their home office that they were gaining ground with Suard, "de qui nous espérons tirer meilleur parti que de son beau frère, homme avantageux, décisif, brusque même et impatient . . . Le ton que prend notre homme ici est de nier et contredire tout ce qui n'est pas selon ses idées et son plan.'"[67] Two days later, Suard had drifted toward the STN's camp and Panckoucke was faltering: "M. Suard blâme hautement son beau frère d'avoir souscrit à un si long renvoi et croit avec raison que le travail de la refonte en souffrira. Il persiste cependant à désirer d'avoir un intérêt dans l'entreprise, et cela répondrait de son assiduité. Panckoucke nous paraît embarrassé et piqué de ce nous voyons clairement à quel point il s'est laissé mené par Duplain.'"[68] On March 14, Ostervald and Bosset reported that Panckoucke had "l'air pensif, un peu embarrassé" when they dined with him. They sensed that he was giving ground. He had agreed to let them appeal to Duplain themselves, and they sent a tough letter. It demanded that they print half the quarto and that the publication of the re-

66. Panckoucke to Duplain, March 10, 1777, Bibliothèque publique et universitaire de Genève, ms. suppl. 148.

67. Ostervald and Bosset to STN, March 10, 1777.

68. Ostervald and Bosset to STN, March 12, 1777.

vised edition begin by the end of 1777. They would never have consented to a delay in the revised edition, they told Duplain; and if he did not make concessions, they could always publish the prospectus for it, which would ruin the market for the quarto.

The Basis of a *Bonne Affaire*

While waiting for Duplain to reply, Ostervald and Bosset made a quick trip to Rouen, where they talked business with seven of the town's thirty booksellers. The exposure to one of the most active centers of the provincial book trade changed their perspective because they learned that the Rouennais had subscribed to the quarto in droves and that the subscription boom seemed to extend throughout France. When they returned to Paris, they joined forces with Plomteux, their *Encyclopédie* associate from Liège, who had arrived to protect his own stake in the negotiations with Panckoucke. "Nous ne pouvons que remercier la Providence de nous avoir envoyé d'aussi bonnes troupes auxiliaires," they wrote home. "Il paraît que ce libraire, qui est homme de grand [sang froid], fait un peu baisser le verbe à notre homme."[69] But meanwhile, Panckoucke had received the two critical letters that he had solicited from Duplain.

In the first, Duplain reported on the subscription rate. He could not provide an exact figure, but he assured Panckoucke that it was phenomenal: "Tout ce dont nous pouvons vous assurer, c'est que calculant d'après toutes les lettres que nous recevons, nous en placerons plus de 4,000; et si vous nous promettiez de nous donner du temps, nous en placerions le double. Nous avons entre nos mains de quoi faire le plus beau coup du monde, mais le projet de la deuxième édition [that is, the revised edition] et le temps trop borné que vous nous donnez nous empêchent d'en profiter. Nos voyageurs [that is, traveling salesmen] récoltent partout. Il n'y a pas de village où il ne trouve [*sic*] des souscripteurs, pas de petite ville qui ne présente jusques à 36 engagés. Valence en Dauphiné en a fait ce nombre, Grenoble davantage, Montpellier plus de 60, Nîmes autant, Dijon nous promet 200. En un

---

69. Ostervald and Bosset to STN, March 20, 1777.

mot, jamais projet n'a été accueilli de cette maniere, et cependant votre diable lettre de défense avait fait une furieuse impression, mais on revient.''[70]

This information confirmed what Ostervald and Bosset had learned in Rouen, and the second letter went further: ''Je ne saurais vous peindre l'enthousiasme du public pour notre projet. Dans le moment que je vous écris, je reçois de Robiquet de Rennes 50 souscriptions, de Catry du Havre 32, d'Aber d'Autun 26 avec assurances d'un cent, d'un avocat d'Aurillac 13. Il n'y a pas de courrier qui n'en réunisse des nombres. Je puis vous assurer que nous placerons nos quatre mille et que si nous avions du temps, je ne craindrais pas d'en tirer six. Au nom de Dieu, mon ami, ne vous inquiétez pas davantage et profitons d'un événement qui ne se représentera jamais. D'ailleurs vous sentez bien que si l'Europe allait encore retentir de nouvelles annonces pour une autre édition, le clergé averti formerait des oppositions, le ministre retirerait sa protection, nous ferions la petite guerre, et enfin les uns par rapport aux autres nous échouerions. Je vous invite à faire entendre raison à Messieurs de Neuchâtel. Ce sont des gens instruits, et la perspective d'un bénéfice immense doit leur faire ouvrir les yeux et leur faire abandonner le projet d'imprimer, ce qui au bout du compte ne peut leur donner qu'un bénéfice qui ne convient qu'à des ouvriers par sa modicité. Si au reste ils veulent absolument faire quelques volumes, s'engager à exécuter comme moi, ils peuvent se procurer une Philosophie neuve et je leur remettrai quand ils l'auront trois volumes.''[71]

Duplain's letters are revealing in four ways. First, although they were written at the instigation of Panckoucke, they suggest Duplain's attitude toward his enterprise: he considered it the most spectacular speculation of his career and thought that the campaign to exploit it should take precedence over everything else. This attitude would prove to be crucial in the final crisis of the enterprise three years later. Secondly, Duplain's remark about the government showed how he understood Panckoucke's ''protection'' from their discussions in Dijon: Panckoucke did indeed have strong backers in Versailles, but they would not act openly. As long as he

70. Duplain to Panckoucke, March 10, 1777, in Panckoucke's dossier in the STN papers.
71. Duplain to Panckoucke, March 16, 1777.

went about his business discreetly, they would pull strings for him behind the scenes. They might abandon him, however, if he aroused the well-entrenched enemies of the Enlightenment. Thirdly, Duplain would only deal with the STN through Panckoucke, and in dealing with them he adopted Panckoucke's line: they should recognize a good thing when they saw it; they should speculate imaginatively, instead of snatching at petty profits and thinking like small-town shopkeepers. And fourthly Duplain made, a small concession: he would let the STN print three volumes. He could not do more, he explained, because he already had contracted the bulk of the job to four printers, who had had special fonts of *Philosophie* made and would have thirty presses at work within a week. The STN would not be able to get the requisite type cast for six or eight months. It would be better to commission them to print some other work in order to keep their plant busy. But if they absolutely insisted, he would give them the three volumes.

That was enough to bring around Ostervald and Bosset. On March 24, they wrote home triumphantly, "Enfin nous avons le plaisir de vous annoncer, Messieurs, que la grande affaire qui nous occupe désagréablement depuis si longtemps est terminée et, ce nous semble, avec autant d'avantage que possible. L'affaire de Duplain réussit étonnamment."[72] But the formal settlement, a contract that they signed with Panckoucke on March 28, did not really represent a triumph for the STN. (This Accession and related documents are reprinted in Appendix A. VII–VIII.) It merely bound the Neuchâtelois to accept the Traité de Dijon in exchange for being allowed to print three volumes according to the specifications of the Traité. Duplain later ratified this agreement by an Engagement of May 28, which also reserved the entire printing job of the revised edition for the STN, as it had demanded. This proviso made it easier for the Neuchâtelois to renounce their earlier demand for half the printing of the quarto, especially as Panckoucke assured them that work on the revised edition would continue, though at a slower pace, and that it would eventually be produced in both folio and quarto format at a total pressrun of 3,500. The continuation of the revision also mollified Suard, who was further compensated by a

72. Ostervald and Bosset to STN, March 24, 1777.

gift of a one-twelfth share in the enterprise. The gift came from the STN's holdings of six one-twelfth shares, but Panckoucke paid for it as part of a general refunding of the STN's debt to him for its original investment.

The refunding was a complicated business because each reversal in Panckoucke's policy had entailed an adjustment in his financial arrangements. On July 3, 1776, the STN had acquired its half interest in Panckoucke's original speculation. (So later, when the shares were divided into twelfths, the STN owned six shares worth 18,000 livres apiece). It promised to pay this sum in sixteen notes, which matured every three months from April 1, 1777. By January 3, 1777, the new arrangements for the revised edition had made it necessary for Panckoucke to agree to a first refunding of this debt. The STN took back its sixteen old promissory notes and issued thirty-six new ones, which came to 110,400 livres in all and matured later: at monthly intervals for three years, beginning on January 1, 1778. On March 28, 1777, the STN's acceptance of the Traité de Dijon required a new financial arrangement. Panckoucke now reduced its debt to 92,000 livres, which compensated it for ceding a one-twelfth share to Suard. The STN bound itself to pay that sum by forty-eight *billets à ordre*. These replaced its second set of notes and were to mature over a four-year period beginning on January 1, 1778.[73]

By making its payments smaller and spreading them out over a longer period, the STN eased the strain on its own finances and could feel somewhat reconciled to the loss of a large share in the printing operation. It could also find solace in contemplating the return on its 5/24 share in the quarto (after ceding one-twelfth of its half interest in Panckoucke's half interest in the quarto, its share in Duplain's enterprise came to 5/24, though it still had a 5/12 interest in the *droits et privilège* and the revised edition of the *Encyclopédie*). Now that "cette affaire est devenue la nôtre," as they put it, Ostervald and Bosset completely changed the tone of their remarks about the quarto. Its pressrun could easily be in-

---

73. See "Troisième addition à l'acte du 3e juillet 1776," Appendix A.VIII. The contract of July 3, 1776, also bound the STN to pay 35,400 livres in six installments between Aug. 1, 1777, and Nov. 1, 1778, to cover half the value of the three volumes of text from the Geneva edition, which Panckoucke had recovered from the Bastille. Although the subsequent agreements made those volumes almost worthless, they did not cancel that debt.

creased to 6,000, they exulted. "Il y aurait 100,000 livres de bénéfice [for their 5/24th] . . . C'est un profit certain."[74] And they fired off instructions about spreading prospectuses, gathering subscriptions, and procuring paper, type, and workers. Their enthusiasm waned for the revised edition as it waxed for the quarto—a process of affective adjustment, which may be a common aftereffect of decision-making. But the Neuchâtelois could hardly deny that Panckoucke had defeated them once again. And this fourth round of negotiations proved to be the most important of all, because it determined the character of a consortium that produced most of the *Encyclopédies* in circulation under the Old Regime in France.

74. Ostervald and Bosset to STN, March 23 and March 24, 1777. See also the similar remarks in their letter to the STN of April 4, 1777.

III

YYYYYYYYYYYY

JUGGLING EDITIONS

After the settlement of the "grande affaire" in Paris, the enterprise shifted from policymaking to manufacturing. But policy continued to be an important element in the efforts of the new quarto associates to guide their speculation to a successful conclusion. In fact the very success of the quarto created problems because it whipped up the profit motive throughout the publishing world, especially among the quarto's own publishers, who faced a crisis in self-government each time the subscriptions broke through a new ceiling, requiring a new agreement on terms for the expansion of production. The story of how the quarto associates fought their way from the first to the third edition shows exactly how the entrepreneurs of the Enlightenment conducted their business.

The "Second Edition"

Throughout 1777 the subscriptions continued to pour in. Traveling salesmen and bookdealers throughout the country reported spectacular sales, and Panckoucke grew more and more excited about the boom. By June 1777, when Ostervald and Bosset had returned to Neuchâtel, he was ready to drop everything in order to exploit this unprecedented success: "Tout ce que je sais très certainement par le rapport de nombre de libraires de province, c'est que l'édition a un prodigieux succès et qu'il faut nous y livrer tout entier, parce qu'un bénéfice tout venu vaut mieux qu'un bénéfice incertain. Il est certain que si cette édition est bien exécutée, qu'on en peut vendre 10 mille.'' In early July, he learned that one of Duplain's agents had sold 395 sets on a recent tour: "Le succès de cet ouvrage m'étonne de plus en plus.'' His astonishment kept growing because Duplain's pressrun of 4,000 copies represented a very ambitious goal for a work that eventually ran to thirty-six enormous quarto volumes in an era when printings of single-volume books normally came to 1,000 or 1,500 copies. In mid-August Duplain reported that the subscription for the 4,000 sets would soon be filled and that he planned to open another one. On August 27 he told the STN, which had begun work on the first of its three volumes, to increase its pressrun from 4,000 to 6,000.[1]

Duplain's letter provides the solution to the mystery of the missing second quarto edition, which has baffled bibliographers for some time. *Encyclopédie* scholars have been able to identify only a first edition of the quarto, whose title page proclaims it to be a "nouvelle édition . . . à Genève chez Pellet,'' and a later edition, described on its title page as "troisième édition . . . à Genève, chez Jean-Léonard Pellet, Imprimeur de la République, à Neufchatel chez la Société Typographique.'' What became of the second edition?[2]

Duplain's letters indicate that by the end of August thirty-two presses, aside from those of the STN, were working on the quarto at a run of 4,000 and that all or part of the first five volumes had been printed. On the last two or three days of the month, each press increased its output to 6,000. But the unfinished volumes had reached different stages of completion, so there was no uniform cutting-off point. The STN had reached sheet T of volume 6 when it received Duplain's order to increase the printing. It therefore reset and reprinted the preceding sheets at a run of 2,000 and continued

1. Panckoucke to STN, June 26 and July 8, 1777. See also the similar remarks in Panckoucke's letters to the STN of May 13 and June 16, 1777, and in Duplain to STN, Aug. 18, 1777.

2. In ''The Swiss Editions of the *Encyclopédie*,'' *Harvard Library Bulletin* IX (1955), 228, George B. Watts made a good guess as to the explanation of the ''second'' edition, although like other scholars he assumed that Pellet was behind the whole affair. Lough agrees with Watts's version of this complicated question (*Essays*, pp. 36–38). For a full discussion see Robert Darnton, ''True and False Editions of the *Encyclopédie*, a Bibliographical Imbroglio,'' forthcoming in the proceedings of the Colloque international sur l'histoire de l'imprimerie et du livre à Genève.

thenceforth at 6,000. The other printers did likewise. But they had reached different points in the production of the other volumes. At the moment that the STN changed gears in Neuchâtel, Pellet in Geneva could have been near the end of volume 5, while J. F. Bassompierre, also in Geneva, could have been at the beginning of volume 4 and the Périsse brothers of Lyons in the middle of volume 3. As there was no uniform order in which the sheets were assembled into volumes and the volumes into sets, there is no standard section in every set that can be identified with some second or intermediary stage of the printing. Each set must be different from all the others, and no second edition ever existed. It does not make much sense, in any case, to speak of editions, because more than half the type of the Pellet quarto was not reset. Instead, the work went through three different "states," corresponding roughly to the pressruns of 4,000, 2,000, and 6,000. But its publishers talked loosely about two editions. In order to avoid confusion, their usage will be followed in this account, despite its inaccuracy according to the tenets of modern bibliography.

Duplain's instructions also provide more specific information about the size of the printing. The Traité de Dijon called for an edition of 4,000 sets but stipulated that 4,250 copies of each sheet would be printed. The 250 extra sheets were intended to be mostly or entirely *chaperon,* to replace those spoiled by the printer. But printers commonly calculated in reams, quires, and sheets (*rames, mains,* and *feuilles;* in eighteenth-century France, 25 sheets made a quire, and 20 quires made a ream, which thus contained 500 sheets). Duplain actually directed the STN to use 3 reams, 10 quires more in the printing of every sheet's worth of text (that is, of every eight pages), making an output of 12 reams, 6 quires, or 6,150 copies. The increase therefore went as follows:[3]

3. In his letter to the STN of Aug. 27, 1777, Duplain phrased his directions as follows: ''Nous nous sommes déterminés à tirer trois rames dix mains de plus. Vous voudrez bien en conséquence, Messieurs, tirer sur chaque feuille que dorénavant vous mettrez sous presse en tout douze rames et six mains, et lorsque vous aurez fini votre volume, vous réimprimerez s.v.p. tout ce qui est fait et tirerez trois rames dix mains seulement.'' In a letter of Aug. 18, he said that the printing of the eighth volume had just begun in Lyons and that the first ten volumes should be completed by September, but his letters do not provide enough information for one to know exactly which proportion of the early volumes were reset and run off at 1,750 copies or who printed them.

original pressrun	8 reams 16 quires or	4,400 copies
increase	3 reams 10 quires or	1,750 copies
total	12 reams 6 quires or	6,150 copies

So many copies of so huge a work seemed staggering to Panckoucke: "Il est certain que le succès de cette édition in-quarto passe toute croyance." He agreed in principle to the increase in the pressrun, but he did not want to accept any proposals for enlarging the enterprise until he had made a personal inspection of Duplain's operation in Lyons, for he had more faith in the success of the quarto than in Duplain's management of it: "Je veux par moi-même m'assurer de la vérité," he wrote to the STN. "Et comme je pars lundi pour Lyon, je verrai alors tout par moi-même, et je ne ferai rien que pour le bien commun. Je pense, Messieurs, que vous vous en rapporterez dans tout ceci à l'habitude que j'ai de traiter les grandes affaires.'"[4] The Neuchâtelois had learned to beware of Panckoucke's grand style of doing business, but they were willing to turn it against Duplain. So the fifth round of negotiations began with the usual conspiratorial preparations, though it mainly concerned technical questions of adjusting the Traité de Dijon to the new dimensions of the quarto.

A great deal of money hung on those technicalities. For example, the Traité de Dijon allotted Duplain nine livres for every ream of the requisite paper that he procured of the *Encyclopédie*. By increasing the printing by three reams, ten quires per sheet, the quarto associates committed themselves to purchasing approximately 11,165 additional reams of paper for 100,485 livres. Such enormous demand was certain to force up the price of paper. In fact, the price had already risen so markedly that in May Duplain had persuaded

---

4. Panckoucke to STN, Sept. 9, 1777. Panckoucke added that he really did believe Duplain's reports: "On doit espérer d'en placer à Paris au moins 1,000. Le débit même, s'il répond aux provinces, peut en être le double. Duplain m'a écrit pour une augmentation de tirage que nous n'aurions pas le droit d'empêcher, quand bien même nous n'aurions pas les raisons de le vouloir. Ainsi que lui, je suis bien sûr que l'on placera ces 6,000 exemplaires, et cette assurance doit vous convaincre, Messieurs, que je vous ai engagé dans une excellente affaire, puisqu'à ce nombre nous devons doubler nos fonds et au-delà.''

Panckoucke to allow five additional sous per ream.[5] How much more should Panckoucke allow in the contract for the second edition? He knew that Duplain would jump at the possibility of raking off the difference between the real costs and the allotted sum. And that difference could be enormous —2,761 livres for an extra five sous in the price of the extra 11,165 reams. Duplain's allotment for printing costs raised the same problems, although in this case Panckoucke might argue for a reduction. The Traité de Dijon permitted Duplain to contract the printing for whatever prices he could get and allowed him 30 livres for the composition and printing of each sheet at a pressrun of 1,000 and an increase of 8 livres for every additional 1,000. Since the Traité provided for an edition of 4,000 copies, it allotted Duplain 54 livres per sheet. Panckoucke evidently believed that the labor involved in printing an additional 2,000 copies would not cost another 16 livres per sheet. He therefore wanted to reduce Duplain's printing allotment. The Traité de Dijon also provided 600 livres per volume for the work of a "rédacteur," who was to incorporate the supplements into the text and probably also to do some copy editing. Duplain had hired a minor littérateur in Lyons called the abbé Laserre, and Laserre wanted more money.

Finally, Panckoucke and Duplain would have to iron out some problems about marketing. Duplain wanted to strike a bargain with a Lyonnais dealer called Rosset, who promised to buy up to five or six hundred subscriptions if he were given special terms. The Traité de Dijon did not permit any deviation from the fixed wholesale price. But in the interest of increasing sales, Duplain had offered to give Rosset a secret rebate, provided that Panckoucke concurred. Panckoucke was suspicious of secret bargains and thought the demand for the book too great for them to be necessary. So he asked the STN to write a contrived letter ("lettre ostensible") to him, which he would show to Duplain and Rosset in order to strengthen his hand in the bargaining. He virtually dictated it, stressing all the arguments against modifying the price policy; and he warned that if the STN wanted to

5. The new price of 9 livres 5 sous per ream was set by an Addition to the Traité de Dijon dated May 15, 1777. This document is missing from the STN papers, but its contents are clearly indicated by the fourth paragraph of the Panckoucke–Duplain agreement of Sept. 30, 1777 (see Appendix A.XI).

say anything confidential to him while he was in Lyons, they should confine it to a separate sheet, "cachetée avec de la cire et sous double enveloppe," because he would be staying with Rosset.[6]

The Lyons conference of September 1777 added yet another contract to the structure of *actes* and *traités* that Panckoucke had built around the *Encyclopédie*. The agreement, which Panckoucke and Duplain signed on September 30 (see Appendix A. XI), regulated the terms for expanding the printing by 1,750 copies. Panckoucke testified that an inspection of the subscription register had convinced him that 4,407 subscriptions had been sold, making the increased pressrun desirable. He consented to an increase of five sous in the set price for the paper, and he got a reduction of three livres per sheet in the printing price of the extra 1,750 copies (Duplain was to get 33 livres per sheet instead of 36 livres, as he might have expected, according to the rates set in Dijon). The abbé Laserre received an increase of 250 livres per volume. The additional salary would permit him to hire *de nouveaux aides* (presumably copyists) and to complete all the work on the copy by the end of 1779. Duplain evidently failed in his attempt to get a special concession for Rosset, but he was compensated by a side-speculation on the *Table analytique,* which he and Panckoucke arranged by a contract dated September 29.[7]

After the STN had refused his offer to buy the *Table,* Panckoucke decided to go ahead with its printing, using the presses of his former Parisian associate, J. G. A. Stoupe. As he had explained to the STN, he expected many owners of the first two folio editions of the *Encyclopédie* to buy the *Table,* which would serve as an index and summary of Diderot's text. The success of the quarto *Encyclopédie* meant there would be a parallel demand for a quarto edition of the *Table.* Panckoucke and Duplain agreed to produce one, splitting costs and profits. Panckoucke would supply Duplain with the sheets of the folio edition as they came off the press. Laserre would adapt them to the quarto format for a fee of 2,400

6. Panckoucke to STN, Sept. 9, 1777.
7. The text of this "Copie de traité pour la *Table analytique* entre M. Duplain et M. Panckoucke" is in the STN papers, ms. 1233. A letter from the STN to Panckoucke of May 3, 1778, shows that the STN did not then know about the secret arrangements for the quarto *Table* and was prepared to make a similar deal with Panckoucke behind Duplain's back.

livres. And Duplain would handle the printing and marketing. The enterprise would begin after the quarto *Encyclopédie* had been printed, and it would be kept secret until then —even from the STN and the other quarto associates.

So Panckoucke and Duplain ended the conference in Lyons on good terms with one another. Panckoucke had arrived ready to do battle and left feeling reconciled and even jubilant about Duplain's handling of the quarto. As he reported in sending a copy of the new contract to the STN: "J'ai eu beaucoup de peine à obtenir une remise de trois livres sur l'impression. Je me suis assuré que l'augmentation sur le papier était nécessaire. Ils n'emploient que de l'Auvergne du poids de 20 à 22 livres. Ils n'y gagnent pas, et je crains bien qu'ils ne se trouvent dans l'embarras cet hiver. L'abbé de Laserre était payé comme un croucheteur. Il avait des titres pour obtenir sa demande . . . J'ai bien vu qu'on m'en avait imposé à Dijon, mais tout cela n'est plus un mal, puisque le succès passe nos espérances. J'ai vu les presses Genevoises. Tout m'a paru bien monté et en bon train. Le nombre de 4,407 est bien réel. Un seul relieur de Toulouse en a fait 200. Les souscriptions viennent tous les jours. Je suis témoin qu'on en a fait 150 en 8 jours. Il ne peut point y avoir de rentrées avant la fin de 1778, puisqu'on est obligé à des achats immenses de papier qu'il faut payer d'avance. Au reste, on donnera le compte tous les six mois. Duplain a à Lyon des associés intelligents qui ont mis plus de 400,000 livres dans son commerce et qui mettent le plus grand ordre dans cette affaire. Les registres sont bien tenus, et il est impossible d'en imposer . . . Enfin, cette affaire, si le gouvernement ne la croise pas, offre les plus grandes espérances . . . La faveur du public est sans exemple."[8]

The Origins of the "Third Edition"

So great was the flood of subscriptions that Panckoucke and Duplain laid plans for a "third edition" while they settled the terms for the "second" (that is, for the increased press-run). On the very day that he signed the Lyons contract, Duplain wrote to the STN that he expected to arrange a new printing of 2,000 but that it would have to be a separate and

---

8. Panckoucke to STN, from Lyons, Oct. 9, 1777.

distinct edition in order to prevent delays in the production of the first 6,000 sets.[9] Launching a new edition was no casual matter, however. Before they could agree on its terms, Panckoucke, Duplain, and the STN spent a year in bargaining and bickering. The unprecedented size of the quarto had already made it almost unmanageable. To increase it by a third strained the publishers' resources and their tempers to the breaking point. Every modification of the old arrangements shifted the budget of the book by thousands of livres, and every attempt to increase profits increased the danger of profiteering.

Duplain sounded the market carefully before committing himself to such a major expansion of the enterprise. The subscription rate continued to be strong—so strong, according to one of his agents, that a third subscription might soon be filled with the surplus from the second.[10] But Duplain had only announced two subscriptions. By mid-January he thought it necessary to announce a third in order to see whether the demand would be sufficient for a new edition. This technique of "taking the pulse of the public" was a form of fraud, which gave subscriptions a bad name, but it helped minimize risk. So Duplain was following the rules of the game rather than breaking them when he asked the STN to place the following notice in various journals:

Les deux premières éditions de l'Encyclopédie in-quarto, annoncées chez Pellet à Genève, se sont écoulées avec une rapidité qui prouve que le public a goûté le projet de cette impression et qu'il est content de la manière dont il est exécuté. Les éditeurs, flattés d'un accueil qui a surpassé leurs espérances, proposent une troisième souscription aux mêmes conditions que les précédentes. Au moyen d'un plus grand nombre de presses qu'on fera monter, ceux qui voudront souscrire auront l'ouvrage complet en même temps que les premiers souscripteurs . . . La souscription est ouverte jusques au premier mars, et la première livraison se fera en mai 1778. On peut souscrire chez les principaux libraires de chaque ville.[11]

9. Duplain to STN, Sept. 30, 1777.
10. The agent, Merlino de Giverdy, told Panckoucke in November that there might be enough subscriptions for a third edition within three months. Panckoucke passed this news on to Neuchâtel with a jublilant remark of the kind that now filled all his letters about the quarto: "C'est un succès incroyable" (Panckoucke to STN, Nov. 8, 1777).
11. Duplain to STN, Jan. 16, 1778. The text is missing from Duplain's letter and is quoted from the *Gazette de Leyde* of Feb. 6, 1778, where the STN had it printed.

Duplain took this step without consulting his associates. Because he administered the subscriptions, he alone knew how feasible a new edition would be, and he dominated the administration of the quarto so completely that he often made such policy decisions by himself. This tendency worried the Neuchâtelois, who had reason to hesitate before plunging into a new speculation on the quarto because they wanted to avoid further postponements of the revised edition, they knew Duplain too well to trust him, and they considered the third edition too important to be left completely in his control. They therefore filled their letters with anxious queries: Had Duplain informed Panckoucke of his decision to make a trial announcement? Was there any sign that the subscription rate had slackened? Did Panckoucke realize that Duplain was drawing them into a major recommitment of capital? They no longer doubted the success of the quarto—"une chose fort extraordinaire"—but they worried that success itself might overexcite Duplain's appetite for gain at their own expense. "Nous voyons clairement que plus l'entreprise prospère et plus il est jaloux de la part que nous y avons," they confided to Panckoucke.[12]

Panckoucke remained unperturbably optimistic. He wanted to subordinate everything, including the revised edition, to the exploitation of the quarto's sensational selling power. The STN deferred to his judgment, "connaissant combien vous êtes expert en ces sortes d'affaires," but pressed him on two points: first, the preparations for the revised edition should continue unabated, so that they could issue a prospectus for it at the end of the year (to issue one earlier might spoil the sales of the third quarto edition); and second, they should make sure that the size of the third edition not exceed the number of subscriptions.[13]

The prospective size of the edition proved to be a sticky point because Duplain resisted the STN's attempts to know what it would be. This information mattered to the Neuchâtelois because they sought to increase their share of the printing. Not only did they want to get new volumes from the third edition to print, but they also hoped to increase the pressrun on the old ones. They could earn far more by producing 8,000

12. STN to Panckoucke, Jan. 25 and 29, 1778.
13. STN to Panckoucke, Feb. 22, 1778.

copies of a volume for all three editions than 6,000 for two. But Duplain had committed himself only to giving the STN three volumes to print at the prescribed rate (see the contract of May 28, 1777, in Appendix A. IX). He could gain more by contracting the work at lower rates to other printers. These and other issues would have to be resolved by the contract for the third edition, if in fact the edition were to take place. Meanwhile, Duplain and the STN played a curious game of probing and parrying in their correspondence: the STN kept trying to pry information and commitments out of him and he replied with elliptical or evasive remarks.

Imbroglios

On March 4, 1778, three days after the deadline for the third subscription had expired, the STN wrote to ask whether the public's response had been sufficient to go ahead with the printing and remarked casually that it expected to produce the same volumes for the third edition as for the first two. Instead of giving a straight answer, Duplain wrote that some customers who had recently subscribed through the STN would have to wait for the third edition to be served. Indirect as it was, his reply indicated that he had decided to proceed with the third edition, but what was to be its size? At the end of March, the STN reminded Duplain that it needed to know what its work load would be in order to make advance plans for its printing operations. It had heard a rumor, it added, that he was now having some volumes run off at 15 reams (7,500 copies). That was an oblique way of sounding Duplain on his strategy: Had he set the size of the third edition at 1,500 or 2,000 copies? And would he have the text recomposed for it or would he print the remaining volumes of all three editions together at a pressrun of 7,500 or 8,000?[14] Duplain seemed to give a forthright answer on April 5: "Nous montons douze presses qui seront uniquement employées à cette édition . . . Nous avons près de 500 souscripteurs sur 1500 que nous tirons. Nous ne ferons point augmenter le nombre de l'autre édition. On recomposera jusqu'à la fin." But the STN did not believe him because it had learned from secret informants that Duplain was printing entire volumes at

14. STN to Duplain, March 29, 1778.

8,000 and because the letter that it had received from Duplain contradicted a letter that Panckoucke had received from him and had forwarded on to Neuchâtel. The STN explained these inconsistencies to Panckoucke, concluding, "Cette petite observation [a report from Lyons that Duplain was printing some volumes at 8,000] et d'autres que l'on pourrait [faire], nous vous le disons dans la confidence de l'amitié, n'inspirent pas une confiance entière et exigent de votre part comme de la nôtre une attention bien entretenue."[15]

Was Duplain printing more copies than he would admit to his associates and raking off excessive profits from his management of the printing? Those questions seemed particularly pressing in April 1778, when Duplain and Panckoucke made a first attempt to agree on a contract for the third edition. It is hard to know what happened in these negotiations, which were conducted by mail, because most of Panckoucke's letters are missing from 1778. But Duplain evidently stressed the enormous increase and expense of his administrative tasks and asked to receive far more than the 8,000 livres in four annual installments provided by the Traité de Dijon. Panckoucke informed the STN that he was trying to hold Duplain to an increase of 16,000 livres, and the STN sent back a statement of support: "Vous avez sagemment répondu aux prétentions de Duplain. Votre offre nous paraît équitable et son calcul enflé toujours à l'extrême. Il faut convenir qu'il est chargé d'un rude détail, mais 16,000 livres font un dédommagement honnête pour quelqu'un qui d'ailleurs partage les bénéfices. Nous vous abandonnons confidemment la suite de cette négociation."[16] The STN also warned Panckoucke that Duplain might be cheating on the printing and took steps to investigate this matter itself. On April 8, it instructed Jacques François d'Arnal, a Lyonnais banker and son-in-law of Bosset, "de vous informer sous main à combien d'exemplaires Duplain et compagnie font tirer l'*Encyclopédie*." The mistrust and intrigue had grown so thick that Duplains' partners actually spied on him in order to know how many copies he planned to print of the book that they were publishing together.

On April 12, d'Arnal reported, "Nous avons su adroitement

15. STN to Panckoucke, April 9, 1778.
16. Ibid.

par deux personnes différentes, qui sont bien instruites, que la nouvelle édition de l'*Encyclopédie* in-quarto sera de 1500.'' Not very incriminating. But the Neuchâtelois suspected that Duplain had set his pressrun at 2,000, perhaps with the intention of selling the extra copies on the sly. They therefore kept d'Arnal's report to themselves and tried to draw Duplain into exposing his true design by keeping up a friendly exchange of letters with him throughout the spring of 1778. In one particularly amicable letter, they said that they had just received some good news from Panckoucke: the third edition was to be printed at 2,000, and the subscription rate was strong enough to justify an even larger pressrun. They assumed, of course, that Duplain would favor the obvious, budget-cutting device of printing the remaining volumes at 8,000, while pretending that the third edition had been reset, as promised in their sales campaign.[17]

Duplain felt that such matters belonged strictly to his domain and would not be provoked into making any revelations about them. The Traité de Dijon only required him to give account to Panckoucke. Their contractual relationship excluded the Neuchâtelois, who were Panckoucke's associates, not Duplain's. Duplain knew they were hungry for the commissions that he was providing to other printers, who lined his pockets by doing the work for far less than the rate set by the contracts for the first two editions. If he gave more printing to the Neuchâtelois, he would have to pay them at the official rate. And as subassociates, they might pry into his management of the enterprise. So he tried to keep them in the dark; and instead of giving them grounds for feeling hopeful about the prospects for the third edition, he sent them a terse and gloomy reply. Far from printing the third edition at 8,000, he wrote, he had not yet decided whether to print it at all. He had received only 500 subscriptions, mainly because of Panckoucke's failure to tap the rich Parisian market. Duplain himself was making every possible effort to drum up sales. He had sent a circular letter to a great many bookdealers and would wait for their response before deciding on the fate of the third edition.[18]

This reply sounded suspicious to the Neuchâtelois. They hid

17. STN to Duplain, April 15, 1778.
18. Duplain to STN, April 21, 1778.

their doubts in their next letter to Duplain, while venting them to Panckoucke. A short time ago Duplain had claimed that the third subscription was a sure sellout, they observed to Panckoucke. Now he doubted that it would produce enough to warrant an increase in the printing. Why had he changed his tone so completely? The Neuchâtelois could guess at the answer to that disturbing question, but they would conceal their suspicions from Duplain and would insist on a close inspection of his accounts. They now appreciated the importance of articles 13 and 14 of the Traité de Dijon, which bound Duplain to give a report on the subscriptions and bookkeeping; and they rejoiced at Panckoucke's announcement that he would come to Lyons to examine Duplain's accounts in person. They agreed with Panckoucke's suggestion that they postpone the revised edition in order to concentrate exclusively on the quarto. Once they had drained all the profits out of it and had closed their accounts with Duplain, they could proceed with other projects, which they could prepare behind Duplain's back.[19]

The relations among the quarto associates had become so conspiratorial that only an external threat prevented an internal rift. They had to drop everything in May 1778 in order to conduct emergency negotiations with other publishers who were attempting to cut into their market with other *Encyclopédies*. On June 22, Panckoucke aligned the quarto group with a consortium from Liège, which had begun to produce an *Encyclopédie* arranged by subject matter instead of by alphabet. This project ultimately developed into the *Encyclopédie méthodique* and put an end to the plans for a revised edition. On June 24, Duplain bought out a Lyonnais group that had begun to produce a pirated quarto edition. And throughout the summer of 1778 the STN was attempting to settle a trade war with the sociétés typographiques of Lausanne and Bern, who were marketing an octavo *Encyclopédie*. These crises made it necessary for the quarto publishers to suspend their negotiations on a contract for the third edition.

But they could not afford any delay in the printing of the third edition. On the contrary, they needed to get their quartos on the market before their competitors could spoil it. Sub-

19. STN to Panckoucke, May 3, 1778, and STN to Duplain, May 2, 1778.

scribers were wary of putting money on books that did not yet exist. They might switch to a second, more attractive subscription, if their money had not been collected for the first. The quarto subscribers were to pay for each volume after they received it, and their payments were to finance the printing of the later volumes. Duplain therefore accelerated his production schedule to an almost unbearable speed. He set presses to work in Lyons, Grenoble, and Trévoux as well as Geneva and Neuchâtel. He had the printed sheets assembled into volumes and stored in Lyons. He arranged for their transportation over thousands of miles of complicated routes. And he tried to keep track of the subscriptions and collections, while keeping the accounts in order, disentangling snarls, and undoing errors.

The problems of administering such a complex operation strained Duplain's temper and his relations with his partners to the breaking point. After receiving a badly printed volume from the STN in January, he exploded in a fit of uncontrollable rage. Two weeks later he was still angry enough to tell his partners what he thought of them in the following terms: "Vous faites mal un volume et M. Panckoucke écrit à tous nos souscripteurs qu'il faut que nous accordions plus de terme. En un mot, nous travaillons jour et nuit pour la réussite de l'affaire, et il semble, Messieurs, que vous fassiez tout ce que vous pouvez pour la détruire. Lorsque nous aurons amoncelé par des crédits des dettes en province, qui nous payera? La majeure partie n'en vaut rien. Voilà où conduisent les discours de M. Panckoucke. Nous vous dirons en passant que nous avons plus de 50,000 écus dehors et que cela a de quoi effrayer et faire de terribles réflexions. Joignez à cela un travail affreux et continuel, et voyez comment vous auriez envisagé un diable de volume qui en vérité est affreux, quoique vous en disiez''[20]

The pressure on Duplain further complicated the situation in which the third edition came into being. While his associates secretly spied on him, he raged against them for making him bear almost the entire burden of the enterprise. Their unwillingness to ease the financial strain made him especially angry. He had delivered the books faster than the subscribers could pay for them. Most of the subscribers were

20. Duplain to STN, Feb. 9, 1778.

booksellers who had sold dozens of sets and needed time to collect the money from their customers. But Duplain had to advance enormous sums for the paper and printing. When his receipts did not even approach his expenses, he began to feel desperate: thus his "terribles réflexions" at the thought of 50,000 écus outstanding and his fury at Panckoucke's willingness to give the booksellers more time to make their payments.

Duplain got some relief from the financial pressure by delaying the payment of his own bills, especially those he owed to the STN for its share of the printing. By mid-June the STN had printed volumes 6 and 15 of the first two editions and had begun work on volume 24. Each volume cost it thousands of livres to produce; and after it finished each one, it billed Duplain according to the rates fixed in the Traité de Dijon and the subsequent contracts. This billing took place in the usual manner of eighteenth-century commerce: the STN sent a statement of its charges to Duplain and then normally wrote bills of exchange on him made out to its own creditors or to d'Arnal, who handled its financial affairs in Lyons. D'Arnal was continually paying out large sums on the STN's behalf, mainly for paper. He therefore needed to cash its notes on Duplain in order to keep its account out of the red. But when those notes became due, Duplain refused to pay them, arguing that his own debtors—the booksellers who had subscribed for the quarto—had failed to pay him on time and he therefore should be able to delay his payments to the STN. As associates in the enterprise, the Neuchâtelois ought to carry their share of its financial difficulties, he maintained. They retorted that as printers they had to be paid. Not only did the laws of commerce entitle them to their wages but also they could not be expected to advance their own capital for their own work without receiving some of the money that must surely be flowing to Duplain from the subscribers. This quarrel broke out in June 1778, just when the quarto group's negotiations with the rival *Encyclopédie* publisher had reached their most critical phase, and it continued intermittently throughout the rest of the year.[21]

At the same time, Duplain and the STN sparred over the

21. This account of the STN's disputes with Duplain is based on d'Arnal's thick dossier in the STN papers as well as on the correspondence between the STN and Duplain.

printing of the third edition. Duplain knew that the Neuchâtelois needed to print more volumes if they were to avoid
firing workers and dismantling their huge shop. He therefore
played on this need in order to postpone the payment of his
debt. On June 2, he asked the STN to extend the maturation
date of some bills of exchange which were then due by three
months. And to show that he could be tractable on his end, he
stopped evading its demands for information about the third
edition: "La troisième édition est commencée, et nous l'avons
donnée exclusivement à deux imprimeurs qui ont monté 18
presses, se sont engagés à trois épreuves de chaque feuille,
et nous voulons faire une belle édition afin que s'il reste
quelques exemplaires, ils ne soient pas à charge. On tire trois
rames dix mains (that is, 1,750 copies)." So Duplain had
begun to print a quite large edition, and he might well exclude
the STN from it.

Duplain let the Neuchâtelois make that last reflection themselves, expecting them to become more flexible about the payment of their bills. They reacted by firing off an urgent letter
to Panckoucke. They had eleven presses to keep occupied,
they lamented; yet Duplain was attempting to cut them out
of the printing of the third edition, and in doing so he was
violating his contractual obligation to give them three volumes.
To be sure, they had received three volumes for the first two
editions, but they were also entitled to three volumes of the
third, as could be proved by a logical extension of the Traité
de Dijon. They would settle, however, for a fourth volume to
print, hopefully at 8,000. They urged Panckoucke to press this
demand on Duplain and also to come to Lyons to check his
accounts because "il doit lui avoir passé de fortes sommes
par les mains." They were worried about how Duplain had
handled this money, and they also felt perplexed about a
discrepancy between the last letters they had received from
Duplain and Panckoucke. Panckoucke's letter reported that
he and Duplain had agreed to simplify the marketing of the
third edition by taking 500 sets apiece and selling them in
their own territories. That agreement seemed to imply that
the edition would consist of 1,000 sets, but Duplain's letter
spoke of 1,750 sets. Moreover, d'Arnal's report set the pressrun at 1,500. The Neuchâtelois still did not know what to believe about the mysterious third edition. But it was now clear

that the disputes over its financing and its printing had become interlocked, stalling the settlement of the contract, even though Duplain's men had begun setting type and printing sheets at a pressrun somewhere between 1,000 and 1,750.[22]

While the STN tried to get Panckoucke to apply pressure on Duplain, Duplain continued to suffer from the strain on his finances. "Nous avons bien eu l'honneur de vous observer que l'argent est ici d'une rareté affreuse, que nos libraires demandent du temps, et qu'enfin nous ne pouvons pas en faire sortir des pierres," he wrote to the STN on June 9. "Le train que nous menons l'ouvrage exige une mise dehors à laquelle nous ne comptions point." He simply could not meet his June payments to the STN. But he could retreat from his adamant stand on the printing—a stand that he had probably taken, in any case, in order to improve his bargaining position on the financial question. The STN showed its willingness to play this game with him by instructing d'Arnal to grant Duplain a delay on his payments in exchange for obtaining a fourth volume. The bargain worked perfectly, d'Arnal reported, despite Duplain's reluctance to sacrifice any of his profitable business as a middleman. In essence, therefore, Duplain bartered a loss in his rake-offs on the printing for impunity in failing to pay his bills on time.

But he snapped up d'Arnal's proposal too quickly. The STN interpreted this alacrity as a sign of weakness and answered d'Arnal with instructions to raise its bid to three volumes of the third edition. It then wrote directly to Duplain, saying that it was glad to help him with his financial difficulties and that it would soon send him some "remonstrances" that would explain its case for the printing of the third edition. Meanwhile it would like to do a fourth volume for the first two editions. It had plenty of paper and workers ready for the job. This tactic backfired. Duplain told d'Arnal that after reconsidering the question, he thought the STN had enough work to keep it busy without anything from the third edition. And in his letters to the STN he merely continued to insist on the need to postpone the payment of his bills. The STN then had to fall back on a strategy of dunning. In late June it warned Duplain that he had accumulated 16,980 livres in debts, that it would insist on being paid, and that it

22. STN to Panckoucke, June 7, 1778.

would charge interest for any delay. In early July, Duplain refused to honor two bills of exchange worth 2,019 livres. The STN then put him on notice that d'Arnal would present the bills once more and that it could no longer accept any postponement of their payment. The quarto associates had reached the brink of a schism, and they still had not settled on a contract for the third edition.[23]

The Neuchâtel Imprint

The STN considered the situation so serious that it sent its most trusted agent, Jean-François Favarger, on a special mission to Lyons. Actually, Favarger was to make a complete tour de France, selling *Encyclopédies* and other books and settling accounts with bookdealers throughout the country. But the most important purpose of his journey was to do some general reconnoitering in Lyons without letting Duplain realize it. By appearing as a traveling salesman, Favarger might be able to discover Duplain's true motives and intentions, for by now the STN's relations with Duplain had become so entangled in bidding and bluffing that the Neuchâtelois no longer knew what his game was. They therefore planned Favarger's interview with Duplain in minute detail and even wrote a scenario for it in Favarger's diary. Favarger consulted the instructions that Ostervald and Bosset wrote in his diary before meeting with the STN's customers along his route, and he recorded the results of each meeting afterward. Most of the entries consisted of a few phrases, but in Duplain's case, the instructions ran on for two-and-a-half pages and contained remarks such as the following:

Voir M. Duplain et tâcher de savoir, mais sans témoigner trop de curiosité, à quoi on en est pour l'impression des volumes de l'*Encyclopédie* quarto, combien de presses y travaillent à Lyon ou ailleurs; si l'on a commencé la troisième édition, à combien on la tire . . .

Vous écouterez attentivement tout ce que M. J. D. [Joseph Duplain] pourra vous dire touchant notre *Encyclopédie,* et vous éviterez de faire aucune ouverture . . .

Vous parlerez à J.D. du désir que nous avons d'imprimer encore un volume à 6,000. Vous le prierez de nous en écrire. Vous lui direz

23. The most important in this exchange of letters are d'Arnal to STN, June 12; STN to d'Arnal, June 17; STN to Duplain, June 24; STN to Duplain, July 8; and STN to d'Arnal. July 8, 1778.

qu'il y a beaucoup d'apparence qu'aucun des volumes n'est corrigé avec plus de soin et par des gens plus instruits, qu'on donne toute l'attention possible à l'exécution, que notre imprimerie est mieux montée à tous égards qu'aucune de celles qu'on emploie à cette entreprise, que nous avons fort papiers supérieurs en beauté à ceux de Lyon, que nous avons monté notre fabrique exprès pour réimprimer l'in-folio que son édition in-quarto a retardée, qu'il est donc juste que nous soyons indemnisés de quelque manière. S'il est impossible d'obtenir de lui un volume à 6,000, dites que nous avons droit par le traité d'imprimer trois volumes à 2,000 [that is, of the third edition] et que nous espérons que cet article ne souffrira aucune difficulté . . . NB Vous nous rendrez compte en détail de ce que vous aurez fait à cet égard.

If he played his part skillfully, Favarger might win the STN's case for an increase in its printing allotment. But he was also to snoop around Duplain's shop, to collaborate with d'Arnal and the abbé La Serre, and to sound the other Lyonnais dealers in order to discover what Duplain's general *Encyclopédie* policy really was.[24]

Farvarger arrived in Lyons on July 13 or 14, just in time for some last-minute instructions from his home office. The STN warned him that Duplain had recently refused to honor its two bills of exchange and had defended his conduct with specious arguments, which it refuted point by point so that Favarger could confound him in their discussions. It concluded with an exhortation: "Relisez bien toutes vos notes pour Lyon avant de faire aucune visite afin d'avoir balle en bouche en traitant."[25] One can imagine Favarger's reaction, as he sat in his inn, rereading his instructions, rehearsing his lines, and fortifying himself for the confrontation with Duplain. But when the great moment came, he found his man surprisingly affable. Duplain talked business for hours with apparent openness and sincerity, though he did not reveal anything about the higher diplomacy of his operations. Favarger held firmly to his appointed role throughout the flood of words, though it was not easy: "je . . . me suis conformé, Dieu merci, en tous points à vos instructions, ce qui n'est pas aisé avec lui quand l'on y est aussi longtemps." And he sent a happy report back to Neuchâtel: Duplain was printing the

24. Favarger's diary, labeled "Instructions et renseignements pour J. F. Favarger," is in the STN papers, ms. 1059.
25. STN to Favarger, July 11, 1778.

third edition at 4 reams 15 quires (2,375 copies); he would give the STN a fourth volume to print for the first two editions (that is, at 6,150) and three volumes for the third; and he would pay the next set of bills when they became due in August. Duplain stated frankly that he made 1,500 livres for every volume that he had printed in Lyons instead of in Neuchâtel, but he had recovered from his fits of temper and wanted to maintain good relations with the STN. He justified his refusal to pay its bills for the printing of volume 15 in June by explaining that the STN's shipment of that volume had arrived late, making it impossible for him to collect from the subscribers in time to make his own payments. He expected the STN to come to his aid in difficult moments because it was a partner as well as a printer. Moreover, specie had been unusually scarce in Lyons, and he thought that as Bosset's son-in-law, d'Arnal ought to agree to a three-month delay in the payment of the bills of exchange. All this sounded disconcertingly reasonable to Favarger. Duplain seemed to be charming and to have conceded the most contested points without a struggle. But what had caused this abrupt change of behavior on his part?[26]

In his report, Favarger mentioned that Duplain "m'a dit vous avoir écrit pour permettre que cette troisième édition parût sous votre nom, que cela lui donnerait plus de relief." This letter arrived in Neuchâtel just in time to take the pressure off Favarger's encounter with Duplain in Lyons because it was Duplain's way of replying to the STN's intransigent demand that he pay his bills: "Nous nous sommes déterminés à réimprimer la troisième édition à 4 rames 15 mains," Duplain wrote "Elle est sous presse, et nous espérons délivrer deux à trois volumes en août. Comme nous voulons que cette édition (entre nous soit dit) soit supérieure à l'autre pour l'exécution, la correction etc. afin que s'il en reste quelques exemplaires ils ne nous soient pas à charge, nous avons pensé que pour qu'elle se distinguât, elle parût sous un autre nom. Nous vous prions en conséquence de nous permettre de nous servir du vôtre. Vous paraîtrez avoir acheté de Pellet etc. Envoyez-nous à cet égard votre consentement s.v.p."[27] After

26. Favarger to STN, July 15, 1778.
27. Duplain to STN, July 10, 1778. This letter was a reply to a letter from the STN of July 8, which was virtually an ultimatum on the payment of Duplain's bill.

waiting three days for this attractive offer to sink in, Duplain sent another plea for the STN to delay the collection of its two bills of exchange until August. This gambit succeeded. The STN replied that it would extend the debt and that it would gladly lend its name to the new edition, for which it expected to print "plusieurs volumes."[28] So when Favarger walked into Duplain's shop girded for battle, the quarrel was being settled above his head. And soon after their anti-climactic confrontation, Duplain handed him a freshly printed prospectus which announced that the STN was sponsoring the third edition.[29]

So it was that the third edition came to appear under the STN imprint. But Duplain was moved by more than a desire to make peace with one of his creditors. As his letters to the STN indicated, the first two editions had received a bad name, owing to sloppy printing and poor quality paper. By changing the typographical false front from "à Genève chez Pellet" to "à Neuchâtel chez la Société typographique," he could attract more subscriptions. He could also get relief from another of the typographical troubles that had plagued the quarto throughout the first year of its existence. When Duplain first offered the quarto for sale, he committed himself to provide the entire text of the original edition and the *Supplément* for a retail price of 344 livres—10 livres for each of the twenty-nine volumes of text and 18 livres for each of the three volumes of plates. The foreman of a printing shop had advised him that the twenty-nine folio volumes of the original text and the *Supplément* would come to twenty-nine volumes in quarto format. Soon after the printing had begun, however, this estimate proved to be far too low. The text of the quarto eventually ran to thirty-six unusually thick volumes. Could Duplain expect his subscribers to pay 70 livres more than the contracted price in order to get the text that they had been promised? Panckoucke thought not: "Le public a souscrit pour 32 volumes [that is, twenty-nine volumes of text and three of plates]. Si l'ouvrage en avait un plus grand nombre, je ne sais trop comment on pourrait les lui faire payer."[30] To make matters worse, Linguet had spotted this mistake and had proclaimed it to be a swindle in his widely

28. STN to Duplain, July 15, 1778.
29. Favarger to STN, July 23, 1778.
30. Panckoucke to STN, June 26, 1777.

read *Annales.* So Duplain could expect to stir up a storm of protests and even lawsuits if he raised the price or cut the text. He eventually muddled through this dilemma by levying a charge on only four of the seven extra volumes. In this way he fraudently raised the retail price to 384 livres—just enough to pay for the additional printing without producing an uproar among his customers.

This was a risky policy, which Duplain tried to foist on the public without being detected. Therefore, when he announced the second edition, he avoided mentioning the number of volumes and the total price, and he merely said that it would be sold on the same terms as the first. This crisis had passed by the time Duplain decided to launch the third edition. He therefore tried to protect himself by changing the terms of the subscription and by attributing it to the STN rather than to Pellet. The new subscribers would not be able to hold him to the promises that he had broken in dealing with the old ones, and the new quarto would appear as if it were the old one under new management. For these reasons, the prospectus that Duplain gave to Favarger announced "une troisième édition de l'*Encyclopédie,* qui contiendra 36 volumes in-4° à deux colonnes, proposée par souscription chez la Société Typographique de Neuchâtel."[31] It did not explain how the STN had come to replace Pellet as the publisher of the quarto. Innocent readers might even conclude that the STN was pirating him, for the prospectus gave no indication that Duplain was behind both enterprises, switching his straw men. Only an astute observer would notice the crucial difference between this quarto and the others: this one was explicitly offered as a thirty-six volume set (plus three volumes of plates) costing 384 livres. Duplain even promised that if the new edition should contain more than thirty-six volumes, they would be provided free of charge. Actually, he had freed himself from the obligation to fit the whole work into twenty-nine of them.

The "Neuchâtel" edition therefore represented a clever attempt to cover up a foolish mistake in marketing—an effort to obscure one fraud by perpetrating another, although "fraud" is too strong a term for the common eighteenth-century practice of publishing books under false imprints.

31. Favarger included a manuscript text of the prospectus in a letter to the STN of July 23, 1778.

This strategy pleased the Neuchâtelois because it seemed to assure them of getting at least three new volumes to print. They told Duplain that they were delighted with his prospectus, which they reprinted and circulated among their correspondents. By this time Duplain and Panckoucke had also made peace with the publishers of the rival *Encyclopédies*. So at last the way seemed to be clear for a settlement of the quarto contract.

Opening Gambits of the Final Negotiations

The external and internal problems of the quarto had forced Panckoucke and Duplain to suspend their initial contract negotiations in April. When they resumed their bargaining in July, the situation had been transformed: the agreement with the consortium from Liège had forced Panckoucke to scuttle the plan for the revised edition, and the STN's quarrel with Duplain had raised suspicions about his handling of the third edition. He had actually produced a portion of that edition by July, so it was too late to prevent him from arranging the production so that he could siphon off thousands of livres by exploiting his role as middleman and manager. Panckoucke could only hope to hold him in check by insisting on a strict contract. The final round of negotiations therefore proved to be even tougher and more complicated than its predecessors.

Originally, Panckoucke favored, a conciliatory strategy. He phrased an early draft of the contract to make it seem as though he were rewarding Duplain for his successful stewardship by giving him a free half interest in the full "privilège, droits et totalité des cuivres" of the *Encyclopédie*. The Traité de Dijon had limited Duplain's half interest to the speculation on the quarto. By extending their partnership to all other *Encyclopédie* projects, Panckoucke apparently hoped to win concessions from Duplain in the bargaining over the contract for the third edition. And to make the prospect tempting, his draft proposal specified that one future project would be a quarto edition of all the original illustrations—a clever idea, as Panckoucke was convinced that many of the quarto subscribers wanted the entire collection of plates, rather than three volumes of them, in order to make their sets of the *Encyclopédie* as complete as possible. Nothing would

be easier than to satisfy this ready-made market, for Panckoucke and the STN owned all the original plates, which
could be retouched, in some cases reduced in scale, and then
used to produce the number of copies ordered—at an enormous profit, with a minimum of effort.[32] Panckoucke considered this gambit seriously enough to submit it to the STN
for approval in late April. But it seems very unlikely that he
ever tried it out on Duplain, for by May the success of the
third edition had become clouded by Duplain's quarrel with
the STN and the threat of the rival *Encyclopédies*. By July,
when those difficulties had subsided, it no longer seemed
appropriate to reward Duplain for anything. Panckoucke had
dropped his plan for a supplementary edition of the plates in
order to pursue other quarry, and he was ready to do some
hard bargaining on the third edition.

The bargaining reopened with an exchange of letters
between Duplain and Panckoucke. As the contracting parties
of the Traité de Dijon, they handled the contract negotiations
for the third edition, but they consulted their own associates,
principally the STN in Panckoucke's case and Merlino de
Giverdy in Duplain's. Most of the surviving documents come
from the Panckoucke–STN consultations, so one must follow
the negotiations from the perspective of Neuchâtel, which was
biased but broad enough to afford an excellent view of the
contest.

On July 7, Panckoucke sent to the STN a draft contract he
had received from Duplain and his own point-by-point commentary on it. A week later, the STN replied in kind. These
three documents therefore show how the issues appeared at
all three points of the Lyon-Paris-Neuchâtel triangle.

First, Duplain set the edition at 3 reams 16 quires or
1,900 copies and asked to receive 38 livres for every sheet
printed at that rate as well as 10 livres for every ream of
paper used. Was he putting his costs too high in order to rake
off the difference between the expenses allotted him and his
actual expenditure? Panckoucke addressed the question
squarely and answered it with a "no." Duplain was only asking for a slight increase over the rates set by the Traité de Dijon. He should be granted it; "il ne faut pas chicaner." Panc-

---

32. Panckoucke's draft proposal, in his handwriting, is in the STN papers, ms.
1233, entitled "Addition à l'Acte du quatorze janvier mil sept cent soixante et
dix-sept."

koucke, who liked to do business in a grand manner without quibbling over small sums, was actually treating Duplain generously, for the increase, which seemed small enough when measured by the ream, came to about 8,190 livres in all—the equivalent of at least eleven years wages for a Swiss printer, though somewhat less than 1 percent of the anticipated revenue of the third edition.[33]

The Neuchâtelois agreed that Duplain deserved this increase in the allotted costs, especially for paper, whose price had risen rapidly in response to the demand created by the quarto. They objected only to Duplain's phrasing of this clause because it gave him complete control over the choice of the printers without obligating him to give any volumes to the STN. They therefore asked Panckoucke to get Duplain to reserve at least three volumes for their presses. But they had no other criticism of Duplain's proposals on pricing. Like Panckoucke, they assumed that he would commission the work at cheaper rates than he received from the quarto Association and that he would pocket the difference. They seemed to consider the money he would make in this manner as legitimate remuneration for his activities as middleman, provided it were kept within bounds by the rates set in the contract. Far from hiding this source of income, Duplain had told Favarger openly that it brought in 1,500 livres per vol-

---

33. In his "Observations sur l'Acte de M. Duplain," which he sent with Duplain's draft in his letter to the STN of July 7, 1778, Panckoucke explained his calculations as follows: "Le prix de 38 livres pour 3 rames 16 mains est conforme à l'Acte de Dijon—30 livres le premier mil, 8 livres le deuxième. Il n'y a que 4 mains de différence . . . Dix livres le papier; il est très recherché, et ils ne gagnent pas cinq sols à ce prix." To follow his reasoning, one should remember that eighteenth-century publishers commonly thought in reams rather than numbers of copies. Duplain wanted to print each sheet 1,900 times for the equivalent of what it would have cost at a pressrun of 2,000, according to the rate set by the Traité de Dijon. The difference of 100 sheets or 4 quires in the printing of each sheet of the text came to 16 sous or 104 livres per volume (the volumes normally contained about 130 sheets) or 3,744 livres for the entire printing. The cost of increasing the paper allotment by 5 sous per ream can be calculated thus: since there were about 130 sheets per volume and each sheet was to be printed 1,900 times, each volume would contain 247,000 sheets or 494 reams, and the entire printing of all the thirty-six-volume sets would contain 17,784 reams. At an additional 5 sous per ream, the increase in the paper costs would therefore come to 4,446 livres. The costs were ultimately higher because Duplain later increased the printing by another ream per sheet. But these calculations illustrate the way eighteenth-century publishers reasoned, beginning always with standard costs per sheet of text or *feuille d'édition*.

ume for him.[34] What would be condemned as rake-offs today passed as a normal business practice among eighteenth-century entrepreneurs, who had to cope with manufacturing techniques that belonged somewhere in the middle of the evolutionary scale between the putting out system and the factory.

Panckoucke and the STN also went along with Duplain's demand for a *réviseur* who would correct the copy for the third edition for 1,000 livres a year. But they balked at a long, complex clause of his draft, which seemed to give him unlimited opportunities to exaggerate his costs at their expense. The clause read: "Et moi C. Panckoucke consens . . . qu'ils [Joseph Duplain & Compagnie] en feront faire l'impression à Genève et dans les différentes villes de la Suisse à leur choix; et dans les cas où ils imprimeraient quelques volumes en France, ils seront expédiés à Genève pour y être mis dans un magasin commun; dans le cas où ils s'y refusassent, les risques en saisie du gouvernement seraient pour leur compte, comme dans nos conventions premières, à la charge par nous d'acquitter les frais; qu'il leur sera remboursé les frais qu'ils feront hors ceux compris dans les frais d'impression, comme séchage et étendage seulement, les autres déboursés, comme commis, magasin, ports de lettres et tous autres relatifs, leur seront remboursés."[35]

Duplain's phrasing implied that he could print, store, and distribute the volumes from Lyons, running the risk of confiscation by the government and charging the Association for what it would have cost him to have the volumes transported from Lyons to Geneva and stored there before distribution. He also seemed to ask for a blank check to cover his operating expenses. Panckoucke countered this proposal with the suggestion that Duplain receive, for all of his expenses, the fixed sum of 16,000 livres, or twice the amount provided by the Traité de Dijon (2,000 livres a year for four years). This sum seemed sufficient to Panckoucke because the output of quartos

34. Favarger to STN, July 15, 1778: "Il me dit qu'il y avait environ 1500 livres à gagner pour lui de faire imprimer ici plutôt que chez nous." For its printing, the STN was paid according to the contract price rather than the price Duplain paid his other printers.

35. "Copie du projet d'Acte proposé par M. Duplain pour la troisième impression" dated July 1, 1778 and sent by Panckoucke to the STN in his letter of July 7, 1778.

had doubled since the signing of the Traité. He explained his reasoning in some confidential remarks to the STN: "Vous savez les prétentions excessives de Duplain sur ces frais de magasin, commis. Il faut bien prendre garde de nous laisser entamer à ce sujet. Ecrivez moi de la manière la plus ferme là-dessus, et j'enverrai cette partie de votre lettre à Lyon."[36]

Duel by *Lettre Ostensible*

The STN complied with a letter that reproduced Panckoucke's arguments exactly and also pushed its claim to print at least three volumes. Publishers frequently used such fake letters—*lettres ostensibles* they called them—as weapons in their bargaining with each other and with the French authorities. The STN had written one for Duplain in 1775, and Duplain now parried Panckoucke's maneuver by using a variant of the same technique. On July 19 he wrote a letter to his own associate, Merlino de Giverdy, who submitted it to Panckoucke as evidence for Duplain's argument on the question of expenses. Merlino was then negotiating in Duplain's name with Panckoucke in Paris; and the letter certainly strengthened his hand, for it defended Duplain's position with extraordinary vehemence.

Duplain made it seem as though his backers in Lyons were exerting pressure on him to resist Panckoucke's terms and that he also had to fight off a far more intimidating pressure group, the local clergy. To be sure, he said, Panckoucke had received a *permission* to market the *Encyclopédie,* but the *permission* said nothing about printing it in France. Panckoucke knew full well that Duplain ran enormous risks by doing part of the printing in Lyons. Duplain had had to guarantee his Lyonnais printers against any loss that might be incurred if the authorities cracked down on them. He had set up secret warehouses to store the volumes and had hired agents to ship them out secretly by night. He had even had to take measures to appease the clergy, "ce corps redoutable qui commençait à gronder." All of those operations cost money, a great deal of it; yet the quarto Association was providing him only 2,000 livres a year for expenses. Very well, he had found a way to cut his expenses, by printing in

36. Panckoucke, ''Observations'' in his letter to the STN of July 7, 1778.

Lyons instead of Geneva. But he expected compensation for running the risks that were necessary to make those savings. He wanted to be paid the amount it would have cost to have the volumes transported and stored in Geneva, even though they would really be distributed from Lyons. That sum would constitute a "droit d'assurance," for Duplain was actually functioning as an insurer who assumed the risk of reimbursing the injured parties should the merchandise be lost. Duplain was also a fighter, battling for the common good of the enterprise, while Panckoucke remained idle on the home front, complaining about costs. Panckoucke's attitude was infuriating, and Duplain was ready to do battle against him, too, if he did not agree to an equitable contract: "Comme je ne suis pas . . . d'humeur . . . à faire la guerre à mes dépenses et que toutes ces lenteurs me font bouillir le sang, je te préviens que si dans neuf jours je ne reçois pas le traité signé, je me mets sur le champ en justice. Il est ridicule qu'après les peines que je me donne, les succès merveilleux que je procure, on me conteste une chose juste et que certainement les tribunaux ne me refuseront pas."[37]

One can never be certain about Duplain's motives, but his letter was probably a bluff. He tried to intimidate his associates several times with angry outbursts. His later letters suggest that he did not really fear the clergy, though he might claim to do so in order to extract concessions from Panckoucke. And he had no grounds for bringing his dispute with Panckoucke before a court. Panckoucke evidently read Duplain's letter in this way and let the nine-day ultimatum expire without giving ground. He wrote to the STN that he would stand by the Traité de Dijon—that is, he would set a limit of 16,000 livres on Duplain's expenses. The Neuchâtelois replied that they hoped he would settle with Duplain "de manière à nous garantir de toute tracasserie," and that he would insist on inspecting Duplain's accounts "pour notre commune sûreté."[38] Although they maintained a friendly tone in their letters to Duplain at this time, their distrust of him had grown so great that they now considered it crucial to force a tough contract on him in order to protect themselves from embezzlement.

37. Duplain to Merlino, July 19, 1778, from the copy kept by the STN in Duplain's dossier. Duplain used the "tu" because Merlino was his cousin.
38. Panckoucke to STN, July 21, 1778, and STN to Panckoucke, July 28, 1778.

They had an opportunity to unbosom their worries in August, when Panckoucke made a quick trip to Switzerland. The "Atlas of the book trade" was carrying several speculations besides the *Encyclopédie* in the summer of 1778. He acquired the *Mercure,* began selling off his stock of books, and laid plans to produce gigantic editions of the works of Rousseau and Voltaire, who had recently died within two months of each other, touching off terrific intrigues among the publishers specializing in Enlightenment. Panckoucke, who led the field in the intriguing, traveled to Switzerland in order to stake out a claim to the manuscripts of the two philosophes, and on his journey he stopped by Neuchâtel to discuss the Duplain negotiations with the STN. The discussions resulted in an agreement on the terms to be demanded and in another *lettre ostensible,* in which the STN outlined those demands.

On August 25 the Neuchâtelois sent the letter to Panckoucke, who had returned to Paris via Montbard, where he had discussed still more speculations with Buffon. They added a covering note that showed how closely they were coordinating their publishing policies: "Nous vous envoyons par ce courrier, suivant nos conventions verbales, une lettre ostensible pour M. Duplain pour en faire usage suivant votre prudence ordinaire, et vous prions de nous informer du suivi, en même temps que vous pourrez nous dire quelque chose positif sur les oeuvres de Voltaire, continuant toujours à nous occuper de celles de J. Jacques." By this time, Duplain had increased the size of the third edition by a ream, making it 4 reams 15 quires or 2,375 copies and raising once again the problem of how much to allot him for his printing costs. This was no trivial matter, for Duplain was already skimming off about 50,000 livres from the printing costs allocated for the first two editions, and he now was pressing Panckoucke and the STN to increase the rate that they had tentatively agreed upon for the third edition. At the same time, he clearly planned to print the new edition as cheaply as possible, for he was not only taking competing bids from the local printers but also had set up some presses of his own and was even asking the quarto Association to pay for them. In the *lettre ostensible,* the STN said it would go as high as 44 livres per sheet, though it thought 42 livres more reasonable. At the higher rate, Duplain would make 13,104 livres more than he would at the rate set by the Traité de Dijon. Since he was also to make another 4,446 livres from the increased allotment of 5 sous per ream on the paper, he would do very well by Panckoucke's proposed contract, the STN wrote, phrasing its comments in order to make the desired effect on Duplain: "Nous devons rendre justice avec vous à l'habileté et à l'intelligence avec laquelle M. Duplain a traité cette affaire là dès son commencement. Nous avons nous-mêmes été témoins de toutes les peines qu'il s'est données à Lyon pour les livraisons; et s'il a un bénéfice comme imprimeur qui ne laisse pas d'être considérable sur la quantité, il lui est bien acquis. Mais nous voyons, Monsieur, trop d'inconvénients à nous écarter du Traité fait à Dijon pour les frais à allouer à M. Duplain; nous avons trop d'envie en notre particulier de nous conserver son amitié et nos relations avec lui, pour ne pas devoir écarter tout ce qui pourrait donner matière à discussions tels que les frais de voyage, magasinage etc."

Considering that the STN had virtually called Duplain a bandit in its confidential correspondence with Panckoucke, its *lettre ostensible* represented its true feelings about as accurately as the diplomatic notes exchanged between sovereign states. It adopted a tone that would flatter Duplain without making any concessions to him. In fact, the STN was both more flattering and more intransigent than in its previous fake letter. It stated explicitly that it had no objection to Duplain's making a profit from his role as middleman, but it would not give an inch in its refusal to let him set his own ceiling on his expenses. It dismissed his argument about running great risks with the comment that it would gladly set up twenty presses and print the whole work in Neuchâtel if the danger from the authorities in Lyons were really so great. It also brought up the embarrassing issue of Duplain's miscalculation on the number of volumes necessary to incorporate the entire folio text. A similar *bévue* had discredited Felice's *Encyclopédie d'Yverdon,* it complained, and it noted critically that Duplain had been forced by this mistake to pacify the subscribers and give away the fifteenth volume free. Yet he had not even informed his partners about this deviation from the policy agreed on by all of them in the Traité de Dijon. But the STN said it would not dwell on such unpleasant matters, nor would it insist on "la somme considérable que nous vous avons payée pour ces cuivres, privilèges etc." —a sum that had been spared Duplain. It merely wanted to avoid future quarrels. So it would hold firm to its determination to have all Duplain's expenses covered by explicit allotments in the contract. In this way, harmony would continue to reign among the partners, and they could fully share in the glory of having produced "cette entreprise, qui est en effet la plus belle qui ait été faite en librairie."

The Last Turn of the Screw

Whether or not he expected Duplain to be moved by such dulcet phrases, Panckoucke could produce them as evidence that his hands were tied: he could make no more concessions because his Neuchâtel allies would not stand for it. He and the STN evidently had agreed on this strategy in Neuchâtel because on September 1 the STN wrote to him saying that it hoped he had arrived safely back in Paris, where he would find the *lettre ostensible* and its endorsement of a draft contract with which he proposed to confront Duplain. They added that they would accept whatever arrangements Panckoucke should make and they had complete faith in his ability to overcome Duplain's *prétentions peu fondées*.

Meanwhile, the STN kept up the amicable tone of the letters that it sent directly to Duplain. On August 26 it wrote disingenuously that it hoped he had settled his differences with Panckoucke, since it believed that Panckoucke, like the STN itself, wanted to make all possible concessions to his wishes; and it closed by expressing its own wish for the copy of the fourth volume Duplain had promised. A week earlier it had informed him that it was nearing the end of volume 24, its third; so its need for new work had become critical. At the same time, Duplain's next set of payments were about to become due. On August 29, the STN notified him that his debt had mounted to 23,723 livres and that this time it expected its bills of exchange to be acquitted without any trouble. It was still waiting for its fourth volume, it added. Duplain replied that he only owed 2,957 livres and that he would not pay a penny more. He produced his own version of the STN's printing account, which differed greatly from the version he had received from Neuchâtel and seemed to provide pretexts for endless haggling. And he showed his determination to haggle rather than pay by arguing that the STN was committed to finance all its purchases of paper and to go without reimbursement until it had delivered the last printed sheet of each volume to him in Lyons. To support this argument, Duplain cited an irrelevant passage from his agreement with the STN of May 28, 1777, and then said that the strain on his finances was so great that the STN should not expect to be paid in any case.[39]

The STN reacted with predictable fury. Duplain was re-opening questions that they had settled in July, when the Neuchâtelois had agreed to extend his debt until the end of August and he had promised to let them print a fourth volume of the first two editions. Now he had produced another specious argument in order to hold back both the payment and the volume. The STN warned him that it could no longer make concessions on either issue, that failure to honor its bills of exchange would throw its finances into chaos, and that his conduct raised grave questions as to his motives.[40] Duplain had made his motives clear enough in a letter of August 21: "Nous sommes en difficulté avec M. Panckoucke pour nos frais. Dès que cela sera terminé, nous vous enverrons un volume." But the Neuchâtelois had ignored that remark, and so he repeated it in his answer to their protest. He would stick to his refusal, he wrote on September 15; and "quant au nouveau volume promis, nous attendons la réponse de M. Panckoucke sur un objet qu'il nous conteste, quoique promis, et duquel nous ne nous départerons point."

Actually, the Neuchâtelois had got the point long ago; and while Duplain intensified the pressure on them, they tried to get Panckoucke to relieve it. On September 15 they inquired anxiously about the negotiations, recounted their latest difficulties with Duplain, and concluded, "En vérité, Monsieur, de tels procédés multipliés nous deviennent très désagréables. Nous en sentons d'autant plus la nécessité de régler compte avec cet associé-là et aspirons à l'époque convenue entre nous pour travailler à cette importante opération." After another week had gone by without any word from Panckoucke about the negotiations, the Neuchâtelois began to feel desperate and filled another letter to him with more urgent complaints about Duplain's tactics, which they now described as extortion. He was holding back the money and the copy to force them to

39. Duplain to STN, Sept. 2, 1778.
40. STN to Duplain, Sept. 9, 1778.

grant better terms in the contract, they said. But vehement as they were in condemning his conduct, they virtually asked Panckoucke to give in to it. Someone had to bend, they explained, and the pressure on them was becoming unbearable: "Vous concevez, Monsieur, dans quel embarras son opiniâtreté doit nécessairement nous jeter . . . Tout cela en vérité est très désagréable, et nous vous prions instamment d'aviser au moyen d'y mettre fin le plutôt qu'il sera possible." But if they were willing to weaken their position on the clause about Duplain's expenses, they were more determined than ever to insist that all the quarto associates meet in Lyons to make a close inspection of his financial management: "Une conduite si déraisonnable démasque ses vues. Elles ne peuvent tendre qu'à extorquer notre acquiescement à ses prétentions, et vous saurez comme nous apprécier ce plan-là . . . Tout cela nous porte à conclure qu'il serait fort bien à souhaiter pour le bien de la chose que l'assemblée des intéressés à Lyon . . . pût être anticipée et avoir lieu le plutôt possible."[41]

The pressure hurt the Neuchâtelois most because the passage of each day brought them closer to the maturation date of their bills of exchange and to the end of the printing of volume 24, the last of the three volumes Duplain had given them. They had written seventeen different bills of exchange on Duplain, worth 19,380 livres in all, and if he refused to acquit them, their bearers would descend on d'Arnal, demanding immediate payment. Should d'Arnal fail to provide the money, the STN would face not merely lawsuits but also the collapse of its credit. Of course it could sue Duplain, but a long and bitter battle in the courts could fatally damage the whole enterprise. In any case Duplain had not given it time to make the threat of litigation effective because it had to produce 19,000 livres at once or suspend its payments—that is, suffer irreparable damage to an excellent commercial reputation, which it had built up by nine years of hard work. Duplain never backed down from his refusal to honor the bills of exchange, despite the entreaties of the STN and d'Arnal. By emergency action on the Bourse, d'Arnal came up with the 19,000 livres. But this heroic *intervention* strained his own capital resources badly, cost a great deal of money in brokerage fees and short-term loans, and only postponed the

41. STN to Panckoucke, Sept. 27, 1778.

reckoning with Duplain until the next series of notes matured.[42]

In a letter of October 3, 1778, the Neuchâtelois complained bitterly to Duplain about the financial crisis he had inflicted on them. They meant to collect their entire debt next time, they warned, and were ready to defend their case "devant tous les tribunaux du monde." But they had to contain their anger, because they needed to persuade him somehow to send them the copy for the fourth volume. By this time two volumes of the third edition had already been printed and distributed and the STN's workers had come within a few days of finishing volume 24. "Ce volume fini, nous aurons 20 ouvriers sur les bras qu'il faudra renvoyer ou occuper à d'autres objets, toujours avec une grande perte pour nous, à moins que vous n'ayez l'honnêteté de nous expédier, à lettre vue, un fragment assez fort de copie," the STN implored. And then it confessed its willingness to play his game against Panckoucke: "Nous l'avons sollicité à vous accommoder. Nous ne faisons aucun doute qu'il ne s'y prête et que vous ne consentiez de votre côté à prendre des arrangements propres à prévoir entre nous toute sorte de difficultés. Vous sentez comme nous que cet écueil serait aussi funeste qu'une contrefaçon ou autre malencontre typographique." No contract, no copy, Duplain replied.[43] There was nothing the Neuchâtelois could do but suffer on the sidelines while their two powerful partners fought it out. The bluffing, grappling, and crushing must have been fierce; but no record of it has survived. All one can know is the result: on October 10 in Paris Panckoucke and Duplain's agent Merlino de Giverdy signed the contract for the third edition of the quarto *Encyclopédie*.

The Contract

The contract represented a compromise, but Panckoucke carried most of the contested points (see Appendix A. XIV).

42. The STN's financial difficulties with Duplain are explained in its correspondence with d'Arnal, notably in d'Arnal's letter to the STN of Sept. 24, 1778.

43. Duplain to STN, Oct. 9, 1778: "Nous attendons que M. Panckoucke ait fini le traité nouveau à signer pour vous envoyer un nouveau volume, et il attend, dit-il, votre ratification. Cela ne nous regarde pas. Tout ce que nous pouvons dire à M. Panckoucke, c'est que nos frais sont immenses et que nous n'aurions jamais pensé qu'ils fussent si considérables. Sitôt donc que nous saurons que M. Panckoucke a consenti, le nouveau volume partira."

As he desired, it set the printing rate at 44 livres per sheet and increased the price of paper by 6 sous to 10 livres per ream. Duplain could not have put up much of an argument against these rates because they would bring him 17,550 livres more than those set by the Traité de Dijon, which were already providing him with an enormous profit on the printing. But the new contract kept strictly to the old allotment for Duplain's general expenses, giving him 16,000 livres for managing the production and distribution of 8,000 *Encyclopédies* in place of his previous allowance of 8,000 livres for 4,000 *Encyclopédies*. Duplain therefore had conceded defeat on the issue that had produced the fiercest fighting. He found some compensation, however, in articles 3 and 5, which authorized him to set up a printing operation in France and to charge the quarto Association for the equivalent of what it would have cost to have those volumes transported to Geneva for storage and distribution. Far from mentioning any obligation to give some of the printing to the STN, this contract for the "Neuchâtel" edition indicated that the printing was to take place "à Lyon et autres villes de France" and that the Association would pay the Geneva–Lyons transport costs in case Duplain considered it "convenable" to have some of the work done in Switzerland. It even committed the Association to pay Duplain's costs for buying his own presses and for bringing workers to Lyons. So it signified a shift from Geneva to Lyons in the production as well as the distribution of the book. Duplain promised to reimburse his associates for any losses in case the French authorities cracked down on his operation, but he demanded payment for assuming this risk: thus the fictitious transport allotment. This "insurance" also compensated him for receiving only 16,000 livres "pour . . . un objet de dépense annuelle infiniment plus considérable." By writing this phrase into the contract, Duplain indicated that he had not taken defeat gracefully in the controversy over his expenses. Panckoucke tried to protect himself from further squabbling on this issue by two other provisions. First, the allocation for the nonexistent transport of the merchandise from Lyons to Geneva should not be effective if the STN opposed it. Second, should the associates disagree over this or any other aspect of the enterprise, they were to choose arbitrators and were to be bound by the contract to accept the arbitrators' verdict, without appealing to the courts. This provision showed the businessmen's distrust of the expensive and inequitable judicial system in France, but it also indicated their distrust of one another. They had not been able to agree without doing battle, and their agreement established a truce, not peace.

The bitterness and belligerency left over from the contract dispute had poisoned relations between Neuchâtel and Lyons, but it had not prevented the STN's shop from continuing work. As soon as it heard that the contract had been signed, the STN shot off a letter to Duplain, requesting its copy in a tone of somewhat strained bantering: "Au cas que vous eussiez encore cette copie si ardemment postulée et si bravement défendue [that is, in case he had not sent it already], veuillez le faire partir sans délai et par la route la plus courte . . . Au reste, nous sommes remplis de confiance dans vos habiletés. Un pilote aussi actif et aussi expérimenté doit infailliblement conduire la barque au port."[44] Unamused and unflattered, Duplain replied that he would not provide the copy until the Neuchâtelois had accepted the contract because clause 5 required their ratification. The STN received a copy of the contract on October 18 and immediately notified Duplain that they accepted it, despite some misgivings about clause five. They wished that it had set a fixed price for the fictitious transport allowance, but they would not refuse to endorse it because they yearned for an end to "la longue contestation." Now, at last, they hoped he would send the fourth volume and perhaps even would let them print it at a pressrun of 8,000—that is, for all three "editions." On the same day, they sent a franker letter to Panckoucke. They worried that clause 5 might give Duplain a loophole for inflating his "insurance" expenses. But they congratulated Panckoucke for his victory on two key points: the limitation of Duplain's general expenses to 16,000 livres and the prevention of further disputes by the provision for arbitration. If they could only give Duplain's accounts a thorough going-over, the Neuchâtelois concluded, they might emerge unscathed from the whole affair. After one more exchange of letters, Duplain finally surrendered the copy. It was volume 35, and the STN was to print it at 6,000 for the first and second editions. So it did not have to fire its work force. It had

44. STN to Duplain, Oct. 14, 1778.

won a fourth volume for the first two editions. But it had not yet received anything to print for "its" edition, which was still far from being what eighteenth-century publishers called *une affaire bouclée.*

The quarto associates had a contract, however. The last round of their long and painful negotiations had developed into a duel by *lettre ostensible.* The duel had ended in a draw, and Duplain had resorted to a more powerful weapon: extortion. He had inflicted wounds on the vulnerable, Neuchâtel flank of his adversary, but he had not forced Panckoucke to sue for peace in Paris. Instead, Duplain himself had made the crucial concessions in the final compromise. And far from feeling pacified by the peace of October 10, he began to look for other ways to trim off fat from the overgrown profits of the quarto, while his partners hardened in their resolve to make a closer examination of his suspicious conduct. So in agreeing on terms for the expansion of his enterprise—"the most beautiful ever in the history of publishing," according to the STN—the quarto associates infused it with so much duplicity, hatred, and greed that it threatened to blow up in their faces. Its very success had increased its explosiveness. But before recounting the struggle that led to the final explosion, it is necessary to backtrack through 1777 and 1778 in order to pick up the story of the quarto's external battles.

I V

PIRACY AND TRADE WAR

While the quarto associates quarreled over their domestic affairs, they had to defend themselves from outside attack. Only a militant foreign policy could protect their market from raids by rival publishers who wanted to profit from the extraordinary demand for relatively inexpensive *Encyclopédies*. That demand had remained hidden until the success of the quarto stunned the publishing world, revealing a market for Encyclopedism throughout the length and breadth of the land. As the STN put it in a letter to Panckoucke about the need to coordinate their defense against *Encyclopédies* being produced in Lyons, Lausanne, and Liège, ''C'est une chose admirable de voir à quel point et en combien de sens on s'occupe de l'*Encyclopédie* depuis notre quarto. Il semble que nous ayons électrifié tout le monde.''[1]

Pirate Raids

Having originally floated his quarto as a pirated edition, Duplain had no way of burying his treasure after he turned legitimate. Other pirates got wind of it and raced to the attack. In August 1777 he heard that some Geneva publishers were planning to counterfeit the quarto and to undersell it by 100 livres a set. He rushed to Switzerland and confronted

1. STN to Panckoucke, June 7, 1778. Similarly, the STN wrote to Belin, a bookseller in Paris, on April 18, 1778: ''On dirait, comme vous l'observez, que le succès de l'*Encyclopédie* in-quarto a tellement électrisé tout le monde que chacun veut avoir part à ce grand dictionnaire.''

them with a counterthreat: he and his associates would not only defend their market with the full force of the French state but also ruin it for their competitors, for they could liquidate their quarto and produce their revised *Encyclopédie* in time to destroy the demand for any other quartos. The project for the revised edition, Suard's *refonte,* therefore served as a weapon in the defense of the quarto; and it proved to be effective because the Genevans abandoned their plan.[2] Meanwhile, however, a group of booksellers from Toulouse plotted to print a counterfeit edition in the shop of Gaude père et fils in Nîmes. This project also foundered, probably because Panckoucke's privilege could not be attacked openly within the kingdom.[3] But as it sank from sight, rumors of another counterfeit quarto began to circulate, this time from Lausanne. Ostervald investigated them and reported that the Lausannois only intended to pirate a Brussels edition of the thirty-volume *Encyclopédie de jurisprudence.*[4] But even false alarms were disquieting. The industry of pirate publishing was too aggressive and too well developed everywhere around France's borders for the quarto to be free from attack.

Having skirmished with pirates for years, Panckoucke knew that they would try to capture the market for the *Encyclopédie.* But he felt confident that he could ward them off by relying on his privilege and protections, the strategy that he had originally intended to use against Duplain. When informed of the danger from Geneva, he replied that he already knew about it, and about raids being prepared from other quarters as well, but that he had erected a solid defense

2. Duplain explained his strategy to the STN in a letter of Aug. 18, 1777: "L'unique but de notre dernier voyage à Genève a été d'empêcher la contrefaçon dont nous étions menacés. La seule man:ère de s'en garantir c'est d'annoncer à ces enragés que d'ici en février 1779 nous aurons tout livré et que de suite après nous annoncions une édition augmentée par les premiers éditeurs. A peine auront-ils fait quatre à cinq volumes dans l'intervale qu'il y aura jusqu'à notre entière livraison. Qu'en feront-ils? Comment leur édition entrera-t-elle en France? Elle trouvera de grandes oppositions. Toutes ces raisons dites verbalement et non par lettre arrêteront les plus obstinés. Faites-en part à la Société de Lausanne s.v.p.''

3. The only information on this project is a passing remark in a letter to the STN on April 6, 1778, from a binder and bookdealer in Toulouse named Gaston, who had collected a huge number of subscriptions for the quarto *Encyclopédie* and would have collected still more, he claimed, had his campaign not been undercut by an edition ''que les libraires de Toulouse font imprimer à Nîmes chez M. Gaude.''

4. STN to Duplain, Aug. 23, 1777.

against them: "On a voulu m'inspirer des craintes sur une pareille entreprise à Avignon. Au reste, j'ai ici tout disposé de manière que dans aucun cas aucune de ces éditions n'entrerait en France, et sans la France, point de succès."[5] Moreover, pirates bluffed as often as they attacked. So Panckoucke advised his partners to ride out the threats, trusting in his ability to defend the French market. At first this strategy worked well: the *Encyclopédies* of Geneva and Avignon proved to be attempts to frighten the quarto Association into paying protection money. But this was a dangerous game. The next assailant might mean business, and the quarto publishers would not be able to tell whether he was faking until he had launched his attack.

After the pirates from Geneva, Toulouse, and Avignon had disappeared over the horizon, Duplain was assaulted from within Lyons by two men of his own breed, Jean Marie Barret and Joseph Sulpice Grabit. Both dealers had built up a large wholesale trade, much of it in prohibited books, which they smuggled in from Switzerland or occasionally had printed in Lyons. Barret, for example, secretly produced an edition of Rousseau's works in 1772, which illustrates the way he and his colleagues did business. Regnault, the future Lyonnais associate of the quarto *Encyclopédie,* learned of the printing and demanded to be given a chance to buy a half interest in it. Fearing that a refusal would result in a denunciation to the police, Barret reluctantly complied. Soon afterward, however, he discovered that Benoît Duplain, the father of Joseph, was trying to beat him to the market by producing a rival clandestine edition in two separate shops. Instead of informing Regnault, Barret let slip a remark that commercial difficulties now made him willing to sell the other half of his Rousseau. Regnault eagerly bought him out, only to become trapped in a trade war with Duplain, which Barret took pleasure in watching from a safe distance. "La méchanceté de mes confrères m'est si bien connue que je crois ne pouvoir prendre trop de précautions," was the conclusion that he drew from the experience.[6]

Grabit lived by the same code. In 1773 he tried to force Saillant and Nyon of Paris to buy a thousand volumes 17–22

5. Panckoucke to STN, Sept. 9, 1777. See also the report on the piracy threats in STN to Panckoucke, Aug. 31, 1777.
6. Barret to STN, Dec. 28, 1772.

from a pirated edition of their own *Histoire de France* by Velly. He threatened to counterfeit volumes 1–16 and to undersell them unmercifully if they refused to collaborate in the undermining of their own privilege. And he sent his threats anonymously through the STN, as if he were a Swiss pirate instead of a Frenchman, who might be expected to defend the law of his land against foreign interlopers. His plot failed, but it illustrates the character of the book trade in Lyons, where the only law that counted was the iron law of the marketplace: maximize profits.[7]

Barret and Grabit naturally jumped at the chance to cash in on the most profitable enterprise that any bookdealer of their generation had ever seen in Lyons: Duplain's quarto *Encyclopédie*. They decided to produce their own quarto—or at least to produce enough of it to convince Duplain that they would go ahead with the printing, presumably using presses in Switzerland, if he did not buy them off. As veteran pirates, they knew that they would have to make a considerable investment for their threat to be credible. Perhaps they even meant to go through with it: it is as difficult to know whether they were bluffing today as it was two hundred years ago. In any case, they printed at least six sheets (48 pages) at a pressrun of four reams eight quires (2,200 copies) and then told Duplain that it would cost him 27,000 livres to get them to stop. That was an enormous ransom, the rough equivalent of a lifetime's wages for one of their printers. Would Duplain pay it?

The STN favored the hard-line Panckoucke policy. "Nous n'avons pas ignoré le projet de Barret," they wrote to Panckoucke. "Mais il nous a paru que son but n'était que d'obliger les entrepreneurs de l'in-quarto à lui donner quelque part au gâteau, et nous avons pensé qu'il vous serait encore plus facile de faire valoir votre privilège contre un libraire du royaume que contre un étranger." But Panckoucke himself had doubts. He later confided that of all the threats to the quarto, "Celle-là m'inquiétait véritablement." So he left the decision up to Duplain, who was best qualified to make it, thanks to his own experience as a buccaneer. Duplain decided to capitulate. In fact he surrendered formally, by contract,

7. This complicated plot may be followed from the correspondence between Grabit and the STN on the one hand and the STN and Saillant and Nyon on the other in late 1773 and early 1774.

and the treaty that he signed with Barret and Grabit is one of the most extraordinary in the string of documents attached to the quarto enterprise (see Appendix A. XIII).[8]

In weighty, legal language, it created a pirates' nonaggression pact. Barret and Grabit solemnly promised to abandon their edition and to refuse any connection with any other *Encyclopédie*. As an indication of their earnestness, they were to deliver 2,200 copies of their first six sheets to Duplain. And in return, Duplain paid them 3,000 livres immediately and promised to pay another 24,000 livres in a year. But he deposited his promissory note with a notary, who was not to release it unless a final condition had been fulfilled: the great bulk of Duplain's quartos had to be sold. If he could produce 500 unsold sets in June 1779, he would not have to pay the 24,000 livres. In that case, however, Barret and Grabit would have the option of buying the 500 sets at half the wholesale price and then collecting the 24,000 livres. This complicated formula would permit them to buy the books for 49,500 livres instead of 147,000 livres, and it protected Duplain by canceling the ransom—or *indemnité* as the contract put it—in case his market collapsed, owing to some unforeseen calamity, like another pirate raid.

This clause seemed reasonable enough, but Duplain may have had a secret motive for inserting it. In discussing the third edition with the STN, Panckoucke later explained that Duplain had decided to increase it by a ream "pour nous dédommager des arrangements qu'on a été obligé de prendre avec Barret.'"[9] One ream per printed sheet meant 500 extra copies, just the number that the Barret-Grabit contract set in the escape clause through which Duplain hoped to rescue his 24,000 livres. Perhaps he planned to print an extra 500 sets on the sly, to produce them as an excuse for refusing the ransom, and to dump them on Barret and Grabit for 48,700 livres. That reprisal would have made an appropriate ending to an unusually tricky episode. But Duplain never executed it. His accounts later showed that Barret and Grabit collected all their money and so succeeded in one of the biggest and boldest raids known to have occurred in the history of pirate publishing. As Panckoucke later observed to Beaumarchais during some negotiations over the Voltaire manu-

8. STN to Panckoucke, June 7, 1778, and Panckoucke to STN, July 7, 1778.
9. Panckoucke to STN, July 21, 1778.

scripts, "Quand on ne peut détruire les corsaires, la bonne politique veut qu'on compose avec eux. C'est la loi de la nécessité."[10]

The Octavo Publishers and Their *Encyclopédie*

While Duplain pursued this policy in Lyons, the STN was sparring with a consortium from Lausanne and Bern, which planned to produce an even smaller, even cheaper *Encyclopédie* than the quarto. It would be an exact copy of the quarto, reduced to the octavo format and marketed at little more than half the price (225 instead of 384 livres). This threat looked far more dangerous than any of the other attacks. But after fending off pirates from Geneva, Toulouse, and Avignon and surrendering 27,000 livres in ransom to the Lyonnais, Panckoucke and his partners decided it was time to stand and fight; they were soon engulfed in a full-scale commercial war.

The publisher of the octavo, the sociétés typographiques of Lausanne and Bern, were formidable opponents because they had years of experience in the front ranks of the pirates who preyed on the French book trade. They operated like the STN, counterfeiting French books from the safety of Switzerland and underselling French publishers everywhere in Europe, including France. The Société typographique de Berne had been supplying French readers with counterfeit books for twenty years when it announced the publication of the octavo *Encyclopédie*. True, it had only begun to print its own works; and the heaviest portion of its wholesale trade went to Northern Europe, especially Germany. It always sent wagonloads of French books to the fairs at Leipzig and Frankfurt. But it also commissioned other printers to produce books that it had smuggled into France. And it speculated increasingly on the French market in the late 1770s, when it came under the influence of a shrewd young man called Pfaehler, who rose from clerk to codirector of the Society and who had a talent for piracy. Pfaehler seems to have been cut from the same cloth as Jean-Pierre Heubach, the director of the Société typographique de Lausanne. Heubach had moved

10. Panckoucke to Beaumarchais, March 10, 1781, ms. Fr. d.31, Bodleian Library, Oxford.

from Mainbernheim, Anspach, where he had worked as a binder and printer, to a partnership in Geneva with Jean Samuel Cailler, a publisher who specialized in the most extreme kind of prohibited French books. Having acquired valuable experience in the underground trade, Heubach then set up his own publishing business in Lausanne. By 1771 he had three presses and a work force of fifteen men. He expanded his shop and took on new partners in 1773 and again in 1774, when he reorganized his company and called it the Société typographique de Lausanne. The business prospered. In 1785 Heubach had seven presses, a stock of books worth 27,388 livres, a town house, 15,000 livres' worth of land in the country, and total assets of 133,190 livres. He probably provided the original impetus for the octavo *Encyclopédie,* but Pfaehler enthusiastically backed him, and the combined support of their sociétés typographiques made it a serious threat to the quarto.[11]

The threat had not seemed grave at first. In March 1777 rumors of an *Encyclopédie* project reached the STN from Lausanne, but the Neuchâtelois dismissed them with a joke: "On fait donc cette année des *Encyclopédies* comme des brochures."[12] The rumors had linked the project with the Société typographique de Lausanne by August, when Ostervald looked into them in Lausanne and was told that they had come from the Society's plan to counterfeit the *Encyclopédie de jurisprudence.* But that report, which apparently came from the Society itself, could have been a *false* false alarm—an attempt to mislead Ostervald while completing the preparations for an attack on the quarto.

The attack came two months later. On November 4, the *Gazette de Berne,* published by the Société typographique de Berne, announced the opening of a subscription for an octavo *Encyclopédie* "exactement conforme à l'édition quarto qui s'imprime chez Pellet à Genève." In its next issue, the *Ga-*

11. This information has been culled from the huge dossiers of the two publishing firms in the papers of the STN. Each société typographique evolved in its own way, attracting many different partners and investors en route. When the Société typographique de Berne was founded in 1758, it had some of the characteristics of a literary club. The Société typographique de Lausanne was always more commercial and more oriented toward France. Its founding is described in a circular letter issued on Feb. 22, 1774.

12. STN to François Grasset of Lausanne, March 17, 1777.

zette explained the terms of the subscription in a way that showed how it was meant to undercut the quarto:

Les 39 [a slip for 29] volumes de discours de l'*Encyclopédie* annoncée par la Société typographique de Lausanne seront imprimés page pour page sur l'édition quarto de Genève, et le 30ème sera un Supplément très intéressant dans lequel on trouvera les découvertes faites dans les sciences et dans les arts pendant le cours de cette édition, avec un catalogue des meilleurs ouvrages sur ces sciences et ces arts publiés chez toutes les nations avec un succès mérité. Ces 30 volumes coûteront 150 livres de France, et les 3 volumes de planches, pour ceux qui les souhaiteront, coûteront 45 livres; en tout 195 livres de France. Somme très modique pour un pareil ouvrage.[13]

The octavo might be smaller in size than the quarto, but it would be larger in scope; for it would contain a bonus Supplément in addition to the old four-volume text of *Suppléments,* which "Pellet" had blended into his edition. And most important, the octavo would be far cheaper—44 percent of the price of the quarto for those who took the option of doing without the three volumes of plates.

It was a brilliant sales plan, which employed the strategy that Bosset of the STN had tried to get his partners to adopt in 1776: it offered the fullest possible text on the cheapest possible terms to the broadest possible audience. Of course the octavo would look puny next to the magnificent folio editions and the magisterial quartos. The quarto publishers later referred to it scornfully as a "miniature" and an *Encyclopédie de poche,* which would blind its readers by its small type.[14] But the pirates had always prospered by purging their editions of *luxe typographique,* as they called it. In this way, they undersold their French competitors and kept their profits high. Duplain had adopted this policy in the first place by reducing the text to the quarto format and eliminating most of the original plates But the octavo publishers took it much further, and by pursuing their course as pirates, they unwittingly furthered a general cultural movement: the popularization of the Enlightenment.

13. *Gazette de Berne,* Nov. 19, 1777. See also the similar notice in the *Journal encyclopédique* of Jan. 1778. The price eventually came to 225 livres because the text of the quarto ran to thirty-six instead of twenty-nine volumes, and the octavo was compelled to do likewise.

14. STN to Panckoucke, Dec. 18, 1777, and Feb. 22, 1778.

The Origins of the Quarto-Octavo War

Publishers frequently used subscription announcements as trial balloons. If they failed to elicit much response, they would drop the project, having lost only a pittance for the publication of the prospectus or having gained a ransom from some competitor who took them seriously. But the publishers of the octavo *Encyclopédie* meant what they said in their announcement, for in early November Duplain made a distressing discovery: "Nous avons des avis certains que la Société typographique de Lausanne fait fondre une fonte petit texte gros oeil et se propose de faire notre *Encyclopédie* in octavo. Vous sentez que c'est une entreprise ridicule mais qui néanmoins nous portera le plus grand coup par rapport aux annonces qu'elle ne manquera pas de faire." Duplain shot this warning off to the STN with a request that it send an agent on an emergency mission to Lausanne. Although the order for the special font of type could mean that the Lausannois had passed the point of no return, they might retreat if confronted with the arguments that Duplain had used to bring around the Genevans two months earlier. "Il faut s.v.p. à lettre vue dépêcher un ambassadeur pour parer le coup . . . Il n'y a pas une minute à perdre pour vous rendre à Lausanne . . . Vous pouvez dire à Messieurs de Lausanne qu'à compter du huitième volume tous nos volumes auront 120 feuilles, qu'ils ne pourront pas entrer en France, et que sitôt notre édition finie, ce qui sera à la fin de 1778, nous en publierons une augmentée par les d'Alembert et Diderot. Ne perdez pas un instant s.v.p."[15] Then, without informing his associates, Duplain prepared a counterattack of his own. He printed a prospectus announcing that the quarto group would produce a rival octavo edition at a still cheaper price. If threatened by him on one side and cajoled by the STN on the other, the Lausannois might capitulate or accept an unfavorable compromise.

The Neuchâtelois could be expected to take a conciliatory line toward the Lausannois because they were friends. The two firms had cooperated closely for years and even had collaborated on pirated editions, sharing costs, profits, and trade

15. Duplain to STN, Nov. 11, 1777.

secrets. They were also united by personal ties, especially the friendship between Ostervald and Jean-Pierre Bérenger, who handled the literary aspect of the Lausanne society's business. Bérenger was an anomaly in the publishing world. A gentle person and a man of principle as well as a man of letters, he had been caught up in the agitated politics of Geneva in the 1760s. The oligarchic Petit Conseil drove him out of the town and burned four of his works, including his six-volume *Histoire de Genève,* for championing the cause of the underprivileged Natifs. For a while he drifted from project to project, fixing his hopes on a professorship in Germany, a tutoring job in Poland, and a pension near Lausanne. Finally, with 4,000 livres from the sale of his library, he acquired a one-tenth interest in the Société typographique de Lausanne and accepted a position as its *homme de lettres,* for 50 louis a year, which was enough to keep his family alive. At each step he had turned for aid and comfort to his good friend Ostervald. So it was natural for Ostervald to turn to him after receiving Duplain's SOS and rushing off to Lausanne.[16]

Ostervald presented Duplain's arguments to Bérenger, Bérenger put them to the directors of the Société typographique de Lausanne, and the directors replied through Bérenger to Ostervald that they would not retreat an inch. They had planned to produce a quarto *Encyclopédie* even before Duplain launched his first subscription, they said. He had got his announcement out first; so they had postponed their project and had offered to buy an interest in his, but they had never even received an answer to their proposal. They therefore had decided to produce an octavo edition, and they meant to go through with it. If Panckoucke had influence in high places, they could rely on protectors of their own, and they could always fall back on smuggling. They dismissed all talk about the revised edition and the rival octavo as bluffing. They would not be intimidated, they assured Ostervald, and they would never abandon their octavo. But they wanted to maintain good relations with the STN, perhaps even to form ''une

16. This information comes primarily from Bérenger's dossier in the STN papers and also from the article on him in the *Dictionnaire historique et biographique de la Suisse* (Neuchâtel, 1924).

espèce de ligue'' among the three Swiss sociétés typographiques against the French. They would gladly help the STN market its quarter share in the quarto if it would do the same for them with the octavo, and they were ready to listen to any other propositions it wanted to make.[17]

The combination of intransigence and appeasement in this reply probably represented an attempt to drive a wedge between the STN and its French associates. The Lausannois could assume that some split had opened up between Lyons and Neuchâtel because Ostervald had not heard of Duplain's octavo prospectus when he arrived in Lausanne. Either the quarto group was unable to coordinate its defense or it was falling apart. And none of the principals had heard from Panckoucke.

The STN had informed Panckoucke of the octavo threat on November 16, the day of Ostervald's departure for Lausanne. About seven days later, it received a vehement reply in which Panckoucke reviewed all his experiences with the *Encyclopédie:*

Je ne suis point surpris de la concurrence que nous éprouvons. Il y a près de 8 ans que cette affaire a été pour moi une occasion de supplices. Dom Félice n'est-il pas venu nous barrer par son in-quarto [that is, the *Encyclopédie d'Yverdon*] au moment de la publication de l'in-folio? N'ai-je pas été mis à la Bastille? 6,000 volumes in-folio n'y sont-ils pas restés six ans? Les portes de la France n'ont-elles pas été fermées deux fois? L'impôt enfin de 60 livres n'a-t-il pas mis le comble à tout ce que nous éprouvions? Notre constance est venue à bout de tous ces obstacles. L'affaire de Genève a réussi. Il en sera de même, Monsieur, de notre entreprise. Cette édition in-octavo peut donner quelques alarmes, mais elle ne nous nuira pas. Je doute même qu'elle s'exécute. On pourra faire 2 à 4 volumes, mais on en restera là. Vous verrez si je ne suis pas bon prophète. Il est fou d'imprimer

17. Bérenger to Ostervald, Nov. 21, 1777. After discussing the situation at length, Bérenger summarized his associates' position as follows: ''Qu'au reste, ils ne croyaient point à cette nouvelle édition de l'*Encyclopédie* sur un meilleur plan, ni à la défense d'entrer la leur en France; que si M. Panckoucke avait du crédit, il n'était pas le seul; qu'ils avaient prévu cet inconvénient, qui lors même qu'il existerait ne les arrêterait pas, parce qu'on fait entrer en France tous les jours un grand nombre de livres défendus; qu'ils ne croyaient pas non plus à cette édition octavo qu'on annonce à Genève pour les intimider; qu'enfin ils étaient résolus de poursuivre, d'autant plus qu'il dépendait d'eux de se mettre en sûreté, puisque s'ils voulaient, ils ne resteraient chargés que d'un sixième de l'entreprise.''

l'*Encyclopédie* en petit texte. Au reste, nous serons ici défendus. J'attends le retour du magistrat pour mettre tout sous ses yeux. Je vous promets bien que cette *Encyclopédie* n'entrera point en France.[18]

Although he felt confident of his ability to defend France's borders, Panckoucke thought that something might be gained by luring the octavo publishers into negotiations. If they would drop the octavo for twenty-five free sets of the quarto, he would gladly buy them off. Perhaps they could be persuaded to postpone it for two years in return for a partnership in some future association that might market the octavo and the revised edition together. But they might take willingness to negotiate as a sign of weakness. It was crucial to prevent them from detecting any fear, though Panckoucke sounded apprehensive enough as he vacillated between a policy of negotiation and one of uncompromising hostility. In the end, he recommended that the STN take the toughest possible line and trust to his protections.[19]

A week later Panckoucke had regained his composure and rethought his strategy. It now seemed to him that the Lausannois could not be stopped: their *Encyclopédie* was certain to succeed, and it could damage the market for the quarto. He therefore advised the STN to try to win them over by offering a compromise: they would have to suspend publication for two years, but in return they would be able to sell their books legally, under the cover of Panckoucke's privilege. He and the STN would join them in a three-way partnership, which would wring every last sou out of the octavo market. To be sure, this arrangement excluded Duplain, but Duplain did not have to know about it, and it would take effect after the expiration of the Traité de Dijon, when he would no longer

18. Panckoucke to Bosset, Nov. 19, 1777. On Sept. 11, 1771, France put a duty of 60 livres per quintal on imports of French and Latin books. This tariff was a terrible blow for the marketing of the Geneva folio *Encyclopédie*.

19. Panckoucke to STN, Nov. 19, 1777. Panckoucke's phrasing revealed his agitated state of mind as he confronted a major policy decision: ''Peut-être pourriez-vous les bercer d'une association dans quelques années, dire que nous prendrions part à leurs entreprises, et qu'en la [the octavo *Encyclopédie*] suspendant cette année et les deux suivantes elle aura sûrement lieu . . . Peut-être ne faudrait-il pas la craindre. Vous leur inspirerez de la confiance en marquant de la crainte. Nos six mille sont placés: que pourrions-nous espérer de plus? Il n'est pas probable que les souscripteurs se retirent. On préférera toujours donner 4 à 5 louis de plus et avoir une encyclopédie lisible et d'un format convenable. Je vous assure que je vois plus de danger à tenter un accommodement qu'à paraître indifférents. Au reste, j'attends le magistrat pour lui mettre le prospectus [of the octavo] sous les yeux et voir les arrangements qu'il y aura à prendre.''

own a share in the *Encyclopédie:* "Vous sentez que si un pareil arrangement avait lieu, il faudrait que Duplain n'en sût rien et qu'on gardât le plus profond secret."[20]

So the octavo attack did divide the quarto group. In publishing his own octavo prospectus, Duplain launched a counterattack without support from his allies. And in proposing a secret truce, Panckoucke attempted to collaborate with the enemy behind Duplain's back. Their contradictory courses of action showed how deeply they distrusted one another. Duplain may have forced the battle in order to prevent the very sort of underhand accommodation that Panckoucke was attempting to arrange at his expense. And Panckoucke was ready to jump ship because he had been waylaid in the first place and would gladly pit one pirate against the other. This division left the STN in the middle, a position made even more difficult by its ties with Lausanne and Bern, which were closer than either of its partners suspected.

At first, the Neuchâtelois tended to favor Duplain's policy, perhaps because they thought that his antioctavo would provide them with the huge printing job they had been deprived of by the quarto. They took his octavo proposal seriously enough to make some discreet soundings about the likelihood of its success.[21] And they did not warm to Panckoucke's suggestion that they make a secret deal with the octavo group, though they agreed to open negotiations if they could argue from a position of strength and could get the Lausannois to propose terms. "Il ne paraît pas que ce fût à nous à leur faire des propositions, tenant, comme vous nous en assurez les clefs du royaume."[22] In short, the Neuchâtelois adopted the position that Panckoucke had abandoned a week earlier, but they remained flexible enough to adjust to the outcome of the battle between Lausanne and Lyons.

20. Panckoucke to STN, Nov. 27, 1777.
21. The STN followed the common practice of announcing a project as a fait accompli and waiting for the reaction from potential customers before embarking on it. Thus after Serini of Basel wrote that he would not order any quartos because his clients preferred the octavo, it answered, in a letter of Nov. 24, 1777, that it would produce an octavo of its own: "Nous voyons que vos commettants préfèrent l'octavo. Nous en entreprenons aussi une en ce format dont le mérite ne sera pas inférieur mais bien le prix. Nous travaillons aux arrangements, et le prospectus en paraîtra avant la quinzaine. Nous comptons que cette seconde édition entreprise pour cet objet aura plus de succès entre vos mains que la première. Nous le désirons pour votre avantage autant que pour le nôtre." See similar remarks in STN to Claudet of Lyons, Nov. 22, 1777.
22. STN to Panckoucke, Dec. 7, 1777.

Duplain had stockpiled his octavo prospectuses in Geneva, waiting to release them until he had heard the Lausannois's reaction to his threat. In late November they replied with an ultimatum that outdid his. Withdraw your prospectuses, they demanded, or we will lower the price of our octavo to the level of yours, and we will undermine your quarto by producing a still cheaper quarto of our own. "Vous serez obligés de nous céder ou de baisser votre prix vous-mêmes. C'est ainsi qu'on se coupe la gorge les uns les autres, mais vous nous en donnez l'exemple et nous en imposez la nécessité. Et ne croyez pas que nous vous faisions une vaine menace. Les prospectus sont prêts, et nous avons les mêmes caractères, presses nécessaires etc. à Yverdon à notre disposition."[23]

The Lausannois, who directed the strategy of the octavo publishers, gave the quarto group fifteen days to reply to the ultimatum. Propositions then arrived in such haste and confusion from both Lyons and Neuchâtel that they suspended the deadline and negotiated on both fronts throughout December 1777. They reserved their toughest language for Duplain, telling him flatly that he would have to withdraw his octavo prospectus or they would sink his quarto. They were more conciliatory to the STN, which had indicated its own desire to reach an amicable settlement by proposing a compromise on December 10. In a personal letter to Ostervald, Bérenger replied that although his associates could not accept the proposal, they wanted to avoid a war. They were suspicious of the propositions they had received from Duplain, which they interpreted as an attempt to deceive them and to play for time. But they would like to offer Ostervald a counterproposal. They had to move carefully, however, because they owned only a third of the octavo enterprise and needed to clear everything with their two partners, one of whom was the Société typographique de Berne. Perhaps the quarto group would let its plates be used for the engraving of the octavo plates. If so, the octavo publishers would cede it half of their 3,000 *Encyclopédies* for only 100 livres a set.[24]

Ostervald answered that he, too, faced the problem of get-

23. The Société typographique de Lausanne addressed this ultimatum to Pellet for forwarding to "vos commettants," and it sent a copy, without comment, in a letter to the STN dated Nov. 20, 1777.

24. Bérenger to Ostervald, Dec. 15, 1777. The Lausannois described their negotiations with Duplain in a letter to the STN of Dec. 23, 1777.

ting his associates to agree. He would try to persuade them to accept a reasonable compromise and trusted Bérenger to do the same. Above all, the octavo group should avoid doing anything rash and should consider the two advantages that their rivals could provide: open entry into France and the use of the plates. If Bérenger could get his group to agree on a serious proposal—and Ostervald indicated that he did not take the 100-livre sales offer seriously—they should be able to prevent a trade war.[25] At the same time, Ostervald communicated Bérenger's offer to Panckoucke, explaining that he thought it should not be taken as a basis for negotiation but as a signal that the octavo group was willing to compromise. What counterproposal should he make? Although he felt nothing but scorn for "cette miniature," he feared that it would sell very well, "tant le bon marché a d'attrait pour le plus grand nombre, c'est à dire pour les sots." He had heard that the subscription was filling rapidly. One bookseller reportedly had ordered 300 sets. But Panckoucke could get more reliable information on the octavo's sales from his contacts in southern France, where Jacob-François Bornand, an agent for the Lausannois, was currently conducting a sales trip. Duplain had embroiled things by negotiating independently and by alienating the Lausannois with "propositions . . . captieuses et contradictoires;" but the quarto group ought to be able to bargain from a position of strength and to reach a favorable settlement. Panckoucke agreed that the octavo was certain to sell, "à cause du bas prix et du goût constant du public pour cet ouvrage." But he would only authorize the STN to offer harsh terms: the octavo group would have to cede a half interest in its *Encyclopédie* and delay its publication for two years; in return, Panckoucke and his partners would open up the French market and provide their plates. If the octavo publishers felt unable to accept those terms, they should renounce the French market; for Panckoucke had complete faith in his ability to close France's borders and to exclude any offensive book from her domestic channels of trade.[26]

Ostervald immediately relayed the message on to Bérenger—and also to Duplain, who had been operating as though

25. Ostervald to Bérenger, Dec. 18, 1777.
26. Ostervald to Panckoucke, Dec. 18, 1777, and Panckoucke to STN, Dec. 22, 1777.

he were a separate power. Duplain had not even kept his associates informed of what terms he had offered to the octavo group. The STN had felt uncomfortable about knowing less than the Lausannois about his end of the negotiations and had asked him what propositions he had made. He would only say that he did not take the octavo threat too seriously: "Leur édition au reste paraît déplaire partout."[27] He clearly wanted to keep the negotiations to himself and he reacted with predictable hostility when the STN informed him that Panckouke had authorized it to try to reach a compromise in Lausanne. "Nous ne prendrons absolument aucun intérêt ni direct ni indirect dans le projet de l'édition octavo méprisé et bafoué partout. Nous vous prions pour notre repos et le vôtre de ne pas commettre sur cet objet notre ami de Paris, qui est un visionnaire et qui change d'opinion comme le temps de vents. Laissez-nous agir, conduire la barque, et soyez sûrs de votre pilote. Nous avons répondu à Lausanne que nous leur abandonnions ce terrain, que nous ne prétendions pas y labourer, et tenons-nous-en là, de grâce. Ne les inquiétons point, et ils ne feront rien contre nos intérêts."[28] Duplain would have used much stronger language had he known about Panckoucke's secret collaboration plot, but he did not need to be apprehensive about the STN's offer to collaborate openly because the Lausannois rejected it out of hand.

Bérenger wrote that his associates considered Panckoucke's proposal so unfair that there was no point in continuing the negotiations. They already had wasted enough time parrying propositions that came from different sources and that contradicted one another. First, they complained, Duplain had asked them to suspend the octavo for eleven months, then he had dropped that demand, and now the STN wanted them to postpone their edition for two years. They would rather be outlawed from France than suffer any delay at all. They meant to proceed with publication; and they would go ahead with their counterquarto if Duplain did not withdraw the prospectus for his counteroctavo within ten days.[29]

It looked as though this new ultimatum would lead to war,

27. Duplain to STN, Dec. 22, 1777. On Dec. 23, the STN warned Duplain that the octavo really was a serious threat. Lausanne and Bern had already sent three or four salesmen to the field, and they reportedly had gathered a rich harvest of subscriptions. See also STN to Duplain, Dec. 28, 1777.

28. Duplain to STN, Jan. 3, 1778.

29. Bérenger to Ostervald, Jan. 2, 1778.

the STN wrote unhappily to Panckoucke. It was Duplain's fault. First he had attacked the octavo group and then he had alienated them by duplicitous negotiating. "Il est facile de conclure que c'est Duplain qui a aigri ces gens-là en lâchant à l'improviste son peu honorable prospectus octavo, et qui de plus les a rendus un peu fiers par les tentatives faites de sa part auprès d'eux. Et comme il a jugé à propos de faire le tout à notre insu, nous sommes fondés à lui en porter plainte." All they could do now was tell the Lausannois that they disavowed Duplain's maneuvers, exert pressure on him for the withdrawal of his octavo prospectus, and print and market the quarto as fast as possible, while leaving the octavo to Lausanne and Bern.[30]

A few days later, word came through that Duplain had renounced his prospectus. The STN tried to construe his retreat as a victory for peace in announcing it to Lausanne, "Chacun travaillera de son côté et fera de son mieux sans donner au public une scène peu convenable au bien de la chose."[31] Actually, Duplain's decision only postponed the battle between the quarto and octavo until the time when they would clash on the market place. It also sent the antioctavo and antiquarto to the limbo of unpublished *Encyclopédies,* where they kept company with the Neuchâtel folio and the Geneva, Toulouse, Avignon, and Lyons quartos—soon to be joined by the Suard *refonte*. The escalation of the projects for editions had even inspired the STN at one point to propose three *refontes:* "C'est à dire qu'indépendamment de l'édition in-folio et in-quarto comme le porte notre traité nous en contrefassions nous mêmes [in-octavo], et cela au prix à faire perdre l'envie de nous contrefaire." That project was even less substantial than the antieditions, but it illustrates the extremes to which publishers would go in their bluffing and bargaining.[32]

The Final Failure of Diplomacy

The cancellation of the antieditions left time for one more round of negotiations before the first volumes of the octavo were shipped to the battlefield in France. This time Panc-

30. STN to Panckoucke, Jan. 4, 1778.

31. STN to Société typographique de Lausanne, Jan. 8, 1778. The Lausannois replied in a similar vein on Jan. 13.

32. STN to Duplain, Dec. 28, 1777.

koucke and the STN maintained control of the quarto Association's policy, and most of the bargaining went on between Neuchâtel and Bern. But this last attempt to avert a trade war put the Neuchâtelois in an awkward position because they owed allegiance to two triple alliances. A penchant for one could look like treason to the other. So the STN tried to keep its diplomacy secret and relatively equitable, mediating between its rival sets of partners and playing them off against one another without betraying either side.

The three Swiss sociétés typographiques did business in the same way and had everything to gain by cooperating instead of competing. Nothing irked a pirate more than to arrive on the scene after some other buccaneer had run off with the booty, yet he had no way of knowing whether he could beat his competitors to the market. When the Société typographique de Lausanne began setting type for a counterfeit edition of Raynal's *Histoire philosophique,* the Société typographique de Neuchâtel might be pulling the last sheets of another counterfeit edition, and the Société typographique de Berne might have reached Paris with shipments of yet another. But if they divided the printing into thirds, they could do it faster than all the other pirates; and by sharing the costs, they would also minimize the risks.

These advantages drew the Swiss into a natural alliance against the French, sometimes even against their best friends in France. In February 1778, for example, the Société typographique de Lausanne proposed to the STN that they cooperate in counterfeiting Robertson's *History of America,* which was being translated by Suard, published with a privilege by Panckoucke, and marketed by Panckoucke and Regnault. Although this project had something to offend almost all its *Encyclopédie* associates, the STN replied, "Nous nous ferons un vrai plaisir de la [*L'Histoire de l'Amérique*] partager avec vous." On February 24, after detailed discussion about paper, type, and profit-sharing, the two firms signed a contract for a joint edition. Each printed half the text, following a common sample sheet and work schedule; and they finished the job at the end of March, just when the antagonism between the quarto and octavo *Encyclopédies* was at its hottest.[33]

33. Quotation from STN to Société typographique de Lausanne, Feb. 9, 1778. Because Suard was slow, the Swiss used another translation and so beat Panc-

The conditions of the trade virtually forced the pirates into such pacts, especially after the edicts of August 30, 1777, indicated a new determination on the part of the French government to destroy the circulation of counterfeit books. Faced with this threat, the three Swiss sociétés typographiques decided to move from occasional collaboration to a formal alliance or "Confédération helvétique," as they called it. In May 1778 they signed a compact, which committed them to pool some of their stock, to hire an agent in Paris, who would provide a steady supply of material for pirating, and to cooperate in producing and marketing counterfeit editions on a large scale. Each firm continued an autonomous line of business but reserved two of its presses for the work of the Confederation, which soon became an important force in French-language publishing. It took at least eight months of conferences and correspondence for the Swiss to reach this agreement, the principal difficulty having been their commitment to rival *Encyclopédies*. In the end, they decided to let that rivalry take its course and to concentrate on the more important business of cooperative piracy. The STN therefore became entangled in two opposed alliances but was in a position to mediate between them.

The mediation involved a certain amount of duplicity, although it might be anachronistic to apply present standards to men who adopted the practices of their time and who had enough experience of their trade to expect some crossing of alignments and enmities. Thus on January 21, 1778, the STN agreed to circulate eighty octavo prospectuses for its ally in Lausanne, and on January 25 it informed its ally in Paris of the points on the French border where customs agents would have most success in seizing octavo shipments. But the Sociétés typographiques of Lausanne and Bern knew that the STN would try to keep their *Encyclopédie* out of France, and the STN made a genuine effort to prevent a trade war. On January 29 Ostervald informed Panckoucke that he had close contacts in Bern and would see whether negotiations could be opened there as well as in Lausanne

Panckoucke had thought that a *lettre ostensible* might

---

koucke to the French market. In 1777 and 1779 the STN planned counterfeit editions of two of Panckoucke's favorite works, Cook's *Second voyage dans la mer du sud* and Buffon's *Histoire naturelle*. It eventually dropped both projects, but not because of any loyalty to Panckoucke.

touch off another round of bargaining after the withdrawal of Duplain's octavo prospectus, and the STN did present some propositions to the octavo group. But the Bernois preferred to deal directly with Panckoucke and dispatched Pfaehler to Paris in February. Those *pourparlers* did not lead anywhere, however. Panckoucke merely stressed that he had the determination and power to drive the octavo from the French market, while Pfaehler replied that he and his associates did not mind because they were concentrating on the market outside France. The STN warned Panckoucke that Pfaehler might be bluffing: he probably had come to Paris to arrange for the smuggling of his *Encyclopédies*. Another octavo agent, Jacob-François Bornand of Lausanne, was also touring France, selling *Encyclopédies* and perhaps setting up smuggling routes. If he showed up in Paris, Panckoucke should take a hard line with him, too. The important thing was to convince the octavo publishers that the quarto group could keep them out of France, thanks to Panckoucke's privilege and protections. If they intimidated their opponents sufficiently, Panckoucke and the STN could wait for them to come forward with peace proposals.[34]

By May the waiting had become difficult. Pfaehler had returned to Bern and Bornand to Lausanne; the STN knew that three presses were churning out *Encyclopédies* in Lausanne alone; and it had received word that a great many French booksellers had taken out octavo subscriptions. On the quarto side, Panckoucke got the French authorities to issue special orders for the confiscation of octavos, and the STN kept up *la guerre à l'oeil*.[35] War seemed only weeks away.

On May 5 the octavo associates held a council in Bern and decided to make a last attempt at a diplomatic settlement. Pfaehler announced their final offer in a letter to Panckoucke: they would cede a half interest in their French sales for 15,000 livres and free access to the French market.[36] Panckoucke pre-

34. STN to Panckoucke, March 10, 1778. Panckoucke and the STN discussed the situation in a series of letters between Jan. and June 1778.

35. STN to Panckoucke, May 3, 1778.

36. In his letter to Panckoucke of May 5, 1778, Pfaehler made the offer seem as attractive and congenial as possible: ''Il n'est pas douteux que vous jouirez par là d'un bénéfice très considérable sans beaucoup de peine et de risque, tandis que nous veillons à l'impression, à l'expédition, à la correspondance, enfin à tout le mercantil. Vous retirerez au bout d'un certain temps le bénéfice sans autres soins ni embarras que d'avoir fait un fonds modique.'' Bibliothèque publique et universitaire de Genève, ms. suppl. 148.

ferred to let the STN handle the negotiations. He passed Pfaehler's letter on to Neuchâtel and informed Bern that the STN had received his *pleins pouvoirs* to arrange a settlement. This maneuver, however, made the Neuchâtelois uncomfortable—not so much because they had compunctions about bargaining with a secret ally as because they felt perplexed about what line to take and what objectives to seek. The offer of half the French octavos aroused their suspicion because they feared that Lausanne and Bern might lie about their sales and falsify their accounts. A cash payment would be preferable. But how much should they demand? Even if they forced a large sum on the octavo group, they would sacrifice an opportunity to own a proportionate share in a market that seemed capable of unlimited expansion. For there was no underestimating the demand of the French public, "entiché comme il est du Grand Dictionnaire et séduit par le bas prix." In the end the STN merely resolved to wait for proposals to arrive from Bern and to bargain hard.[37]

The Bernois produced a draft contract that they wanted to serve as a basis for the bargaining. It gave the STN a choice between a half interest in the French sales and a quarter interest in the entire enterprise In either case, the quarto group would have to pay 15,000 livres for its share and would have to guarantee the shipments against confiscation by the French authorities. If the quarto group wanted to collaborate on those terms, the octavo group would increase production from 3,000 to 5,000 copies. If not, it would fight. The Bernois said that they did not desire a battle, and especially wanted to avoid any harm to the Confederation. But they would not make any more concessions: "Nous romprons plutôt toute négociation que de souscrire à des conditions onéreuses."[38]

The Bernois seemed so determined to hold this position that the STN doubted that the two sides could be brought together. But it did not want to give ground either. It thought that the octavo group still failed to appreciate the likelihood of suffering severe casualties in any attempt to break through the quarto's defenses in France. So it sent an intransigent reply, and then each of the negotiators paused to consult his constituents. The STN sent a full report to Panckoucke, with copies of the correspondence, and also dispatched a few lines

37. STN to Panckoucke, June 7, 1778.
38. Société typographique de Berne to STN, June 14, 1778.

to Duplain—his first notification, apparently, that fresh negotiations had begun: "Les gens de Berne travaillent et marchandent. Ils offrent le quart, et nous insistons sur la moitié."[39]

At this point, Bérenger tried to break the stalemate with a personal letter to Ostervald. The quarto *Encyclopédie* had been produced "en France et pour la France," he argued. But the octavo was a non-French affair. Four-fifths of its subscribers lived outside the kingdom. So Lausanne and Bern had little to gain from Panckoucke's privilege, even if they increased their pressrun to 6,000. It was not reasonable to ask them to sacrifice half of all the subscriptions they already had in hand for the mere possibility of increasing their relatively small number of French subscriptions. Two days later the Société typographique de Berne wrote a letter indicating that there was some flexibility on the octavo side. Instead of giving the quarto publishers a percentage of the sales, the octavo group might consider a cash payment; but then it needed hard information about "la somme qu'il faudrait payer pour une et toute fois et quels seraient les sûretés qu'on pourrait nous donner pour nous mettre à l'abri des événements que nous pourrions encore courir, non obstant de l'entrée accordée? La variabilité du gouvernement et tant d'autres choses rendent cette précaution nécessaire, puisque la mort ou la disgrâce d'une seule personne peuvent amener des changements inattendus." If Panckoucke wanted protection money, he would have to provide some assurance of his protectors' effectiveness. The publishers of Lausanne and Bern had had enough experience with the French system of government to know that influence could be bought and also that it could evaporate overnight.[40]

The STN's reply is missing, but it clearly did not satisfy the Bernois, for on July 1 they abandoned the negotiations. They did so rather tentatively, however, dangling the bait while reeling in their line for the last time. Although they

---

39. STN to Duplain, June 24, 1778. The STN also warned Duplain that he had underestimated the selling-power of the octavo: "Cette édition menée avec l'activité que vous avez déployée aurait un succès aussi brillant que l'in-quarto."

40. Bérenger to STN, June 23, 1778, and Société typographique de Berne to STN, June 25, 1778.

could rely entirely on the non-French market, they said, they still would give up half their French sales in exchange for free access to the kingdom.[41] This offer, which was tantamount to a gift of at least 300 octavos worth 58,500 livres, must have tempted the STN; but it did not bite, probably because it did not want to flood the market with cheap *Encyclopédies* before it had sold the 2,000 sets of the third quarto edition. To the octavo group, however, the offer seemed generous. It demonstrated a genuine desire to keep the peace, and the rejection of it smacked of a kind of commercial tyranny, which Bérenger unhesitatingly attributed to "l'Atlas de la librairie française." In a last, bitter letter to Ostervald, he explained that the publishers of the octavo had gone as far as they could to prevent a conflict, but they could not possibly turn over half their *Encyclopédies* to the quarto group. If they increased their printing to 5,000, they would only get to keep 2,500; and they had already sold that many. Moreover, "ils auraient eu la peine de former et préparer l'enterprise et de l'exécuter. En l'augmentant ils multiplieraient leurs frais, leurs presses, leurs embarras, et cela pourquoi? Pour avoir l'honneur d'en donner gratis à M. Panckoucke, qui les a regardé faire les bras croisés. Oh non, cela ne se peut. Ils ne sont ni sots, ni insensés. Ils aiment mieux laisser libres leurs souscripteurs français. S'ils veulent des *Encyclopédies* de ce format, ils viendront les prendre sur les frontières. Et M. Panckoucke pourrait bien être faché un jour d'avoir poussé trop loin ses prétentions. Les deux sociétés me paraissent disposées à lui rendre le change, non pas de la même manière—elles ne le peuvent pas—mais par des moyens qui ne lui seront pas moins défavorables." Panckoucke and company wanted war; very well, war they would get.[42]

41. Société typographique de Berne to STN, July 1, 1778: "Après avoir consulté Messieurs nos intéressés à l'*Encyclopédie* et après avoir tout balancé de côté et d'autre, ils ont trouvé qu'il faudra nous en tenir à notre édition, qui est presque toute placée hors la France, que nous abandonnons entièrement. Il y aurait peut-être eu moyen de s'arranger encore, si vous nous eussiez pu faire entrer quelque sûreté en cas de changement dans le ministère, ou si vous n'auriez pas rejeté notre proposition d'association pour le quart du tout en faisant les fonds nécessaires ou pour la moitié des exemplaires qui entreront en France . . . Nous avisons aujourd'hui M. Panckoucke d'avoir rompu la négociation pour l'*Encyclopédie*."
42. Bérenger to Ostervald, July 1, 1778.

Open War

Trade wars among eighteenth-century publishers did not involve pitched battles and clear lines of conflict. Each side set out to capture the market, of course; but they campaigned surreptitiously and their strategies differed. A publisher might try to undersell the enemy, to malign the quality of his product, to cut his supply lines, to intimidate his distributors, to deter potential customers by impugning his honesty or his ability to satisfy them, or finally to get his books seized by customs agents, guild officials, and the police. All or most of these techniques were used in the war between the octavo and the quarto *Encyclopédies,* but essentially the conflict developed into a contest between a strategy of smuggling and a strategy of protection and privilege.

Panckoucke's ability to defend his privilege by invoking his protections was more important than the ownership of the privilege in itself. He could crush an interloper by setting wheels in motion in Versailles. Duplain would not have surrendered half his quarto and the Sociétés typographiques of Lausanne and Bern would not have offered half the French sales of their octavo if they had not recognized the strength of Panckoucke's unique position in the book trade. He had complete confidence in his power to carry out the threats that he directed at the octavo group through the STN: "Je réponds bien qu'ils n'entreront pas en France. Le magistrat me l'a promis, et les nouveaux règlements offrent les moyens les plus faciles pour les en empêcher quand il y a un tiers opposant. Vous sentez, Messieurs, qu'étant munis d'un privilège, vous ne devez point, ainsi que moi, vous relâcher de vos droits. Duplain, en vertu de nos actes, de notre privilège, est venu composer avec nous. Il faut que les Lausannois en fassent de même."[43]

Panckoucke's capacity to fight his battles with the weapons of the state accounts for the severity of the terms he offered to Lausanne and Bern and also for their willingness to persevere with the negotiations, despite his intransigence. At first he refused even to reveal the full extent of his influence

43. Panckoucke to STN, Dec. 22, 1777.

in Versailles, and the STN did likewise in its early negotiations with Lausanne; but its letters indicate that Panckoucke had received a definite commitment of support from a powerful source: "Nous n' avons pas dit un mot aux Lausannois ni de vive voix ni par écrit de la protection que l'on nous accorde, et ils n'avaient que faire de le savoir. Nous nous sommes bornés à leur parler du privilège pour les cuivres et de la refonte projetée."[44] Because thousands of livres hung on the effectiveness of this support, the octavo group had asked for assurances—an unconditional guarantee that Panckoucke would remove all obstacles to the circulation of their *Encyclopédie* in France and also "quelque sûreté en cas de changement dans le ministère."[45] The STN had advised Panckoucke to let their opponents know how powerful his protectors were in order to strengthen its hand in the bargaining.[46] He must have done so, considering the persistence with which Lausanne and Bern sought to place their *Encyclopédie* under the cover of his protection. The whole character of the negotiations indicates the importance that each side attributed to Panckoucke's influence among the French authorities. So Panckoucke's remarks about his conferences with Le Camus de Néville, the director of the book trade, should not be interpreted as name dropping. As the STN put it, he held "les clefs du royaume":[47] he meant to open the doors of France to the quarto and to close them to the octavo.

It is difficult to know the details of Panckoucke's tactics because he kept them secret, but he certainly got Néville to issue a special alert against the octavo, just as he had originally done in the case of the quarto. Judging from allusions in the correspondence of Panckoucke and the STN, Néville ordered his subordinates, the inspecteurs de la librairie attached to the booksellers' guilds throughout the kingdom, to

44. STN to Panckoucke, Jan. 25, 1778.

45. Société typographique de Berne to STN, June 14 and July 1, 1778.

46. STN to Panckoucke, Feb. 22, 1778: "Nous croyons qu'il est très à propos que vous fassiez connaître, soit directement soit par personne tierce, mais d'une manière nette et claire, au Sieur Bornand, commis voyageur des Lausannois, la résolution où vous êtes d'user de vos droits et les mesures déjà prises pour rendre efficace votre opposition. C'est un moyen assuré de l'arrêter tout court, s'il en est un."

47. STN to Panckoucke, Dec. 7, 1777.

confiscate any octavos circulating in their territory.[48] The territory of La Tourette, the inspector in Lyons, covered most of the eastern border of France, including the main arteries of the trade from Geneva and Lausanne. Since the quarto shipments were to follow the Geneva–Lyons route and the octavo might do likewise, Panckoucke and Duplain made a special effort to enlist La Tourette on their side. They gave him a set of their *Encyclopédie* and provided free copies to de Flesselles, the intendant; Prost de Royer, the lieutenant de police; and the local academy. Panckoucke also gave quartos to his most important protectors in the capital: Néville, Vergennes, and probably also Miromesnil, the Garde des Sceaux, and Lenoir, the lieutenant-général de la police. Altogether the Association distributed almost two dozen complimentary quartos—gifts that were valuable enough to reinforce the proquarto, antioctavo bias on the local and national levels of the administration.[49]

Panckoucke could also draw on a legacy of bribes and gratifications; and once he had oiled the machinery of the state, he set its wheels in motion. He conferred with Néville, got authorization to take special measures in Lyons, and then issued orders to La Tourette, writing as if he himself were a minister: "J'ai eu l'honneur de vous mander que je m'étais arrangé avec M. Duplain au sujet de la nouvelle édition inquarto du *Dictionnaire des Arts*. Il me mande dans ce moment qu'il attend de Genève les deux premiers volumes. Je vous serai obligé de donner vos ordres pour que ces volumes passent sans difficulté et d'accorder toute votre protection à cet ouvrage. Monsieur de Néville est prévenu de tout ce que j'ai fait à ce sujet."[50]

48. Because most of Panckoucke's letters are missing from this period, his activities are difficult to follow, but the letters to him from the STN sometimes allude to a promise from Néville that special orders would be issued against the octavo. On May 3, 1778, for example, the STN wrote, "Il sera très bien que les ordres soient adressés aux inspecteurs dans les différentes chambres syndicales, vu l'autorité que les nouveaux règlements leur attribuent."

49. Under the rubric "gratis," the subscription list (see Appendix B) included twenty-five copies intended to smooth the quarto's way into France. At least ten of them went to Lyonnais, and four went to Panckoucke's protectors in Paris and Versailles.

50. Panckoucke to La Tourette, July 18, 1777, from a copy made by Panckoucke, which is in the Bibliothèque publique et universitaire de Genève, ms. suppl. 148. Although Panckoucke may not have actually sent this letter, the following document, which accompanies it in the Genevan archives, leaves no doubt

The protection given the quarto proved to be so effective that the shipments from Switzerland to Lyons served as a cover for a smuggling operation mounted by Revol et Compagnie, shipping agents in Lyons. The "company" actually consisted of nothing more than Jacques Revol and his wife, and Revol had nothing more than know-how. But he knew a great deal. He had a wide acquaintance among wagoners, warehousemen, innkeepers, and customs inspectors—the human material out of which a smuggling business could be built—and he understood the weak points in the system for controlling the importation of books. Crates full of books were supposed to enter France through border stations, where customs officials tied them with rope, sealed the rope with lead, and made out a customs permit called an *acquit à caution*. Accompanied by the *acquit,* the crates then journeyed to the nearest town with a booksellers' guild whose officials (*syndic et adjoints*) were authorized to make inspections and discharge *acquits*. Swiss shipments almost always had to go through the chambre syndicale of the guild in Lyons. There, under the surveillance of the local inspecteur de la librairie, the officials broke the seals, made sure the crates contained nothing illegal, and discharged the *acquit* by signing it and returning it to the border station from which it had originated. The books then could be forwarded by a shipping agent to their ultimate destination within the kingdom. At this point they traveled as a domestic shipment and usually were left alone by the authorities, unless Néville issued a special alert or unless they were bound for certain cities, like Marseilles, Toulouse, and especially Paris, where additional inspection was common.

This system minimized fraud by enlisting both administrative officials and established bookdealers in the policing of the

---

that Néville cooperated with the campaign to protect the quarto long before the octavo was a threat. Perrin (Néville's secretary) to Panckoucke, July 19, 1777: "M. de Néville me charge, Monsieur, de vous renvoyer la lettre ci-jointe que vous lui avez communiquée. Ce magistrat ne voit point d'inconvénient à la faire partir, mais M. de La Tourette n'est pas actuellement à Lyon. Il va arriver au premier jour à Paris, et M. de Néville ne croit pas qu'il soit à propos d'écrire aux officiers de la chambre syndicale. Si vous voulez venir demain matin ou mardi matin à dix heures, vous verrez avec M. de Néville à prendre un autre parti. J'ai reçu ce soir les trois exemplaires [that is, the free copies of the first volumes of the quarto] de la chambre syndicale. J'en ai fait partir un aussitôt pour M. le Comte de Vergennes."

trade, but it could be circumvented by the techniques used by Revol and other marginal middlemen. Revol instructed the wagoners to meet him at inns in the outskirts of Lyons, notably the establishment of M. Boutarry, a half-league outside the Faubourg Sainte Claire, where he kept a supply of legal books. He would open the crates, substitute the legal for the prohibited works, and close the crates up again, counterfeiting the lead seal, which was easier than falsifying the discharge of the *acquit*. He then would store the contraband goods in a secret warehouse until they could be forwarded safely as a domestic shipment, and he would process the innocent merchandise through Lyons's chambre syndicale.

Of course Revol could cut corners and save expenses if he could persuade a guild official to make a fraudulent or even a negligent inspection while the inspector turned his back. Inspection was long, dull work, involving a good deal of red tape and pawing about in piles of loose sheets (books were normally shipped in sheets rather than stitched and bound). Many syndics and inspectors did a cursory job: they checked the top sheets in a crate and glanced at whatever could be seen from the side. The sheets of prohibited books might pass unnoticed if stuffed in the middle of a crate full of inoffensive sheets. This practice of "larding," as it was called, involved more risks than the detour through inns and secret warehouses, but it was cheaper and easier; and the risks could be reduced if the officials knew that the shipping agent was receiving large, regular shipments of legitimate books.

The quarto *Encyclopédie* fit this requirement perfectly. The inspector, La Tourette, had received special orders to facilitate its passage through Lyons; huge crateloads of it came almost every week from Switzerland; and the agent whom Duplain had hired to clear the crates through the guild was a boyhood friend of his, Jacques Revol. Revol was also the STN's first-string smuggler in Lyons. So the *Encyclopédie* provided him and the Neuchâtelois with a golden opportunity for cut-rate contraband, as he explained in a letter of July 5, 1778: "Nous vous prévenons qu'avons des entrepôts sûrs; et si vous voulez profiter de l'occasion qu'avez de l'*Encyclopédie,* vous pouvez d'une balle en faire passer quatre, en les masquant sur les bords et aux têtes. Nous vous les ferons passer avec facilité, sans que personne s'en apperçoive, d'autant plus que c'est nous qui retirons toutes les balles de M. Duplain et qu'elles sont entreposées dans nos magasins. Nous lui les envoyons déballées par ballot. C'est fort rare quand on les visite à la Chambre; ou si on les visite, on ne les regarde jamais en dedans ou ne visite que les bords. Vous devez juger qu'il nous serait facile de mettre en sûreté tout ce que vous voudrez.'' The STN funneled its prohibited and pirated books through Lyons in this manner for a year and a half. Far from drawing the fire of the French authorities, as it had done in the 1750s, the *Encyclopédie* circulated under their protection and circulated so openly that it served as camouflage for the diffusion of works that they really did want to suppress.[51]

Panckoucke and Duplain had no idea that Revol had grafted a small smuggling business onto their enterprise. They wanted to prevent smuggling, at least in the case of the octavo; for after the failures of the negotiations with Lausanne and Bern, they knew that the octavo group would try to penetrate France by clandestine shipments. As Revol's operations illustrate, contraband was particularly vulnerable at two points: the border stations, where customs agents could inspect the crates instead of sending them on under *acquit à caution,* and the guild towns, where the crates normally underwent inspection. Néville's orders spread the antioctavo alert throughout the network of guilds, while the STN helped put the border agents on their guard. Writing to Panckoucke in the expectation that he would pass the word on to the French officials, it revealed the locations on the border where the smugglers were most likely to pass.[52]

The Neuchâtelois could speak with some authority on this subject because they had done business for almost a decade with the shipping agents and smugglers of the area, particularly those who operated around the border station of Frambourg, which dominated the route between Neuchâtel and Pontarlier, France They normally dealt with Jean-François

---

51. This account is based on Revol's dossier in the STN papers. The quotation comes from his letter to the STN of July 5, 1778.

52. STN to Panckoucke, Jan. 25, 1778: ''Il s'agit de mettre en oeuvre tous les moyens que nous avons en main . . . L'un des plus essentials est de veiller avec le plus grand soin à ce qu'il ne s'introduise en France aucun exemplaire de l'octavo, car les Lausannois posent toujours en fait qu'il en entrera en dépit de nous. Les bureaux de *Frambourg,* de *Jougne, Moret* et *Coulonge* sont les principaux de ceux qui répondent à nos frontières. Nous nous persuadons que vous saurez prendre les mesures les plus efficaces pour empêcher cette contrebande-là.''

Pion of Pontarlier, who commanded a large team of wagoners. But Pion was lazy and unreliable, and the STN tended to entrust more delicate shipments to the smaller but surer agency of Meuron et Philippin in Saint Sulpice, a village on the Swiss side of the border. Meuron et Philippin specialized in border crossings, both legitimate and clandestine. They wanted to prove their prowess in order to steal the STN's business from Pion. And they were proficient in various techniques, ranging from backpacking to bribery, which had evolved over the years in the Jura valleys, where smuggling was an important industry. It was particularly important to the Swiss sociétés typographiques, which spent a great deal of time and money on the maintenance of their clandestine routes to the rich French market for prohibited books. So the STN knew that a vast effort of secret route building lay behind the willingness of Lausanne and Bern to break off negotiations concerning the *Encyclopédie*. "Nous avons eu occasion de voir ici nos Encyclopédistes in-octavo," it informed Panckoucke. "Ils ne nous ont fait aucune ouverture relative à cet objet, et nous avons lieu de croire ou qu'ils veulent traiter avec vous ou qu'ils croient pouvoir s'arranger sans cela. Ils tirent actuellement à 3,000 et ils ont dit dans la conversation qu'ils s'engagaient à rendre les exemplaires dans tout le royaume à leurs périls et risques. Nous croyons devoir vous rendre ce propos pour l'expliquer par leur conduite. Si la correspondance de Pfaehler avec vous n'a point de suite, il faut en cas conclure qu'ils ont ou croient avoir une route sûre."[53]

But where did that route pass? If it could be discovered, the quarto group could cut the octavo's supply lines, perhaps with such devastating effect that Lausanne and Bern would abandon France. Favarger kept his eye out for such a discovery when he left for his confrontation with Duplain and his sales trip through southern and central France. He also intended to inspect the STN's own supply lines, and his first stop was Saint Sulpice, in the office of Meuron et Philippin. There the Meuron brothers, Théodore Abram and Pierre Frédéric, dropped a remark about crossing the border with five 800-pound crates bound for Paris containing volume 1 of the octavo *Encyclopédie*. This was a capital revelation,

53. STN to Panckoucke, May 21, 1778.

which the Meurons slipped into the conversation with more than a touch of professional pride, because they knew that it would help them in their competition with Pion. Favarger tried to draw them into further disclosures, but "leurs réponses ont toujour été, 'Quand nous avons de vrais amis, nous savons aussi dans le besoin leur donner un coup d'épaule.' "[54]

Favarger then hurried on to Pontarlier in order to confront Pion with this information. Pion had gone to Besançon on business, but his son, who was minding the shop, came up with a crucial piece of missing information: he had seen the *acquits à caution* for the shipment. They had been discharged by Capel, syndic of the guild in Dijon. The agents in the station at Frambourg, who usually only delivered *acquits* for Lyons, had just begun to permit shipments for Paris to pass directly through Dijon. So Capel was collaborating with the octavo group by discharging its *acquits* and relaying its crates on to their destination instead of confiscating them.

Favarger sent a triumphant letter to his home office about this discovery, and the Neuchâtelois immediately reported it to Panckoucke, taking care, however, to avoid mentioning Capel and insisting that Panckoucke not mention them. They knew that he would get his protectors to confiscate the shipment, and they did not want to be associated, even as informers, with a smuggling operation—or rather they did not want the French authorities to associate them with it. In the reply that they sent to Favarger in Lyons, they emphasized the importance of concealing Capel's part in the affair, because "nous espérons que pour de l'argent il nous rendra le même service." If Capel would discharge the STN's *acquits* as well as those of its neighbors, he might well replace Revol as its principal underground agent, and Dijon could become the main entrepôt of its illegal books. The conditions of the import trade in France had forced the Neuchâtelois to route most of their shipments through Lyons, an enormous and expensive detour in the case of books bound for the rich markets of northern France. Capel's collaboration would open up a northwest passage to the capital, which they had been seeking in vain for years. Lyons would still be the gateway to the Midi, and the STN wanted to encourage Revol's enterprise, in case the northwestern route should collapse. So it also instructed

54. Favarger to STN from Pontarlier, July 8, 1778.

Favarger to reinforce its ties with Revol: "Plus aurons de cordes à notre arc et mieux la chose ira." Of course Duplain should be informed of the octavo breakthrough but should be kept in the dark about Revol's activities. The STN wanted to play both sides of the smuggling game.[55]

Favarger succeeded in feeding the prescribed amount of information to Duplain, which was no small accomplishment, given the vivacity of Duplain's conversation and the complexity of the *Encyclopédie* intrigues: "M. Duplain n'a pas peu été surpris de l'adresse des entrepreneurs de cette édition pour faire passer leurs balles . . . Il est enchanté de cette découverte et dit, 'Il faut trémousser, c'est ce que je ferai; ils ne seront pas peu capots quand ils en verront quelques magots d'arrêtés.' Je me suis bien gardé de compromettre Capel et me suis conformé, Dieu merci, en tous points à vos instructions, ce qui n'est pas aisé avec lui quand l'on y est aussi longtemps que j'y ai été." In its next letter, the STN told Favarger that Capel had refused its overtures and so his name could be mentioned to Duplain. It produced quite an effect: "Il a été étonné des démarches de Capel, d'autant plus qu'il a souscrit pour un grand nombre de l'in-quarto," Favarger reported. In fact, Capel eventually ordered 131 quartos, an enormous number but not enough for him to resist the temptation to make some extra money by working with the octavo group. His unwillingness to discharge *acquits* for the STN did not surprise Favarger, who reminded the Neuchâtelois that they had failed to persuade him to cheat at his job two years earlier. But they would get surer service from Revol, who had made splendid use of the quarto shipments (Favarger's investigations confirmed that "lorsqu'il s'agit de l'*Encyclopédie* quarto, on n'en ouvre qu'une ou deux [balles] d'un envoi, et l'acquit est déchargé") and now had perfected his measures for smuggling books without them.[56]

The octavo publishers therefore had got off to a good start in their smuggling, but the quarto group was on their track. The octavo crates went from Lausanne and Bern right past Neuchâtel and up the Val de Travers to Saint Sulpice. Next, thanks to Meuron et Philippin, they made the border crossing to Pontarlier and, with Capel's help, traveled directly to Paris

55. STN to Favarger, July 11, 1778.
56. Favarger to STN, July 15 and 21, 1778.

via Dijon. The STN believed that they were stored in a secret entrepôt outside the city walls and then were smuggled in small bundles past the customs to retailers and subscribers in the capital.[57] Provincial subscribers might have received their shipments from Dijon, too, but the octavo publishers also used other routes. Favarger learned that Robert et Gauthier, booksellers of Bourg en Bresse, had smuggled in 50 octavos for their own customers and that the octavo group had asked Jaquenod, an influential colleague of Duplain's, to undertake ''l'opération périlleuse'' for them in Lyons but that he had refused.[58] So the quarto publishers had uncovered a good deal of their enemy's distribution system. How effective were they in destroying it?

After learning that the four tons of the octavo's first volume were bearing down on the gates of Paris, Panckoucke reassured the STN that he had prepared for their reception: ''Le magistrat a donné des ordres.''[59] Too few of his letters survive from this period for one to know whether his protectors captured this shipment or any of the cartloads of volumes 2–36 that were to follow it. But it seems unlikely that the octavo agents could keep such huge shipments hidden from officials who were primed to discover them and who knew the route they would take.

One indication of the casualty rate among the octavos comes from the diary and letters written by Favarger during his *tour de France*. Word reached him in Nîmes that the authorities of Toulouse had seized an important shipment of octavos. In Marseilles he crossed paths with Duplain's main assistant, Amable Le Roy, who had been selling quartos throughout the south. Le Roy confirmed the reports of the Toulouse confiscation and added that it was doubly disastrous for Lausanne and Bern because it had destroyed their customers' faith in their ability to deliver the merchandise: ''Les souscripteurs, déjà lassés des entraves et désagréments qu'a essuyés le premier volume, se dégoûtent et prennent de l'in-quarto, qui va son train. Il est certain qu'ici [Marseilles, which eventually took 228 quartos] ils n'en ont pas placé quatre exemplaires.''[60] In subsequent discussions with bookdealers, Favarger continued

57. STN to Favarger, July 11, 1778.
58. Favarger to STN, July 15, 1778.
59. Panckoucke to STN, July 21, 1778.
60. Favarger to STN, Aug. 23, 1778.

to hear of desertions among the octavo subscribers: Odezenes et fils of Morbillon had not received a single volume of the six octavos they had ordered and were about to convert to the quarto; Fuzier of Pézenas had persuaded his clients to switch to the quarto because of the uncertainty of receiving the octavo; and Hérisson of Carcassone had done the same. Similar information arrived from several correspondents of the STN in France. Panckoucke reported the confiscation of a shipment in Caen.[61] Champmorin of Saint Dizier wrote of another in Sedan.[62] Chaurou of Toulouse said the authorities there would not release the octavos they had seized until the quarto had been sold out.[63] And the Société typographique de Bouillon was stopped dead in its attempts to smuggle octavos to customers in northern France.[64] It took time for the state to organize the extermination of the octavo; so some early shipments reached their customers. In October 1778 Duplain complained, "L'édition de Lausanne in-octavo se répand en foule partout. Il faut donc que M. de N. (Néville) envoie à chacun de ses inspecteurs un prospectus de cette édition in-octavo et défende qu'on en vende . . . Il est bien singulier que Panckoucke fasse si fort sonner sa protection et qu'elle nous soit si inutile."[65] But a year later, Duplain's lieutenant, Amable Le Roy, reported that 1,400 copies of the octavo had been confiscated by the French authorities.[66] Unable to sustain such losses any longer, the octavo publishers retreated from France in 1779. Panckoucke had won the war in the western sector.

But Panckoucke's interdiction did not extend to the rest of

61. Bosset, reporting a conversation with Panckoucke, to STN from Paris, May 22, 1780.

62. Champmorin to STN, Nov. 26, 1780.

63. Chaurou to STN, Jan. 15, 1779.

64. Jean-Pierre-Louis Trécourt to Pierre Rousseau, Feb. 23, 1780. Trécourt had taken over the daily management of the Société typographique de Bouillon from Rousseau, who had withdrawn in semiretirement to Paris. In this letter, he reminded Rousseau that they had failed to provide their French subscribers with octavos and had abandoned the marketing of it. In a letter of Aug. 18, 1780, he mentioned the confiscation of a large shipment of octavos by the chambre syndicale of Nancy. Both letters are in the Archives Weissenbruch of the Bibliothèque du musée ducal in Bouillon, and they were kindly communicated by Dr. Fernand Clément of Bouillon.

65. Duplain to Merlino de Giverdy, Oct. 14, 1778, from a copy sent to the STN by Panckoucke.

66. Le Roy mentioned the confiscations in a conversation with Mallet Dupan. Mallet to STN, Oct. 1, 1779.

Europe, where Lausanne and Bern had extensive contacts. In Germany, for example, the Société typographique de Berne reputedly did business with every important bookseller, and Pfaehler traded regularly at the book fairs of Frankfurt and Leipzig. When the quarto group tried to penetrate Germany, it found the territory occupied everywhere by the octavo. The STN got Serini of Basel to distribute quarto prospectuses among his large German clientele and at the two great fairs, but he found the situation hopeless.[67] The Neuchâtelois had a great many German correspondents of their own, who wrote "de tous les côtés" that the octavo had captured the market.[68] Similar reports arrived from other parts of Europe, while the quarto associates concentrated on reaping their three harvests from the denser market in France. The STN estimated that at least seven-eighths of the quartos had been sold in France by mid-1780, and it noted that they had circulated openly everywhere, without the slightest difficulty from the French authorities.[69] So the trade war developed into a stalemate: the quarto group had driven the octavos from France, and the octavo forces had prevented the quartos from penetrating extensively beyond the Rhine and the Rhône. Except for occasional sniping and border raids, each side stayed behind its lines. The great European market for *Encyclopédies* had been divided into two spheres of influence.

Pourparlers for Peace

The stalemate did not please any of the Swiss sociétés typographiques, who were used to selling books on both sides of the divide. In the spring of 1779 the STN wrote ruefully to Panckoucke and Duplain that the octavo's success in Germany proved that they should have made peace in 1778—and it en-

67. Serini to STN, March 27, 1779.

68. STN to Duplain, April 7, 1779. See also STN to Panckoucke, June 24, 1779: ''En Allemagne, l'édition octavo que vous avez si fort méprisée, nous y a fait le plus grand tort.''

69. STN to Dufour et Roux of Maestricht, Aug. 14, 1780. In a letter to Champmorin of Saint Dizier dated Nov. 12, 1780, the STN emphasized that the quarto association had marketed its *Encyclopédies* ''dans toutes les villes de province où il y a chambre syndicale et même à Paris, sans aucune précaution, et jamais il n' y a eu aucun empêchement.'' This statement is borne out by the record of book confiscations kept by the Chambre syndicale of Paris. Bibliothèque nationale, ms. Fr. 21933–21934.

tered into secret negotiations with its octavo neighbors. At first they merely discussed a marketing agreement: the STN would sell octavos in its territory, while Lausanne and Bern sold quartos in theirs, and each side would supply the other with its *Encyclopédies* at a 25 percent discount. The octavo publishers even offered to provide the STN with a list of the subscribers whom they had not been able to reach in France, owing to the effectiveness of Panckoucke's embargo. But this proposal probably would have involved the Neuchâtelois in a campaign to smuggle octavos past their own anti-smuggling defenses; after serious consideration, they rejected it. They did cooperate informally with Bern in May and June 1779, but these attempts at mutual marketing came to nothing. After offering the octavo in its commercial correspondence and spreading around some prospectuses, the STN sold only one set. Pfaehler failed to sell any quartos at all, although he proposed them to his customers on a sales trip through Germany.[70]

By this time, however, few quartos remained. Panckoucke's Association had sold almost all the sets of its three editions. The quarto group may therefore be considered the victor of the *Encyclopédie* war, especially as the octavo publishers were the first to sue for peace. They had most to gain from a cessation of hostilities because the more expensive quarto had never been a threat in their territory, while there was still a demand for the octavo in France, judging from the reports of their French correspondents. They sounded the STN on the possibility of a truce at a meeting of the Swiss Confederation in October 1779, and soon afterwards the Lausannois proposed terms: the STN would arrange for the octavo's free entry into France and then would join Lausanne and Bern in a new edition of it, a tripartite edition, like the other productions of the Confederation. With the rich French market open to them, they could sell 2,000 or 3,000 more sets, for they had once had 1,200 French subscribers "et serons sûrs de placer un beaucoup plus grand nombre en France dès que cet ouvrage aura un cours libre en France, dont il sera aisé de l'obtenir, puisque la moisson des Encyclopédistes in-quarto est faite."[71]

70. Société typographique de Berne to STN, March 14, June 15, and Dec. 13, 1779; and STN to Société typographique de Berne, June 12, 1779.
71. Société typographique de Lausanne to STN, Oct. 9, 1779.

Tempting as it was, this proposal might badly entangle the STN's two alliances; so Ostervald wrote a lukewarm reply. The whole business stirred up painful memories about the abortive negotiations of June 1778, he said. That had been the time to come forth with concessions. He could see what the octavo publishers were after now: the lifting of the French embargo. But that would be more difficult than they realized, and he would not even consider it unless he had a realistic idea of the profits to be expected from the deal. He needed information that they kept secret—about costs, subscriptions, and the like—then he would decide whether "la chose vaudra la peine que nous y employons ce que nous avons d'amis et de crédit à Paris."[72]

The Lausannois replied that they did not want to rake up old troubles but rather to give the STN an opportunity to cash in on a new speculation, which would surely succeed if they could break through to the French market. They had made a success of their first edition without France. Only 100 of its 3,200 sets remained to be sold, and they had reached the twentieth volume in their printing. They had received word from Lyons that the quarto publishers had sold out almost all their *Encyclopédies*, too: "leur moisson est faite; ils paraissent inclinés à nous permettre de glaner après eux." The time had come to expand across the Jura. They could reprint volumes 1 through 20 at 3,000 and continue with volumes 21 through 36 at 6,000. The increase in output would permit them to save about 20,000 livres in the costs of composition and correction for the second "edition." They could use that sum to buy their way into France from the quarto group. And once in France, they could make a killing, for they were convinced that the quarto had not exhausted the demand for inexpensive *Encyclopédies* among French readers: "Le public a bien accueilli la première édition [octavo], et l'on nous fait de toutes parts des instances en France pour cet ouvrage, qui nous sont un bon garant de l'écoulement de la seconde. L'un de nous pourra faire le voyage pour renouveller les anciennes souscriptions et en acquérir des nouvelles." They would give the STN at least fifteen volumes to print; and in order to satisfy Ostevald's demand for information about costs and

72. STN to Société typographique de Lausanne, Oct. 11, 1779.

profits, they produced the following "Tableau de l'entreprise":

Pour la réimpression de 3,000 exemplaires en 36 volumes de discours et 3 volumes de planches, les frais d'impression, gravure, etc., total environ .... 280,000 livres

Ajoutons pour faux frais 20,000

300,000

Le produit des 3,000 exemplaires, déduction faite du 25% [for the booksellers' discount] et le 13eme gratis [for bakers' dozens] sera .... 450,000 livres

Par conséquent on aura 150,000 livres de France de bénéfice.

The STN could therefore expect to make 50,000 livres from its one-third share in the enterprise, aside from its portion (presumably 4,167 livres) of the protection money to be paid the quarto group and its profits from printing the fifteen volumes.[73]

It was a tempting prospect; to make it more so, Bérenger reinforced the octavo argument in a personal letter to Ostervald. He was a literary man, not a merchant, he said, but he had been amazed at the rapid sales of the octavo, especially as it was "bien inférieur" from a literary point of view. Heubach had assured him that only a few sets remained and that they could easily sell another 3,000 copies, for they were already sure of 500 sales in Paris and 700 in other French cities. But Ostervald knew that he could not make a deal without the support of Panckoucke, who had ultimate control of the French border. So he sent a noncommittal reply: he would have to consult the quarto associates.[74]

Bérenger sent back a more insistent letter. The octavo publishers had reached an important turning point. They would either continue their *Encyclopédie* as it was (and it was profitable enough in its present form), or they would double it. To delay the decision would be to increase their expenses, and they did not want to set type now for volumes that they would have to recompose if they were to print at 6,000 instead of 3,000.[75] But Ostervald could not get a commitment out of Panckoucke, who was then absorbed in his negotiations with

73. Société typographique de Lausanne to STN, Oct. 21, 1779.
74. Bérenger to Ostervald, Oct. 15, 1779, and Ostervald to Bérenger, Oct. 21, 1779.
75. Bérenger to Ostervald, Nov. 5, 1779.

the Liégeois, and left Bérenger's letter unanswered. The Lausannois then began to suspect that the quarto group was playing for time, while it prepared some new move—perhaps an octavo of its own. After all, the information supplied by Lausanne and Bern proved the profitability of their project, which the quarto group could simply appropriate. Panckoucke and his partners had enclosed the French market within an impenetrable trade barrier, and they could harvest their own octavo in it without sharing the yield. On November 11, the Lausannois demanded a categorical reply. They could wait no longer; the quarto group would have to commit itself.

A week later the Neuchâtelois answered that they could not answer because their partners had not yet decided. If the octavo group forced the issue, they would have to say no—not that they found the proposal unappealing; they simply could not commit their associates. The quarto Association was to meet in Lyons in a month, and the STN would try to get a decision from it at that time, "notre délicatesse ne nous permettant pas de prendre aucun parti à ce sujet avant cette époque."[76] That reply sounded so evasive to the Lausannois that it confirmed their suspicions. The quarto group must surely be preparing a surprise raid on the octavo market that they believed belonged to them. Worse, their own confederate, the STN, had lured them into revealing confidential information, which was the crucial element in the decision to attempt this stab in the back. Now that the quarto publishers knew that there were at least 1,200 potential customers for another octavo in France, and that the demand for it had continued to make itself felt in Lausanne and Bern, they would produce the book themselves, extracting one more victory from a war they had already won.

This was the theme of a bitter letter which Bérenger sent to Ostervald, a letter whose bitterness was especially acute because of the closeness between the two men. Only three months earlier, Ostervald had made a surprise visit to the Bérenger household during a trip to Lausanne and had spent some delightful hours with Madame Bérenger and her children. Bérenger had written later about how sorry he was to have been away at the time and about the warm feelings that Ostervald had left behind: "Si je pouvais être jaloux, je le

76. STN to Société typographique de Lausanne, Nov. 17, 1779.

serais de vous: il n'est pas jusqu'à nos petits enfants qui n'estropient un peu votre nom pour parler de vous, et le petit homme se vante d'avoir gagné aux quilles le grand Monsieur qui était tant bon.'"[77] Now Bérenger wrote in a different tone, virtually accusing Ostervald of stealing the octavo group's plans for a second edition. "C'est notre Société qui vous y a fait penser; c'est elle qui vous l'a proposée; c'est à elle que vous avez demandé des détails pour voir si l'entreprise était faisable. Elle se confie à vous comme à un allié; elle vous détaille son projet, le gain qu'elle en espère, les frais qu'il exige, etc.; et dès que vous le savez, vous faites de sa confiance un piège. Elle ne pense pas que vous en ayez agi avec franchise et comme un associé." This treachery hurt Bérenger more than anyone else, not merely because of his friendship with Ostervald but also because the Société typographique de Lausanne had allotted him one-fourth of its one-third interest in the octavo. He had pinned his hopes on it. He had expected it to bring him 9,000 livres, perhaps even enough to retire on. "Déjà je voyais dans le lointain le moment où j'acheterais une cabane solitaire, un verger, un petit champ, où j'irais m'y égayer avec ma famille. Puis vous soufflez sur ce rêve de bonheur. Ah! ce n'est pas vous qui deviez le faire évanouir!" The STN's apostasy would also damage the Confederation and would even undermine Bérenger's position within his own société typographique. For he had always advocated closer relations with Neuchâtel, and now the Neuchâtelois had made that policy look foolish—unless Ostervald would change his mind: "Voyez, pesez, jugez, je ne dirai rien de plus; je voudrais oublier cette affaire désagréable."[78]

Ostervald never expected to receive such an indictment because, in fact, he had not abused the Lausannois' confidence and the quarto group had not planned to produce its own octavo, although Bérenger's misconception suggests that such foul play was not unthinkable. No holds were barred in the commercial warfare of the eighteenth century. In this case, however, Ostervald was able to reassure Bérenger that the STN had not betrayed its Swiss allies and that it still would argue their case at the meeting of its French alliance. As

77. Bérenger to Ostervald, Aug. 1779 (no precise date given).
78. Bérenger to Ostervald, Nov. 23, 1779.

that meeting had been called for the final settling of the quarto's accounts, it would provide an appropriate occasion for a settlement with the octavo publishers. Bérenger sent a retraction in reply: he had been misled by a false rumor about the STN's *Encyclopédie* plans. In subsequent letters, the two friends tried to cover up the chasm that had opened between them. Ostervald explained the difficulty of being forced to play both "l'enclume et le marteau." And Bérenger expressed some sympathy for "votre situation . . . fâcheuse."[79]

The Lausannois then went directly to Duplain, who proved ready and willing to sell them the entry into France for 24,000 livres, as he had almost liquidated the quarto. But he, too, could not make any settlement without the agreement of Panckoucke, whose main concern by the end of 1779 was the launching of the *Encyclopédie méthodique*. Fearing damage to the market for his new *Encyclopédie*, Panckoucke vetoed the proposal, reinforced his defenses, and reported to the STN that Néville had sounded the antioctavo alert once more: "Soyez sûrs que l'in-octavo . . . n'entrera pas en France. Le Magistrat me l'a renouvellé de nouveau."[80]

A Drôle de Paix

The attempt to bring the quarto and octavo groups together had almost split the Swiss Confederation. So much was at stake in so many different intrigues by 1780 that fissures had developed everywhere in the publishing alliances. The octavo publishers had tried to use the STN to divide their conquerors, or at least to extract easy terms from them; for 20,000 livres was not an excessive price for the liberation of France, espe-

79. Ostervald to Bérenger, Nov. 29 and Dec. 20, 1779, and Jan. 3, 1780; Bérenger to Ostervald, Dec. 10 and Dec. 31, 1779, and Jan. 15, 1780.
80. Panckoucke to STN, Jan. 6, 1780. Panckoucke was referring to Néville's order directing the inspecteurs de la librairie to seize all copies of the octavo. Panckoucke also indicated, in a letter to the STN of Dec. 2, 1779, that he felt great concern about Duplain's negotiations with the Lausannois: "Duplain vient de m'écrire qu'il était à la veille de traiter avec les Lausannois, qui lui offraient mille louis pour l'entrée en France. Je lui ai mandé sur le champ de ne rien faire, puisque pour une somme modique ce serait nous enlever toutes les espérances de l'*Encyclopédie méthodique*, à laquelle on travaille à force et dont la partie physique est même sous presse. Comme M. Duplain est allé très en avant avec ces Messieurs de Lausanne, que sa lettre semble même annoncer un rendez-vous, je vous prie en grâce, Messieurs, de veiller sur tout ce qui passera."

cially if compared with the 27,000 livres that the quarto group had paid to Barret and Grabit. But for reasons that will become apparent later, the Neuchâtelois could not afford the slightest falling out with Panckoucke at this time. Even if they wanted to, they could not make a separate peace. They owned half the privilege for the *Encyclopédie,* but Panckoucke controlled the protections. Duplain suffered from the same weakness, though he followed an independent line of secret diplomacy. Ultimately, therefore, the octavo publishers bowed to necessity and laid their peace proposals before the imposing Atlas de la librairie.

The pacification took place, along with all the other *règlements de compte,* at the Lyons meeting of January 1780. Unfortunately, there are only fragmentary reports of that event, so one cannot follow the last maneuvers that finally brought the octavo–quarto war to a close. But the main result is clear: the two sides signed a *traité,* which gave the octavo publishers the right to market their *Encyclopédie* in France for 24,000 livres.

Rather than confront Panckoucke directly, the Lausannois revived the preliminaries that they had agreed upon with Duplain in December. Duplain received their bills of exchange, made out to him, for payment of 24,000 livres at some future date, and then negotiated the final contract with Panckoucke. This proved to be a difficult task, however. In order to insure themselves against the disappearance or malfunctioning of Panckoucke's much-vaunted protections, the octavo publishers insisted that he guarantee to get 500 sets into the capital. For his part, Panckoucke continued to worry that the flood of octavos would damage the market for the *Encyclopédie méthodique,* and he also considered the sum insufficient. But it represented a badly needed asset at a moment when the quarto associates were haggling over their own debits and credits, and Panckoucke hit upon an idea for translating it into something of immediate value. He allowed Duplain to endorse the notes over to him as part of Duplain's general financial settlement with the associates, and then he gave them back to the Lausannois in return for the equivalent value in octavo *Encyclopédies,* presumably 150 sets of the 500 he had guaranteed to get into France. This arrangement seemed to be mutually beneficial: Panckoucke acquired a valuable commodity in place of a paper asset, and the octavo publishers compensated him from their stock without depleting their capital.[81]

They used the capital for the new edition they had planned to produce with the STN. Secure in the knowledge that Panckoucke had finally sanctioned their entry into France, they immediately doubled their output. They printed volumes 21 through 36 at approximately 6,000 and reset the earlier volumes for a supplementary pressrun of approximately 3,000. Hence the answer to another bibliographical riddle, for librarians and book collectors have searched in vain for a missing octavo edition like the missing quarto.[82] In fact, the unfindable second "edition" of the octavo was only an extension of the first. But there were two distinct octavo subscriptions. Lausanne and Bern opened the second one in April 1780 with high hopes for capitalizing at last on the French sales that had eluded them for two and a half years. But just when those hopes seemed certain to be fulfilled, Panckoucke obliterated them. He slashed the price of his octavos, dumped them on the market, and then compounded the damage by publicizing his plans to produce an *Encyclopédie* to end all *Encyclopédies*—the *Encyclopédie méthodique*.

How extensive was the damage to the second octavo edition? How dastardly was the deed? And why did Panckoucke do it? Irresistible but unanswerable questions, given the sparseness of the documentation after February 1780, when the quarto Association came to an end. One can approach them, however, by comparing the terms offered by the octavo publishers and Panckoucke in their advertising:

Lausanne and Bern, announcing their new edition in the *Gazette de Berne* of April 8, 1780:

L'*Encyclopédie* in-octavo qu'impriment les Sociétés typographiques de Berne et de Lausanne, page pour page après l'édition in-quarto par Pellet à Genève, ayant reçu l'accueil le plus favorable, elles ouvrent une nouvelle souscription, à raison de 5 livres de France le volume de discours (il y en aura 36) et de 15 livres chacun des 3 volumes de

81. The main provisions of the settlement are described in notes made by Ostervald and Bosset at the Lyons meeting (STN papers, ms. 1220) and in their letter to their home office from Lyons of Jan. 29, 1780. See also the complementary remarks in Panckoucke to STN, March 31, 1781.

82. See George B. Watts, "The Swiss Editions of the *Encyclopédie*," *Harvard Library Bulletin* IX (1955), 230–232 and Lough, *Essays*, pp. 40–41.

planches, [making 225 livres in all] comme on peut voir plus au long dans leurs Prospectus. Pour cette deuxième édition, on délivrera les volumes de discours depuis 1 à 20 par 5 volumes à la fois, et depuis 21 à 36 par 2 volumes. On paye en souscrivant chez lesdites sociétés et les principaux libraires 6 livres de France et 25 livres en recevant les 5 premiers volumes, qui paraîtront au plus tard au mois d'août prochain, après quoi la souscription sera fermée et chaque volume se vendra 6 livres de France. L'ouvrage entier sera terminé dans le courant de juin 1781.

Panckoucke, in a printed circular to booksellers, dated February 27, 1781:

L'ENCYCLOPEDIE, in-octavo, *édition de Lausanne, en trente-six volumes de Discours et trois de Planches.* Cette édition aussi complète que celle connue sous le nom de Pellet, est pour vous, Monsieur, du prix de 168 livres 15 sols, et le treizième exemplaire *gratis,* en feuilles. Le prix pour le particulier est de 225 livres, de sorte que votre remise est plus d'un quart. Il en paraît actuellement 26 volumes de discours et un de planches. Toute cette édition sera finie au mois d'avril prochain. Pour ce seul article, je tirerai en expédiant à un an et quinze mois. Le prix de chaque volume de discours est de 3 livres 15 sols et le volume de planches 11 livres 15 sols. Je ne ferai que deux livraisons pour éviter l'embarras . . . Il n'y a jamais eu de livre donné à meilleur compte.[83]

Panckoucke was hardly exaggerating: an *Encyclopédie* for 168 livres, 15 sous, and an additional 8⅓ percent discount for large orders, with a year's credit for the payments represented an incomparable bargain, one that would have driven any competitor to despair—and the despair must have been heavy in Lausanne and Bern. The octavo publishers may have dropped their price (their subscription notice apparently gave the *prix de particulier* or retail price, and Panckoucke's circular the wholesale *prix de libraire*) but they could not cut it down to Panckoucke's level. He got his octavos free, and they needed to clear some profit after covering publishing costs, which had gone up by 8 percent, owing to the now use-

83. The circular is in a letter from Panckoucke to the STN of March 31, 1781. Panckoucke issued these printed circular letters every year to advertise his wares among a large clientele of retailers. They show how he came to function as a wholesaler-impresario, for they differed from the catalogues printed and circulated by other wholesalers in that they offered only a few works of massive size, like Buffon's *Histoire naturelle* and La Harpe's *Abrégé général des voyages,* instead of a general stock. By studying the series of *lettres circulaires* in Panckoucke's dossier in Neuchâtel, a biographer would be able to follow the evolution of his enterprises and his marketing strategy.

less purchase of the right to sell their merchandise in France. Even by the standards of the time, Panckoucke seemed to have struck below the belt; and even his former partner, the STN, protested that he had committed a foul. But Panckoucke replied that his conscience was clear: "Vous vous plaignez de l'in-octavo de Lausanne, mais vous avez partagé les mille louis qu'ils ont donné pour entrer en France. Vous ignorez donc que dans l'acte passé avec Duplain il est convenu que j'en ferais entrer cinq cent à Paris; et puisqu'il fallait remplir cet engagement, j'ai cru devoir en prendre en payement et les servir. Si je ne l'eusse pas fait, un autre l'eût fait à ma place, et nous n'y eussions rien gagné." He still viewed Lausanne and Bern as interlopers in his territory. True, he had opened it up to them, but he felt no moral obligation to honor their deal with Duplain. He executed his part of the bargain and then turned it against them—a *coup de théâtre,* perhaps, but not a *sale coup.* If anyone were to skim the cream off the market, it should be he. And he could spoil it at the same time—that seems to have been his main motive. For the Lyons agreement committed him to allow 500 octavos into Paris, not to make sure that they got sold. He could prevent their sale, if he sold 150 of his own, at a drastic reduction. In that way, he would minimize the damage to the *Encyclopédie méthodique,* which had become his supreme speculation by 1780: better to dump 150 octavos on the market than to swamp it with 500.[84]

It seems unlikely, however, that 150 cut-rate octavos could permanently destroy the demand for the 3,000 that Lausanne and Bern expected to sell. Panckoucke probably calculated on a short-term victory, one that would drive the bulk of the second octavo edition off the market until he could produce the first volumes of the *Encyclopédie méthodique,* which would be so superb as to make the public forget the primitive, imperfect work of Diderot. In the long run, however, Lausanne and Bern probably sold a great many of their books in France because Panckoucke's ultimate *Encyclopédie* became mired in difficulties and delays. Thus the last phase in the diffusion of Diderot's text and the first phase in the production of Panckoucke's overlapped and intertwined. In fact all the speculations on the *Encyclopédie* were connected, for all of

84. Panckoucke to STN, March 31, 1781.

them represented attempts to satisfy the seemingly insatiable appetite for Encyclopedism among the readers of the Old Regime. But before they can be traced to their denouement, it is important to take a closer look at the process by which the *Encyclopédies* reached the readers.

V

YYYYYYYYYYYY

BOOKMAKING

The contract disputes, pirate raids, and trade wars all point to one central fact: the *Encyclopédie* had become a best seller—the biggest best seller anyone had ever heard of, a publisher's dream, "la plus belle [entreprise] qui ait été faite en librairie."[1] Whether it really was the most lucrative speculation in the entire history of publishing before 1789 cannot be determined because almost nothing is known about the sales of other early-modern books. But the papers of the STN reveal almost everything about the production and diffusion of the quarto *Encyclopédie*. Each of those aspects of the book's biography deserves a chapter to itself, for each leads to unexplored areas in the past—areas where publishing history borders on the history of economics and technology, of work and the working class, of management and advertising, and of communications and idea diffusion. By following this best seller from producer to consumer, one can investigate the literary market place from a dozen different angles. In the end one should be able to chart the course of a phenomenon that has eluded earlier investigation, although it has shaped much of modern history: the spread of the Enlightenment.

Strains on the Production System

The extent of the quarto's success can be appreciated from the enormous effort that was necessary to manufacture the

1. STN to Panckoucke, Aug. 20, 1778.

book. After collecting more than 8,000 subscriptions, Duplain faced the problem of producing 306,900 huge quarto volumes.[2] He was not a master printer himself, and in any case, the task exceeded the capacity of a single printing shop; so he contracted the work out to two dozen shops scattered about eastern France and western Switzerland. Jean-Léonard Pellet, under whose name the first two editions appeared, had only a peripheral connection with the enterprise. The quarto partners, who paid him 3,000 livres for the use of his name on the prospectuses and title pages, referred to him disparagingly as their "prête-nom" and "commissionnaire" in their confidential correspondence. Judging from colophons, Pellet printed only four volumes of "his" edition—no more than those done by the STN, which also printed four, though it produced only one of the volumes of its own "Neuchâtel" edition.[3]

In order to make the most of his role as contractor, Duplain played the printing houses off against one another and bargained them down as far as they could go. They had to compete not only for commissions but also for supplies—type, ink, paper, and workers—and the competition got rougher as the enterprise increased in scale. The original nucleus of production was in Geneva, in the shops of Pellet, Bassompierre, and Bonant. By March 1777, thirty Genevan presses were working on the quarto, and a year later the whole city was printing it, according to a Genevan bookseller.[4]

Meanwhile, Duplain had shifted the center of production to Lyons. Once he had covered himself with Panckoucke's privi-

2. This figure is based on the total pressrun of 8,525, stipulated by the contracts for the three editions. The number of complete sets must have been smaller, owing to spoilage. Panckoucke saw to the production of the three volumes of plates in Paris.

3. See STN to Ranson of La Rochelle, May 24, 1778: "Pellet, qui n'est qu'un simple imprimeur, est un prête-nom pour nous"; and STN to Graffenried of Avrenches, March 6, 1780: "Celui-ci [Pellet] n'était que notre commissionnaire, chargé de notre part d'imprimer quelques volumes . . . Pellet avait commission aussi de collecter des souscriptions, et c'est là à quoi tout son intérêt a été réduit." Pellet's fee for the use of his name on the title page appears in Bosset's notes, entitled "Dépenses," on the expenses of the enterprise, in the STN papers, ms. 1220. And Pellet's colophon appears at the end of volumes 2, 7, 11, and 31 of the first edition of the quarto in the Bibliothèque de la ville de Neuchâtel. The only other volumes with colophons in that set are volumes 3 and 8 (Bassompierre of Geneva) and 14 (Société typographique de Genève.)

4. Duplain to Panckoucke, March 16, 1777 (from a copy sent to the STN by Panckoucke) and Barthélemy Chirol of Geneva to STN, July 17, 1778.

lege, he was able to operate legally in France. The French authorities permitted him—informally and without acknowledging it—to have the quarto printed in Lyons, and the contract of October 10, 1778, virtually made the third edition a Lyonnais product, despite the "Neuchâtel" on its title page.[5] When Favarger inspected Duplain's operation in July 1778, he found that the quarto was dominating the printing industry of the whole region: "Il y a a environ 40 presses qui travaillent à cet ouvrage, tant ici [Lyons]qu'à Grenoble et Trévoux . . . A la réserve de quelques usages, l'on n'imprime ici autre chose, et dans toutes les imprimeries, que l'*Encyclopédie* quarto . . . Quinconque avait un certain argent à mettre tous les mois ou tous les ans sur des livres, l'a placé sur l'*Encyclopédie* quarto."[6] Duplain himself listed fifty-three presses working on the third edition in January 1779.[7] Considering that the first two editions were almost three times as big as the third, it seems likely that about one hundred presses in twenty different shops worked on the quarto between 1777 and 1780. At the same time, the sociétés typographiques of Lausanne and Bern were putting out the octavo *Encyclopédie* at a pressrun of 6,000, and Félice was printing 1,600 sets of his version of the *Encyclopédie* in Yverdon. Diderot's book was being produced on such a scale that it strained the capacity of the printing industry throughout a vast stretch of France and Switzerland.

The strain showed in every segment of the economy with any connection to the book trade. It is difficult to appreciate today, however, when books, as physical objects, do not have the same importance they possessed in the eighteenth century. Before buying a book, the readers of the Old Regime inspected the merchandise carefully, rubbing the pages between their fingers, holding them up to the light, scrutinizing the shape of the characters, the clarity of the impression, the

5. At first Panckoucke had doubted that Duplain could persuade the authorities to let him print in France: "Il sollicite pour en obtenir la permission à Lyon, et je sais qu'il ne l'obtiendra pas." Panckoucke to STN, July 4, 1777. But on Aug. 5, 1777, Panckoucke reported to the STN, "A force de sollicitations on a obtenu d'imprimer quelques volumes à Lyon."

6. Favarger to STN, July 21, 1778.

7. Duplain to STN, Jan. 21, 1779: "Voici l'énumération des presses que nous avons sur la troisième édition: 6 chez Bélivre, 4 chez Labbe (?), 4 chez Chavanne, 6 chez Vatan, 8 à Trévoux, 4 chez Goeri, 3 chez Dégoutte, 6 chez Pellet sous 15 jours, 3 chez Cuty, 9 chez Cuchet."

width of the margins, and the overall elegance of the design. When they found faults, they protested loud and clear. "Vous aviez garanti, Messieurs, pour l'impression du *Dictionnaire encyclopédique* un beau papier, un caractère neuf," an indignant subscriber wrote to the STN. "Cette promesse, permettez-moi de vous le dire, n'a pas eu son entière exécution, car le papier est généralement défectueux et la caractère presqu'-éteint, ce qui fatigue beaucoup les yeux du lecteur. Des ouvrages de ce genre, faits pour vivre éternellement, méritent qu'on y apporte un peu plus d'attention. La majeure partie des feuilles sont maculées ou déchirées. Vous sentez, Messieurs, que ces négligences de la part de vos ouvriers ne peuvent être que préjudiciables à vos intérêts, en dégoûtant le public des nouvelles souscriptions à ouvrir."[8]

As this reader pointed out, the quarto publishers had emphasized the physical qualities of the book throughout their advertising. In their main prospectus, they insisted that all the volumes would be printed on the best quality paper with handsome type, called appropriately *philosophie* (small pica). Individual retailers embroidered on this theme in their own sales campaigns. Thus Téron of Geneva assured the readers of the *Gazette de Leyde* that "tous les papiers sont tirés d'Auvergne, et on n'emploie que des caractères de France, qui seront renouvellés après le tirage de chaque cinquième volume."[9] These remarks would seem out of place in the advertising for a modern, machine-made book, but they were standard fare in an age when books were made by hand, when sheets of paper were manufactured individually through months of careful handling in remote mills, and when an army of ragpickers was required to gather the raw material for paper—cast-off linens, whose threads can still be seen in the fabric of the *Encyclopédie*. It took more than a million sheets to produce just one of the thirty-six quarto volumes, in all three editions. And it took five months of hard labor by five compositors and twenty pressmen for the STN to transform those sheets into one volume of printed pages.[10] Although they represent only

8. Champmorin of Saint Dizier to STN, July 17, 1780.
9. *Gazette de Leyde*, Oct. 7, 1777. See also similar remarks in the notices of Jan. 3 and Feb. 11, 1777.
10. According to its account books, the STN used 1,762 reams for volume 24 of the first and second editions and 669 reams for volume 19 of the third edition. By concentrating exclusively on this job, a team of five compositors managed to set the type for volume 24 between June 6 and Nov. 7, 1778. The pressmen, whose

a small fraction of the entire printing process, the STN's operations illustrate the complexity of producing a book on a mass scale before the advent of mass production.

Plant expansion proved to be the least of the STN's problems. It bought a new house for its printing shop and purchased six fully equipped, secondhand presses, doubling its printing capacity. The presses came from the Lyonnais printer Aimé de la Roche Valtar and cost only 250 livres apiece, the equivalent of four months' wages for an ordinary journeyman printer.[11] Getting type was another matter. In an attempt to maintain some harmony in the physical appearance of the volumes, Duplain directed all his printers to buy their fonts of *philosophie* from a Lyonnais founder named Louis Vernange, but Vernange was soon overwhelmed with work. On April 20, 1777, he signed a contract with the STN, promising to supply a font weighing 1,800 *livres,* half on June 1 and half on July 1, and to cut his price by 20 percent if he failed to make the deadline. On June 2, he asked the STN to give him another two weeks. It agreed reluctantly, because it had planned to put six presses on the *Encyclopédie* from the beginning of the month. Still without type on June 18, it warned him that it needed the shipment urgently. And on June 26, it threatened to apply the penalty clause because its presses and workers were standing idle. The first shipment finally arrived on July 8, the rest at the end of July and the end of August. Even then, the STN had to order another 500 *livres* and various *assortiments,* which did not arrive until the end of the year. The main font, weighing 1,471 *livres,* had taken twice as long as Vernange had promised and had cost 1,852 livres tournois. Vernange justified the delay by recounting his difficulties: his workers had fallen ill and had gotten into trouble by their "caprices" (the Lyonnais authorities had

---

number varied, printed it concurrently with four other large-scale projects and various small jobs, all of which were composed by the half dozen other compositors in the shop.

11. On the STN's purchase of "la maison Brun" in order to expand its shop for the printing of the *Encyclopédie,* see STN to Pettavel, July 22, 1776; Mme. Bertrand of the STN to Bosset, May 21, 1780; and Bosset to Mme. Bertrand, May 29, 1780. On the purchase of the presses see the STN's letters to la Roche Valtar of Sept. 8, Sept. 24, Oct. 6, Oct. 12, and Nov. 24, 1776, and Sept. 17, 1777. Favarger inspected the presses during a trip to Lyons in 1776. He found them to be better value than new presses, which would cost 300 livres, without copper platens, and would take a month each to be built by Tardy, the best pressmaker in town. Favarger to STN, Aug. 25, 1776.

exiled one of them for misbehavior); he had not been able to find a wagoner for the route through the Franche-Comté, owing to the harvest; and he simply could not cope with all his orders. Actually, the STN was fortunate to get its type with only two months' delay. Bonnant of Geneva had to suspend operations in November 1777 because his type failed to arrive—not from Lyons, for Vernange could no longer supply all the quarto printers, but from Avignon, where another foundry was already swamped.[12]

Demand also smothered supply in the ink trade, which was monopolized by two Parisian firms, Langlois and Prévost. The Neuchâtelois dealt with Langlois, who charged 22 sous per *livre*, 2 sous less than Prévost. But they nearly ran dry in October 1777; and when Langlois finally came to their rescue, after several urgent appeals, he was able to ship only one of the two 250-*livre* barrels that they needed. He took three months to supply the second barrel; and when it arrived in April 1778, it turned out to be faulty, forcing the STN to suspend its printing for a while. The STN extracted two barrels out of Langlois in May, another in October, and two final barrels in February and August 1779. Meanwhile he had inched up his price—from 22 to 28 sous per *livre*. He blamed the suppliers of his own raw materials—walnut gatherers and resin merchants from the Midi, traders in turpentine and linseed oil in Paris, and even the American revolutionaries, who, he claimed, upset commerce so badly that he had to push his price up another 2 sous in 1782. It may be that walnuts and revolutionaries were bound up in a world economic system, but Langlois probably was cashing in on the boom in *Encyclopédies*—and doing very well from it, too, for one of his barrels of ink cost more than a fully equipped printing press.[13]

12. The transactions about type can be studied in detail in the dossier of Vernange, which contains twenty-one letters, and in the STN's replies, particularly its letters of June 26, July 8, and Sept. 4, 1777. See also STN to Duplain, April 26, 1777, and Bonnant to STN, Nov. 14, 1777. In general, the STN ordered its type both by weight and by sets or *feuilles,* and it paid about 25 sous per Lyonnais *livre* of 14 *onces.* It provided Vernange with two sample letter *m*'s so that he would cast the type in conformity with its height to paper.

13. The most important letters in Langlois's dossier are Langlois to STN, Oct. 27, 1777; Jan. 22, Feb. 5, May 2, and Sept. 17, 1778. Duplain insisted that his printers use top-quality Parisian ink: ''Employez-vous de l'encre de Paris? Il n'en faut pas d'autre et vous adresser à Messieurs Prévost-Langlois, l'un et l'autre marchand d'encre à Paris, sans autre adresse.'' Duplain to STN, Feb. 9, 1778.

The millroom of a paper mill as represented in the original *Encyclopédie* ("PAPETTERIE," plate IV). The waterwheel (E) turns an axle or camshaft (B). As they rotate, the cams raise the ends of twelve mallets arranged in groups of fours (a, b, c) and then release them so that the heads of the mallets drop into three troughs, where rotted rags have been dumped. The pounding and rinsing transforms the rags into stuff.

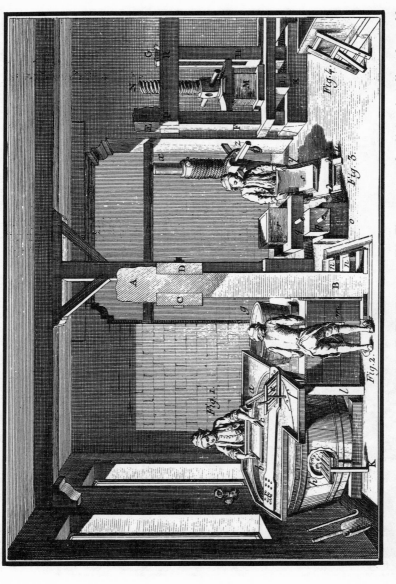

The vat room of a paper mill ("PAPETTERIE," plate X). The *ouvreur* (fig. 1) has just dipped a mold of woven wire into a vat full of heated stuff and is letting the water drain out before tilting the mold in such a way as to "lock" the fibers of the stuff together and form a sheet. Meanwhile the *coucheur* (fig. 2) separates a newly formed sheet from its mold by flipping it onto a piece of felt. Then the *leveur* (fig. 3) presses the sheets in piles of 260 called *porses* and separates them from the felts so that they can be dried and sized.

Procuring Paper

The most costly element in book production was paper. Paper obsessed eighteenth-century printers and determined many of their calculations. When they discussed pressruns, they often talked in reams and quires rather than in thousands and tokens. And when they made budgets for books, they figured in *feuilles d'édition,* that is, the cost of producing all the copies of one sheet, including composition, presswork, and paper. Those three elements varied with the size of the printing: on the one hand, the cost of setting the type remained the same, while the cost of the presswork increased with the number of copies printed; on the other, the cost of paper went up at a faster rate than that of composition and presswork combined. The *Encyclopédie* contracts made room for the first variation by setting a flat rate of 30 livres per *feuille d'édition* for the composition and presswork of the first thousand copies and 8 livres for every additional thousand. So according to the first contract (the Traité de Dijon of January 14, 1777), Duplain would receive 54 livres to cover the labor costs of producing one sheet of the first edition at a pressrun of 4,000. According to the contract of September 30, 1777, he would receive 71 livres 4 sous for the labor on the first and second editions, which were printed together at a run of 6,150 (that is, 12 reams 6 quires). The presswork had gone up from 59 percent to 69 percent of the labor costs. But the cost of paper had increased by even more—from 72 livres per *feuille d'édition,* according to the first contract, to 110 livres 14 sous, according to the second. The contracts also recognized the critical importance of paper as a variable by special clauses that set a fixed rate per ream for Duplain's provisioning. After doing some arithmetic, therefore, one can compare the proportions in the "budgets" of a *feuille d'édition* from the contract for the first edition and a *feuille d'édition* in the contract for the combined first and second editions.

First edition (*pressrun of 4,000*)		First and second "Editions" (*pressrun of 6,150*)	
Typesetting	22 livres	22 livres	
Presswork	32 "	49 "	4 sous
Paper	72 "	110 "	14 sous
	126 livres	181 livres 18 sous	

The role of the paper in the calculations of the publishers had expanded from 57 percent of the first budget to 61 percent of the second. Of course the actual printing costs differed considerably from the standard costs set by the contracts, but the differences only magnified the importance of paper. For example, the STN provided all the paper for volume 24, which it printed at a pressrun of 6,150. The typesetting and presswork came to 4,828 livres and the paper to 13,897 livres, or three-quarters of the production costs, excluding overhead. Moreover, the quarto association reimbursed the STN for its paper expenses at the rates fixed in the contracts, which were higher than the rates it had paid; so the STN got back 15,875 livres—a profit of 1,978 livres on the paper for volume 24 alone. In general, the expense of paper and the cheapness of labor made the proportions in the budgets of eighteenth-century printers look like the opposite of those in modern printing, for in the nineteenth century labor costs soared, and the price of paper—mass-produced paper made by machines from wood pulp—plummeted. With the spread of offset printing and the devastation of forests, costs may now be shifting back toward the eighteenth-century pattern. Looking backward from the 1970s, however, one can hardly overestimate the importance of paper for the publishing industry two hundred years ago.[14]

Duplain had to produce 36 million sheets of paper for his printers. Just how he managed this feat of engineering is not clear, because one can see only the managerial aspects of the enterprise from the perspective of the archives in Neuchâtel. But the rising pressure of demand can be followed from the quarto's contracts. On January 14, 1777, Duplain's paper costs were fixed at 9 livres per ream, on May 15 at 9 livres 5 sous, on September 30 at 9 livres 10 sous, and on October 10, 1778, at 10 livres. After investigating the situation in Lyons in September 1777, Panckoucke reported that the increases could not be avoided: the papermakers were forcing up their prices, and they might not be able to supply enough to get Duplain

14. These remarks on paper are based on a close study of the dossiers of twenty-three *papetiers* and *marchands papetiers* in Neuchâtel, but they are intended only as a sketch of a subject that will be treated in full in the forthcoming thesis of Jacques Rychner. For a good account of all aspects of book production and references to the literature in analytical bibliography see Philip Gaskell, *A New Introduction to Bibliography* (New York and Oxford, 1972).

through the winter at any price.[15] Duplain himself finally told the STN to find its own paper: "Ne comtez point sur nous pour des papiers. C'est la chose impossible."[16] Although the STN protested vehemently, arguing that the responsibility for provisioning lay with Duplain, there was little he could do. The paper market of Lyons, and the mills of the Massif Central which fed it, had run dry. Reports from Lyonnais paper merchants confirmed Duplain's account of the crisis, and so did the STN's banker, Jacques-François d'Arnal, who scoured the city for paper in December 1777: 'Nous avons été visiter tous nos marchands de papier . . . mais nous n'avons pas trouvé une seule rame de papier de la qualité que vous demandez. M. Duplain enlève tout.'[17]

Finally, the STN gave up on Duplain and the great paper mills of Auvergne and the Lyonnais, and it tried to piece together its own supply system. It had bought paper from the millers of its region for years, but few of them could make the heavy, white *carré fin* required for the *Encyclopédie*. Duplain insisted that every sheet conform to the samples he sent to his suppliers and to the STN, and that every ream weigh at least 20 *livres* (in Lyonnais *livres* of 14 *onces*), as stipulated in the contract. He had to enforce strict standards because each volume required at least 1,000 reams; the paper came from many different mills; and if the millers' shipments did not conform well enough to be "married" or blended by the printers, the book would take on a piebald appearance and customers would cancel subscriptions. The provisioning problems were compounded by the delicate and rather primitive character of papermaking as a craft. Despite the introduction of some modern machinery (cylindrical pulpers called Hollanders), it remained tied to the rhythms of an agrarian economy. Ragpickers took to the road after working the autumn harvests. Millers stocked rags and prepared stuff (a watery paste from which sheets were made) in the winter and manufactured most of the paper in the spring and

15. Panckoucke to STN, Oct. 9, 1777.
16. Duplain to STN, Jan. 3, 1778.
17. D'Arnal to STN, Dec. 21, 1777. The STN tried to get paper from six of Lyons's thirteen *marchands papetiers* without success. In a typical reply, one of them, Dumond, wrote on Dec. 16, 1777: "Ces papiers sont très rares actuellement et augmentés par la recherche que M. Duplain en fait faire dans toutes les fabriques. Vous vous y prenez trop tard, n'étant pas dans la saison."

summer, when the weather turned warm enough for sizing (applying a delicate finish, which easily spoiled while drying). They often operated by *campagnes* or large batches, which they sold in advance after hard bargaining during the off season. Having contracted for a *campagne,* they would blend their stuff from different grades of rags according to the quality desired. So they could not switch to *Encyclopédie* paper in the late summer and winter of 1777, when the printers needed it most. Duplain had not allowed for the inflexibility of this system when he increased production, and therefore the STN nearly had to fire its workers and to close its shop in the winter of 1777-1778 for lack of supplies.

The STN got through the winter by combing a vast area of France and Switzerland for every last ream of 20-*livre carré*. It wrote dozens of letters, even to millers in the southernmost stretches of Duplain's raked-over territory southwest of Lyons. It haggled with merchants in far-off corners of Switzerland and Alsace. It sent Favarger by horse on a paper hunt through the remotest valleys of the outer Jura. And in the end, it built up a network of suppliers who kept it going until March 8, 1779, when it finished its final volume. The paper, 5,828 reams in addition to some earlier shipments from Duplain, came from thirteen millers and merchants, scattered around an axis that ran for 450 kilometers from Strasbourg to Ambert.

Thanks to the STN's account books, one can follow almost every ream in this flood of paper as it passed from the mills, through the presses, and into actual copies of the quarto sitting on the shelves of libraries today. The table illustrates the evolution of the shipments.

At first the STN depended entirely on Duplain and the Lyonnais merchants who worked for him. Favarger's paper scouting opened up some new sources in mid-1777, but the millers he enlisted, Gurdat of Bassecourt and Morel of Meslières, could not produce much *Encylopédie* paper until the next spring. So when Duplain's provisions gave out in December, just when the STN began printing volume 15, the Neuchâtelois had to patch together enough supplies from the market in Lyons to avoid suspending production in the winter. Relief came with the first big shipment from Morel in March 1778. By May the paper was flowing from Alsace, Switzerland, and the Franche-Comté. And by the end of 1778, the STN had accumulated enough stock to print almost all of its last two volumes.

Drawing supplies from such scattered sources raised the problem of "marrying" the sheets so that their different colors and consistencies would not offend the eye of the customer.[18] To see how the marriages were arranged, one can compare the entries in the STN's accounts with the pattern of watermarks in a copy of the quarto. According to an entry dated November 6, 1778, in the account book called Brouillard B, the STN constructed volume 24 out of paper from five suppliers in the following amounts:

Schertz (Strasbourg)	149 reams 10 quires
Vimal (Ambert)	431 reams 1 quire 3 sheets
Gurdat (Bassecourt)	90 reams
Morel (Meslières)	930 reams 9 quires
Fontaine (Fribourg)	44 reams 11 quires
	1645 reams 11 quires 3 sheets

This information indicates the proportions of different suppliers' paper used in the entire printing of 6,150 copies of volume 24. By converting it into *feuilles d'édition*, one can construct a model volume 24, which can then be compared with an actual copy—in this case the copy in the Bibliothèque de la Ville de Neuchâtel.[19] Volume 24 of the Neuchâtel quarto contains three sorts of paper that are fairly easy to identify: thirty sheets (or gatherings) of Vimal, which have nearly complete watermarks and countermarks; twelve sheets of Schertz, which lack marks but contain a "DV" mentioned in

18. The expression "marrying" comes from the rich slang of eighteenth-century papermakers. For example, in a letter to the STN of Sept. 7, 1778, Jean-Georges Schertz, a paper merchant from Strasbourg, promised to maintain the same degree of whiteness in the paper from all the mills that worked for him: "S'il y manquait un brin, la différence sera si peu de chose qu'on ne s'en appercevra pas. Je sais qu'il faut pouvoir les marier."

19. The conversion can be done by dividing the amount of paper needed for one *feuille d'édition* into the total amount provided by each supplier. As each sheet was printed 6,150 times, one *feuille d'édition* required 12.3 reams of paper. When divided by 12.3, Schertz's 149.5 reams come out as 12.1 *feuilles d'édition*. An archetypical or model copy of volume 24 would therefore contain 12.1 sheets of Schertz, although actual copies might have considerably more or less, depending on how the paper was doled out to the pressmen. The long and difficult task of inspecting the Neuchâtel quarto was done with the help of Jacques Rychner, whose expertise in reading watermarks proved to be crucial for its success.

The Flow of *Encyclopédie* Paper to Neuchâtel (in reams)

				Supplier			
	Date	Girard merchant, Lyons	Tavernier merchant, Lyons	Duplain merchant, Lyons	Gurdat miller, Bassecourt	Morel miller, Meslières	Claudet merchant Lyons
1777	April	160					
	May		10				
	June			48			
July 19–Dec. 13 printing volume 6	July			226	70		
	Aug.				50		
	Sept.			100		10	
	Oct.						
	Nov.			42			
				100			
				50			
	Dec.			214			
1778	Jan.		54				100
	Feb.		168				
			27				
			86				
Dec. 13–June 13 printing volume 15	March					108	
	April						
	May	100			95	234	
	June					270	
June 6–Nov. 7 printing volume 24	July					270	
Nov. 7–	Aug.					310	
	Sept.				90		
	Oct.					102	
Feb. 27	Nov.				70		
1779 printing volume 35	Dec.				116		
	Jan.						
	Feb.						
Feb. 27–	March						
May 8	April						
printing volume 19	May						
	Totals	260	345	780	491	1,304	100
	Total						

By reading this table diagonally from upper left to lower right, one can trace the shipmen of *Encyclopédie* paper from thirteen suppliers, both millers and merchants (*marchands pap tiers*) to the STN. Each number represents a shipment. Thus Tavernier sent the ST three shipments in February 1778, one for 168 reams, one for 27 reams, and one for 86 ream The information comes from two of the STN's account books, Brouillards B and C, whi cover the period from January 1778 to December 1789. A missing account book, Brouilla

			Supplier			
Joannin merchant, Lyons	Fontaine miller, Fribourg	Vimal miller, Ambert	Petitpierre miller, La Motte	Planche miller, Vuillafans	Schertz merchant, Strasbourg	Desgrange miller, Luxeuil
58						
	2					
		213				
		96	40			
	52	304	120		150	
	54				120	
	132					
	28					
				120	200	
	20		327			125
	55½					
						132
						200
58	343½	613	487	120	470	457
					5,828½ reams	

., contained all the credits for paper shipments in 1777, when the STN printed its first *Ency-*
clopédie volume (volume 6). Most of the missing information can be compiled from the correspondence of the suppliers, whose *lettres d'avis* and *factures* for the latter period
correspond precisely with the entries in Brouillards B and C. But a few paper transactions
were settled verbally and some commercial correspondence is missing from the STN's papers.

Schertz's correspondence as countermarks; and four sheets of Fontaine, which have grapes as marks and "MF" for Maurice Fontaine as countermarks. The rest of the paper consists of eighty sheets with no mark or countermark and seven sheets with a tiny cross. The former must have come from Morel, who told the STN that he would keep all identifying signs off the molds of his *bâtard,* which did not conform to the quality standards of the French authorities;[20] and the latter must have come from Gurdat, who provided seven sheets, according to the model. Thus the actual copy and the model correspond quite closely:

	The model	*The Neuchâtel quarto*
Schertz	12.1	12
Vimal	35.0	30
Gurdat	7.3	7
Morel	75.6	80
Fontaine	3.6	4
	133.6	133

This correspondence can be seen only on an abstract level, not in the physical examination of a particular copy—that was the point of the marrying. As Figure 1 shows, the STN shuffled the sheets in an irregular pattern throughout the volume instead of exhausting one lot of paper and then moving on to the next. In June 1778 when it began to print volume 24, the STN had plenty of paper from all five suppliers in stock and probably also had a fair amount from two others, Girard and Petitpierre. Neglecting the latter, it concentrated on the Morel, which was flowing into Neuchâtel at the greatest rate; it also drew heavily on the Vimal, which ranked second in the stockroom, judging from its printing and paper accounts. These two sorts accounted for 83 percent of the paper used in volume 24. Rather than using them en bloc, the STN scattered them throughout the volume in runs that usually varied between two and twelve sheets, except for the two large runs of Morel toward the end. The Gurdat served as a filler,

20. Using an overworn paper-miller's joke, Morel informed the STN in a letter of May 2, 1778, ''Les papiers que je ferai pour vous n'auront point de marque. Ils seront bâtards de nom et de fait.''

in single-sheet units that would not be noticed. The Schertz came first, probably because it was the most beautiful. Then the STN zigzagged between the Morel and the Gurdat, the Morel and the Vimal, the Morel and the Gurdat again; finally it ended with the Fontaine. The pattern could have been more complex and the mixing more complete, but the STN did not want to undertake the back-breaking job of blending hundreds of reams sheet by sheet. It only needed to produce a mixture that would pass muster with a paper-conscious public.

That task required enormous effort, not because marrying was difficult but because buying paper demanded a great deal of time, energy, and craftiness. Each purchase had to be arranged through elaborate bargaining, and each supplier did business in his own way. The famous firms of Johannot and Montgolfier in Annonay barely condescended to sell their medium-quality (*moyen*) paper and quoted prices on a take-it-or-leave-it basis. The little millers of the Franche-Comté, like Planche of Vuillafans and Sette of Chardon, had to scramble hard to sell any of their *fin,* and they often hauled their wares to Neuchâtel themselves so that they could barter in person (many of them found it difficult to write even primitive letters) over a bottle of wine, giving full rein to their Comtois palaver. Merchants like Girard of Lyons and Schertz of Strasbourg filled their commercial correspondence with sophisticated talk about bills of exchange and interest rates, while ordinary millers like Morel of Meslières and Desgranges of Luxeuil hustled for quick cash. When specie was scarce, Morel asked for payment in barrels of wine—he needed vintage Neuchâtel, he explained, as medication for an ailing son—and he mixed his sales talk with garbled quotations from Saint Paul and proposals to cut prices behind Duplain's back by cheating on the weight of the reams and by adding quicklime to the stuff.[21]

The trade was built on such tricks, for the millers rarely had enough good rags to satisfy all their customers and therefore slipped inferior sheets into their *fin* or stuffed extra sheets into substandard reams in order to meet weight requirements. The *Encyclopédie* touched off battles for rags, especially the well-washed linens that came out of Burgundy. Thus Jean-Baptiste Gurdat, a miller from the mountain ham-

21. Morel to STN, Nov. 30, 1777, and May 2, May 16, and July 1, 1778.

Shertz	Morel	Gurdat	Vimal	Fontaine

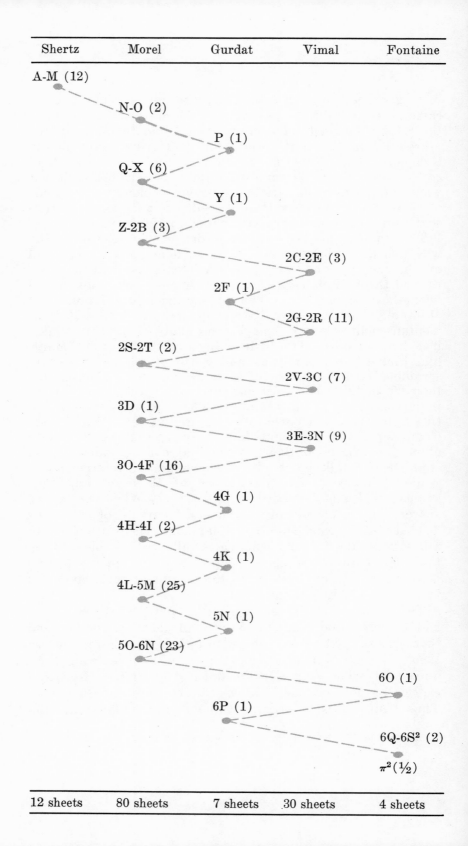

A-M (12)

N-O (2)

P (1)

Q-X (6)

Y (1)

Z-2B (3)

2C-2E (3)

2F (1)

2G-2R (11)

2S-2T (2)

2V-3C (7)

3D (1)

3E-3N (9)

3O-4F (16)

4G (1)

4H-4I (2)

4K (1)

4L-5M (25)

5N (1)

5O-6N (23)

6O (1)

6P (1)

6Q-6S^2 (2)

π^2(½)

12 sheets	80 sheets	7 sheets	30 sheets	4 sheets

Bookmaking

Figure 1. The Distribution of Paper in the Neuchâtel Quarto:
Volume 24

This figure, based on the inspection of watermarks in the copy of the quarto in the Bibliothèque de la Ville de Neuchâtel, shows how the STN distributed sheets while printing volume 24. The letters correspond to the signatures at the bottom of pages, the numbers in parentheses to the number of sheets in a run of paper from a single source. Thus the printer began with twelve sheets of Schertz, switched to Morel when he reached sheet N, moved on to Gurdat at P, went back to Morel for sheets Q through X, and so on. The notations correspond to standard bibliographical descriptions. Thus 2B stands for the sheet signed Bb, π stands for the preliminary leaves, and the calculations fit the 23-letter alphabet used by the printers, which eliminated I or J, U or V, and W. The formula for describing this volume would be as follows: *carré au raisin* 4°: π^2A-6R^{4}6S^2; 532 leaves, pp. [4], *1* 2-1060. (For details about this system of bibliographical description see Philip Gaskell, *A New Introduction to Bibliography* [Oxford, 1972], especially pp. 328–332.) 6S^2 and π^2 were half sheets, which were cut from the same sheet to make the "prelims" at the beginning of the book and to avoid waste at the end. This interpretation is confirmed by the pay book (Banque des ouvriers) of the foreman in the STN's shop, which shows that 6S was paid for as a whole sheet and that there was no special payment for the composition of the title page and half-title, as occurred in cases where the prelims were composed separately.

let of Bassecourt near Porrentruy, demanded quick payment for his *Encyclopédie* paper because he wanted to exploit a coup against the papermakers of Basel: "je conte de me transporté ches vous en viron le 3 ou 4 avris prochain pour à voir mon argen en samble car je en aÿ de grand besoin pour le present car jaÿ oppetenus de notre prince un ordonance quil desfant de lesser passer des guenille pour bale et il men vien baucout de la bourgogne presentement."[22] Gurdat may not have had a perfect mastery of written French, but he knew how to wring specie out of his customers, to lobby with his protectors, and to divert the rag traffic from his competitors.

For its part, the STN used all the devices in its power to manipulate the millers. It demanded a reduction on almost every ream that it received for the *Encyclopédie,* even when paper was acceptable. By finding fault, it often knocked a few sous off the price or got better treatment in the next shipment or forced a miller to accept inferior bills of exchange, which usually bounced and then spent months being passed from bill collector to bailiff to attorney until at last the original sig-

22. Gurdat to STN, March 2, 1778.

natory agreed to a settlement, normally at a reduced rate, or fled town, leaving his creditors to argue over the remains of a bankrupt business. The millers fought back by playing their customers off against one another. This strategy worked beautifuly in 1777, when demand outstripped supply and prices soared. But in 1778 the ragpickers decided that it was time for them, too, to take a cut in the *Encyclopédie* bonanza. Desgranges claimed that the cost of top-quality rags shot up by 25 percent in a little more than a year.[23] Meanwhile, more mills switched to *Encyclopédie* paper and more printers built up their stock; so the pressure shifted back to the producers of paper. By the spring of 1779, the price of 20-*livre carré* had leveled off, and the supply system had adjusted to the *Encyclopédie*. But the adjustment had been slow and painful because the system responded poorly to short-term fluctuations. It moved at a pace that was set by an ancient style of market-place bargaining and by the more fundamental seasonality of nature. It worked well enough, however, to supply the raw material for 8,000 thirty-six-volume sets of Diderot's great work.

Copy

Diderot and his collaborators had done their share of the work many years ago, but that was only the beginning of a long process which culminated around 1780 with the reproduction and distribution of their copy on a mass scale throughout Europe. The text that reached the general reading public, if not the masses, differed somewhat from theirs, however, because it, too, suffered from the strains of the production process. In his prospectus, Duplain had promised not only to reprint the original text in its entirety but also to improve it in three ways: to correct its numerous typographical and factual errors; to add a great deal of new material; and to blend the four folio volumes of the *Supplément* into it. He never intended to produce a literal copy of the first folio edition but rather to create a superior version of it—or at least to persuade the public that he had done so. The correcting, augmenting, and blending would require a great deal of editorial work, so the contracts between Duplain and Panckoucke provided for a *rédacteur,* who was to receive 600 livres per

23. Desgranges to STN, Jan. 9, 1779.

volume, later increased to 850 livres with an additional 3,000 livres for further work on the third "edition." Duplain gave this job to the abbé Jean-Antoine de Laserre, an Oratorian and a minor literary figure in Lyons. Laserre therefore became the successor of Diderot and the intermediary through which Diderot's text reached most of its readers in the eighteenth century.

Laserre's main qualification for improving on the work of Diderot seems to have been friendship with Duplain. He did not worry about tampering with the text or adapting his changes to Diderot's style because he had other things to worry about—promoting his own career, for example, and courting his superiors in the church. He replaced the original article APOLOGUE by the abbé Edme Mallet with a selection from his own *Poétique élémentaire;* he added an excerpt from his *Discours de réception* in the Academy of Lyons to the article NATUREL (Belles-Lettres) in volume 22; and at TESTAMENT in volume 33 he included an edifying extract from a *pastorale* by his archbishop, which began, "Tout l'Ancien Testament n'est, dans le dessein de Dieu, qu'un grand et magnifique tableau, où sa main a tracé d'avance tout ce qui devait arriver au libérateur promis."

For the most part, however, *le saint homme,* as the Neuchâtelois sardonically called him, left the text alone—not because he respected it but because he did not have time to make changes. He worked at a furious pace, cutting out references to the eight volumes of plates that were not to be included in the quatro, attaching snippets from the *Suppléments* to the main body of the book by bits of his own prose, and reading over the final amalgam of printed and manuscript copy that was to be mailed out to the printers. As a half dozen printing shops were working on different volumes at the same time, he could hardly keep up with their demand for copy. He supplied the STN in small batches, and it kept urging him to work faster and to send larger quantities, so that it could maintain its rhythm of production: "Copie et papiers, c'est toujours notre refrain."[24] It also objected to his tendency to slip his own writing into the text and to overlook errors. "La copie

24. STN to Laserre, Oct. 19, 1777. See also STN to Duplain, Sept. 20, 1777: "Sur l'article de la copie, qui va nous manquer, si nous n'en recevons entre ce et 8 jours, nous vous demandons la grâce d'y pourvoir et de faire en sorte que nous en ayons toujours une certaine provision."

que vous nous avez envoyée va être épuisée dans quelques jours,'' it wrote to Duplain in July 1777. ''Faites-nous en passer de la nouvelle s.v.p. . . . Vous ferez bien de prier l'abbé de Laserre de lire avec soin la copie avant de l'envoyer, ayant trouvé de fautes contre le sens que nous avons redressées et qui ont donné lieu à tout le sarcasme des Encyclopédiques.''[25]

These criticisms stung the abbé, who replied that he could not do an adequate job of cleaning up the text while turning out six volumes in three months. He counted on the printers to make editorial as well as typographical corrections and not to snipe at him behind his back. True, he had not kept up with the polemics surrounding the *Encyclopédie,* so he could not disarm its enemies by correcting the text in places where they had concentrated their attacks, but the book was big enough to withstand criticism and to contain contradictions.[26]

Comme la critique de Fréron ne m'est parvenue que depuis l'envoi des premières feuilles du 6ème volume, j'ai laissé subsisté quelques unes des fautes que le journaliste y avait trouvées. Celle qui pourrait le plus choqué est à l'article Canathous, où le mot de divinité substitué à celui de virginité forme une absurdité. Peut-être regardez-vous aussi comme une faute d'attention d'avoir accepté des articles qui semblent se contredire. Mais les gens de lettres que j'ai consultés m'ont confirmé dans l'idée que l'*Encyclopédie* étant un répertoire des différentes opinions et non un ouvrage systématique, il fallait y insérer le pour et le contre . . . c'est du choc des opinions que sort la lumière, et notre dictionnaire doit avoir l'avantage des académies de recueillir tous les systèmes sans les adopter.

It may seem odd that a contemporary editor of the *Encyclopédie* should have been so badly informed about the contemporary criticism of it, but Laserre and his fellow ''gens de lettres,'' that is, his colleagues in the academy of Lyons, had watched the *Encyclopédie* controversies from a distance instead of engaging in the thick of them. Consequently, the quarto took on a certain provincial flavor. It was a Lyonnais product, edited, printed, and managed for the most part from Duplain's circle of acquaintances. Diderot and Panckoucke

25. STN to Duplain, July 30, 1777.
26. Laserre to STN, Aug. 4, 1777. In a letter to the STN of Oct. 1, 1777, Laserre again complained about being criticized behind his back and asserted that he would provide better copy in the future, for he had hired additional *aides* (evidently copyists and copyreaders) and would make sure that the entire text was read three times before it went to the printers.

might know what was necessary to satisfy Parisians, but Laserre and Duplain knew what the provincials wanted, or at least what they would buy.

For its part, the STN wanted to avoid damaging the market anywhere. It saw Laserre more as a liability than an asset, and when possible did some fairly extensive copy-editing of its own, explaining to Panckoucke: "Nous nous attachons aussi à la correction, non seulement pour les fautes typographiques, mais encore pour les fautes de sens, qui se trouvent dans la copie même qu'on nous envoie Il faut prier M. l'abbé de Laserre notre rédacteur d'y faire attention." "L'écrivain de la présente [Bertrand] se joint à M. Ostervald pour vous recommander de ne pas permettre [que] M. l'abbé de Laserre mette de sa prose dans l'*Encyclopédie,* mais qu'il y joigne seulement les *Suppléments.*"[27] Panckoucke intervened, but rather gently. "Les auteurs sont un peu plus vains que les autres hommes,"[28] he explained to the STN. The Neuchâtelois —who had dealt with Rousseau and Voltaire as well as a great many pretentious, second-rate authors—agreed, writing as one publisher to another: "Sans doute que les auteurs sont vains; la science réelle ou prétendue enfle, et l'abbé encyclopédiste n'est pas le seul qui ne sache pas recevoir un bon conseil"[29] But when they wrote to Laserre, they took a different tack. Abandoning their earlier demands that he respect the integrity of the text, they tried to make the most of his willingness to manipulate it. They indicated that they could produce some propaganda for his books in their literary review, the *Journal helvétique,* if he could promote their edition of the *Description des arts et métiers* in the cross references of the quarto. This formula restored peace between editor and printer. From volume 15 onward, Laserre directed the quarto's readers to seek further information on the *arts* in the STN's book, and the STN reviewed Laserre's works as so many "livres classiques" in its journal.[30]

Unfortunately, Duplain had to deal with editorial problems that could not be solved simply by tampering with the text and

27. STN to Panckoucke, July 27, 1777, and May 1, 1777.
28. Panckoucke to STN, Nov. 8, 1777.
29. STN to Panckoucke, Nov. 16, 1777.
30. STN to Laserre, Oct. 19, 1777; Laserre to STN, Jan. 28, April 6, June 10, and Oct. 24, 1778; and *Nouveau journal helvétique,* July 1778, pp. 38–42. Judging from the STN's papers, trumped-up book reviews were as common in the eighteenth century as Balzac said they were in the nineteenth.

tickling the ego of his editor. As explained in Chapter III, he had based his subscription campaign on a disastrous miscalculation about the size of his *Encyclopédie:* it would have to contain thirty-six volumes instead of the twenty-nine for which the subscribers had contracted if it were to include the whole text. (The subscribers paid by the volume, but Duplain had promised to keep the overall retail price down to 344 livres.) He hoped to skirt this difficulty by announcing ostentatiously that he would give away three extra volumes free, while quietly billing the subscribers for the other four. He also made some subtle changes in the terms for the second subscription and marketed the third as if it were a new enterprise —in thirty-six volumes—by the STN. Even then, he had to increase the thickness of the volumes in order to restrict their number. And in order to keep this obesity within bounds, he instructed Laserre to do some discreet abridging. By dropping sections from the main text and condensing articles from the *Supplément,* Laserre kept the first eight volumes down to 800 pages each. But this maneuver did not escape the sharp eyes of some rival Swiss publishers. In April 1778 at the height of the quarto–octavo conflict, H. A. Gosse of Geneva sent the following letter to the Société typographique de Lausanne:

Nous venons d'apprendre une nouvelle trop importante pour vous, Messieurs, concernant l'*Encyclopédie* pour ne pas vous en donner avis. M. Cramer, libraire, vient de sortir de chez nous, nous a assuré qu'ayant confronté l'édition in-quarto qui se fait ici avec la sienne in-folio, il a trouvé qu'outre les retranchements qu'occasionnent naturellement l'omission des figures il y en a beaucoup d'autres, que cela venait sûrement de ce que les libraires avaient promis l'édition in-quarto en 33 volumes et que chassant davantage qu'ils n'avaient cru d'abord ils se sont crus obligés de faire plus de volumes ou d'y remédier en tronquant leur édition. Ils ont choisi le dernier parti. Vous ferez, Messieurs, le cas que vous jugerez à propos de cet avis. Nous l'avons cru de trop de conséquence pour ne pas vous le marquer.[31]

What the Lausannois did with this explosive bit of information is not clear. They may have used it in their attempt to ex-

31. H. A. Gosse to Société typographique de Lausanne, April 11, 1778, in Geneva, Archives d'Etat, Commerce F 62. By thirty-three volumes, Gosse evidently meant the original twenty-nine and the four extra volumes for which Duplain planned to charge the subscribers, or he may have made a slip for thirty-two—the twenty-nine volumes of text and the three volumes of plates that Duplain had originally announced.

tort favorable terms from the quarto publishers, although
both sides stood to gain by suppressing information about the
deficiencies of their common text. In any case, word soon
spread around publishers' circles that Duplain was cheating
on the length of his *Encyclopédie,* and Duplain had to order
Laserre to stop. The volumes grew to 1,000 pages from volume
9 onward. Volume 11 contained an ''Avis des éditeurs,'' which
indignantly denied that any cuts had taken place. The editors
had merely rearranged some material, it explained. For example, PSEUDO-ACACIA would appear under P instead of A. The
last volume would contain a generous section of additions and
corrections; and if any subscriber discovered any genuine cuts
in the text, the editors would publish them in a free supplement. Laserre restored the material he had amputated between ABATARDIR and HORN as some ''Additions,'' appended
rather awkwardly to the end of volume 16, and he remained
faithful to the original text throughout the second half of the
quarto, although he confessed in private to the STN that the
whole business disgusted him:

On nous force de gâter l'ouvrage. Depuis le 20ème volume à l'article
du *Dictionnaire,* nous y joignons celui des *Suppléments,* qui disent
souvent la même chose. Mais ''chat échaudé craint l'eau froide.'' Il
est essentiel selon M. Duplain et ses associés qu'on ne trouve aucune
suppression. Avec ce système nous nous serions évité bien des peines.
On ne voulait d'abord que 32 volumes de 100 feuilles. Il fallait donc
supprimer. L'on m'a fait un crime de n'avoir point tenté l'impossible.
Aujourd'hui que l'on donnera de 39 à 40 volumes, il faut autant qu'il
sera possible tout conserver . . . Je sens comme vous combien cette
marche est vicieuse, mais nous sommes forcés de la suivre.[32]

Although the Neuchâtelois replied sympathetically to the
abbé, they berated him behind his back in their letters to Panckoucke. They had cause to complain, for Laserre had compounded his previous errors instead of correcting them when

---

32. Laserre to STN, Aug. 4, 1778. Laserre acknowledged his *suppressions* in
similar terms in a letter to the STN of June 10, 1778. And in his letter of Aug. 4,
he revealed some of the details of his copyediting operation: ''Je vous proteste
que depuis le 9ème volume, il n'en est aucun qui n'ait été prélu par quatre personnes différentes, dont une est entièrement occupée à vérifier les citations, presque
toujours défectueuses. Il échappe des fautes malgré ces précautions. Mais c'est
une suite nécessaire de la précipitation avec laquelle on est obligé de travailler
. . . J'ai renvoyé à la *Description [des arts]* de M. Bertrand dans presque tous
les articles d'arts et métiers, et c'est encore une des choses qui nous a été re-
proché par les censeurs que l'envie a armés contre nous.''

he reworked the copy for the third edition. This final episode of editorial bungling became clear to the STN after it received the copy for volume 19 of the third edition and compared it with volume 19 in the previous editions. To make matters worse, Duplain refused to let the STN print more than one volume of the ''Neuchâtel'' quarto, even though he issued all thirty-six volumes under its imprint. Thus, as the STN explained to Panckoucke, it was being made to bear the onus of the slipshod work turned out in its name by rival printers whom Duplain favored.

A mesure que notre 19ème volume 3ème édition avance, nous y remarquons un si grand nombre de fautes et de fautes si lourdes qu'il ne nous est pas possible de les digérer de [sang] froid. D'abord ce volume ne renferme aucune correction nouvelle de ce qui se trouvait de fautif dans les précédentes éditions, ensorte que les 1,000 écus alloués en supplément de salaire à notre abbé sont à pure perte; mais ce qui est pire encore, nous voyons des fautes nouvelles que nous ne pouvons rétablir qu'à l'aide de la première édition folio. Voilà de quoi faire épanouir la rate aux journalistes anti-Encyclopédistes. Cependant ce bel ouvrage porte en titre le nom de notre Société. On ne manquera pas de nous en faire des reproches. Il faudra nous justifier aux dépens de qui il appartiendra.[33]

It is impossible to acquire more precise information about Laserre's operations—how he organized his copyists, how he processed the copy, and so on. But the above account should make it clear that neither he nor the publishers considered the text of the *Encyclopédie* as sacred. On the contrary, they stuffed it with extraneous material, squeezed it out of shape, cut it apart, and reassembled it as they liked, without the slightest consideration for the promises of their prospectus or the intentions of Diderot. Of course, Diderot himself, having suffered through more than his share of editorial difficulties, had described the *Encyclopédie* as a monstrosity that needed to be completely redone. This casual and critical attitude toward the text runs through all the projects for reproducing it, from the initial proposal for a *refonte* to the *Encyclopédie méthodique*. The publishers also treated other books in the same manner. They worked capriciously with texts, for it never occurred to them that they ought to feel

33. STN to Panckoucke, March 14, 1779.

religious respect for the written word. The age of "scientific" editing had not yet dawned.

Recruiting Workers

Having sent for copy, presses, type, ink, paper, ink-ball leathers, candles, quills, imposing stones, galleys, chases, and a hundred other articles, the STN needed men to put the matter in motion.[34] It ordered the workers pretty much as it ordered the equipment and ran into the same problems of supply and demand. But it also had to cope with the peculiarities of printers as human beings. They had no notion of joining a firm. Instead, they worked by the job, coming and going according to the availability of tasks and their own inclination. Although some stayed several years in one shop and a few settled down and raised families, most printers seem to have lived from job to job and to have spent much of their lives on the road. For printing was a tramping trade. Men went where they could find work, even if they had to hike hundreds of miles. When work abounded, they sometimes changed jobs in order to collect travel money or simply by "caprice," as they put it.[35] They came and went at a furious pace during the *Encyclopédie* boom, which sent repercussions throughout the migratory circuits of France, Switzerland, and parts of Germany, producing as much competition for workers as for paper.

The masters of the Swiss printing houses could not draw on a large supply of local labor, so they secretly tapped each other's shops and tried to siphon off printers from the larger labor pools of France. The STN used a colorful cast of re-

34. The following account is meant only to describe the main aspects of bookmaking at the STN, not to serve as a substitute for the thorough analysis that will appear in the thesis of Jacques Rychner. On the two other early-modern publishers whose papers are comparable in richness to those of the STN see D. F. McKenzie, *The Cambridge University Press, 1696–1712* (Cambridge, Eng., 1966), 2 vols. and Leon Voet, *The Golden Compasses* (Amsterdam, 1969–72), 2 vols., an account of the Plantinian press in sixteenth- and seventeenth-century Antwerp.

35. Offray, a compositor working for Félice in Yverdon, to his friend Ducret, a compositor with the STN, Dec. 20, 1770, quoted in Jacques Rychner, "A l'ombre des Lumières: coup d'oeil sur la main-d'oeuvre de quelques imprimeries du XVIIIème siècle," *Studies on Voltaire and the Eighteenth Century*, CLV (1976), 1949.

cruiting agents: a pub-crawling bookseller in Basel, a book-loving magistrate in Strasbourg, a sometime smuggler from the booksellers' guild in Dijon, a marginal bookdealer in Paris, a shipper and a type founder in Lyons, a down-and-out watchmaker in Geneva, and a Genevan pressman with the alluring address, "chez la veuve Joly, rue des belles filles." These agents dispatched printers and discussed them with the STN in a stream of letters, which reveal a great deal about attitudes toward work and workers at the level of the "bourgeois," as the men called the masters.

The STN ordered printers in batches, like paper. "Il serait bon qu'ils fussent assortis, c'est-à-dire tant de compositeurs, tant de pressiers,"[36] it instructed an agent in Lyons. But it worried that the *assortiments* might get out of hand. Thus the safety provisions in its instructions to Perregaux, who was to send a servant around the printing shops of Paris in search of recruits:

Nous cherchons des ouvriers par mer et par terre. Si M. Boniface, qui va furetant partout, pouvait nous procurer, enrôler et expédier diligemment 3 compositeurs et 3 pressiers, aussi jolis garçons que possible, nous lui en serions vraiment fort obligés, fussent-ils même au nombre de 4 de chaque espèce. Il pourrait leur promettre à chacun 1 louis pour le voyage. Nous avons accoûtumé [sic] de ne le payer qu'après un mois de résidence, mais il n'est pas nécessaire de leur dire cette particularité-là. Il peut encore les assurer que s'ils travaillent ici de suite pendant un an, et mieux encore jusqu'à la fin de l'impression de notre *Encyclopédie,* qui doit aller à 2 or 3 ans, nous leur allouerons très certainement une récompense dont ils auront lieu d'être contents.[37]

Evidently the STN did not expect the workers to remain in Neuchâtel for as much as a year and would not trust them with as little as a louis. It hoped to keep them in the shop by means of bonuses, but the bonuses had strings attached, which it would not reveal until the men had arrived, 500 kilometers later. Thus it tempted them with *le voyage*—a sum of money roughly equivalent to the amount they would have made during the time it took to walk to Neuchâtel—but refused to pay it until they had worked at least a month. Sometimes the STN kept the men's belongings (*hardes*),

36. STN to Claudet, May 8, 1777.
37. STN to Perregaux, June 24, 1777.

which the recruiters sent separately, as a kind of security deposit; for it feared that they might take another job en route or disappear, after collecting *voyages* from several masters.[38]

On their end, the recruiters put the STN on guard against drunkenness and sloth. A Genevan agent gave a qualified endorsement to a compositor but warned that he should be kept on piece rates: "Il est diligent et habile, à ce qu'on dit; mais ce qu'il y a de sûr, c'est qu'il est fainéant et ivrogne."[39] And a Parisian recruiter hedged a recommendation of a pressman with similar reservations: "Et par rapport à ce dernier, qu'on m'a assuré être bon ouvrier, je vous prie de ne lui donner jamais rien d'avance. Son pere m'a dit qu'il était un peu fainéant."[40] All the letters about the workers struck a note of basic mistrust. At best a printer would knock off for irregular bouts of *débauche;* at worst he would sell sheets to pirate publishers or spy for the police.[41] The employers and their agents wrote about working men as if they were children (*garçons*) or things (*assortiments*)[42] or an alien species. At one point, the STN complained to Duplain that a recruiter had sent off some men without inspecting them: "Il nous en a adressé une couple en si mauvais état que nous avons été obligés de les renvoyer."[43] And at another, it begged to be given an additional volume to print so that it would not have to break up its shop—not that it objected to firing the men; it wanted to avoid the necessity of rebuilding its labor force for a later job. Duplain replied, "Qu'y a-t-il de plus simple que de choisir 6 bons pressiers sur votre nombre et renvoyer les

38. The STN discussed these stratagems with its recruiting agents, notably the Parisian bookseller Pyre. See Pyre to STN, June 16, 1777, and STN to Pyre, July 1, 1777.

39. Marcinhes to STN, July 11, 1777.

40. Pyre to STN, June 16, 1777.

41. While on a business trip in France, Ostervald warned the STN that one of their apprentices might be an agent for the Société typographique de Berne. Ostervald to STN, April 25, 1780. He also tried to ingratiate himself with Beaumarchais by informing him that the workers in Beaumarchais's printing house at Kehl were likely to accept bribes from publishers who wanted advance copies of the Kehl Voltaire so they could pirate it. Ostervald to Bosset, May 3, 1780. At the same time, the STN received warnings that its own shop had been infiltrated by the Parisian police. See J.-P. Brissot to STN, April 23, 1781; July 26, 1781; and Jan. 12, 1782, and the discussion of this correspondence in Robert Darnton, "The Grub Street Style of Revolution: J.-P. Brissot, Police Spy," *Journal of Modern History*, XL (1968), 322–324.

42. STN to Pyre, July 1, 1777, and STN to Bosset, Aug. 30, 1779.

43. STN to Duplain, July 2, 1777.

autres, qui refluant ailleurs nous procureront plus de pouvoir sur une race effrénée et indisciplinable dont nous ne pouvons jouir ?''[44]

The workers exchanged letters of recommendation about bosses, just as the bosses corresponded about them. The few worker-to-worker letters that survive—some of them so crudely written and misspelled that they have to be read aloud to be understood—show the same set of concerns.[45] The men wanted to find out where the supply of work was plentiful, the pay good, the company congenial, and the foreman a soft touch. This information also traveled by word of mouth, when printers crossed paths on the road or in taverns frequented by their trade. The printers' taverns in Paris, notably Le Panier Fleury, rue de la Huchette, became important clearinghouses for reports about jobs and wages. They even served as centers for collective action, as can be appreciated from the following account in an unpublished autobiography of a Parisian foreman:

Ces Messieurs choisissent un cabaret qui leur sert de tripot. Il y a dans ce tripot toujours compagnie, on y débite toutes les nouvelles de l'imprimerie. On sait l'état des prix, on prend des mesures pour ne les point laisser tomber, on parle du gain excessif des maîtres, et l'on peut juger comme on les habille. Il y a de quoi faire de bonnes copies. Aussi ne les ménage-t-on pas beaucoup. On y apprend les places vacantes. Il y a, dit-on, un ouvrage qui va commencer; il faut tant de compositeurs.

On endoctrine les nouveaux venus sur l'état des prix, et on leur recommande surtout d'être fidèles à la société et de défendre les prix. Quelques uns en ont un état écrit. En voici copie fidèle.[46]

Whenever possible, the masters tried to manipulate the workers' grapevine. In the summer of 1777, for example, the STN had its printers write to friends in other shops, urging

44. Duplain to STN, Dec. 10, 1778.

45. The archives of the STN contain a half dozen of these letters, which may be the only specimen of worker-to-worker correspondence from the early modern period. They will be published in the thesis of Jacques Rychner.

46. Nicolas Contat (dit Le Brun), *Anecdotes typographiques d'un garçon imprimeur*, ed. Giles Barber (forthcoming, Oxford Bibliographical Society, 1979), part II, chap. 2. This passage goes on to quote the written report on wages, which circulated in the Panier Fleury and which provided the workers with detailed information on payments per sheet for composition. For example: ''Premièrement le gros Romain in 4°, in 8° et in 12 se paie . . . 3 [livres] 10 [sous]. Le même avec des notes et des additions . . . 4 [livres] 10 [sous].''

them to come to Neuchâtel. In this way an STN pressman called Meyer persuaded his brother in a printing house in Strasbourg to emigrate with five comrades at the Nativity of the Virgin (September 8), when the Germans traditionally changed jobs. Before then, however, the Strasbourg printers heard a bad account of the STN from a journeyman who was passing through town, and they canceled their trip. Evidently they had more faith in an oral report from a fellow worker than in a letter from a relative, which might be inspired by a *bourgeois,* as indeed it was.[47]

Thus the recruiting business was swept by cross currents and conflict. One side dangled propositions, the other played with the bait. Employers might dock wages, hold back belongings, or fire the unruly; but they could not reel in men at will. And if the workers sometimes played one *bourgeois* off against another, extracting *voyages* and bonuses, they lost their room for maneuver as soon as demand declined. The tactics on each side can be seen most clearly in the case of some recruits from Lyons and Paris, who did most of the work on the *Encyclopédie* for the STN in the summer of 1777.

Duplain had warned the STN not to go fishing in his labor pool: "Les ouvriers sont ici d'une rareté extrême. Tâchez de vous en procurer de vos côtés."[48] But it secretly arranged for shipments of workers with Claudet, its shipping agent, and Vernange, its type founder. Claudet entered into negotiations with several printers, but he found that they asked sophisticated questions, which he could not answer satisfactorily: What formats did the STN favor? Did it pay by time as well as by the piece? How long would the jobs under way in its shop last? And above all could it guarantee them some marginal advantage by its wages? "C'est précisément sur cet appointement qu'ils insistent, parce qu'ils ne veulent pas

47. STN to Turkheim of Strasbourg, Sept. 4, 1777. The importance of maintaining a good reputation among the workers is also clear from the entry for May 27, 1777, in the diary of J.-F. Favarger, the STN's traveling agent. Favarger had just visited the Société typographique de Berne, hoping to secretly recruit some workers, when he ran into workers that the Bernois had recruited secretly from the STN. "Vu Pfaeler le cadet qui m'a accompagné dans leur imprimerie, où j'ai trouvé Christ et Brosé de ceux qui nous ont quitté avec lui. Il ne s'est pu empêcher de divulguer [notre] imprimerie par devant moi. Je l'ai fait taire, mais il n'en faut pas davantage pour engager les autres à ne point y venir." Favarger, Carnet de voyage, STN papers, ms. 1150.

48. Duplain to STN, May 28, 1777.

quitter un endroit où ils sont bien, si ce n'est pour trouver mieux."[49]

The STN refused to be drawn into such elaborate commitments and fell back on Vernange, who came up with two compositors and two pressmen. They needed travel money, however, and the STN refused to advance it; so they remained in Lyons. After another month of prowling about taverns, Vernange finally found three men who were willing to tramp to Neuchâtel without a prepaid *voyage*. But when they arrived, the STN rejected two of them: "Deux de ceux que vous nous avez addressés nous sont arrivés mais malades au point d'infecter, en sorte que nous n'avons pu les occuper. Personne n'a voulu les loger. Ils sont repartis et ont pris la route de Besançon pour se rendre à l'hôpital et y chercher du soulagement."[50] Vernange admitted that he had reached the bottom of the barrel, but he could not come up with anything better unless he paid an advance, and even then he would have difficulty because good workers were getting scarcer and scarcer. At last the STN authorized him to pay his recruits 18 livres and to promise that they would get 6 livres more upon their arrival.

Two of them, "un nommé La France et son compagnon dont on a été très content en cette ville,"[51] arrived in time to begin the printing of the *Encyclopédie* in July, but they did not hurry. They took two and a half weeks to walk the 300 kilometers and stopped off in Geneva, where they borrowed 12 livres in the STN's name. Two others never arrived at all. One had left without insisting on his *voyage* and probably got snapped up by a rival printing house en route. The other, a certain "Jean Maron," pocketed 12 livres and disappeared. "Nous craignons qu'il ne se serve de cet argent pour aller ailleurs, où il exigera peut-être encore des frais de voyage,"[52] the STN concluded. In printers' slang "marron" meant a forbidden book, "marronner" and "marronage" involvement in the illegal trade. Evidently "Jean Maron" had learned some tricks in the shadier side of the book business. But whatever the etymology of his nickname, he taught the STN to revert to

49. Claudet to STN, June 18, 1777. See also Claudet's letters of June 6 and July 30, 1777.
50. STN to Vernange, June 26, 1777.
51. Vernange to STN, June 3, 1777.
52. STN to Vernange, July 8, 1777.

the safer practice of paying the men after their arrival—and from then on it failed to dredge any more workers out of Lyons.

The STN had more success in the larger labor market of Paris, thanks to a bookseller called Pyre, who, as a friend of the STN's foreman and an enemy of the Parisian guild, was well placed and willing to help.[53] On June 16, 1777, six of Pyre's recruits set out for Neuchâtel, bearing a letter from him which attested to the terms of their employment: 24 livres in *voyage* upon their arrival and a 24 livres bonus if they remained until the end of the year. In another letter, sent directly to the STN, Pyre explained that he had not told the men about the requirement of a month's labor for the payment of the *voyage* because he feared that the proviso would prevent them from leaving. He also thought it best to hold back their *hardes* until he had learned of their arrival. They arrived on July 1, having walked the 500 kilometers across France in two weeks at an average of 36 kilometers a day. Their 24 livres in travel money was good pay for such a summer hike because it equaled what they would have earned from two weeks of hard labor in the STN's shop. When they reached Neuchâtel, however, they learned that they could not collect it for at least a month. They had no choice but to set to work on the *Encyclopédie* and the other items being printed by the STN.

Their progress through the shop can be followed, week by week and job by job, from the wage book kept by the foreman, Barthélemy Spineux.[54] Three compositors, Maltête, Poiré, and Chaix, set type for lottery tickets, the STN's catalogue, and other small jobs for two weeks. Then in the week of July 14–19, Spineux gave them regular assignments. Poiré and Chaix worked together on a tract of the Enlightenment, *Instruction donnée par Cathérine II, Impératrice de toutes les Russies, à la Commission établie par cette souveraine, pour travailler à la rédaction d'un nouveau code de loix*, and Maltête com-

---

53. In a letter to the STN of Sept. 17, 1777, Pyre described himself as an ''ennemi juré du corps de la librairie.'' He was a small and somewhat marginal bookseller, who suffered from the oligarchy of the great merchants in the guild. The most important of the letters about his activity as a recruiter are STN to Pyre, June 1, 1777; Pyre to STN, June 15, June 16, and July 1, 1777; and STN to Pyre, July 1, 1777.

54. The wage books, called *Banque des ouvriers,* in the STN papers, ms. 1051 provide information on the composition and printing of every sheet produced by the STN from 1770 to 1782. They are the principal source for the following discussion of work and workers.

posed the first formes of the STN's first volume of the *Encyclo-pédie*. The three men continued to labor on their assignments until August 23, when Poiré and Chaix disappeared. They had worked eight weeks, long enough to collect their *voyage* and their *hardes,* and had moved on—presumably to places in another shop, which they had reserved through the workers' grapevine. Maltête left two weeks later. This time, Spineux refused to pay the travel money. "Je ne donne point le voyage à M. Maltête, et comme je n'ai rien payé, je ne payerai rien," he scribbled in an angry note at the bottom of his entry in the wage book for September 6, which was the last to mention the Parisian threesome but not the last that was heard of them. Two months later Duplain reported that the men who had quarreled with the STN were blackening its name in the shops of Geneva, where they were still working on the *Encyclo-pédie.*[55]

Pyre's other recruits scattered in other directions. A pressman called Jean left for Félice's shop in Yverdon after eight weeks with the STN. He returned fifteen weeks later, but his second stint lasted only three weeks; so Spineux never let him have his *voyage.*[56] Another pressman, Gaillard, remained in the shop from July 5 to December 20—a fairly common stint and a satisfactory one, as far as one can tell from the wage book. But in July 1778, a Parisian tradesman, who supplied the STN with ink-ball leathers, wrote that Gaillard was back in Paris and ready to set out for Neuchâtel again—for the third time. "Il fait [*sic*] tout plein d'éloge de vous et s'attribue à lui-même toutes les fautes qu'il a faites."[57] Evidently Gaillard had quit after some kind of dispute. His experience and that of his fellow travelers confirms a theme that runs through all of the STN's correspondence about workers: employment in the printing trade of the Old Regime tended to be stormy and short.

How often the storms burst is imposible to say, but the wage book shows a very rapid rate of turnover in the STN's work

55. Duplain to STN, Oct. 31, 1777. The payments of *voyage* did not always take place one month after a recruit's arrival. In a letter to Vernange of May 24, 1777, the STN said it would pay for the travel after three months of work in its shop.

56. In a note following the entry of Sept. 13, 1777, in the wage book, Spineux wrote, "J'ai payé à M. Erb [probably the "Bergue" whom Pyre had mentioned as Jean's traveling companion in a letter of July 2] pour voyage 42 batz. Je les avais retirés de M. Jean, venu de Paris et parti pour Yverdon."

57. Thomas to STN, July 19, 1778.

force. Few workers remained in Neuchâtel as long as a year, although a half dozen veterans, who stayed for two- or three-year hitches, maintained some continuity in the shop. Judging from letters about the "caprice," "curiosité," and quarrelsomeness of the workers, they often left of their own accord.[58] But there must have been much more firing than quitting after the *Encyclopédie* boom subsided. Once they had finished printing the quarto, Ostervald and Bosset decided to reduce their shop from twelve to two presses. Mme. Bertrand, who was handling the firm's correspondence while they were away on a business trip, wrote to them about a problem with this policy: you could not close down presses without dismissing pressmen. "On ne saurait mettre sur la rue du jour au lendemain des gens qui ont femme et enfants."[59] This objection apparently had not occurred to the directors. They brushed it aside with a lecture about profitability, and soon afterward the STN was operating at two presses.

There is no way of knowing what became of the workers after the STN turned them out. They must have found it difficult to get jobs elsewhere because other printing houses also seemed to be cutting back and the book business in general went into a slump during the 1780s.[60] Some of the men and their families—for the *garçons* were not necessarily youths enjoying their Wanderjahre—probably disappeared into the "floating population" of the poor, which surged through the roads and flooded the poorhouses (*hôpitaux*) of western Eu-

58. Although the letters about recruitment indicate that the workers changed jobs to get better pay, they sometimes allude to more flighty motivation. In a letter to Vernange of May 24, 1777, the STN indicated that the men might come in order to taste the local wine, and in a letter to Pyre of Oct. 14, 1777, it said they might leave Paris because they were "curieux" to see Switzerland. The compositor Offray said that he and his comrades changed jobs by "caprice" in the letter cited above, note 35. And caprice was not limited to printers. On Aug. 17, 1777, Vernange wrote to the STN, "Vous connaissez aussi bien que moi ce que c'est que d'être exposé aux caprices des ouvriers. J'en ai un qui est tombé malade et un autre qui, par son inconduite, a été obligé de s'exiler de la ville." Similar remarks appear in some of the letters of the paper millers.

59. Mme. Bertrand to Ostervald and Bosset in Paris, Feb. 12, 1780.

60. The slump shows up clearly in the letters and orders that the STN received from booksellers all over France in the 1780s. For example, on March 31, 1780, Pierre Machuel of Rouen wrote, "La vente . . . est totalement morte et de nouveaux correspondants—car la majeure partie [of the old ones] a fait banqueroute—ne payent rien, et les bons ne demandent plus rien . . . Les magasins sont plus que remplis, et l'on meurt de faim auprès . . . Les seuls heureux sont ceux qui se seront trouvés les éditeurs de l'*Encyclopédie*. Mais le temps en est passé actuellement pour les autres."

rope on the eve of the Revolution. The jobless had often drifted through the STN's shop, begging for pennies, which Spineux doled out from time to time and noted in the wage book: "7 batz d'aumône au relieur"; "aumône 3 batz 2 creuzer"; "pour aumône à un ouvrier allemand 7 batz" "à un pauvre Allemand imprimeur 7 batz."[61] Printers could easily sink into indigence because they rarely accumulated savings. According to Pyre, few of the potential recruits in Paris had saved enough to support themselves for two weeks on the road while tramping to Neuchâtel. "La difficulté ne laisse pas que d'être assez grande, parce que la plupart n'ont point d'argent, même ceux qui sont assez rangés. Il s'en est présenté plus de vingt, qui tous n'avaient pas un sou pour partir."[62]

Although they received relatively good wages when they found work, printers had no protection against unemployment, disease, and old age. Illness had transformed Vernange's two substandard recruits from skilled artisans into beggars; and if they ever made it across the Jura Mountains to the pestilential *hôpital* of Besançon after the STN rejected them, they may well have spent their last days with other workers whom they had known in the Swiss printing shops—men like the compositor from the Société typographique de Berne who had come close to his final tour when he was recommended to the STN: "C'est un bon ouvrier, qui a travaillé du temps de M. Droz assez longtemps à Neuchâtel, mais il vous faut dire que la vue et l'ouie commence à lui manquer et que par sa viellesse il n'a plus la vitesse en composant qu'un jeune homme robuste. Mais comme vous ne lui payez que ce qu'il gagne [that is, piece rates], je vous supplie de le garder aussi longtemps, étant réduit par la grande misère . . . à un état pitoyable."[63]

Setting Wages

Cruel as it was, the labor market favored the workers for a few years at the height of the *Encyclopédie* fever. They reacted by attempting to force up wages, but the attempt provoked a counteroffensive by the masters, who closed ranks

61. The references, in the order of their appearance, come from the wage book, entries for Feb. 14 and July 25, 1778 and Jan. 16 and Feb. 20, 1779.

62. Pyre to STN, June 16, 1777.

63. Pfaehler of the Société typographique de Berne to STN, March 3, 1772.

after a great deal of sniping at one another. As this episode provides some rare insight into labor relations during the pre-industrial era, it deserves to be analyzed and documented in detail.

None of the twenty or so Swiss and French houses which were printing the quarto on contract for Duplain knew precisely what all the others received from him or what they paid their workers in wages. This information could influence the bidding for commissions and the competition for labor, so the master printers kept their cards close to their chests—and instructed their agents to look over their competitors' shoulders. When Favarger toured around printing shops for the STN, he was sometimes received as a spy—and rightly so. After visiting the shop of Cuchet in Grenoble, he sent the following report to Neuchâtel:

Cuchet imprime l'*Encyclopédie*. Le pauvre garçon a fait des frais immenses pour s'assortir. Il a fait faire 5 presses neuves, qui lui reviennent à 480 livres, dit-il, une fonte neuve. Il a déjà fait le 7ème volume. Il fait à présent le 19ème et a chez lui la copie pour le 27ème. Il croit qu'après il sera encore à temps pour en faire un 4ème, ce que je doute, quoique je ne lui aie pas dit. Il a 9 presses dessus. Je n'ai pu savoir combien il paie les ouvriers, mais ce qu'il m'a dit, c'est qu'ayant fait des prix si bas avec M. Duplain, s'il n'a pas un 4ème volume à faire au bout de ses trois, il se trouvera avec beaucoup de presses qu'il ne pourra pas occuper, une fonte usée et point de bénéfice que la gloire d'avoir fait l'*Encyclopédie* in-quarto. Je ne sais pas quels sont ses prix. Ci-joint un échantillon de son impression, qui en général n'est pas absolument mal, mais vous remarquerez qu'il y a un côté mieux fait que l'autre. Au reste, c'est pris d'une feuille à l'aventure, sans qu'il s'en soit apperçu. J'ai été fort bien gardé chez lui. J'y ai reconnu quelques ouvriers qui ont travaillé chez nous, auxquels je n'ai pas pu parler. Ses papiers me paraissent assez beaux et bien tirés. Mais il ne peut rien gagner dessus.[64]

Louis Marcinhes, the STN's man in Geneva, sent similar information about the shops of Pellet and Bassompierre, which the STN had requested "pour ne pas être dupes de ces gens là."[65] Marcinhes also infiltrated the Genevan shops in order to procure workers for the STN:

64. Favarger to STN, July 26, 1778.
65. STN to Marcinhes, July 7, 1777. The STN continued, "Il nous importe fort de savoir ce que l'on paie pour le 1000 de tirage et composition de chaque feuille de l'*Encyclopédie* chez Nouffer et Pellet. Veuillez nous l'apprendre. Il nous conviendrait de savoir en quel nombre on la tire."

J'ai revu quelques ouvriers, mais je ne sais qui les a si fort prévenus contre votre typographie. Vous avez sans doute des concurrents ennemis et jaloux; et il est, j'en conviens, fort disgracieux d'avoir à traiter avec une canaille dépourvue de tout sentiment d'honneur. Cette semaine pourtant il part un compositeur entendant fort bien cette partie et le grec. Il est long [that is, lent] mais assidu, travaillant avec beaucoup de soin, ne rassemblant point aux barbouillons, qui m'avaient tous promis et qui se dédisent sans donner d'autres raisons que la médiocrité des prix. Pellet et Bassompierre, qui ont sous de fortes promesses séduit plusieurs de ces ouvriers et dégarni les imprimeries des environs, ne veulent pourtant leur donner que quinze florins neuf sols de notre monnaie par feuille, le florin de 12 creuzer. Aussi une bonne partie veulent quitter, parce qu'ils demandent 17 florins la feuille. Celui qui part cette semaine en est un. Il se nomme Caisle. Il doit aussi partir deux pressiers, qui ont promis de venir me parler mais que je n'ai pas encore vu. On paie ici 4 florins le 1000 à la presse . . . Je ne perdrai pas de vue les occasions de vous adresser les mécontents de Messieurs Pellet et Bassompierre et Nouffer.[66]

As the Genevans also had drained off workers from Neuchâtel, the reciprocal raiding threatened to develop into an open war. But before things got out of hand, Duplain's agent in Geneva, Amable Le Roy, intervened by sending the following directive to the STN:

Vos ouvriers ont répandu plusieurs lettres dans les imprimeries de Genève tendantes à engager les nôtres à se tourner de vos côtés; et pour mieux y réussir, ils leur ont donner l'espoir d'un voyage payé et de six sols par 1000 de plus qu'ils n'ont ici.

Nous avons arrêté toutes les presses des principales imprimeries de Genève, et même elles rouleraient toutes dès à présent, si le besoin d'ouvriers ne nous avait pas réduit à 12. Si vous mettez une augmentation dans les prix, elle contribuera à les rendre plus rares dans un pays où il est très intéressant qu'ils soient abondants pour la plus grande célérité des choses; parce que c'est là où il est le plus possible d'avoir un aussi grand nombre de presses qu'on le désire. Votre augmentation de prix d'ailleurs pourrait les porter à cabaler pour l'obtenir ici. Vous jugez, Messieurs, combien une telle cabale nuirait à la célérité de l'ouvrage et à sa bonne fabrication; car si l'on ne

66. Marcinhes to STN, July 11, 1777. Caisle, or ''Quelle'' as he appeared in the STN's wage book, arrived in Neuchâtel soon afterward—followed, a week later, by a letter from Pellet saying he had left town without paying a debt of 22 florins. Marcinhes got the wages of the Genevan compositors slightly wrong for his figure did not include imposition. He also provided information on the Genevans' pressrun, because the STN suspected some fraud: ''Pellet tire pour sa part 4000 et 14 mains, Nouffer et Bassompierre autant. Voilà ce que Bassompierre m'a assuré et ce que m'ont confirmé tous les ouvriers à qui j'en ai parlé.''

souscrivait pas à leur demande, les places chez vous étant remplies, ils aimeraient mieux courir à Bâle et Yverdon que de revenir ici, et nous aurions beaucoup de peine à en trouver autant qu'il nous en faudrait, la France même étant dépourvue. Il conviendrait beaucoup que votre imprimerie fût fermée à tous ouvriers venant de Genève pour ne l'ouvrir qu'aux Allemands qui sont à votre proximité.[67]

As the danger of losing Duplain's patronage outweighed the difficulty of procuring workers, the STN decided to retreat. It sent a full report on its wages to Nouffer and Pellet and asked them to do likewise, so that the masters could present a united front to the men. The Genevans replied with detailed accounts of their wage scale and expressions of solidarity:

Il est certainement tres essentiel que nous nous conformions pour les prix de cet ouvrage, vu que les ouvriers nous mettent les pieds sur la gorge (wrote Pellet). Votre prote ayant écrit que vous payez 15 batz le mil et une gratification à la fin, voici, Messieurs ce que je paie, de même que dans les autres imprimeries, savoir 15 florins en paquet et corrige [sic] première et seconde. Le prote corrige la troisième. Le prote met en page, ayant trouvé qu'il me revenait mieux à compte, vu que l'on demandait 4 florins de mise en page et même 5 florins. A la presse je paie 4 florins le mille, soit 12 batz, et à chaque feuille finie je leur donne 5 cruches, de sorte que la feuille entière vient 97 batz et demi, le florin toujours à 3 batz.[68]

To make sense of this situation, it is important to take account of some techniques of bookmaking. Early-modern printing shops were divided into two halves, *la casse,* where the compositors set type, and *la presse,* where the pressmen printed sheets. As the pressmen operated downstream to the compositors in the flow of work, they generally ran off whatever formes were ready. In Neuchâtel they printed the *Encyclopédie* along with four or five other books, which the STN also had in production at various times between 1777 and 1780. Instead of receiving special wages for the presswork on each of the books, they were paid a standard rate of 15 batz (a little more than 2 livres tournois) for every thousand impressions they produced, no matter what they worked on.[69] Pellet's let-

67. Amable Le Roy to STN, July 22, 1777.
68. Pellet to STN, July 23, 1777.
69. The only exceptions to this rule concerned unusually large or small press-runs, in which the preparatory labor ("make ready") took up an unusually small or large amount of work in proportion to the printing. So when the STN increased the pressrun of the *Encyclopédie* from 4,000 to 6,000, it increased the pay of the pressmen from 60 to 86 batz instead of 90 batz for the printing of

ter showed that his standard rate was a good deal less than the STN's: only 4 Geneva florins or about 12 Neuchâtel batz per thousand. The STN could not easily lower its rate to his level, however, because it could hardly pay its pressmen less for working on the *Encyclopédie* than on their other jobs, nor could it expect them to accept a reduction in wages across the board. But it could tinker with the wages of its compositors, which it set in the opposition fashion—that is, ad hoc, according to the size of the type and the difficulty of each job.

Several elements could be adjusted in the composition of the *Encyclopédie* because the compositors worked by the companionship or "paquet" system.[70] In the case of the quarto, a *paquet* was a page of type; four pages made a forme; two formes were necessary to print both sides of a sheet; and 100 to 120 sheets went into each of the thirty-six volumes of text. The sheets were identified by their "signatures," letters at the bottom of certain pages. By aligning the sheets according to the alphabetical order of their signatures and by folding them correctly, a binder could transform a loose stack of printed paper into a volume, ready to be read, page by page, after stitching, cropping, and binding. Thus the sheet was the key unit of production. Compositors were paid by the sheet. But in the *paquet* system, four or five men composed nothing but *paquets,* working as a team under a head *paquetier* or *compositeur en chef.* Although he, too, made *paquets,* he spent most of his time on the more skilled labor of imposition. He arranged the *paquets* in the proper manner on the composing stone, added final touches such as headlines and signatures, laid an iron frame or *chassis* (chase in English) around the assemblage and packed it tightly with wood quoins and wedges (*garniture* in French, "furniture" in English), so that the newly constructed "forme" would hold together and give an even impression when inked, covered with paper, and pressed beneath the platen of the printing press.

---

each forme. Similarly, when it did small jobs, it paid the printers 2 batz for 100 impressions, a higher rate than the standard 15 batz per thousand. Thus the entry in the wage book for April 24, 1779:

"Pour le gouvernement { Deux affiches 300 en tout   . . . 6 [batz]
Congés militaires 100   . . . 2 [batz]"

70. For contemporary accounts of the *paquet* system see the article PAQUET in the *Encyclopédie;* A.-F. Momoro, *Traité élémentaire de l'imprimerie, ou le manuel de l'imprimeur* (Paris, 1793), pp. 247–248; and S. Boulard, *Le manuel de l'imprimeur* (Paris, 1791), pp. 96–97.

By dividing labor in this fashion, a group of compositors could move through a text at great speed. They were also paid as a group, once a week, at the Saturday evening "banque." The foreman would record the number of sheets they had composed and the amount they had earned in his wage book ("banque des ouvriers"). Thus the entry for August 30, 1777, "Maltête Encyclopédie, tome VI T.V.X.Y. . . . 236," meant that Maltête had received 236 batz for the composition of four sheets, T through Y, during the week of August 25-30. He then divided that sum among the members of his team according to the amount of work each had done—so much per *paquet* and so much for his own imposition.

Having adopted this system for the composition of the *Encyclopédie,* the STN had to decide how much it would pay per sheet and how it would apportion the piece rates between *paquet* work and imposition. It experimented with different combinations for a month and then set its wages as close as possible to those reported by Pellet and Nouffer. According to Pellet's report, compositors in Geneva would not accept anything less than four florins per sheet for imposition. He considered that too expensive, so he left the imposing to his foreman, who received an annual salary, probably something comparable to the 1200 livres tournois that the STN paid Spineux. But the STN had too many other jobs under way to let Spineux concentrate exclusively on the *Encyclopédie.* It did not even free its time-hand compositor, Ruhr, for the impositions of the *Encyclopédie.*[71] Instead, it followed the practice of Nouffer, who had colluded with the attempt to freeze wages by sending the following report to Neuchâtel:

Pour satisfaire à vos désirs . . . et dans l'espérance que vous vous conformiez exactement à nos prix, nous vous les donnons ci-bas . . .

Pour la composition d'une feuille en paquet	ff.15
Pour la mise en page [that is, imposition]	4
Pour le tirage à 4 [mille] 14 mains	32-6
Ensemble	ff.51-6

Nous donnons le louis d'or neuf à ff. 51 et par conséquent l'écu neuf à ff.12-9.[72]

71. Ruhr received a weekly salary of 105 batz, or 15 livres tournois, for special assignments, such as unusual corrections (*remaniements*) on the *Encyclopédie.* The wage book for Oct. 4, 1777, for example, shows that his work that week had included "remaniement . . . sur six paquets de l'Encyclopédie."

72. Nouffer to STN, July 23, 1777. Although Nouffer expressed himself differently, his figures on presswork tally with those in Pellet's letter of July 23,

Nouffer's letter provided just the information, including a precise exchange rate, that the STN needed. The Neuchâtelois finally decided to pay their compositors 59 batz per sheet: 50 batz for the 8 *paquets* and 9 batz for imposition.[73] This adjustment brought the STN's wages within close range of those in Geneva, as can be seen from the following table:

	Genevan wages in florins	*Equivalent in batz*	*STN wages in batz*
Composition			
8 paquets	15	49-½	50
imposition	4	13-¼	9
total	19	62-¾	59
Presswork	32-½	107	120
Total cost of labor per *feuille d'édition*	51-½	170	179

In the end, the *feuille d'édition* cost 9 batz more in Neuchâtel than in Geneva, but the difference did not amount to much—a matter of 171 livres for all the labor on volume 6, which was trivial in comparison with the 11,545 livres that the STN charged Duplain for producing that volume. The

quoted above. The pressrun of 4,000 required 8,000 impressions for each *feuille d'édition*, which, at four florins per thousand (Pellet's rate) came to 32 florins. The final six sous (half a common Genevan florin) correspond to the six *cruches* mentioned by Pellet and covered the remaining sheets. Thus Nouffer's price of 32½ florins was the same as Pellet's, as is confirmed by Pellet's remark that his presswork came to 97½ batz of Neuchâtel (3 times 32½).

73. The STN's attempt to adjust the components of its wage scale is evident from a note in the wage book for July 26, 1777, written just after the arrival of Nouffer's letter:

Composition de Genève	. . . 49–2
Mettage en page	. . . 8–2
ajouté au prix de Genève	. . . 1–
	batz 59

In deciding how to rearrange these proportions, the STN did not use the crude exchange rate of 3 batz to the Genevan florin mentioned by Pellet. Like Nouffer, it reckoned according to the standard French louis d'or, which was worth 24 livres tournois, 51 Genevan florins, and 168 Neuchâtel batz. See Samuel Ricard, *Traité général du commerce* (Amsterdam, 1781), I, 105–111 and 121–123. Thus the florin was really worth 3.294 batz, and the Genevan wages of 15 florins for *paquet* work on a sheet came to 49 batz 2 creuzer, almost exactly what the STN paid; but the Genevans paid 4 batz 1 creuzer more than the STN did for imposition. Printers normally calculated that imposition should come to one sixth of the costs of composition. See Boulard, *Le manuel de l'imprimeur*, p. 61.

STN paid a little more than the Genevans did for presswork, but it paid a little less for imposition, while matching the Genevan rate for *paquets*. In this way it moved as close to the Genevan rates as it could, without upsetting the wage scale of its pressmen and without penalizing any of its compositors, except Maltête, who received 50 percent less than his counterparts in Geneva for making up pages and imposing formes. Perhaps that was why he quit the STN early in September, before he could collect his *voyage*. By that time, however, the STN had built up a large enough work force to maintain a high level of production, despite temporary ups and downs. It replaced Maltête with a veteran compositor called Bertho, who led the *Encyclopédie* team satisfactorily for the next two years. The masters stopped raiding one another's shops, and the complaints about competition for labor disappeared from their correspondence. They had out-caballed the workers.

Pacing Work and Managing Labor

Even though the masters held the line on wages, the men received comparatively good pay. Comparisons are hard to make because so little is known about wages and output among early-modern workers, but the Swiss printers clearly belonged to the so-called "labor aristocracy."[74] The STN's employees usually made 10-15 livres (70-105 batz) a week, or about 2 livres (40 sous) a day, depending on their productivity. They did not earn as much as their counterparts in Paris, where composition of an ordinary sheet in pica normally fetched about 8 livres as opposed to 5 livres (35 batz) in Neu-

---

74. For information on wages during the last years of the Old Regime in France see C.-E. Labrousse, *Esquisse du mouvement des prix et des revenus en France au XVIIIe siècle* (Paris, 1932), pp. 447–456. The figures in George Rudé, *The Crowd in the French Revolution* (Oxford, 1959), pp. 21–22 and 251 also are helpful, although they do not apply to the period before 1789. Using these and other data, Pierre Léon distinguished three strata of workers in the Old Regime: the poor, who made under 20 sous and often less than 10 sous a day; a middle group, who made 20–30 sous; and the elite, who made more than 30 sous. See Léon's contribution in C.-E. Labrousse and others, *Histoire économique et sociale de la France* (Paris, 1970), II, 670. Although French historians have produced estimates of typical daily wages, they have failed to provide statistical series on earnings over long periods of time. So the labor history of the Old Regime remains too underdeveloped to provide comparisons with the rich material in Neuchâtel. McKenzie and Voet also use the notion of "labor aristocracy" in discussing the wages of the printers in Cambridge and Antwerp. See McKenzie, *The Cambridge University Press*, I, 83 and Voet, *The Golden Compasses*, II, 341.

châtel, and presswork brought in about 2 livres 10 sous per thousand impressions as opposed to 2 livres 3 sous (15 batz).[75] But the cost of living must have been much higher in Paris, and the printers in Switzerland earned more than almost all the other categories of French workers. According to the estimates of C.-E. Labrousse, common laborers made about a livre a day (19–21 sous) in rural France and a little more in towns (23–24 sous). Skilled workmen such as carpenters and masons made about 30 sous, and highly skilled artisans, like locksmiths, made 30–50 sous. Those estimates are not very revealing, however, because many workers received some form of piece rates, and almost nothing is known about variations in their output. So a great deal can be learned from the STN's wage book, if one shifts the question away from the consideration of rates to a more fundamental issue: given their wage scale, how much did the printers work?

A close study of the wage book from June to November 1778 shows an extraordinary variety in the men's productivity. Income and output rose and fell so erratically from week to week that one cannot talk of averages but only of fluctuations ranging for the most part between 70 and 120 batz with peaks of 130 batz or more and low points of about 45 batz. The steadiest compositor in the shop was Maley, who set the type for Cook's *Voyage au pôle austral,* one of the four books, along with a journal and various odd jobs, that the STN was producing at this time. For six weeks in a row, from August 15 to September 26, Maley composed two sheets and received 70 batz, although he could compose three sheets for 105 batz, as he did in the week of October 3. Nicholas and Quelle, who

75. In a letter to the STN of April 15, 1780, Pyre set the typical costs for producing a sheet in pica (*cicéro*) at a pressrun of 1,000, presumably in octavo, as follows:

composition	. . . 8 livres
presswork, at 2 livres 10 sous per thousand	. . . 5
total	. . . 13
"étoffes" (overhead)	. . . 6–10
profit	. . . 5– 5
total cost	. . . 23–15

The printer Emeric David of Aix made almost exactly the same calculations in a diary that he kept during a trip through France in 1787: "Mon voyage de 1787," Bibliothèque de l'Arsénal, ms. 5947, fol. 50. According to Contat, *Anecdotes typographiques,* ed. Barber, part I, chap. 9, time hands in Paris received 50 sous a day, which was the same as the 17 batz 2 creuzer paid for "conscience" work by the STN. Evidently the more skilled compositors could command relatively higher wages than the piece workers in Switzerland.

worked together on the STN's edition of the Bible, went to
the other extreme. They received 60 batz for composing one
sheet in the week of October 10, 150 batz for two and a half
sheets in the following week, and 120 batz for two sheets in
the week after that. Erb worked slowly and erratically. He
often set only a sheet of Millot's *Elémens d'histoire universelle* and came home with 46 batz for his week's work, but he
could double his output, as in the weeks of July 4 and August
15, when he composed two sheets and collected 92 batz.
Champy, who normally set the *Journal helvétique,* worked
fast. He frequently set three sheets of the journal and made
108 batz, although his output declined to half that amount in
the weeks of June 27 and August 1, when he earned only 54
batz. The composition of the *Encyclopédie* proceeded just as
unevenly. Bertho normally worked with a team of four paquetiers, and their output in September and October varied as
follows:

	Sept.				Oct.				
	5	12	19	26	3	10	17	24	31
s								
posed	7	3½	5½	10	6	5½	7	7½	8
s								
batz)	413	206½	324½	590	354	324½	413	442½	472

The press crews worked at even more irregular rates. During three weeks in June, the output of Chambrault and his
companion plummeted from 18,000 to 12,000 to 7,000 impressions and their income from 258 to 172 to 101 batz. During
three weeks in October, the output of Yonicle and his companion soared from 12,525 to 18,000 to 24,000 impressions and
their income from 182 to 258 to 344 batz. It is impossible to
find any consistent pattern in the fluctuations. They were relatively moderate for Roat's crew, enormous for the crews of
Georget and Lyet, erratic at a high level of productivity for
Foraz's crew, and erratic at a low level for Bentzler's. The
weekly output of veterans like Albert and his companion rose
and fell continuously between 12,000 and 19,000 impressions.
Meyer and his companion, also veterans, usually made more
than 200 batz a week and once earned 303 (for 21,000 impressions), but their income sometimes dropped to 165 batz (for
11,200 impressions). Kroemelpen and his companion usually

made about 172 batz (for 12,000 impressions), but they were capable of making 280 (for 19,500 impressions). The heaviest week put in by any press crew occurred between February 16 and 21, 1778, when Chambrault and his companion made 379 batz (about 27 livres each) for 26,250 impressions: 4 formes of the *Encyclopédie* at 6,000, 1 forme of Millot's *Elémens d'histoire* at 2,000, and 250 impressions of the local *Feuille d'avis*.[76]

All these data point to the same conclusion: the men set their own pace. No matter how much one allows for outside factors —holidays, shifting assignments, occasional work at half press, advances from one week to the next—it seems certain that the variations were voluntary. If the men labored less, it was because they wanted to; the drop in their output did not result from irregularities in the supply of work. Compositors almost always worked exclusively on one job, so they could set as much copy as they desired each week. Pressmen ran off formes as the compositors finished them, no matter what the job, so they rarely ran out of material to print— and if they did, they were entitled to compensation, called *temps perdu*. Only once during the two years of work on the *Encyclopédie* does the wage book mention any such payment: "Gaillard . . . une demie journée perdue . . . 8 [batz] 3 creuzer."[77]

Holidays do not account for the variations, because one man's output often increased while another's declined and there was no general drop in production even during the weeks of Good Friday, Easter, Christmas, and New Year's. Although Neuchâtel was Protestant territory, many of the workers were Catholic. They did not enjoy the rich diet of feast days, which may (or may not) have reduced the work year to 250 or 300 days in Catholic Europe.[78] But they gave themselves holidays.

76. The highest output of a press crew at Cambridge in the early eighteenth century came to only 20,700 impressions, but in general the productivity of the compositors and pressmen in Cambridge fluctuated as wildly as that of the men in Neuchâtel. See McKenzie, *The Cambridge University Press*, I, chap. 4.

77. Wage book for Nov. 29, 1778. That sum was the same as the payment for a half day's work *en conscience*.

78. In discussing real or "effective" income, historians have assumed that French laborers did not work during an enormous number of unpaid feast days— 111 per year, according to G. M. Jaffé and George Rudé. Rudé, *The Crowd in the French Revolution*, p. 251. But it seems unlikely that many of those holidays were observed. In his *Art de faire le papier* (Paris, 1761), p. 84, J. J. Lefrançois

Although the wage book does not provide a day-by-day attendance record, it indicates the rate of absenteeism among time-hands or "conscience" workers, who lost 8 batz 3 creuzer for every half day that they missed. Their record is particularly revealing during the summer of 1778, when several pressmen worked for short stretches "en conscience.":[79]

Days worked among time hands

	June				July				Aug.					Sept.		
	6	13	20	27	4	11	18	25	1	8	15	22	29	5	12	19
melpen	4															
ud		5	5	6	6	3										
:			4	6												
r					4	5	6	6	4	6	6	6	6	6	6	4½
c					4½	6	6	6	6	6	6	6				
.							6	6								

If the six-day week prevailed, it was broken almost as often as it was observed. And "conscience" workers, as their name indicates, were considered the most reliable in the shop. The piece workers probably labored more erratically. Indeed, the whole process looks so erratic that one wonders whether the six-day work week and the 300-day work year ever existed, and whether one day was ever the same as another.

The irregularity in the pace of the work compounded the irregularity in the stints of employment and left the STN with enormous problems of labor management. With men moving unexpectedly in and out of the shop and working at idiosyncratic rhythms, Spineux could not possibly turn out the

de Lalande noted, "On suppose 300 jours ouvrables dans l'année, puisqu'on ne chôme dans ces sortes de manufactures que les dimanches et fêtes principales." The wage book of the STN does not provide information on daily output, but the STN's *Copie de lettres* shows that its clerical staff and directors did not slacken their work on Dec. 25 and Jan. 1. The only indications of official festivity in the STN papers are a few unimpressive entries in the account book entitled *Petite Caisse*, ms. 1048. Thus Dec. 27, 1777: "une bouteille aux ouvriers . . . 6 [sous]."

79. As the time rates remained constant, the number of days worked by a time hand can be determined by his weekly pay. Furthermore, Spineux usually noted cases when time hands did not work a full week—for example, in the entry for June 13, 1778: "Pataud 5 jours . . . 87 [batz] 2 [creuzer]." He also recorded their output according to piece rates and then subtracted it as *ouvrage* from the gross total of the week's *banque*, thereby indicating that the men received only their time wages and did not finish incomplete weeks by working at piece rates.

Encyclopédie and the other books at an even rate. But he could minimize the chaos by certain practices. He attached compositors to specific jobs called *labeurs* or *ouvrages*,[80] using his time-hand compositor for specialized work—unusual corrections (*remaniements*) and small commissions like lottery tickets and posters. The piece-work compositors could proceed at their own pace as long as their copy held out. If they reached the end of their *labeur* without quitting the STN and if the STN did not replace it with a new assignment, they would be fired. Thus Erb set nothing but Millot's *Elémens d'histoire* for 32 weeks from January 10, 1778, until August 22, when he quit. Then Tef, a *paquetier* on the *Encyclopédie* took over; when he completed the Millot six weeks later, he was fired. The STN had hired a new man to replace Tef on the *Encyclopédie* and it did not want to start a new book for several weeks, so it let him go. Tef was joined by another compositor, Comte, who had been working parallel to him on the *Description des arts et métiers* and had come within two and a half sheets of the end of that job. The STN was liquidating two of six long-term *labeurs* without renewing them, so Tef and Comte could not be absorbed back into the ranks of the compositors, and they took to the road together. Spineux might have made room for them by dismissing men with less seniority, but he adhered to another principle: consistency in assignments. He almost never shifted men back and forth between *labeurs*. Thus if job security did not exist, at least a man could count on keeping his *labeur* as long as the copy held out.

Although regularity in assignments simplified the allotment of work, it did nothing to mitigate the irregularity in output. On the contrary, the differences in the compositors' productivity meant that the formes of one job would become ready for printing at different intervals from the formes of another. Moreover, the number of characters per forme and the size of the pressruns also differed, so it was impossible for certain compositors to feed certain pressmen with any consistency. Instead, after running off one forme, a press crew would take the next forme that was ready, no matter who had composed it. Every pressman took part in the printing of every book. In this

80. The article LABEUR in the *Encyclopédie* suggests the job-oriented nature of this work: ''Terme en usage parmi les compagnons-imprimeurs; ils appellent ainsi un manuscrit ou une copie imprimée formant une suite d'ouvrage considérable et capable de les entretenir longtemps dans une même imprimerie.''

way the extreme variety of the tasks in presswork compensated for the regularity of the assignments in composition, and total output at the *presse* balanced that of the *casse*. It was up to the foreman to maintain the equilibrium between the two halves of the shop. He had to be careful, above all, to prevent the compositors from falling behind the pressmen, to "ne pas laisser manquer les presses," according to one of the most emphatic themes in the contemporary literature on printing.[81] If a foreman kept a press crew idle for lack of a forme to print, he was expected to compensate it with *temps perdu*. But he could also fire it.

Hiring and firing were the most important ways of maintaining balance in the shop.[82] For example, on October 3, 1778 when Tef and Comte left the STN, they were joined by a third compositor, Mayer, causing the *casse* to shrink from thirteen to ten compositors. Spineux had eighteen pressmen to keep busy, so he fired six of them. Almost overnight the work force had declined by a third, but it had retained its equilibrium. Three weeks later the STN began to increase production again and hired new men off the road. Labor management was a balancing act, performed at a heavy cost, both economic and human.

How frequently and drastically the balance shifted can be appreciated from Figure 2. Except for three weeks in July, the composition of the work force was never the same from one week to another. Men came and went pell-mell; their output rose and fell in terrific fits and starts; and the productivity of the shop as a whole fluctuated as wildly as the behavior of the individuals. Even during the period from June to September, when the turnover of labor was least severe, total output per week zigzagged by factors of 15 percent or more; in September and October it often doubled or halved.

For the masters, this irregularity betokened an inefficient use of resources and some loss in profits. What it meant for

81. See the article PROTE by Brullé, the foreman of Le Breton's shop, in the *Encyclopédie* and the more detailed discussion of the foreman's functions in Contat, *Anecdotes typographiques*, ed. Barber, part I, chap. 4 and part II, chap. 10.

82. It was in order to maintain balance that the STN asked its recruiters to send matching numbers of compositors and pressmen. The need of pressmen outstripped that of compositors in times of high pressruns, so during the printing of the *Encyclopédie* the STN deviated from the conventional formula of two compositors for two pressmen. On July 8, 1777, it instructed Vernange, "Il nous faut 2 pressiers pour 1 compositeur."

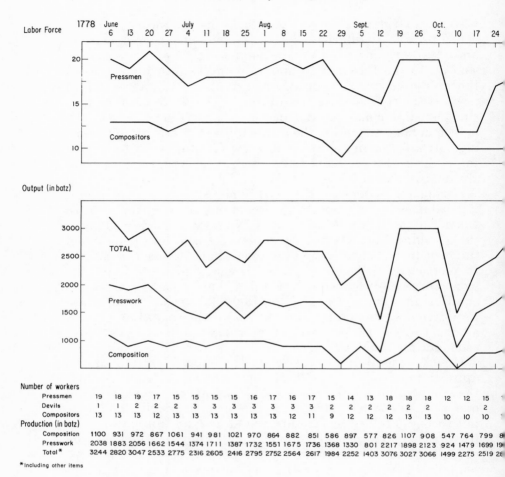

Figure 2. Manpower and Productivity, June–November 1778

the workers is difficult to say, but some speculation may be permissible. After studying the wage book for many weeks, one develops a sense of the rhythms at which work progressed and the units into which it was divided. Press crews, for example, often made 172 or 258 batz a week because those sums represented a full run of two and three formes, respectively, of the *Encyclopédie*. A crew might print two formes (one perfected sheet or 12,000 impressions) and then knock off for the week

—or it might decide to stretch its pay 30 batz by printing a forme of Millot's *Elémens d'histoire* (another 2,000 impressions) or perhaps a forme of the *Description des arts et métiers* (1,000 impressions), if they were available. That is why figures like 12,000, 14,000, and 13,000 recur often in calculations of press-crew output. Despite its irregularity, work involved combinations of various tasks, each of which required a fixed amount of effort. Within limits, the men could combine the tasks as they pleased, taking on more or fewer formes and *paquets*. The units of work shaped their output, and work in general remained task-oriented—that is, men labored to compose so many sheets or to print so many thousands, not to make a standard amount of money or to fill the hours of a standard work week.

Psychologically, nonstandardized work must have differed considerably from the kind of work that was then being imposed on the laboring classes in England.[83] The pace of work in the factories was set by clocks and bells, by the opening and closing of gates, by fines and beatings, and ultimately by the production process itself; for later, in assembly-line production, the men were reduced to "hands," and work streamed past them in an endless, undifferentiated flow. The compositors and pressmen of the STN worked at their own pace. They exerted some control over their production. And when they laid claim to their *banque* at the end of their long, erratic weeks, they may even have looked back on the sheets they had composed and the thousands they had printed with a sense of satisfaction.

Printing: Technology and the Human Element

Lest this interpretation seem romanticized, it should be added that some mastery over the production process did not

---

83. As examples of the historical literature on this subject see E. P. Thompson, "Time, Work-Discipline, and Industrial Capitalism," *Past and Present*, no. 38 (1967), 56–97; Sidney Pollard, "Factory Discipline in the Industrial Revolution," *Economic History Review*, 2d ser., XVI (1963), 254–271; and Neil McKendrick, "Josiah Wedgwood and Factory Discipline," *Historical Journal*, IV (1961), 30–55. And for examples of studies in the sociology of work see Sigmund Nosow and William H. Form, eds., *Man, Work, and Society. A Reader in the Sociology of Occupations* (New York, 1962) and Eugene L. Cass and Frederick G. Zimmer, eds., *Man and Work in Society* (New York, 1975).

mean that the workers developed any special affection for the real masters of it. The *bourgeois* retained most of the power and manipulated it brutally, by hiring and firing, while the workers responded with the few devices at their disposal. They quit; they cheated on their *voyage;* they collected small advances on the next week's work (*salé*) and then disappeared;[84] and sometimes they spied for rival publishers or the police. Although they may have felt some pride in their craft, they took shortcuts and compromised on quality when it made labor easier. The results can be seen in any copy of the *Encyclopédie* today—clear, crisp typography for the most part, but margins askew here, pages misnumbered there, uneven register, unsightly spacing, typographical errors, and smudges —all of them testimony to the activity of anonymous artisans two centuries ago.

Sometimes one can break the anonymity. For example, page 635 in volume 15 of the quarto in the Bibliothèque de la Ville de Neuchâtel contains a vivid thumbprint, which was almost certainly made by one of the STN's printers.[85] The wage book shows that the printer of that page (sheet 4L) was a certain "Bonnemain"—evidently a nickname and a singularly inappropriate one in comparison with Maltête and Maron. A letter from a master printer in Dole called Tonnet reveals how Bonnemain reached the STN's shop. He was a dark-haired Norman, who had had a checkered career in the printing houses of Paris and then tramped to Lyons, where he fell in with the Kindelem family—father, mother, and son, who drifted from shop to shop along the typographical tours de France. After committing *quelques coquineries* in Lyons, the Kindelems and Bonnemain took to the road together and eventually showed up in the little shop in Dole, where Tonnet printed saints' lives for the peddlars of the Franche Comté.

84. For example, the pressman Huché quit the STN after collecting his *banque* on June 6, 1778, for printing one forme of the *Encyclopédie* and two formes of the Bible. But he had not completed five *marques* (1,250 impressions) of the last forme of the Bible, which the STN was printing at 3,000. So that work had to be done in the following week by Pataux, and Spineux noted in the wage book for June 13: "Pataux—Salé fait pour Huché, cinq marques . . . 18 [batz] 3 [creuzer]."

85. There is no doubt that this and similar fingerprints came from the printers rather than from subsequent readers of the *Encyclopédie*. The fingerprints are in ink and they often disappear into the binding—that is, they were made while the pages were being produced as sheets. Also some subscribers complained about receiving copies with printers' fingerprints on them.

Tonnet put Kindelem père and Bonnemain to work at one of his three presses, while Kindelem fils set type. This arrangement did not last long, however, because Tonnet detected some kind of *complot odieux* involving his shop girl, who had a weakness for young Kindelem. He planned to fire the whole group as soon as they had finished printing a *Vie de Sainte Anne*. But they struck first. After collecting a week's pay and an advance on the next week's work, they dumped 1,000 half-printed sheets under Bonnemain's press and fled town by different routes. Young Kindelem ran off with the girl, and his parents took 200 copies of a devotional tract, which they sold along the road—or so Tonnet claimed. He wrote to warn the STN about the group, which reportedly had reunited in its shop, and he came down hard on the theme of solidarity among employers: "J'entre dans le malheur d'un bourgeois qui ne peut tout faire lui-même . . . Nous devons entre nous maîtres imprimeurs nous prévenir de quelques coquins, qui sont parmi nos ouvriers."[86]

The STN may not have taken Tonnet's version of the incident at face value because he was something of a shady character himself. He cheated on the bills for the books he bought, and the books were often obscene and seditious works, which he peddled under the cover of his trade in popular religious literature. But his warning did not matter very much in any case because it arrived in Neuchâtel after the Kindelems had left. The wage book indicates that the father worked for seven weeks during the summer of 1777 as "Rodolphe," an alias he had used earlier in Lyons. The son worked for only three weeks, and he must have left under a cloud because the last entry next to his name, dated September 6, 1777, shows that Spineux had docked him a week's wages. Having got into some kind of trouble, he probably had gone off in search of work in the other Swiss shops and then arranged for his parents to join him. Perhaps Tonnet's shop girl did, too, at least for a few stops along the well-worn route from Neuchâtel to Yverdon, Lausanne, Geneva, and Lyons. But Bonnemain did not. He remained at his press from August 1777 until March 1779, one of the longest stints put in by any of the men who

---

86. Tonnet to STN, Nov. 12, 1777. Judging from Tonnet's dossier in the STN papers, his printing and bookselling business did not amount to much. But in the *Gazette de Berne* of Dec. 23, 1780, he said that it included three presses and a stock of 40,000 volumes and that he would sell the whole thing for 10,000 livres.

printed the *Encyclopédie*. Evidently he liked Neuchâtel well enough to break with his fellow travelers, and the STN liked him well enough to keep him, although it had assured Tonnet that it would purge its shop of such *mauvaise compagnie*.[87]

To see into the life behind a fingerprint in the *Encyclopédie* is to get some sense of how men moved through the obscure channels of working-class history, but Bonnemain's thumbmark also can be studied for its typographical significance. It illustrates a point that is difficult to appreciate in an age of automation: the printers of the Old Regime left their mark on their books—literally, in Bonnemain's case, and figuratively in all the others. For each workman stamped each page with something of his individuality, and the quality of his craftsmanship affected the success of the product.

Bonnemain's fingerprint really resulted from a typographical trick. By smearing the forme excessively with ink, he and his companion did not have to pull so hard at the bar of the press to get an impression. But the extra ink came off on their fingers and smudged the sheets during handling. This maneuver escaped the attention of Spineux but not Duplain, who sent some blistering criticism to the STN.

Il y a plusieurs jours que nous avons avis de Genève que votre volume est mal fait. Nous attribuions cela à la jalousie des ouvriers, qui de chez vous ont passé à Genève. Mais nous avons été bien étonnés à la réception de vos feuilles de voir que malgré qu'elles ont été choisies, elles ont été tirées par des ouvriers qui ne sont bons qu'à tirer de l'eau du puits et non le barreau. Nous voyons des bras énervés ou parasseux, qui distribuent sur leurs formes beaucoup d'encre pour avoir moins de peine à tirer. En un mot, ce que nous pouvons vous dire, Messieurs, c'est que si tout est comme cela notre entreprise en souffrira beaucoup. Votre prote, qui a du goût pour la composition, n'a donc jamais les yeux sur les presses. Nous ne vous dissimulerons pas que cela nous donne de grands regrets et qu'absolument nous ne donnerons aucun autre volume si la fin n'est pas mieux.[88]

Having learned to expect such outbursts from Duplain, the STN turned this attack aside by replying in kind. Not only had it done a creditable job on volume 6, it answered, but also the other printing houses had botched volumes 1 through 4.

87. STN to Tonnet, Nov. 16, 1777. Spineux's entry in the wage book for Sept. 6, 1777, reads: ''Retenir à M. Kindelem toute sa banque.''
88. Duplain to STN, Oct. 31, 1777.

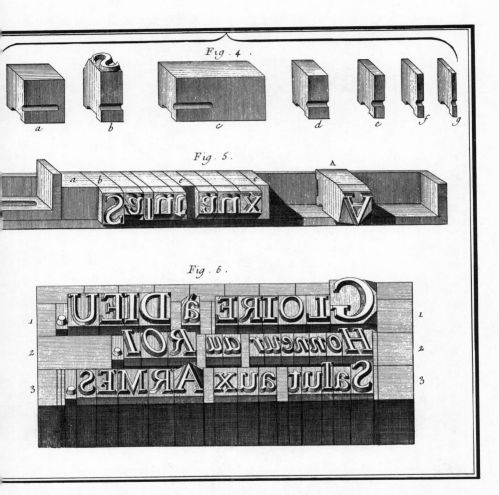

Fig. 4.

Fig. 5.

Fig. 6.

Type ("IMPRIMERIE," plate I), including various spaces, which are inserted in the composing stick (fig. 5) so that the lines are justified—that is, end evenly, as in fig. 6. The letters appear backwards, as though reflected in a mirror, so that the printed text will read correctly—and the text could hardly be more correct if read ideologically: "GLOIRE à DIEU. *Honneur au ROI*. Salut aux ARMES."

The composing room of a printing shop ("IMPRIMERIE," plate I). The first compositor (fig. 1) transfers type from the case to the composing stick, while keeping his eye on the copy, clipped onto a visorium. The second compositor (fig. 2) moves a newly composed line to a galley, where it will form part of a page. The third compositor (fig. 3) has imposed two folio pages inside an iron chase packed with furniture and is planing them with a mallet and wooden block so that the completed forme will present an even surface to the platen of the press. Newly printed sheets—in the folio, quarto, octavo, and duodecimo formats—are drying on cords above the men's heads.

The press room ("IMPRIMERIE", plate XIV). The first pressman (fig. 1) lays a sheet on the tympan of the press below the frisket, while his companion (fig. 2) spreads ink over the type of the forme in the coffin. Next the first pressman will fold the frisket over the tympan and the tympan over the forme. Then, turning a rounce with his left hand, he will winch the forme under the platen and print the sheet by pulling on the bar of the press, as in fig. 3. While he does so, the companion will distribute ink over the surface of his ink balls and inspect the previously printed sheet, as in fig. 4.

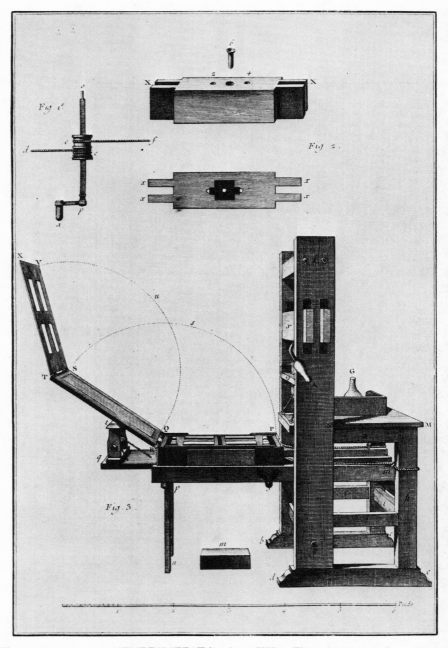

The common press ("IMPRIMERIE," plate XV). This side view shows how the pressman folded the frisket (X V S T) over the tympan (S T R Q) and the tympan over the forme in the coffin (R Q P O) on the horizontal carriage of the press. He ran the coffin under the platen (z) in the vertical part of the press by means of a windlass or rounce, shown in detail in fig. 1. Then by pulling on the bar, he turned a spindle in the head of the press (x, detail in fig. 2), which worked like a screw in a nut, forcing the platen down onto the forme. He printed only half the forme at a time, so it required two hefty *coups* at the bar, using the footstep (m) for purchase, to print one side of each sheet.

Ainsi que nous l'avions prévu, Monsieur, l'examen attentif et détaillé que nous venons de faire des quatre premiers volumes de l'*Encyclopédie* imprimés sous votre direction nous a mis à même d'y découvrir tant de défautes de tant de sortes que nous en sommes profondément affligés et admirons toujours plus le style de votre dernier, en le comparant avec de tels chefs-d'oeuvre. Pages de beaucoup trop noires, d'autres et grand nombre blanches au point de n'être presque pas lisibles, moines multipliés, pâtés, manque de registre, à la première page lettre grise dont on ne se sert plus, titres et réglets différents, au tome 3 pagination fausse, double signature, les lettrines prises de plusieurs corps, l'algèbre très mal traité, toutes les lignes où il y a du grec imprimées de travers etc. etc.[89]

Meanwhile the STN confided to Panckoucke and to d'Arnal that its presswork did need improving, though it would not admit any imperfection to Duplain. With him, it explained, the best defense was a good offense.[90] The typographical debate continued to rage through the mail for several weeks. Duplain wrote that the STN's volume was so "abominable" that it would ruin his sales campaign, and he threatened to reject it.[91]

89. STN to Duplain, Feb. 7, 1778. The STN then continued its criticism as follows: "Mais ce qui nous a singulièrement frappé, c'est que les caractères ne sont point des mêmes que ceux dont nous nous sommes servis pour l'impression du 6ème volume, desquels nous étions convenus. Fidèles à nos engagements sur ce point, nous avons préféré de différer notre travail de plusieurs mois plutôt que d'y manquer en employant une fonte Philosophie toute neuve et très ample dont nous étions pourvus. Cet article capital mérite la plus grande attention de votre part, Monsieur; et s'il en résulte quelque dommage pour l'entreprise, bien plus que de nos prétendues feuilles trop noires, vous vous rapellerez, s'il vous plaît, que nous vous en faisons aujourd'hui la remarque.

"Quant aux défets, en voici une ample note, procédant principalement des feuilles déchirées pour avoir été l'emballement mal fait, d'autres dont le papier est devenu jaune, qui n'ont pas été remplacées convenablement. Tout cela présente matière à bien des réflections, mais en attendant nous ne recevons toujours point la copie promise et nécessaire, et ne pouvons que vous représenter encore qu'il est très fort de vos intérêts de ne pas laisser chômer nos 11 presses, qui ne vous déplaise travaillent pour le moins aussi bien que beaucoup d'autres, et nous sommes d'ailleurs pourvus de fort beaux papiers . . . P.S. Les vignettes sont affreuses et mal tirées."

90. STN to Panckoucke, Feb. 8, 1778, and STN to d'Arnal, Feb. 8, 1778.

91. Duplain to STN, Jan. 21, 1778: "Il faut en vérité, Messieurs, que vous soyez aveugles pour qu'un ouvrage fait sous vos yeux sorte de vos mains avec autant d'imperfections. Votre tome 6 est abominable, et il l'est d'une manière à ruiner la société entière si nous le donnons. Mais comme nous ne voulons pas qu'elle [that is, the quarto association] souffre de votre négligence et du peu de connaissance qu'ont les personnes à qui vous confiez la direction de votre imprimerie, nous vous prévenons qu'il est ici pour votre compte et qu'à aucun prix nous ne le prendrons. Quand votre caractère aurait servi dix ans, il ne sortirait pas aussi plein et aussi charbonné. Les ouvriers que vous avez employés n'ont cherché qu'à faire des feuilles, et pour aller plus vite et s'épargner la peine de tirer, ils

The STN retorted that it had done a better job than his other printers and that he was inventing a pretext to avoid paying his bill. Ultimately he did pay, and the quarrel blew over.[92] But it was a significant episode because it showed how much importance publishers attributed to the physical character of their books.

Their concern was economic, not esthetic. They assumed that a badly made book would not sell—and they were right. Not only did customers complain but also potential subscribers refused to buy the quarto, merely because of the way it was printed. The work habits of Bonnemain and his companions had a direct effect on the literary market place, as the STN learned from letters like the following, from a subscriber in Saint-Dizier: "Je connais plusieurs personnes qui se proposaient de souscrire pour l'*Encyclopédie* et qui ont été arrêtées à la vue des négligences multipliées de vos pressiers, dont les doigts sont imprimés sur presque toutes les feuilles."[93]

Thus in the era of the handmade book there existed a typographical consciousness that disappeared sometime after the advent of automatic typesetting and printing.[94] To anyone still interested in typography, the imperfections that outraged producers and consumers in the eighteenth century constitute some of the charm of old books today. Every page, every line has its individuality. Each character bears the imprint of a gesture made by someone like Bonnemain. It would be misleading, however, to represent bookmaking under the Old Regime as idiosyncrasy run wild because the printers operated within a system of technological constraints, and the technol-

---

chargent leur forme d'encre de manière que sans effort la lettre sort—mais comment?—comme un pâté, sans régularité, sans traits . . . Nous vous prions de surseoir absolument tout ce que vous faites, de nous renvoyer la copie, et nous continuerons. Il ne fallait pas vous charger d'une opération que vous n'étiez pas en état de faire et nous exposer à une ruine inévitable."

92. For further details on the dispute and its resolution see Duplain to STN, Feb. 9, 1778; d'Arnal to STN, Feb. 4, 1778; and STN to d'Arnal, Feb. 8, 1778. "Puisque l'affaire est en règle, nous oublierons volontiers ce que sa vivacité a pu lui faire écrire de trop peu ménagé dans l'espérance qu'il n'oubliera plus à l'avenir les égards que les honnêtes gens se doivent entr'eux. Nous lui avons écrit consécutivement deux lettres qui lui feront comprendre que nous savons aussi bien que lui juger le travail des imprimeries."

93. Champmorin to STN, Nov. 26, 1780.

94. Thus a typical remark on the commercial importance of good typography in a letter to the STN of July 22, 1780, from Gaspard Storti, a bookseller in Venice: "C'est le pressier qui peut donner beaucoup d'aide au bon succès."

ogy of bookmaking remained fixed in essentials between the sixteenth and the nineteenth century.[95]

Like their forerunners of the Renaissance, the compositors of the STN made lines by transferring type from cases to composing sticks; they made pages by moving lines from the composing sticks to galleys; and they made formes by imposing the pages in a chase. Their work on the *Encyclopédie* probably differed from that of the compositors in sixteenth-century Antwerp—and perhaps even fifteenth-century Mainz —in only one way: it was organized according to the *paquet* system. But *paquet* production involved the division of labor rather than technology. It increased the speed with which a text could be set, because the *paquetiers* worked simultaneously, composing separate segments of cast-off copy.

As the directors of the STN wanted to print as many volumes of the *Encyclopédie* as possible, speed was very important to them. They even pulled compositors off other jobs in order to form several teams of *paquetiers* in the spring of 1779, when the printing houses were racing to win Duplain's last commissions. In the week ending April 24, for example, the STN had three teams and one lone compositor, Albert, working on volume 19 of the third edition, and they composed fourteen and a half sheets, an extraordinary output for the era of hand production (see Figure 3). Different versions of this system, called companionships, prevailed in the larger printing houses of the nineteenth century. Compositorial practices could not go further in the direction of speed and efficiency until the introduction of mechanization, in the form of cold-metal machines during the 1820s, Linotype in the 1880s, and electronic composition today.

Presswork, too, did not advance by any great leaps forward in technology before the adoption of the cylinder press from 1814 and steam power in the 1830s. The *Encyclopédie* was printed on the venerable common press, in essentially the same way that books had been printed for the previous two or three hundred years. The two-man press crews worked from heaps of wet paper, which they had soaked and stacked on the preceding day. After a good deal of preparatory labor—working up ink, stuffing the ink balls, and make-ready at the press— the crews began "pulling" and "beating." The beater or "sec-

---

95. For a good summary of the bibliographical literature on which the following account is based see Gaskell, *A New Introduction to Bibliography.*

ond'' would distribute ink over the surface of the ink balls by rubbing them together. Then he would ink or "beat" the forme, which had been locked in a movable box or "coffin" on the horizontal carriage of the open press. Meanwhile, the puller or "premier" would position a sheet on a parchment-covered frame, the "tympan," suspended over the forme by a hinge. He would close the press by lowering another frame called the "frisket" over the sheet and by folding frisket, sheet, and tympan together onto the forme. Then he would winch half the forme under the platen, a flat block suspended from a spindle in the vertical part of the press. By pulling the bar of the press, he would turn the spindle like a bolt in a nut, forcing the platen down on the back of the tympan and making an impression on the paper between the tympan and the type. After winching the other half of the forme under the platen, he would print it; winch the forme out again; unfold the tympan and frisket and remove the freshly printed sheet onto a new pile. After so many tokens or *marques* (250 sheets), the men would switch roles. And after one side of a sheet had been printed, the other would be run off from the second forme, usually by another crew.

Printing was hard work. To winch the forme forward and backward with one hand while pulling at the bar with the other required strength and endurance—hence the nickname for pressmen, *ours,* as opposed to the nimble-fingered compositors, *singes.*[96] The presswork on the quarto *Encyclopédie* must have been especially grueling because the run was so long—6,000, plus 150 *chaperon* to cover spoilage, in the case of the combined first and second editions. Faced with so much hoisting and heaving, the pressmen sometimes ran off only one or two formes a week. But they never followed any consistent pattern because they worked at an irregular pace and took formes as the compositors finished them, turning out the *Encyclopédie*

---

96. According to the printers' lore, the term ''ours'' was coined by one of the original Encyclopedists, a certain Richelet, during the printing of Diderot's first edition. While Richelet was examining some newly printed sheets on a heap near a press, one of the pressmen let go of the bar, which sprang back toward its resting position and knocked Richelet off his feet. He retorted with the epithet ''ours,'' and the pressmen then devised ''singe'' as a corresponding term for compositor. See Momoro, *Traité élémentaire de l'imprimerie,* pp. 308–309 and Honoré de Balzac, *Illusions perdues* (Paris, Editions Garnier, 1961), p. 4. Richelet's name does not appear in any list of Diderot's contributors, but the lists are far from complete.

along with the Bible and several other jobs.[97] Therefore the flow of work between the two halves of the shop could become extremely complex, especially when several teams of *paquetiers* were composing the *Encyclopédie,* as in the week of April 24, 1779 (Figure 3). By organizing work in this way—*paquet* production at the *casse* and fluid assignments at the *presse*— the STN stretched the production of the *Encyclopédie* to the outermost limits of a technology that had remained unchanged in essentials for perhaps three centuries.

In contrast to Figure 3, which illustrates the work-flow on the *Encyclopédie* for one week, Figure 4 shows how the production of an entire volume evolved over five months. This was the STN's first volume, the one whose pressrun it increased according to Duplain's directions for a second "edition." So the figure also shows how one edition was blended into another in mid-production. The STN switched from 4,000 to 6,000 copies in the first week of September, as soon as it received Duplain's order of August 27. At that point, Maltête's team of *paquetiers* had reached the end of the first alphabet, and the pressmen, working a few sheets behind them, had reached the end of sheet V. The compositors continued as usual, setting sheets Z through 2D. But the pressmen printed sheets T and V once again (there was no U or W in the printer's twenty-three letter alphabet), this time at 2,000, and then continued with sheet X at the increased run of 6,000. They could not do R and S from the formes they had printed the previous week because the type from those formes had been distributed before the arrival of Duplain's letter. Thus one can see the precise point at which one edition shaded off into another.

From that point onward, the composition and presswork continued at its usual erratic pace, the shop as a whole producing between 4 and 8 sheets a week, until December 13, when the last sheet, 5I, was pulled. But during the last six weeks the compositors had to reset sheets A through S, which had been

97. The output of three crews in the week of April 24–29, 1778, provides a final example of variations in work at the *presse.* Georget and his *second* printed only one forme of the *Encyclopédie* (6,000 impressions) that week, while Albert and his *second* pulled three formes of it (18,000 impressions) in addition to 525 impressions of the *Journal helvétique.* Gaspard and his *second* completed a more varied set of jobs: one forme of the *Encyclopédie* (6,000 impressions), one of Cook's *Voyage* (1,500 impressions), one of Millot's *Elémens d'histoire* (2,000 impressions), and a sheet of the local *Feuille d'avis* (250 impressions).

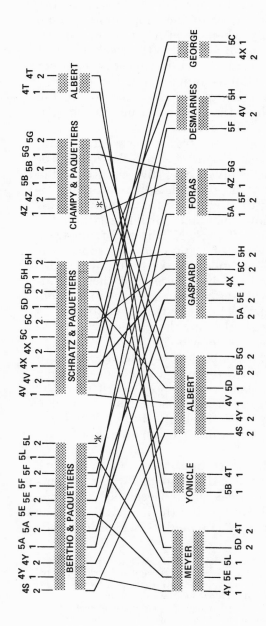

Figure 3. Work-Flow in the Shop, April 19–24, 1779

This figure shows the flow of work between the compositors, at the top, and the pressmen, at the bottom, during the week of April 19 to April 24, 1779, when the STN was nearing the end of volume 19. The notations correspond to the signatures on the sheets: thus 4S over 2 refers to the second forme of sheet Ssss, and 4Y over 1 refers to the first forme of sheet Yyyy (one quarto forme contained four pages or *paquets*, and it took two formes to print one sheet). As the press crews usually kept a little behind the compositors, some of the sheets were not run off until the week of May 1. Two of them, the second formes of sheets 5L and 4Z, which are marked by asterisks, were not printed until the week of May 8. It is impossible to know which press crews ran them off because at that time the subforeman had temporarily taken charge of the shop and he did not record the names of the pressmen next to the sheets they pulled. At the same time, the press crews also worked on three other books that the STN was producing concurrently with the *Encyclopédie*, which were being composed by six other compositors. Thus the work pattern was even more complicated than it appears here. But the figure does demonstrate how complex the flow of work on one book could become. It also indicates that the compositors worked from east-off copy, moving from one eight-page unit to another without following the sequence of the text. And it shows that they fed their formes to whatever press crew was available, instead of working regularly with one crew or another. To operate on such a scale required careful management as well as a huge font of type. Given the complexity and the speed of the job, it does not seem surprising that the final product aroused so

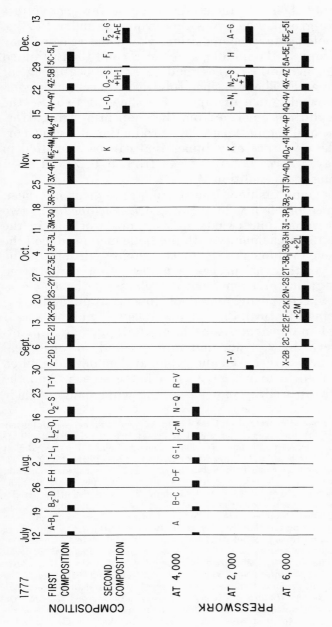

Figure 4. The Production of Volume 6, July–December 1777

The horizontal bars represent the number of sheets composed and printed each week, according to the entries in the foreman's wage book. The notations correspond to the signatures of the sheets, as explained in the caption to Figure 3.

printed at only 4,000. Spineux detached a few *paquetiers* from the main group, which was then working under Bertho, to do the supplementary composition. For some reason, they attacked the copy in the middle, at sheet K, and then rejoined Bertho's team in the last week, ending at the beginning, at sheets A through E. The pressmen followed wherever the compositors led, usually at a distance of a sheet or two. So the *Encyclopédie* did not emerge from the shop in a straightforward manner, and the pattern of its production does not correspond to the structure of volumes that can be consulted in libraries today, where signatures run from A to Z and a set seems to belong to an "edition."

In this laborious fashion, line by line and sheet by sheet, the STN turned out five volumes of the *Encyclopédie* in two years. At the same time twenty other shops working in the same manner and occasionally with the same men produced the other sixty-seven volumes of the three quarto "editions." The human reality of this vast process cannot be appreciated from work-flow diagrams or even from the plates of the *Encyclopédie* itself, where the printers look like wind-up dolls: expressionless and identical, they turn cranks and pull bars as if they inhabited an immaculate, mechanical utopia.[98] Real printing shops were dirty, loud, and unruly—and so were real printers. The presses creaked and groaned. The ink balls, filled with wool soaked in urine, gave off a fierce stench. And the men waded about in filthy paper, swilling wine, banging their

---

98. Roland Barthes interprets the plates in the opposite manner—as intensely human scenes, a "sorte de légende dorée de l'artisanat," in which men dominate machines—but he provides no evidence for his argument. Roland Barthes, Robert Mauzi, and Jean-Pierre Seguin, *L'Univers de l'Encyclopédie* (Paris, 1964), p. 11. Of course, Diderot himself came from a family of skilled artisans, and the main article on printing in the *Encyclopédie*, IMPRIMERIE, was written by Brullé, the foreman of Le Breton's shop. But the text of the article, and the other articles on crafts, fails to say much about the craftsmen as human beings and does not mention anything about their ceremonies, humor, and lore. There is much to be said for the common opinion that the *Encyclopédie* rehabilitated the dignity of crafts, but there is no reason to link that tendency with an incipient populism or revolutionary egalitarianism, as in C. C. Gillispie, *A Diderot Pictorial Encyclopedia of Trades and Industry: Manufacturing and the Technical Arts in Plates Selected from l'Encyclopédie* (New York, 1959), 2 vols. and Gillispie, "The *Encyclopédie* and the Jacobin Philosophy of Science" in Marshall Clagett, ed., *Critical Problems in the History of Science* (Madison, 1959), pp. 255–289. On the contrary, in stripping artisanal work down to its technological base—or recasting it as it ought to exist according to a more rational technology—the *Encyclopédie* eliminated a fundamental aspect of it: its culture.

composing sticks against the type cases for the sheer joy of making noise, bellowing and brawling as opportunities arose, and tormenting the apprentices with practical jokes. (If a printer's devil took a turn at the press, one could coat the bar with glue or ink, or one could produce bald spots on his sheets by surreptitiously touching the inked surface of his forme with one's fingertips; and a new boy at the *casse* could always be asked to clean the ink out of the eye of a capital P, which was really an eyeless paragraph sign or *pied de mouche*.

It is difficult to estimate the extent of the disorder because the masters did not discuss such things in their correspondence. But sources like printers' manuals contain so many imperatives about what workers should not do—above all, eating, drinking, and fighting in the shop—that one can construct a negative picture of their actual behavior. That picture conforms to a few first-hand accounts, particularly those in the autobiographies of Benjamin Franklin and Restif de la Bretonne and Balzac's *Illusions perdues*. And the literary descriptions correspond to the impressions given by printers' slang, which can be compiled from eighteenth-century glossaries and which stressed themes such as noisiness (*donner la huée*), horseplay (*faire une copie*), pub-crawling (*faire la déroute*), drunkenness (*prendre la barbe*), brawls (*prendre la chèvre*), indebtedness (*faire des loups*), absenteeism (*promener sa chape*), and unemployment (*emporter son Saint Jean*).[99]

However unruly they may or may not have been, the men who made the *Encyclopédie* certainly did not resemble the automatons of its plates. They could have figured in the *Comédie humaine*. Spineux, for example, worried that his brother-in-law was cheating him out of an inheritance in Liège and that his wife was betraying him in Paris, while he oversaw the workers in Neuchâtel. No wonder that he failed to notice the tricks played by the pressmen, for his family problems nearly drove him mad. "Spineux m'a écrit ce matin un billet qui porte sa confession générale et annonce que la tête lui tourne," Bertrand wrote to Ostervald in Paris. "Je crains

---

99. A great deal of information about the atmosphere of eighteenth-century printing shops can be gathered from the contemporary sources cited throughout this chapter, especially Momoro's *Traité élémentaire de l'imprimerie* and the "Anecdotes typographiques," which contain full discussions of printers' slang. See also the glossary in the *Encyclopédie méthodique, Arts et métiers mécaniques* (Paris, 1784), III, 475–636, and the first chapter in *Illusions perdues*.

que nous ne perdions ce pauvre garçon.[100] Spineux finally made
it to the end of the *Encyclopédie,* but he had to be helped by a
subforeman called Colas, who had problems of his own. Having left his wife behind in Geneva, Colas migrated around the
French and Swiss printing houses with a son, who worked as
a pressman and a compositor, and an impressive wardrobe,
which included twenty-three shirts, eleven pairs of silk stock-
ings, three complete suits, four pairs of breeches, and sixteen
handkerchiefs. He must have been the best-dressed man in the
shop, but he never seemed to have any money. Although he received 18 livres a week from the STN, he remained in Neuchâtel only three months; and he had to borrow from his land-
lady, leaving a gold watch behind as security, in order to pay
his way to his next job, in the printing shop of Cellot, rue
Dauphine, Paris. He had grown weary of living out of board-
ing houses, he complained: ''Je voudrais trouver un endroit
stable, pour ne plus être à même d'aller et venir.'' And he also
suffered from ''chagrins cuisants''—the result, apparently, of
bad relations with the other workers. ''Mon amour pour le
travail, le grand emploi du temps, mon attention à ce qu'aucun
ouvrier ne perde son temps et ne fasse mal, n'ont pas toujours
trouvé des approbateurs parmi mes confrères de la pro-
terie,''[101] he had explained in applying for the job with the
STN.

Superiority of this sort did not go down well with the workers, who could respond with the varieties of indiscipline celebrated in their slang: *joberie, copies, romestuques, bais,* and
grattes. Work might be disrupted, too, by a bout in the tavern,
a funeral procession, a wandering beggar, a cockfight, a hang-
ing, or some dramatic event. In July 1781, one of the STN's
former pressmen, Jean Thomas, who had worked on the *En-*

100. Bertrand to Ostervald, Feb. 23, 1777. On the suspicious conduct of Mme.
Spineux, who had remained in Paris after Spineux joined the STN, see Ostervald
and Bosset to STN, Feb. 28, 1777. However, three years earlier Mme. Spineux
had seemed to be full of wifely devotion, when she wrote to the STN about her
husband's failure to arrive on schedule for a visit with her in Paris: ''vous ob-
bligeres une mere de famille affliger et sil vous saves ou est mon mari je vous prist
de vouloire bien me le marque.'' Undated letter, received by the STN in Oct.
1774. Spineux's dossier reveals a great deal about his family affairs and his
character, which seems to have been exacting, upright, and generous.

101. Colas to STN, Aug. 15, 1777; Feb. 27, 1779; and Aug. 25, 1779. Colas
may well have been the foreman ''Colas'' described in Contat, *Anecdotes typographiques,* ed. Barber, part I, chap. 9: ''C'est un original d'ouvrier qui tranche
du grand personnage.''

cyclopédie three years earlier, arrived on the run from Geneva with a new bride and the bride's guardian in hot pursuit. The guardian had tried to prevent the marriage on the grounds that Thomas was a Catholic and the girl a Calvinist; and after their elopement, he asked the STN to flush the girl out of her hiding place, while he chased after Thomas with an arrest warrant. The STN's papers do not reveal what happened next, but the incident must have kept tongues wagging in the shop for weeks.[102]

Although such episodes stand out in the STN's correspondence, the routine letters can be more revealing, for they sometimes mention the workers' ties to friends and family, the ordinary stuff of which ordinary lives were made. In July 1779, for example, the STN received an appeal for news about one of its young compositors, Orres (Oreste?) Champy. Five months earlier, he had informed both his uncle, who lived across the border in Arbois, and his father, who lived somewhere farther away in France, that he was spitting up blood. The uncle had urged him to "se bien ménager" until Easter, when he could come to Arbois for some rest and home cooking. But Orres had not replied; so the uncle wrote to the STN, having received a worried letter himself from the boy's father, "où il me dit être fort inquiet et très en peine de ce jeune homme. Si vous vouliez bien, Messieurs, m'honorer d'un mot de réponse et me marquer ce que vous pensez de cet enfant, s'il est malade ou non, ce qu'il fait ou ce qu'il est devenu, vous obligerez infiniment le père et l'oncle."[103]

These people had families. They came from somewhere—with passions, problems, hopes, and fears. Although most of them have vanished irretrievably into the past, they did not compose some faceless proletariat while they lived. They brought their own cultural world to their work—a world that hardly touched the Enlightenment, except through the composing stick and the bar of the press—and they worked in their own way, imprinting the physical substance of the *Encyclopédie* with some of their individuality, just as Diderot had breathed his spirit into its text.

102. Rocca of Geneva to STN, July 17, 1781, and an undated letter from Rocca received by the STN on July 26, 1781.
103. Champy of Arbois to STN, July 10, 1779. Although the STN's reply to this letter is missing, young Champy probably had not been seriously ill because the wage book shows that he continued to work without interruption.

V I

ʏʏʏʏʏʏʏʏʏʏʏʏ

DIFFUSION

The printed sheets had a long way to go before they reached the shelves of the subscribers. Not only did they have to be assembled, folded, stitched, trimmed, and bound before they took on the appearance of a book, but they had to do a great deal of traveling—from the print shops to Duplain's storehouses in Lyons, from Lyons to the shops of retail booksellers everywhere in Europe, and from the bookshops in the libraries of individual subscribers, with stops along the way in entrepôts, custom offices, guildhalls, and tariff bureaus. The commercial circuit was not completed until the individual subscribers paid the booksellers, the booksellers paid Duplain, and Duplain divided the profits with his partners at a final settling of accounts. All these printed sheets rumbling about the highways of Europe, accompanied by bills of lading, invoices, customs declarations, and tariff forms and crossing paths with bills of exchange, letters of protest, and court summonses—these waves of paper surging across Europe threatened at times to overwhelm the entrepreneurs, who tried to contain them in the channels of commerce and ultimately produced the spread of Englightenment.

Managerial Problems and Polemics

Most of the paper passed through Duplain's headquarters in Lyons, which soon became flooded. In December 1778 after berating the STN for several months about the slowness of its shipments, Duplain suddenly told it to stop them. He had run out of storage space: "Nous ne pouvons pas absolument en loger ni recevoir aucune feuille, à moins de nous chasser de chez nous et en meubler nos appartements."[1] At the same time, he worried that the financial basis of the enterprise might give way, for he had to advance enormous sums in cash to the paper millers and printers while the subscribers and booksellers took their time in paying him. Once caught in this squeeze, he found it impossible to pay several of his bills to the STN: "Nous avons bien eu l'honneur de vous observer que l'argent est ici d'une rareté affreuse, que nos libraires demandent du temps, et qu'enfin nous ne pouvons pas en faire sortir des pierres. Le train que nous menons l'ouvrage exige une mise dehors à laquelle nous ne comptions point."[2] Constant bungling in Duplain's warehouse and shipping operations compounded this problem because customers who received faulty shipments refused to pay their bills. Their refusals touched off exchanges of nasty letters—Duplain was not the gentlest of correspondents—and even lawsuits, which added further to Duplain's burden.

The load had become more than he could bear by March 1778, when one of his clerks complained to the STN about the "embarras dont nous sommes surchargés . . . la multiplicité des envois qui nous excèdent. Il y a bientôt six mois que Monsieur Duplain n'a pu mettre le pied à la rue."[3] A week later Duplain informed the STN that he had turned over all the quarto's bookkeeping to a firm of merchants, Veuve d'Antoine Merlino et fils, which had also bought into his share of the enterprise: "Nous nous débarrassons du détail de banque, et malgré cela nous sommes excédés, malgré 15 personnes que nous avons au moins."[4] Actually, Duplain's staff never amounted to much and could not cope with the work. When Favarger reached Marseille during his sales trip of 1778, he reported to the STN, "Quelques libraires que j'ai vus m'ont dit avoir reçu à bien des reprises des volumes pour des autres et avec cela quantité de feuilles très mal propres avec beaucoup de défectuosité. Il me paraît qu'il [Duplain] a trop peu de monde, car il n'a de commis que les deux frères Le Roy, un teneur de livres, et Le Roy l'aîné est presque toujours en voy-

---

1. Duplain to STN, Dec. 20, 1778.
2. Duplain to STN, June 9, 1778.
3. Duplain et Compagnie to STN, March 17, 1778.
4. Duplain to STN, March 24, 1778.

age. En magasin il a deux assembleurs, trois femmes pour collationner et son domestique pour mettre de côté etc."[5] When he reached Bordeaux six weeks later, Favarger discovered that the local booksellers had banded together and were withholding their payments as a group until Duplain corrected the defects and disorder of his shipments.[6] Duplain responded to these protests with threats of lawsuits; and instead of enlarging and reorganizing his staff, he drove it harder. After the STN objected to the way his warehouse workers assembled and inspected shipments, he replied, "Nous vous dirons qu'il faut faire collationner avant d'envoyer, que notre monde achève les tomes 21, 22, 23 [that is, of the first two editions], vient de finir 3 et 4 [of the third edition], qui sont partis, va se mettre sur 24, 25, 26 et après sur 5, 6, 7, 8. Vous rendez bien peu de justice à notre travail immense, et vous calculez bien peu son étendue. Ignorez-vous qu'il faut collationner près de 1800 rames sur chaque volume?"[7] All the letters about Duplain give the impression of an obsessed man who saw in the *Encyclopédie* the chance of a lifetime and who drove himself to the breaking point in order to exploit it for everything it was worth.

The lack of documentation makes it difficult to say much more about Duplain's role as a manager, except insofar as it concerned public relations. This aspect of his functions involved not only marketing but coping with a basic problem that threatened to spoil the market: how to persuade the subscribers to pay for thirty-six volumes of text after promising to supply them with twenty-nine. As already explained, Duplain tried to maneuver around this difficulty by surreptitious

5. Favarger to STN, Aug. 15, 1778.
6. Favarger to STN, Oct. 1, 1778: "Les libraires de cette ville qui ont pris de cette 3ème édition se récrient très fortement de ce que M. Duplain veut leur expédier indifférement des volumes pris du milieu du corps de l'ouvrage. Ils ne les payeront pas à 90 jours de date, parce qu'il arrivera que le souscripteur qui aura reçu les deux premiers volumes ne se souciera pas [d'avoir] à présent le 12ème ou 14ème sans avoir les entredeux, et le laissera sans le payer chez le libraire, qui ne sera pas moins obligé de le payer, ce qui mettra ledit libraire dans le cas de faire de fortes avances, auxquelles la moitié d'entre eux ne pourront pas suffire. Ils ont déjà mille peines de retirer ou faire retirer ceux qui leur viennent par ordre, car je suis persuadé d'après ce que j'ai vu que la moitié de ce qui a paru jusqu'à présent est encore à retirer du particulier. Le libraire ne l'a pas moins payé. Ils concluent de là que M. Duplain a envie dans cette marche d'aller plus vite. Or ils trouvent qu'il va déjà trop vite, qu'il ferait beaucoup mieux de se presser moins et avoir plus de goût dans le choix de ses imprimeurs, car le public est très malcontent d'une partie des volumes qu'il a déjà livrés."
7. Duplain to STN, Dec. 20, 1778.

cuts in the text and changes in the terms of the subscription. But he had not got far when his entire position was exposed in perhaps the most widely read French-language periodical then in existence, the *Annales politiques, civiles, et littéraires du dix-huitième siècle* of Nicolas-Simon-Henri Linguet. Under the headline "Brigandage typographique d'une nouvelle espèce," Linguet warned the world that the quarto editors could not possibly fit the full text into twenty-nine volumes. They would either have to cut it by half or double the size of their edition, drawing vast sums in additional payments from the subscribers. Linguet did not think that Diderot's *rapsodie Encyclopédique*—a slipshod compilation of bad philosophy and plagiarized treatises—deserved another edition in the first place. If the public insisted on buying one, it should spend its money on the *Encyclopédie d'Yverdon,* a good, cheap work from which the poison of Diderot and d'Alembert had been removed. But at all costs—and the costs would be enormous—it should avoid the quarto. For the quarto was a fraud, and the fraud could be explained in one word: Panckoucke. "Cette manoeuvre n'étonnera pas, quand on saura que c'est le libraire Panckoucke qui se cache sous le masque de l'imprimeur Pellet; il n'y a jamais eu de partisan, en guerre ou en finance, aussi fécond en ruses de cette espèce que le libraire Panckoucke."[8]

As in most of the polemics in eighteenth-century journalism, there was more to this than met the eye. Linguet's hatred of the philosophes went back a long way and had exploded a year earlier in an attack on La Harpe and the rest of the *secte* in the *Journal de politique et de littérature.* Panckoucke was then publishing the journal and Linguet editing it—but not for long, because a cabal of philosophes and academicians persuaded Panckoucke to fire him and replace him with none other than La Harpe. This incident led to the founding of the *Annales,* which Linguet published from London and filled with declamations against Panckoucke and his philosophic friends. The quarrel also had a commercial aspect because Linguet's new journal became a spectacular success and cut into the market for the *Journal de politique et de littérature,* which Panckoucke merged with the *Mercure* in June 1778. Moreover Linguet's main distributor on the Continent was the firm of Pierre Gosse Junior of The Hague, which had also taken over

8. *Annales politiques, civiles, et littéraires du dix-huitième siècle,* no. 15, II (1777), 465.

the marketing of the *Encyclopédie d'Yverdon*. Gosse must have rejoiced in Linguet's attack on the quarto, if he did not actually instigate it, because the quarto was damaging his attempts to sell the last sets of the *Encyclopédie d'Yverdon*, just as, back in 1769, Panckoucke's folio edition had undercut his campaign to sell the first sets.[9]

By combining so many combustible elements—personal, ideological, and commercial—Linguet's outburst sent repercussions throughout the book trade. The STN received several letters from subscribers whose confidence in the quarto had been shaken, and it dealt with them as best it could. Thus it reassured a customer in La Rochelle, who worried about having to pay for extra volumes, that the whole affair could be dismissed as a personal vendetta: ''Auriez-vous adopté le calcul de ce fou de Linguet, qui aveuglé par la haine implacable qu'il porte au pauvre Panckoucke et croyant que cette entreprise lui appartenait uniquement, s'est permis contre lui des propos les plus extravagants, dont les gens du métier se sont moqués et ont bien su le lui dire?''[10] And it tried to rally a bookseller in Naples, who was tormented with second thoughts about a speculation on fifteen subscriptions, with some bandwagon sales talk.

Outre que comme chacun le sait le caractère de cet homme fameux est de dire du mal de tout le monde, nous lui opposons en confiance un fait de notoriété publique. C'est que nous avons actuellement 6,000 exemplaires placés et sommes obligés d'ouvrir un 3ème souscription pour satisfaire à de nouvelles demandes. Il serait bien étrange que parmi un si grand nombre d'amateurs aucun d'eux n'eût eu assez de bon sens pour prévoir l'objection qu'a imaginé cet Aristarque moderne.[11]

These arguments, as the STN realized, did not constitute much of a defense, so it tried to cut off further attacks by go-

9. There is a good deal of information about these polemics in the dossiers of Pierre Gosse Junior et Pinet and of Mallet du Pan in the STN papers and in the *Mémoires secrets* of Bachaumont, entries of Aug. 8, Sept. 6, and Sept. 17, 1777, and Oct. 6, 1778.

10. STN to Ranson of La Rochelle, May 24, 1778.

11. STN to Società letteraria e tipografica di Napoli, Jan. 26, 1778. The STN went on to explain that it would keep down the number of volumes by increasing their size. The effect of Linguet's denunciation on the Neapolitans is clear from the worried tone of their letter to the STN, dated Jan. 6, 1778: ''M. Linguet nei suoi Annali no. 15 avverte il pubblico contro l'edizione dell'*Enciclopedia* in 4° di Ginevra e fa toccar con mano che in 29 vol. non conterrà che la metà. Questa è una cosa che merita ogni seria considerazione, soprattutto perchè non si sono gli editori avvaluti delle correzioni d'Yverdon, che sono buone.''

ing to their source. It did not appeal directly to Linguet but to his main collaborator, Jacques Mallet du Pan, who oversaw a Swiss edition of the *Annales* being printed by the STN's other rival *Encyclopédie* publisher, the Sociéte typographique de Lausane. Because of his personal friendship with Ostervald, Mallet felt more allegiance to the Neuchâtelois than to the Lausannois, and he eventually succeeded in ending Linguet's campaign against the quarto.[12]

Meanwhile, however, Linguet had badly damaged the quarto's market, and Duplain had to devise some way of repairing it. Panckoucke had suggested a round of soft-peddled publicity. "J'ai mandé à Duplain ce qu'il fallait qu'il répondît,"[13] he informed the STN. "Il faut que sa résponse soit générale, sans nommer ni Linguet, ni les *Annales*." Accordingly an "Avis" appeared in volume 11 that replied to Linguet's charges, without naming them or their author, by presenting the quarto's case positively: "Nous avons pris avec nos souscripteurs un engagement solonnel et sacré de leur livrer gratis tous les tomes qui excéderaient le nombre de 36." In other words, what Linguet had denounced as a swindle should be understood as a bonus, especially as the individual volumes would each contain 200 more pages than anticipated. But who was to pay for the seven extra volumes between volumes 29 and 36, a matter of 70 livres per set and more than a half million livres for all three editions? Duplain tried to construe this difficulty as an advantage for the subscribers by announcing

---

12. In a letter of March 14, 1778, the STN assured Duplain "qu'il n'y a aucun danger à courir de la part de Linguet et qu'il s'en tiendra aux sarcasmes qu'il a lâchés dans son journal . . . Le sieur Mallet, avec qui nous sommes liés particulièrement, est parti dernièrement pour aller joindre Linguet à Londres, avec lequel il s'est associé pour son journal; et il nous a donné sa parole qu'il ne paraîtrait plus rien contre notre entreprise, et nous sommes très portés à croire que s'il avait su que nous y fussions intéressés, il n'aurait point fait cette incartade." A close reading of the *Annales* for 1777 and 1778 shows that Linguet did indeed cease his attack on the quarto. Although Mallet's letters from this period are missing, his dossier shows that he was a close friend of Ostervald. He attempted to negotiate a peace in the quarto–octavo war in Jan. 1778. And in June, when he had returned from London to oversee the Lausanne edition of the *Annales*, he tried to get Linguet to plug the STN's edition of the *Description des arts et métiers*. The Neuchâtelois then sent a sycophantic appeal to Linguet himself, while indicating to one of their closest customers that they expected to make some important "arrangements" with Linguet and at the same time assuring Panckoucke that they were "bien résolus de ne former aucune liaison avec un tel homme." See STN to Linguet, May 30, 1778; STN to Charmet of Besançon, May 3, 1778; and Mallet to STN, June 6, 1778.

13. Panckoucke to STN, Nov. 27, 1777.

another "engagement solonnel" in a circular letter and an Avis in the eighteenth volume: the subscribers would have to pay for only thirty-three volumes and would receive the rest as an additional bonus. By this stroke of generosity he actually increased the retail price of the quarto from 344 to 384 livres. He could not expect the increase to go unnoticed, so he went over to the offensive by attacking all the quarto's enemies in a broadside *Mémoire:* "C'est en vain que des personnes mal intentionnées ont voulu tenter de décrier cette édition; on reconnaîtra aisément qu'un motif d'animosité et une basse jalousie ont guidé leurs honteuses démarches, qui ne peuvent que faire avorter leurs iniques projets."[14]

Duplain's tone and tactics had clearly changed: now he denounced Linguet by name, raked over the quarrel with Panckoucke, and refuted the typographical argument in detail (Linguet had got the size of the quarto's type wrong and had underestimated the thickness of its volumes). He returned to the attack in an Avis to the nineteenth volume of the quarto that repeated the same themes while attempting to cover his exposed flank by a new maneuver. Writing as "Pellet," Duplain announced a transfer in the ownership of the third edition: "Dans le désir de répondre à l'empressement de mes anciens souscripteurs, j'ai traité avec la Société typographique de Neuchâtel. En me remplaçant pour cette troisème édition, ils me laissent le temps et les soins que je me fais un devoir sacré de donner aux deux premières." The STN confirmed that it had bought out Pellet in a printed circular dated July 24, 1778. Another circular and a prospectus followed on November 10, while Duplain sent out similar announcements from Lyons. All this propaganda stressed that the STN's quarto would be even better edited, if possible, than Pellet's; that both editions would contain everything in the original text and a good deal more besides (" augmentations" as well as the uncut *Supplément*); and that the subscribers would have to pay for only thirty-two volumes. This stratagem had two purposes: it would promote sales by disassociating the third "edition" from the typographical and editorial faults of the first two, and it would shift the conditions of the sub-

14. Undated two-page "Mémoire pour les éditeurs de l'*Encyclopédie* de Genève en 32 vol. format in-4°."

scription so that no customer could claim that he had only bargained for twenty-nine volumes.[15]

The STN supported Duplain's fraud enthusiastically because it expected to get three more volumes to print by lending its name to the third edition. As already mentioned, however, it printed only one; and despite the claims of its advertising, the new quarto outdid its predecessors in the profusion of typographical errors. Then, while the STN was gnashing its teeth over the deformities of the third edition, Duplain went on to announce a fourth. Still using Pellet's name as a cover, he explained that he would publish one more edition in order to satisfy those who had failed to get their money in on time to buy the first three. He (Pellet) had received almost 400 more orders than he could supply. But he could not produce an edition of 400 sets economically, so he would have to charge his last subscribers 10 livres for each of the three volumes that he was giving away free to the first subscribers, bringing the price of the final quarto up to 414 livres instead of 384.[16] Duplain sent the text of this announcement to the STN with instructions to publish it in Dutch and Swiss journals. He merely meant it as a ruse ''pour enferrer de plus en plus les souscripteurs,'' he explained. The sales campaign for the third edition had really become stalled some 300 subscriptions short of its goal, and he needed to convince the subscribers of the first two editions that they were getting a bargain by receiving three extra volumes free instead of getting swindled by being forced to pay for four.[17]

Duplain's announcement probably had a still more devious purpose: to convince his partners that he was doing all he

15. The *Avis* quoted above all appear on unnumbered pages at the beginning of the quarto volumes. The circular letters come from the STN papers, ms. 1233, and a copy of the STN's prospectus for the third edition can also be found in Case Wing Z 45.18, Newberry Library, ser. 7, Chicago. The underlying reasons for the shifts in Duplain's strategy are clear from the following letters: Duplain to STN, July 10, 1778; STN to Duplain, July 15, 1778; and Favarger to STN, July 21 and July 23, 1778. Favarger wrote the last letter on a copy of Duplain's prospectus for the ''Neuchâtel'' edition, which the STN reprinted and circulated among its own correspondents.

16. Duplain published this announcement in the *Gazette de Leyde* of April 20, 1779.

17. Duplain to STN, March 31, 1779. Duplain timed the announcement to coincide with the period when the subscribers had finished paying for the twenty-nine volumes stipulated in the sales agreements and were about to pay for the extra volumes. See Duplain to STN, April 17, 1779.

could to market the unsellable rump of the third edition. The STN read it that way—as a laudable attempt to stave off lawsuits while stampeding customers into buying the last of the subscriptions.[18] Panckoucke, however, disapproved of it, not because he objected to lying to the public but because he thought it would not work: "La notice de Pellet est une charlatanerie dont je ne crois pas que personne soit la dupe."[19] The success of this maneuver cannot be calculated, but it brought the selling of the quarto to an end in appropriate style, by a final round of false advertising.

Marketing

All this feinting and sparring with journal notices raises two further issues: how did publishers market books in the eighteenth century, and what do their sales campaigns reveal about the literary market place? Admittedly, one must surround those questions with a caveat: the *Encyclopédie* was not a typical book, and its subscription campaign cannot be taken to typify book salesmanship. By 1777, the *Encyclopédie* had come to be seen as the embodiment of the Enlightenment. It had aroused so much controversy that the publishers of the later editions did not have to worry about making its name known. In fact, its name was used as a selling service device by other publishers, who wanted to capitalize on its success by presenting their own books as *Encyclopédies*. The register of applications for privileges kept by the booksellers guild of Paris shows a surge of requests to publish encyclopedic works around 1770, including an *Encyclopédie économique*, an *Enclopédie médicale*, an *Encyclopédie mathématique*, an *Encyclopédie domestique, économique, rurale et marchande*, an *Encyclopédie militaire*, an *Encyclopédie littéraire*, an *Encyclopédie des dieux et des héros*, and an *Encyclopédie des dames*.[20] But far from coasting on this wave of notoriety and

18. STN to Panckoucke, April 5, 1779: "Après avoir lu et examiné cet avis, il nous a paru être un moyen assez adroit d'engager les curieux à souscrire pour la 3èmé édition de l'*Encyclopédie* dans la crainte de payer plus cher s'ils attendaient l'apparition de celle qu'on annonce."

19. Panckoucke to STN, April 25, 1779.

20. Registre des privilèges et permissions simples, Bibliothèque nationale, ms. Fr. 22001, pp. 141, 146, 150, 164, 183, 225, and 266. Of course *Encyclopédie* had appeared in titles before the publication of Diderot's work, but the incidence of its use in the guild's registers shows that Diderot's *Encyclopédie* created a vogue for the term.

letting the book sell itself, the quarto publishers worked furiously to market as many copies as they could. Their sales campaign has some general significance for understanding eighteenth-century publishing, because they used the standard techniques of their trade. They distributed prospectuses and circulars, placed advertisements in journals, slipped sales talk into their commercial correspondence, and sent traveling salesmen far and wide in search of subscriptions.

The publishers adopted these techniques as a matter of course, without coordinating them or calculating their effect. They evidently did not feel a need to work out any special sales strategy but assumed that the whole association would benefit if each partner hit the public with all the propaganda he could muster, firing away in whatever manner suited him. "Il faut répandre avis sur avis et ne pas vous décourager,"[21] Panckoucke exhorted the STN in December 1778, when the demand seemed to be declining. Duplain stressed the same point: "Il ne vient plus de souscriptions, et si malgré 200 lettres que nous venons de faire partir, nous ne trouvons un débouché, nous serons nécessités à une mise en magasin. Ayez s.v.p. la bonté de réitérer des annonces, de faire de nouveaux efforts, de parler dans vos gazettes."[22] The flow of subscriptions provided a rough indication of demand and helped to orient policy. Thus Duplain did not decide to go ahead with the third edition until he received a sufficient response from a circular letter, which announced that Pellet had already begun to print it.[23] But the demand for the quarto had taken the publishers by surprise, and they never had a clear idea of its extent. In fact each of them had only the vaguest notion of what the others were doing to promote sales. They all produced their own prospectuses and placed their own advertisements. Duplain did not consult his partners before announcing his fake fourth edition. The STN printed a long circular letter about the third edition, which it addressed to itself and signed "Pellet," without informing Duplain or Panckoucke, not to mention Pellet himself, who was used to having others put words in his mouth.[24] Panckoucke sent out similar circulars from Paris and then delegated the selling on his end to a Pa-

21. Panckoucke to STN, Dec. 22, 1778.
22. Duplain to STN, Dec. 22, 1778.
23. Duplain to STN, April 21, 1778.
24. *Journal helvétique*, March 1778.

risian bookdealer called Laporte, leaving it two removes away from Duplain's central operation in Lyons. All three of the principal partners dispatched traveling salesmen to the French provinces, but without coordinating their routes, so that the salesmen sometimes got in each other's hair. Favarger sold only a few subscriptions during his tour of southern and central France in 1778 because he unknowingly retraced the steps of Duplain's agent, Amable Le Roy, who had already harvested most of the sales. Le Roy then complained that Favarger spoiled the harvest by telling the subscribers that they could delay payments. And Panckoucke's man came home almost empty-handed from a tour of the north because he became caught in the rivalry between provincial and Parisian dealers:

Mon commis voyageur n'a rien fait du tout. Il sera ici dans quelques jours. Il a vu la Normandie, la Bretagne, m'a dépensé environ six cents livres et m'a donné des mémoires pour quinze cents en totalité. On le regardait presque partout comme un espion. Les libraires de province n'aiment point ces commis voyageurs des libraires de Paris. Les Lyonnais, les Rouennais ont eu soin d'établir des préventions de toutes espèces.[25]

Of course the partners did not always work at cross purposes. Panckoucke applauded the success of one of Duplain's earlier traveling salesmen, who sold 395 subscriptions on one tour, and he also approved of Duplain's decision to bait subscriptions with an offer of free engravings of Diderot and d'Alembert: "C'est une dépense bien vue et qui peut produire un bon effet."[26] By processing all the subscriptions in Lyons, Duplain maintained control over the whole enterprise. And by observing the same terms in their sales, which went almost entirely to retailers rather than to private individuals, the partners worked as collaborators instead of as competitors. The proceeds of the sales did not get dispersed in commissions but went to Duplain, who used some of them for operating expenses and kept the rest to be distributed among the partners, in proportion to each man's share in the association, at the final settling of accounts. The associates also tended to aim their sales campaign at different areas, even though

25. Panckoucke to STN, June 26, 1777. On Favarger's difficulties see Favarger to STN, Aug. 23, 1778, and Duplain to STN, Nov. 10, 1778.
26. Panckoucke to STN, July 8, 1777.

they got their salesmen tangled. Panckoucke concentrated on Paris, Duplain on the provinces and southern Europe, and the STN on northern Europe. Thus when Duplain wanted to make a general announcement, he sent its text to Neuchâtel with a request for the STN to publish it in whatever journals it thought would be effective in "le nord," a term that he applied vaguely to all the territory between England and Russia.[27]

This propaganda cost very little—another indication of the relatively primitive character of eighteenth-century marketing. Favarger wore out a horse during his sales trip in France, but the STN easily wrote the beast off as a loss of 96 livres. Actually the total cost of the five-month journey came to quite a lot—1,289 livres and 240 livres for Favarger's salary—but only a fraction of it can be attributed to the *Encyclopédie;* for Favarger also collected bills, set up smuggling routes, established relations with new booksellers, and sold all the other works in the STN's stock.[28]

The STN charged the quarto association only 18 livres for producing 1,500 copies of a four-page "Grand Prospectus" for the third edition in November 1778.[29] Duplain printed his own version of it in Lyons, and the STN went on to reprint 2,060 more copies on various occasions between November 1778 and January 1780. It also produced a one-page *petit prospectus* to be distributed with the *Gazette de Berne* and 300 *lettres circulaires,* which it sent around its network of booksellers.[30] As Duplain's other printers also turned out prospectuses, it seems likely that they produced enough to supply every bookseller in Europe with a dozen or more and to flood

27. On May 11, 1777, for example, Duplain wrote to the STN, ''Nous vous envoyons ci-inclus le prospectus que vous ferez imprimer tel quel au même nombre que la *Gazette de Berne,* en vous entendant avec le gazetier pour les frais de poste . . . Le même prospectus . . . vous servira pour le nord, l'Allemagne, l'Angleterre, la Hollande et tous les autres pays où vous voulez en envoyer. Faites-en tirer le nombre qu'il vous en faut.''
28. STN papers, Brouillard B, entry for Dec. 1, 1778, and Favarger's ''Carnet de voyage,'' ms. 1059.
29. Brouillard B, entry for Nov. 28, 1778. A copy of this prospectus can be found in Case Wing Z 45.18, ser. 7, Newberry Library, and a copy from Duplain's edition of it, which Favarger sent to the STN from Lyons with his letter of July 23, 1778, is in Favarger's dossier of the STN papers.
30. The information on pressruns is derived from the STN's ''Banque des ouvriers,'' entries for Oct. 31, 1778; Aug. 1, 1778; Nov. 28, 1778; June 5, 1779; July 10, 1779; and Jan. 22, 1780.

the channels of the book trade. With a ream of paper and a day's work by a compositor and a press crew, the STN could easily run off a thousand prospectuses for 12 livres.

Notices in journals were just as cheap. The STN placed a full-page *annonce* in the *Gazette de Berne*—thirty-six lines in small pica—for only 7 livres 10 sous. Similar notices in the journals of Basel and Schaffhausen cost 6 or 7 livres apiece, and almost eight pages of sales talk in the STN's own *Journal helvétique* cost the association only 6 livres. These prices are difficult to evaluate, owing to the scanty state of research on French-language periodicals in the eighteenth century. But by modern standards the circulation of the journals was certainly small. In 1778 pressruns ranged from 7,000 in the case of the *Mercure* to 250 in the case of the *Journal helvétique*. The *Gazette de Leyde* probably had as much influence as any journal in *le nord,* judging from the frequent references to it in the STN's correspondence. In 1779 it charged only 26 livres for a half-page notice consisting of twenty-five lines in small type. The STN also paid about a livre a line for three other advertisements for the quarto, which it placed in the *Gazette de Leyde* in 1777 and 1778.[31] But it had to pay almost twice as much (38 shillings) for two half-page notices in the widely read *Morning-Herald* of London (only two lines of them were devoted to the quarto, the rest concerned the *Description des arts et métiers*). Booksellers who speculated on the quarto also took out their own advertisements. Téron of Geneva, for example, published five notices in the *Gazette de Leyde* and four in the *Gazette de Berne* between 1777 and 1779. Therefore the total cost of advertising for the quarto cannot be established, but it did not amount to much. For its part, the STN placed ten notices in five journals for about 120 livres—less than a third of the price of one *Encyclopédie*. The sum looks insignificant, if seen from the perspective of Madison Avenue, but it produced a heavy dose of publicity by eighteenth-century standards.

Twentieth-century notions of advertising do not fit the prac-

31. The STN mentioned these expenses in a letter to Duplain of June 26, 1779. The text of the half-page notice appeared in the *Gazette de Leyde* of April 20, 1779. And the pressrun of the *Journal helvétique* has been calculated from the STN's "Banque des ouvriers." On the *Mercure* subscriptions, which may have doubled by 1789, see Panckoucke to STN, July 21, 1778, where Panckoucke said they had reached 6,500, and the *Mémoires secrets* of Oct. 6, 1778, where they are estimated at 7,000.

tices of the Old Regime, and it would be anachronistic to comb through the papers of the quarto publishers in the expectation of finding an advertising budget or reports on market research. The publishers could have afforded many more journal notices, but it never occurred to them that they could saturate the "media" with a "campaign." They contented themselves with a few *annonces*—that is, literally, with announcing, in single issues of a half-dozen journals, that their product could be obtained at a certain price and in a certain manner.[32]

They did not, however, compose these announcements casually. Duplain wrote his own copy and passed it on to Neuchâtel with strict orders that it be printed without changes: "Nous vous remettons ci-inclus un avis à joindre tel quel dans les gazettes de Berne ou autres, si vous en connaissez, sans y rien changer."[33] In this case, the text announced the fake fourth edition with the covering explanation that "Pellet" had received nearly 400 more orders than he could supply, whereas in fact the real entrepreneur behind the quarto, Duplain, had just informed his partners, who also remained hidden, that the sales for the third "edition," which was really the second, had fallen short by 300 subscriptions—and even that information was false. Having woven together such a complex pattern of lies, Duplain could not afford to have a single word

---

32. The costs of the STN's advertising can be reconstructed from Brouillard B, entries for Feb. 8, May 3, June 13, and Nov. 28, 1778, and from the STN's "Compte courant" on all of its *Encyclopédie* work for Duplain in m.s. 1220. The entry for Feb. 8, 1778, in the Brouillard provides an example of different costs and account keeping:

Encyclopédie 4° doit aux suivants, prospectus et avis insérés dans les papiers publics concernant la 3ème souscription:

à Banque [des ouvriers], composition et tirage d'avis et de prospectus		. . . 4–10
à *Journal helvétique* 1778, insertion de l'avis		. . . 6–
à Profits et Pertes: Papier de poste pour dit	. . . 3–	
Port et affranchissage d'avis en Hollande, à Bâle, Schaffhausen et Berne		. . . 4–10
Droits aux 3 gazetiers de Suisse	. . . 21	
	28–10	. . . 28–10
		39 (livres)

Jean-Baptiste d'Arnal, a Swiss merchant in London, informed the STN on April 19, 1782, that he had placed its notice in two issues of "le *Morning Herald*, qui est le papier le plus à la mode; cela m'a coûté 38 schellings." He also included a copy of the printed text. The text of the other notices has been established by a study of the *Gazette de Leyde*, the *Gazette de Berne*, and the *Journal helvétique* for the period 1777–1780.

33. Duplain to STN, March 31, 1779.

disrupted. With seeming artlessness, he began the advertisement as follows:

Jean-Léonard Pellet, imprimeur à Genève et éditeur de l'*Encyclopédie* in-quarto, annonce qu l'ouvrage des rédacteurs est entièrement fini et que son édition contiendra en tout 36 volumes de discours et 3 de planches. Il se fait un devoir sacré de tenir les promesses qu'il a consignées à la tête des tomes XI, XIII, et XVII. En conséquence, MM. les souscripteurs recevront gratis trois volumes de discours.[34]

Duplain did not develop his main point—the announcement of the new edition, which was meant as a foil to make the terms of the old editions look good—until well down in the body of the text. This indirectness suited the casual, epistolary style of eighteenth-century journalism. Pellet spoke directly to "MM. les souscripteurs," using the polite form of the third person, as if he were merely imparting information. His voice sounded like that of the journalists themselves, for journals were conceived as open letters. They presented their news in an offhand manner, without headlines or special make-up, as so much letter-writing between ordinary *correspondants* and the *gazetier*. Advertisements, called simply *annonces* or *avis*, could hardly be distinguished from news, although they usually came at the end in French periodicals. Everything about them gave off an air of casualness and confidentiality, just the thing for systematic falsehood.

The slant of the publishers' sales pitch suggests the way they expected the *Encyclopédie* to appeal to the reading public. They could have presented it as a superb reference work or as a manifesto of the Enlightenment. Either approach would have indicated the importance of ideology in their perception of the public. In fact they combined themes in a way that is still more significant, as can be appreciated from the opening sentences of their main prospectus, which they reprinted in most of their *annonces* and *avis*, including the "Avertissement des nouveaux éditeurs" at the head of volume 1 in the quarto.

Les deux écrivains qui conçurent le projet de l'*Encyclopédie* en firent la bibliothèque de l'homme de goût, du philosophe, et du savant. Ce livre nous dispense de lire presque tous les autres. Ses éditeurs, en éclairant l'esprit humain, l'étonnent souvent par l'immense variété de leurs connaissances, et plus souvent encore par la nouveauté, la pro-

---

34. *Gazette de Leyde*, April 20, 1779.

fondeur, et l'ordre systématique de leurs idées. Personne n'a mieux connu qu'eux l'art de monter des conséquences aux principes, de dégager la vérité de l'alliage des erreurs, de prévenir contre l'abus des mots, qui en est la principale source, d'épargner des efforts à la mémoire qui recueille les idées, à la raison qui les combine, à l'imagination qui les embellit. Cette marche vraiment philosophique a dû accélérer les progrès de la raison ; et depuis quelques années l'on court à pas de géant dans une route qu'ils ont applanie et dont ils ont souvent changé les épines en fleurs.

A quarto on the shelf would demonstrate its owner's excellence in three capacities: as a man of taste, as a man of learning, and as a philosophe. Far from being incomptabile, these roles complemented one another; and best of all, they were easy to play. Diderot and d'Alembert had laid out such pleasant paths through the arid expanses of knowledge that one could merely follow their lead, stopping now and then to enjoy the flowers along the way, and still have the satisfaction of belonging to the intellectual vanguard. One did not even have to read very many other books, for the *Encyclopédie* was a library unto itself. The editors did not list the works that it rendered obsolete, but anyone who consulted its *Discours préliminaire* would have no difficulty in distinguishing between the heavy tomes of traditional learning and the streamlined, modern model. Modern learning meant Enlightenment; the prospectus made that point clear, not only by invoking reason and the progressive march of philosophy but also by attributing knowledge to the operation of the three faculties, memory, reason, and imagination, exactly as d'Alembert had done in the *Discours préliminaire*. It made these points gently, however, without any flag-waving about Bacon and Locke or any rhetoric about trampling superstition underfoot.

Instead of emphasizing the *Encyclopédie*'s challenge to accepted values, it stressed the ease with which the subscribers could become learned and progressive at the same time. Diderot and d'Alembert had made those qualities seem inseparable—a neat trick, like turning thorns into flowers, which the publishers adapted to their propaganda. They wanted to sell *Encyclopédies,* not to make them seem difficult, and the book's authors had provided its main selling point: it was both a compendium of knowledge and a vehicle of Enlightenment. To ask whether it appealed to eighteenth-century readers as the one thing or the other is to miss the point, for it was intended to appeal to them in both respects, by the men who wrote it and the men who sold it. Thus in a vulgar and simplified way, the advertisements amplified the message of the book; they did not distort it. And in doing so, they testified to the spread of a certain cultural tone. The publishers calculated that many people would buy the quarto merely in order to appear intellectually fashionable. Not only did the prospectus suggest that anyone who owned an *Encyclopédie* could call himself a philosophe, but Ostervald made the strategy of exploiting intellectual snobbism explicit in a letter to Panckoucke: "Il faut commencer par distinguer et ranger sous deux classes tous ceux qui se sont pourvus chez nous: les uns sont gens de lettres ou curieux de s'instruire à l'aide de cette compilation; les autres n'ont été guidés que par une sorte de vanité, se faisant gloire de posséder un ouvrage si renommé.'"[35]

Intellectual snobbism seems to have been a new phenomenon at this time, perhaps because intellectuals had just begun to get a hearing in the general public. In any case, it seems significant that the snobs sided with the *Encyclopédie,* while the book remained officially illegal. It could not be advertised in journals printed in France, although foreign periodicals like the *Gazette de Leyde,* which circulated widely in the kingdom, carried notices about it. Duplain had to keep his operation hidden behind the Swiss title pages, and even the prospectuses had to seem Swiss. "Il a fallu des courses, des sollicitations, des démarches pour obtenir la circulation *modérée* et *comme venant de Neuchâtel* de ce prospectus,"[36] Panckoucke informed the STN. The STN received several letters from bookdealers who were afraid to sell the quarto, and its replies, though reassuring, indicated the limits to the book's legitimacy: "Quoiqu'il y ait eu une défense de la part du magistrat pour cet ouvrage, nous sommes très assurés que l'effet en est suspendu et qu'il n'y a absolument rien à craindre pour les collecteurs, surtout en mettant, quant à la capitale, quelque prudence dans leur marché.'"[37] The government could not openly renounce its persecution of the *Encyclopédie* without seeming to endorse the ideology of the book. So the publishers had the best of both worlds: they could capitalize on the allure

35. Ostervald to Panckoucke, May 9, 1779.
36. Panckoucke to STN, Dec. 22, 1778. Of course the limited toleration granted to their publicity meant that the publishers had to restrain the ideological element in their prospectuses and notices, but they did not eliminate it entirely.
37. STN to Pyre, May 1, 1777.

of illegal literature while exploiting the unacknowledged protection of the authorities. But the most important point about their marketing concerns the way the *Encyclopédie* was presented and perceived during the late eighteenth century. It had originally made a splash by arousing controversy; it had spread throughout Europe on the waves of its *succès de scandale;* and its enemies kept stirring up the scandal by renewing their attacks. Far from evolving into a neutral reference work, it remained official anathema until the Revolution. Its illegality continued to be so good for business that the STN prayed for more edicts against it.[38]

Booksellers

The publicity and polemics show how the *Encyclopédie* was seen by its contemporaries, when it appeared on the market in a relatively inexpensive form. But the marketing itself involved a great deal more than the production of propaganda. It occurred in two stages: the publishers sold subscriptions to booksellers, and the booksellers sold them to individual customers. This two-step sales process required two systems of communication. The publishers and booksellers broadcast general messages to the public, and they exchanged commercial information through their own network, a trade grapevine that operated through exchanges of letters, personal visits, and traveling agents.

This process put the middleman in a strategic position. Far from maintaining sales organizations that could operate on a national scale, the publishers had only *comptoirs,* tiny home offices composed of a bookkeeper and one or two clerks. But after years of doing business, they had built up extensive contacts everywhere in the retail trade. They therefore aimed most of their sales campaign at the retailers, who handled the selling on the local level. "On peut souscrire chez les principaux libraires de chaque ville"[39] was the formula that closed most of the *annonces* for the quarto and for most other books as well. In order to persuade the booksellers to act energeti-

38. STN to Bosset, Aug. 28, 1779. For further discussion of this point see Chapter IX.
39. In this instance the formula comes from an *annonce* in the *Gazette de Leyde* of Feb. 6, 1778. Sometimes the notices said that the public could subscribe with Pellet or the STN, but they usually directed subscribers to retailers.

cally as *Encyclopédie* salesmen, the publishers allotted them a large share in the profits. The dealers bought their quartos for 294 livres and sold them for 384, and they received a free copy for every twelve that they ordered. If they sold a baker's dozen, they would make 1,464 livres, the equivalent of two years' wages for a journeyman printer and a profit of 41 percent on an investment of 3,528 livres. The importance of the trade discount emerges clearly in their letters. Gaston of Toulouse, for example, told the STN that he had made a great effort to sell subscriptions, because "si la remise est honnête, on ne craint point la dépense pour se mettre en campagne pour tâcher de se procurer de bien loin des souscripteurs."[40] The quarto–octavo war became in large part a struggle over booksellers, and Panckoucke expected to win it because of the profits he shared with them: "Vous sentez, Monsieur, que le très gros bénéfice des libraires les empêchera de favoriser cette entreprise [the octavo]," he explained to Ostervald.[41] "Et il n'y a qu'eux qui puissent faire le succès d'un ouvrage de cette nature. Cette seule raison suffirait pour ne pas vous faire craindre cette concurrence."

The STN also attributed the quarto's success to the support of the booksellers and ascribed their support to the money they made from it.[42] In its commercial correspondence, the STN kept pushing them to sell the book harder. Again and again it returned to the same point: by buying the quarto volumes for 7 livres and 10 sous apiece and selling them for 10 livres, they could make a killing. "Nous avons pour maxime général de faire gagner gros à tous les libraires avec qui nous travaillons,"[43] it explained to Abert of Avallon, adding suggestively that Abert's colleagues in nearby Dijon had sold 100 subscriptions in a few months. Meanwhile it urged the Dijonnais to emulate the booksellers of Besançon, who had sold still more, and it goaded one Bisontin for lagging behind the others: "Personne ne vous demande-t-il de l'*Encyclopédie* quarto? Nous croyons que vous êtes le seul libraire en France et même ailleurs qui n'y ait spéculé. Leur bénéfice est assuré."[44] It was by letters of this sort, dozens of letters every week, that the publishers handled most of their end of the marketing.

40. Gaston to STN, April 6, 1778.
41. Panckoucke to STN, Nov. 19, 1777.
42. STN to Cardinal Valenti of Rome, July 12, 1779.
43. STN to Abert, May 1, 1777.
44. STN to Charmet, June 19, 1777.

Diffusion

The booksellers produced still more propaganda on the other end. They took out advertisements in local periodicals, spread around prospectuses supplied by the publishers, and talked up the quarto when customers wandered into their shops. Tonnet of Dole favored posters: "Vous ne feriez pas mal d'y ajouter un placard en grosses lettres pour afficher devant ma boutique, ainsi que devaient faire tous vos correspondants."[45] Charle of Meaux relied on the *Affiches de la Brie* and on the heavy use of prospectuses: "En les répandant dans nos environs avec soin, je pourrai bien vous faire 24 souscriptions au lieu de 12."[46] And some dealers even produced prospectuses of their own. The Società letteraria e tipografica di Napoli circulated an elegant, two-page prospectus in Italian, which invited subscriptions for both the quarto and the octavo editions.[47] The publicity for the quarto therefore mushroomed all over Europe, and most of the selling was done by local bookdealers, who used whatever devices they could command. But what did the quarto look like from their perspective, and how did they understand their role in the diffusion process?

Those questions can be pursued in two ways, by consulting the booksellers' letters and by following Favarger's reports on retailing during his sales trip of 1778. The letters make somewhat disappointing reading because the booksellers did not expatiate on their notions of literature and their relations with their customers. They maintained a tight-lipped, businesslike tone and kept to the subjects that concerned them and their suppliers: shipping arrangements, payment problems, and the quality of the merchandise. But this emphasis shows how the *Encyclopédie* appeared from their point of view. They saw it primarily as an opportunity to keep their heads above water in a trade where men were constantly going under. This hard-bitten and strictly economic view of the book prevailed especially among the smaller dealers, as is clear from a few examples chosen from the area around Nancy.

Fournier of Saint-Dizier, a town of 5,600 souls in western

45. Tonnet to STN, Nov. 8, 1769. In this case Tonnet was referring to another work, but his remark applies to the *Encyclopédie*. Bookdealers also circulated sample title pages, though they did not mention this common sales technique in their correspondence about the quarto.
46. Charle to STN, Dec. 8, 1777. As his remark indicates, the booksellers strained to get the free thirteenth copy, so the baker's dozen provision stimulated their salesmanship.
47. Società letteraria e tipografica di Napoli to STN, Feb. 17, 1778. This letter includes a copy of the prospectus and a long discussion of it.

Lorraine, counted himself lucky: he had a small nest egg. His wife wanted him to invest it in land, but he thought he might do better in *Encyclopédies*—provided he could get special terms. If he could have a discount of 94 livres a set in addition to the free thirteenths, he told the STN, he would pay cash on the barrel head. It was a tempting offer, for cash payments were rare in a trade that suffered from lack of specie, and Fournier thought he could sell two dozen subscriptions. The STN could not accept it, however, because once one of the quarto associates deviated from the fixed wholesale price, they would be deluged with requests for discounts. So in the end, Fournier's wife had her way, and he sold only three quartos, at the standard price.[48]

Meanwhile in Joinville, a nearby town of 3,000, the local bookseller, a certain de Gaulle, was filling his letters with lamentations about the cost of transport. "Comment voudriez-vous que je puisse souscrire pour un exemplaire de l'*Encyclopédie?*" he asked. "Les ports et les droits m'emporteraient le profit." Although he finally sold one subscription, he considered Joinville a poor market for books. People preferred to spend their money on tapestries, he explained: "J'ai le malheur d'être dans une ville où il n'y a pas beaucoup de curieux."[49]

A little to the north, in Verdun, a small city of 10,300, a bookseller called Mondon worked hard to sell a dozen subscriptions, in order to profit from the free thirteenth copy. The prospectus produced a promising response among his customers, particularly the officers of two regiments stationed in the town, he reported. Within four months, he had sold eight *Encyclopédies;* and after the announcement of the third edition, he sold four more. But then difficulties set in. Duplain's first shipments did not contain enough copies of the first volumes, so that Mondon had to supply some of his customers before others. "Je suis dans l'embarras," he wrote to the STN. "Lequel de mes souscripteurs préférerai-je? Ce sont tous des personnes de la première classe." Next, Duplain lost Mondon's order for the third edition. When he finally found it, he sent twice as many copies as Mondon had requested. And when Mondon inspected the shipment, he discovered that the whole thing was worthless because at least 200 sheets had

48. Fournier to STN, Oct. 5, 1777; May 30, 1780; and June 20, 1780.
49. De Gaulle to STN, May 3 and Aug. 13, 1777.

been spoiled by dampness and mishandling. As Duplain failed to replace the defective sheets and barely condescended to answer his letters, Mondon tried to get the STN to intervene. Duplain had ceased to care about the small booksellers and had failed to keep control of the quarto, once it had become a gigantic enterprise, he complained. And worst of all, Duplain had refused to give him his free thirteenth copy, on the grounds that four of his twelve quartos came from the third edition, which was a separate affair. "Vous devez bien imaginer que dans une petite ville comme celle-ci il était difficile sans soin et sans se déplacer de faire un nombre de souscriptions. Je vous avouerai que c'était dans l'espérance de jouir du 13ème." All these difficulties created problems with Mondon's customers, who refused to pay until they received complete and correct volumes. And finally, the slump in the book trade and the departure of a great many troops for the American war left Mondon near the edge of bankruptcy. "Je ne regarde pas comme mortification de vous faire l'aveu de mon peu de fortune. Comme père de famille, je dois prendre garde à mes engagements et les respecter . . . Le commerce ici est totalement tombé. Nul état ne peut fixer son débit sans troupes, et c'est ce dont nous sommes dégarnis depuis la misérable guerre qui nous ôte toutes ressources . . . Je suis hors d'état de payer : six enfants, peu de commerce ne me laissent entrevoir que beaucoup de peine."[50]

Such were the tribulations of small-town booksellers. The large dealers did not agonize over payments, but they, too, treated the *Encyclopédie* exclusively as an opportunity to make money. Machuel of Rouen appraised the demand for the quarto as "lucrative."[51] Mathieu of Nancy said that he expected to sell a great many subscriptions and then did so, without going into further detail.[52] Bergeret of Bordeaux, Chevrier of Poitiers, Letourmy of Orléans, Rigaud of Montpellier, and Buchet of Nîmes discussed the quarto in similar fashion. They always treated it as a best seller, but never analyzed the reasons for its success. Instead, they maneuvered for favorable treatment—secret discounts or priority in the dispatching of shipments—and filled their letters with complaints about Duplain. Rigaud, for example, tried to play

50. Mondon to STN, March 20, 1778; Sept. 26, 1778; and April 9, 1780.
51. Machuel to STN, April 11, 1779.
52. Mathieu to STN, March 30, 1779.

the STN off against Duplain in order to get an extra free copy as a reward for collecting over a hundred subscriptions. The STN turned his request aside with a remark about what his success had meant for him: "Nous devons aussi vous féliciter de votre spéculation à raison du bénéfice qu'il y a à faire pour vous." And he replied with some fulminating criticism of the first two "editions":

On trouve le papier gris et inégal, le caractère usé, les corrections mal faites, car l'ouvrage fourmille de fautes d'impression. Enfin, le discours préliminaire du premier volume est exécuté d'une manière infâme, c'est-à-dire de caractères usés sur de très mauvais papier etc. Si on continue de même, cela attirera infailliblement des discussions et vraisemblablement des procès avec les souscripteurs, qui ne cessent de se plaindre et qui enfin feront éclat.[53]

Finally, the booksellers' letters illustrate the functioning of their trade grapevine. Faced with the dangers of being caught in the crossfire between warring consortia of publishers or of getting lost in the blizzard of false advertising, the retailers needed to know what was really going on among the publishers. An obscure, marginal dealer like Lair of Blois found it extremely difficult to get a clear view of events. In December 1773, Lair heard (wrongly) that the work on the *Supplément* was floundering, owing to the desertion of Diderot (who had no connection with it); so he decided to order Félice's *Encyclopédie d'Yverdon*. By March 1777, the news of the quarto's existence had reached him, though he believed (wrongly) that it was being published in Bouillon and was a cheap counterfeit of Félice's work. He favored the quarto until he read Linguet's attack on it in the *Annales*, which convinced him that Pellet and Panckoucke meant to swindle their subscribers and that he should remain faithful to Félice after all. But his faith was shaken when he heard rumors about the plan to publish a version of the *Encyclopédie Méthodique* in Liège. Favarger finally clarified the situation when he came through Blois on his sales trip of 1778; but in revealing the true nature of the quarto, he disguised that of the *Méthodique*, because he did not want the new *Encyclopédie* to ruin the market for the old one. He therefore convinced Lair that the project for the *Méthodique* was a pipe dream, even though Panckoucke had

53. STN to Rigaud, Nov. 23, 1777, and Rigaud to STN, March 9, 1778.

already taken it over and was about to execute it in Paris.[54]

Important booksellers located at nodal points along the grapevine succeeded much better in keeping themselves informed. Jean Mossy, a powerful and canny dealer in Marseilles, knew all about the connections between Neuchâtel and Lyons six weeks after the STN had accepted the Traité de Dijon, but he hesitated to collect subscriptions for the quarto because he wanted to be certain about which of the competing editions the government would favor. "Quant à votre nouvelle édition de l'*Encyclopédie* in-quarto, je nè sais encore que vous dire," he wrote to the STN in May 1777. "Nous avions reçu des ordres du gouvernment de ne point nous mêler ni de la vente, ni de la souscription de cet ouvrage. Outre cela, on parle d'une autre que le gouvernement veut favoriser, qui sera aussi in-quarto. Il s'agit d'avoir patience pour voir ce que tout cela deviendra." A month later, Mossy realized that the government had thrown its support behind Duplain's quarto; and three months later he had penetrated the motive behind Duplain: "J'entrevois que cette entreprise est combinée par un français et qu'il a envie d'abandonner le commerce après cette opération."[55]

Above all, the booksellers relied on their information network to avoid being stuck with old editions when new ones threatened to capture the market. As soon as Sens of Toulouse heard about the quarto, he cancelled an order for the second folio edition; and as soon as Carez of Toul heard about the *Méthodique,* he cancelled an order for the quarto.[56] The booksellers also exchanged information about other subjects, such as the solvency of their competitors and the direction of government policy. For example, Chaurou informed the STN that the authorities in Toulouse did not intend to release the shipment of octavos that they had seized: "*L'Encyclopédie*

54. The sequence of the misinformation that reached Lair can be reconstructed from his letters to the STN of Dec. 14, 1773; Sept. 21, 1774; March 23, 1777; May 15, 1777; Feb. 18, 1778; and Nov. 11, 1778.

55. Mossy to STN, May 16 and Aug. 4, 1777. The "ordres" to which Mossy referred probably concerned the *Encyclopédie d'Yverdon,* which was strictly forbidden in France and was also published in the quarto format.

56. Carez to STN, Dec. 17, 1781, and Sens to STN, March 5, 1777: "Quant à l'*Encyclopédie,* la personne pour laquelle nous voulions la [a copy of the folio edition] faire venir, sur les bruits qui courent que l'*Encyclopédie* de Genève en format in-quarto aura lieu et que le libraire Panckoucke de Paris s'est arrangé avec celui qui en a projeté l'édition, nous a dit qu'il donnerait volontiers la préférence à l'édition in-quarto, ce qui est cause également que nous vous prierons de ne pas nous l'envoyer."

in-octavo n'a pas été rendue, et l'on assure même qu'elle ne le sera pas. Je pense que l'on n'obtiendra le privilège tacite de cet ouvrage qu'après que celle in-quarto sera consommée.''[57] This was an important piece of news, which meant in effect that the quarto group had won the trade war; for the more it traveled the more it provoked the octavo subscribers to desert to the quarto, and the STN took care to keep it circulating through the grapevine.

Not surprisingly, the booksellers appear only as business-men when seen through business letters, but they were also cultural agents who operated at the meeting point between literary supply and demand. When a publisher's representa-tive came through town, they often discussed public tastes in literature, and their discussions often influenced the pub-lisher's decision about what works to reprint and what genres to emphasize. Favarger took soundings on literary demand wherever he went, traveling from book shop to book shop during his tour de France in 1778. His diary and letters con-stitute a virtual survey of the *Encyclopédie* market, which one can follow as he progressed around the map of France.

In Lyons, for example, the booksellers talked about nothing but the success of the quarto, although they were careful about how much they revealed in their talk. After a discussion in the powerful firm of Périsse frères, Favarger reported to the STN, ''Je n'ai pu savoir leurs pensées sur l'*Encyclopédie* quarto. Ils sont très réservés, ces Messieurs. Mais ils s'ac-cordent à dire qu'il n'y a rien sous presse dans tout Lyon que quelques misères et l'*Encyclopédie* quarto.'' In Vienne Veuve Vedeilhé was all enthusiasm: ''Elle a placé 48 exemplaires de l'*Encyclopédie* quarto, et la même chose est à Vienne qu'à Lyon: l'on n'achète plus de livre que celui-là.'' Similarly in Grenoble Veuve Giroud said that she had sold twenty-six sub-scriptions and could sell another two dozen, if Félice could be persuaded to take back some incomplete sets of the *Encyclo-pédie d'Yverdon*. After receiving the first fourteen volumes of it, her clients had canceled their subscriptions, and they now wanted to give up Félice's *Encyclopédie* for the STN's. Push-ing farther south, Favarger entered some arid territory.

Dans Valence c'est Aurel qui fait le plus, mais il est chargé de famille et ne gagne guère . . . Il n'y a point de libraire à Viviers. Un

57. Chauron to STN, Dec. 22, 1778.

ambulant du Vivarais dont je n'ai pu savoir le nom ni la demeure y apporte, comme à Montélimar, 3 ou 4 fois pendant l'année des livres à vendre. Orange n'a qu'un nommé Touït, perruquier de profession, mais qui vend des usages et rien d'autre. Calamel, qui est noté sur l'almanach des libraires, est un marchand d'étoffes, qui autrefois a vendu des livres mais qui n'en tient plus.

Carpentras did not contain a single bookseller, but like other small cities in the area it drew its supplies from the pirate publishers in the papal territory of Avignon. The Avignonese treated Favarger as an enemy and a spy, and the going got rougher on the route to Toulon because unemployed silk workers had turned to highway robbery. Favarger made it safely, but he found only three booksellers in Toulon, who told him that their trade had slumped so badly that they could only sell books on navigation. Business was booming, however, in Marseilles, Nîmes, and Montpellier. Hérisson of Carcassone had canceled his subscriptions to the octavo in order to get the quarto. And Fuzier of Pézenas and Odezenes et fils of Morbillon also seemed inclined to desert to the quarto, mainly because the smuggling operation of the octavo publishers had collapsed in the southwest. The quarto had swept everything before it in Toulouse, though Favarger found the city to be a "centre de la bigoterie" and its booksellers a collection of scoundrels, who tried to eliminate one another by mutual denunciation to the police. A bookbinder called Gaston had gathered eighty to eighty-five subscriptions for the quarto by offering free binding as a bonus, but the booksellers' guild had forced him to give them all up on the grounds that a non-guild member had no right to sell books. The quarto also sold well in Bordeaux, despite a slump in the local book trade, which the dealers attributed to the American war. After going north through La Rochelle, where he did not do much business, and Poitiers, "une bien pauvre ville pour tout commerce," Favarger returned home through the Loire Valley, which was also a disappointment. His diary contains a series of entries like "Saumur, rien" and "Chinon, encore moins." The three L'Etourmy brothers dominated the trade of the valley from their strongholds in Tours, Blois, and Orléans; but Favarger did not succeed in capturing their business, and he did not want the business of most of the lesser dealers, who speculated in prohibited books but failed to pay their bills, according to information that he picked up from the local merchants. After inspecting two of the biggest markets for the quarto, Dijon and Besançon, Favarger finally made it back to Neuchâtel, five months and hundreds of hours of shop talk after his departure.[58]

Although one cannot recapture the talk of those shops, one can catch something of its flavor from Favarger's letters. He found the big Lyonnais dealers to be difficult and aloof: ''Ils n'ont jamais le temps de vous écouter. Il semble qu'ils ont des empires à gouverner.'' The booksellers of Toulouse adopted a more casual, southern style of business:

Lorsque vous avez fait vos offres, l'on vous répond que l'on examinera le catalogue etc., vous priant de repasser. Vous repassez donc 3 ou 4 fois, et il arrive que le patron ne s'y trouve pas. Si vous le trouvez, il n' a pas eu le temps d' examiner vos propositions. Il faut donc y retrourner, pourquoi? Pour rien la plupart du temps. Tous sont sur ce ton. Il faut qu'un étranger coure constamment depuis l'un des bouts de la ville à l'autre et cela dans l'avant midi, car après dîner il est rare de trouver quelqu'un de ces Messieurs chez lui.

When he finally collared his clients, Favarger not only made some sales but got some valuable information out of them. For example, he found a strong demand for Rousseau everywhere: ''Chacun me demande les mémoires [that is, *Confessions*] de J.-J. Rousseau. On croit fermement qu'ils existent, non à Paris mais peut-être en Hollande. Ce serait un livre à faire à 3,000 si on l'avait dans la primeur . . . On est impatient partout de savoir des nouvelles de cet auteur. Peut-être, et cela est sûr, qu'une nouvelle édition augmentée de ses oeuvres se vendrait bien.''[59] This enthusiasm was purely commercial. Personal taste and values seem to have had no influence on the booksellers' assessment of books. As André of Versailles put it, ''Je ne néglige pas non plus le débit des livres que je ne saurais lire jamais, et c'est uniquement parce qu'il faut vivre avec la multitude et parce que le meilleur livre pour un marchand de livres est celui qui se vend.''[60] Only once did Favarger run into a bookseller who took an ideological approach to his business: ''Arles. Gaudion vaut de l'or, mais c'est un singulier personnage . . . Quand je lui ai parlé de la Bible et de l'*Encyclopédie,* il m'a répondu qu'il était trop bon Catho-

58. Quotations from Favarger's letters to the STN of July 21, July 26, Aug. 2, Sept. 13, and Oct. 28, 1778.
59. Favarger to STN, July 23, Sept. 13, and Aug. 15, 1778.
60. André to STN, Aug. 22, 1784.

lique pour chercher à répandre deux livres aussi impies, que toutes les encyclopédies lui ont bien éte offertes, mais qu'il se gardera bien d'en placer.''[61] Everywhere else, Favarger found that booksellers had greeted the *Encyclopédie* boom with gusto. The demand varied: it seemed to be weakest in remote inland areas—the mountains of the Midi, Berry, Poitou, and parts of the Vendée—and strongest around a huge semicircle, which ran down the Rhône Valley and back up along the Garonne. In general, then, Favarger's soundings confirmed all the other information that reached the publishers: the *Encyclopédie* had taken the trade by storm.

Prices and Consumers

The booksellers liked the *Encyclopédie* because their customers bought it. But who were those customers, and how far did the book penetrate into the world beyond the trading posts visited by Favarger? Those questions, like so many problems in the sociology of literature, are difficult to resolve. But one can measure the outside boundaries of the *Encyclopédie*'s readership, even if one cannot get inside the minds of the readers. First it is necessary to calculate the economic limits of the consumption pattern; then it should be possible to chart the geographical and social distribution of the quarto editions.

The *Encyclopédie* became smaller in size and cheaper in price as it progressed from edition to edition. While the format shrank from folio to quarto and octavo, the subscription price fell from 980 to 840, 384, and 225 livres. At the same time, the size of the pressruns increased—from 4,225 and 2,200, in the case of the two folio editions, to more than 8,000 quartos and 6,000 octavos.[62] Having satisfied the "quality market," the publishers tried to reach a broader public by producing in quantity. They made this strategy explicit in their commercial correspondence: "Le format in-folio sera pour les grands seigneurs et les bibliothèques, tandis que l'in-quarto sera à la portée des gens de lettres et des amateurs dont la fortune est moins considérable."[63] They also emphasized it in their advertising. Throughout their work they had

61. Favarger to STN, Aug. 15, 1778. Gaudion considered Favarger's Bible impious because it was a Protestant edition.
62. For the sources of this information see Chapter I.
63. STN to Rudiger of Moscow, May 31, 1777.

been guided by "vues économiques," they explained in the prospectuses. They wanted to bring the *Encyclopédie* within the reach of ordinary readers, who could profit most from it and who had been repelled at the *luxe* of the folio editions.[64]

The elimination of typographical luxury was a theme that would appeal to the values as well as the purses of a middle-class public, and it became possible as a policy by the elimination of most of the plates—plates which looked pretty but had no usefulness, the prospectus explained, because they never could be complex enough to improve any artisan's technical skills, and they were often so simple as to be unnecessary. Who needed engravings of everyday objects like hammers and bellows? If compared physically with the folio editions, the quarto stands out by its simplicity and sobriety, its flimsy paper, modest margins, and undistinguished type. Duplain did not go to the magnificent mills of the Johannot and Mont-golfier firms for paper but to second-rank supplies scattered throughout France and Switzerland. He did not order type from Fournier le jeune of Paris, the most elegant founder in Europe, but from Louis Vernange, a capable but pedestrian provincial. Even the printing of the quarto looks sloppy in comparison with that of its predecessors. A close examination of any set will reveal a profusion of workers' finger prints, overinked pages, misfolded sheets, badly made register, and typographical errors. The quarto was a rushed job, done on the cheap; and the octavo was even worse. If stood next to its sister editions, it looks like a poor stepchild, ragged, blotchy, and unkempt. Not only did the publishers cut frills as they cut costs, they tailored their book to fit the plainest provincial library. The *Encyclopédie* had gone from one extreme to another. Its typographical metamorphoses suggest that after having been originally aimed at an audience of seigneurs and sophisticates, it had penetrated into the remotest sectors of the reading public.[65]

The "democratization" of the *Encyclopédie* had limits, however, because even the cheapest edition would have seemed expensive to the common people. One can see how far beyond their reach it remained by translating its price into bread,

64. These phrases, which recurred in all the prospectuses and notices, are quoted from the first "Grand prospectus," published by the STN in the *Journal helvétique* of May 1777, pp. 76, 78.

65. These remarks are based on the examination of several sets in libraries in Switzerland, France, The Netherlands, Great Britain, and the United States.

the basic element of their diet. A first folio was worth 2,450 loaves of bread, a quarto 960 loaves, and an octavo 563 loaves, the standards of measurement being the subscription prices of the book and the "normal" price of 8 sous for a four-pound loaf of ordinary bread in prerevolutionary Paris. An unskilled laborer with a wife and three children would have had to buy at least 12 loaves a week to keep his family alive and would have earned about a livre a day, when he could find work. Even in good times and even when the wife or children had jobs, half the family's income would have gone for bread. A "cheap" octavo represented almost a year of this precarious food budget, a quarto a year and a half, a folio four years. It would have been about as likely for a laborer to buy an *Encyclopédie*—even if he could read it—as for him to purchase a palace. Skilled artisans—locksmiths, carpenters, and compositors—made 15 livres in a good week. Judging from signatures on marriage certificates and from *inventaires après décès,* they often managed not only to read books but also to buy them. But they could not have bought any *Encyclopédies,* for a first folio represented sixty-five weeks of labor, a quarto twenty-six weeks, and an octavo seventeen weeks. Diderot's work remained beyond the purchasing power of the "labor aristocracy," including the men who printed it.[66]

But the men who wrote it, the "Gens de Lettres" invoked on its title page, could have purchased the cheaper editions. Diderot himself made an average of 2,600 livres a year for his thirty years of labor on the *Encyclopédie.*[67] A quarto would have cost him seven and a half weeks of his wages and an octavo four and a half—not an extravagant sum, considering that he had other sources of income. Many lesser writers enjoyed greater wealth than Diderot, thanks to patrons and pensions. B. J. Saurin, a typical figure from the upper ranks

66. The above information on artisans' "budgets" and bread prices comes from the work of C.-E. Labrousse, Pierre Léon, Albert Soboul, and George Rudé. See particularly, Labrousse, *Esquisse du mouvement des prix et des revenus en France au XVIIIe siècle* (Paris, 1932), pp. 447–463, 582–606 and the forthcoming treatise on bread and bakers by Steven L. Kaplan, who gave me a generous preview of it. For estimates on literacy, which are even trickier than those on bread consumption, see Michel Fleury and Pierre Valmary, "Les progrès de l'instruction élémentaire de Louis XIV à Napoléon III d'après l'enquête de Louis Maggiolo (1877–1879)," *Population,* XII (Jan.–March 1957), 71–92 and François Furet and W. Sachs, "La croissance de l'alphabétisation en France XVIIIe–XIXe siècle," *Annales. E.S.C.,* XXIX (May–June 1974), 714–737.

67. Jacques Proust, *Diderot et l'Encyclopédie* (Paris, 1967), pp. 59, 81–116.

of the Republic of Letters, now deservedly forgotten, made 8,600 livres a year in pensions and *gratifications*.[68] He could have treated himself to a quarto, the equivalent of two and a third weeks' income. The octavo was for hack writers like Durey de Morsan, a literary adventurer who lived off the crumbs from Voltaire's table and wrote as "un des souscripteurs zélés" of Lausanne and Bern: "Le nombre des littérateurs pauvres surpasse de beaucoup celui des lecteurs opulents. Je suis charmé, en mon particulier, que cet ouvrage, ci-devant trop cher, n'excède pas les facultés des demi-indulgents tels que moi. Je voudrais que la porte des sciences, des arts, des vérités utiles fût ouverte jour et nuit à tous les humains qui savent lire."[69]

It is impossible to produce typical figures for the wide variety of incomes among the middling classes of the provinces, but the following calculations should give some idea of what the purchase of an *Encyclopédie* would have represented for persons located well below the great noblemen and financiers and well above the common people.[70] Although curés received only 500 livres as their *portion congrue* after 1768, their annual income often amounted to 1,000–2,000 livres. The purchase of a folio *Encyclopédie* would have absorbed twenty-five weeks of a prosperous curé's revenue, a quarto ten weeks. Magistrates of the bailliage courts stood at the top of the legal profession among provincial bourgeois and often earned 2,000–3,000 livres a year: a folio was worth seventeen weeks of their income, a quarto seven weeks. To live *noblement* a bourgeois had to count on at least 3,000–4,000 livres a year in *rentes:* a folio represented thirteen weeks of his income, a quarto five weeks. In each case, the difference between the cost of the folio and the quarto corresponded to the difference between an extravagance and a manageable luxury. Duplain had pitched his price at just the right level for the *Encyclopédie* to penetrate beyond the restricted circles of the wealthy avant-garde in which it had originally been confined and to

68. Robert Darnton, "The High Enlightenment and the Low-Life of Literature in Prerevolutionary France," *Past and Present*, no. 51 (1971), 87.

69. Durey de Morsan to Ostervald, April 17, 1778.

70. The calculations are based on information in Marcel Marion, *Dictionnaire des institutions de la France aux XVIIe et XVIIIe siècles* (Paris, 1923), p. 446; Henri Sée, *La France économique et sociale au XVIIIe siècle* (Paris, 1933), pp. 64–66, 162; and Philip Dawson, *Provincial Magistrates and Revolutionary Politics in France, 1789–1795* (Cambridge, Mass., 1972), chap. 3.

reach a general public of professional people and small-town notables.

The accuracy of that calculation became clear from the response to the subscriptions. As already explained, the orders poured in at such a rate that Duplain was overwhelmed and Panckoucke could hardly believe it: "C'est un succès incroyable."[71] The STN proclaimed the quarto to be the greatest coup in the history of publishing.[72] And reports from booksellers confirmed that view. In Rouen Machuel found the quarto to be "répandu partout"; Rigaud considered it "bien répandu" in Montpellier; from Maestricht DuFour reported, "Il n'est pas d'ouvrage si universellement répandu"; Resplandy described it in Toulouse as a work "dont nos rues sont pavées"; and d'Arnal echoed in Lyons, "Notre ville en est pavée."[73] The octavo sold even better, at least until the confiscations of its shipments discouraged the subscribers. Lair of Blois reported that a traveling salesman for Lausanne and Bern had gathered 3,000 subscriptions in a quick tour of a few provinces and that the octavo's price had made it a dangerous competitor of the quarto.[74] Similarly, Gaston of Toulouse wrote that he could have sold twice as many octavos as quartos if the publishers' intrigues had not undercut his efforts:

J'en plaçai 182 exemplaires [of the quarto] dans trois semaines de temps. Je me trouve souscripteur de 104 exemplaires de cette même *Encyclopédie* qui s'imprime à Lausanne en format in-octavo; et si on n'en avait annoncé deux autres in-octavo et une in-folio—savoir une que les libraires de Toulouse font imprimer à Nîmes chez M. Gaude et les deux autres s'imprimant à Liège sous le titre d'Amsterdam—il est très assuré que j'en aurais placé au moins 400 de celle de Lausanne.[75]

Reports such as these convinced the quarto publishers that they had been wrong to dismiss the octavo as an absurd "*Encyclopédie* de poche." They had failed to foresee that the

71. Panckoucke to STN, Nov. 8, 1777.
72. STN to Panckoucke, Aug. 20, 1778.
73. Machuel to STN, March 31, 1780; Rigaud to STN, Nov. 22, 1779; Dufour to STN, Aug. 2, 1780; Resplandy to STN, Jan. 2, 1778; and d'Arnal to STN, Nov. 12, 1779.
74. Lair to STN, Nov. 11, 1778: "M. Witel, gendre de M. Fauche leur associé, m'a dit qu'il en avait vendu plus de 3,000 dans le peu de provinces de la France qu'il n'a fait que traverser, sans ceux qu'il espérait de placer à Paris, où il s'est rendu en sortant d'ici 14 septembre."
75. Gaston to STN, April 6, 1778. Although none of these *annonces* ever came to anything, they illustrate the extent of the demand for cheap *Encyclopédies* and the scrambling among the booksellers to exploit it.

public would put up with its small type in order to take advantage of its reduced price, they confessed to one another. "Si nous en croyons quelques amis, l'entreprise, quoique ridicule, ne laissera pas que de réussir, tant le bon marché a d'attrait pour le plus grand nombre, c'est-à-dire pour les sots, qui le forment toujours," the STN warned Panckoucke early in the quarto–octavo war. And Panckoucke replied that the octavo would indeed succeed, "à cause du bas prix et du goût constant du public pour cet ouvrage."[76] Pricing therefore proved to be crucial in the process by which the *Encyclopédie* came within the range of ordinary readers. Duplain had suspected that somewhere in the *grand public* a sizeable demand for the book lay latent. He had tried to tap it by offering the quarto at 39 percent of the price of the folio. And he found to his surprise that he had struck a vein of gold, one that ran even deeper and farther than he had suspected, for the octavo publishers outdid him in exploiting it by selling their *Encyclopédie* at 59 percent of the price of his. But where exactly was this public located, and how was it composed?

The Sales Pattern

Grand public is one of those phrases that the French use to indicate that they have subjected unexplored territory to the suzerainty of rational discourse. In fact, very little is known about the extent, composition, and tastes of the audience for books in the early modern era, when mass literacy and market research did not exist. Duplain was shooting in the dark when he sent out his prospectuses and traveling salesmen; but every time he made contact with the demand for *Encyclopédies,* he kept a record of it. His subscription list covers virtually all the quartos (8,010 sets) and about 60 percent of all the *Encyclopédies* that existed in France before 1789. So by transposing the list onto a map (see Appendix B and Figure 5), one can get a clear view of the *Encyclopédie* market in eighteenth-century France.

The picture is not completely accurate, however, owing to four factors. First, it does not do justice to the Parisian market. Panckoucke and Duplain expected to sell a great many quartos in the capital, and when they failed to do so, each

76. STN to Panckoucke, Dec. 18, 1777, and Panckoucke to STN, Dec. 22, 1777.

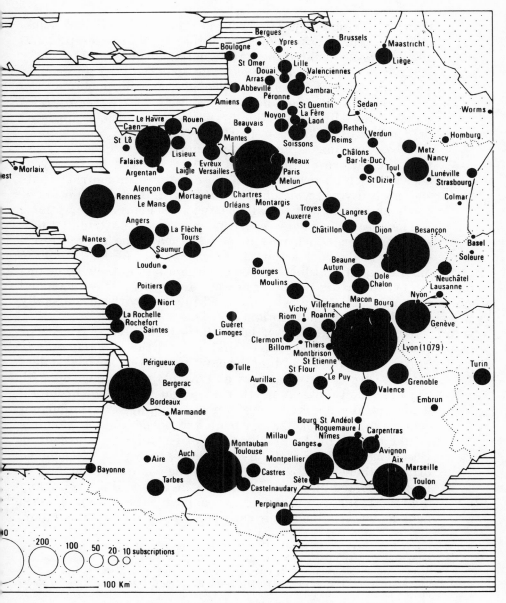

Figure 5. The Diffusion of the Quarto, France and
the French Borderland

blamed the other: Duplain accused Panckoucke of slack sales-manship, and Panckoucke claimed that Duplain's sloppy management had alienated the Parisians, who would not buy badly printed books. Although there was some validity to each of those arguments, Panckoucke probably put his finger on the main explanation when he observed that "Paris regorge des précédentes éditions."[77] The first two folio editions appealed primarily to the luxury market of the court and capital, whereas the quarto suited more modest provincial purses: thus the failure of the quarto in Versailles, a large city of 30,000, which absorbed only 5 subscriptions, in contrast to its success in Lyons, which had about a fifth of the population of Paris and accounted for almost twice as many subscriptions.

The spectacular sales in Lyons may also be attributed to a second factor, which may distort the map: the uneven effectiveness of the bookdealers as *Encyclopédie* salesmen. Duplain and his agents outdid themselves as retailers in their home territory, and so did two other dealers, Lépagnez of Besançon and Gaston of Toulouse, who operated in cities with an unusually high density of subscriptions.

Thirdly, the density of booksellers themselves varied. For example, the Sartine survey of 1764 and the *Almanach de la librairie* of 1780 show that a heavy population of booksellers existed throughout Flanders and Artois in comparison with the Franche-Comté. A northerner might subscribe through a dealer in any one of several towns, including the important centers of the book trade across the border, but a Comtois would not be likely to place his order anywhere except in Besançon or Dole. Actually, however, very few Frenchmen ordered their quartos from outside the country. Plomteux of Liège reported mediocre sales, and the STN did not sell many quartos from its own stock in eastern France. The French bookdealers confined their sales to their local or regional mar-

---

77. Panckoucke to STN, April 25, 1779. Duplain complained about his partners' selling in a letter to the STN of Jan. 21, 1779: "Si vous et M. Panckoucke vous donniez autant de mouvements que nous, vous reconnaîtriez aisément que ce livre est encore le meilleur des livres et se vendra jusqu'au dernier . . . Mais M. Panckoucke ne se donne pas de mouvements." Panckoucke expressed his complaints in a letter to Duplain of Dec. 22, 1778, in the Bibliothèque publique et universitaire de Genève, ms. suppl. 148: "Votre édition est déjà si décriée que je doute très fort que je parvienne à placer les 500 exemplaires. Il paraît que c'est un cri général . . . Je présume bien que j'en pourrai placer ici quelques cents, mais l'édition n'est pas estimée, et on aime ici la grande correction." See similar remarks in Panckoucke to STN, Dec. 24, 1778, and March 18, 1779.

ket and did not pass their *Encyclopédies* on to other retailers, because the quarto partners kept the wholesaling in their own hands.

But the hinterlands served by the retailers varied in size, and those variations should put one on guard against a fourth factor: the map underrepresents rural subscriptions because Duplain entered only the names of booksellers, not their clients, on his subscription list, and the booksellers almost always lived in cities. The circles on the map therefore represent zones of regional diffusion, not precise points of demand. Besançon (338 subscriptions) did not really dwarf Lille (28 subscriptions) by as much as it seems because about a third of the Bisontin subscribers came from outlying towns and villages. Even so, the disparity between the two cities looks enormous, especially if one considers their population—about 28,700 in the case of Besançon and 61,400 in the case of Lille. Why did the book sell so much better in some places than in others?

Any attempt to explain the sales pattern involves some treacherous steps: the argument may collapse where the statistical base is too thin, or it may slide off course in pursuit of spurious correlations. But research on the Old Regime has reached such an advanced state that one can draw on a kind of cultural geography in order to propose some tentative interpretation.[78] In general, it seems clear that the quarto reached every corner of the country, including the remote areas of the Pays Basque and the Massif Central. Its diffusion corresponded fairly well to the density of the population on a national scale, despite important discrepancies from city to city. Sales were concentrated in the great provincial capitals and scattered about smaller cities in secondary zones of dif-

78. The quantitative and geographical study of eighteenth-century French culture goes back to Daniel Mornet, F. de Dainville, and the Maggiolo survey of literacy cited above in note 118. Its more recent variations include research on the book trade (see especially François Furet and others, *Livre et société dans la France du XVIIIe siècle* [Paris and The Hague, 1965–70], 2 vols.), on education (see Roger Chartier, Dominque Julia, and Marie-Madeleine Compère, *L'éducation en France du XVIe au XVIIIe siècle* [Paris, 1976]), and on the intellectual elite of the provinces (see the series of articles by Daniel Roche culminating in his magisterial thesis, *Le siècle des lumières en province. Académies et académiciens provinciaux, 1680–1789* [Paris and The Hague, 1978]). The following account makes use of all this work. For population statistics, the census of 1806, despite its late date, remains the most reliable overall indicator of population around 1780. See René Le Mée, ''Population agglomérée, population éparse au début du XIXe siècle,'' *Annales de démographie historique* (1971), pp. 455–510.

fusion, but they do not demonstrate the existence of a ''Maggiolo line'' of literacy, dividing a semiliterate south-southwest from a progressive north-northeast. Instead, the map shows a fertile crescent of *Encyclopédies,* curving through the Midi from Lyons to Nîmes, Montpellier, Toulouse, and Bordeaux, just where Favarger found the market to be richest. Actually, one would not expect much correlation between minimal literacy—the mere ability to sign a marriage certificate, as measured in the Maggiolo survey—and the sophisticated mastery of reading necessary to make use of the *Encyclopédie.* So it may not be significant that the quarto had only a mild success in northeastern France, where Maggiolo found literacy to be most advanced. But Maggiolo's findings have been confirmed, with some modification, in recent studies of education in the eighteenth century; and those studies show that primary and secondary schools were scarcest in the area where the quarto's sales were weakest—that is, in the southwestern circle formed by the Loire and the Garonne, with Limoges at its center and tangential stretches of cultural desert running through Brittany and the Landes.[79]

It is difficult to correlate sales with cities because urban centers had so many different characteristics under the Old Regime, and it would be arbitrary to attribute the quarto's success to one element in the population rather than another. Bordeaux, for example, was the site of a parlement, an intendancy, an archbishopric, an academy, and a port. As it ranked fourth in population and fourth in sales of the quarto, the incidence of the sales hardly seems surprising; but it cannot be explained by labeling Bordeaux as a legal, administrative, religious, cultural, or commercial center, for it was all of those at once. Certain characteristics predominated in some of the smaller cities, however. In many cases the size of their population does not correspond to the sales of the quarto, even after one makes allowances for the distorting factors mentioned above. If the discrepancies point in the same direction, it might be possible to formulate some hypotheses about the nature of the market for the *Encyclopédie.* In Appendix C the thirty-seven largest cities have been ranked

79. See Chartier, Julia, and Compère, *L'éducation en France,* chaps. 2 and 3 and especially the map of schools run by the Frères des Ecoles Chrétiennes on p. 79, which corresponds fairly well with the map of the diffusion of the quarto *Encyclopédie.*

according to both the number of their subscriptions and the size of their population, and data about the existence of parlements, intendancies, and academies have been compiled. Cases in which sales were disproportionate to population stand out in the following table, which illustrates contrasts between pairs of cities.

	Population	Subscriptions	Parlement	Academy	Capital of Généralité
Bordeaux	92,966	356	x	x	x
Nantes	77,226	38			x
Lille	61,647	28			x
Toulouse	51,689	451	x	x	x
Amiens	39,853	59		x	x
Nancy	30,532	120	x	x	x
Clermont-Ferrand	30,982	13			
Rennes	29,225	218	x		x
Besançon	28,721	338	x	x	x
Toulon	28,170	22			
Brest	22,130	20			
Grenoble	22,129	80	x	x	x
Dijon	22,026	152	x	x	x
Limoges	21,757	3			x

Despite the arbitrariness involved in any set of comparisons, the table suggests two general tendencies : cities in which sales were high in relation to population tended to be primarily administrative and cultural centers, and cities where sales were low in relation to population tended to be primarily commercial and industrial centers.

If one goes over the subscription list with that formula in mind, it seems clear that the quarto sold best in cities with parlements and academies. The only cases that ran counter to this trend are Metz and Aix-en-Provence, parliamentary cities that had an exceptionally thin density of subscriptions; but this aberration may be explained by the unusual character of their book trade. The powerful booksellers' guild of Nancy, led by Mathieu and Babin, had almost destroyed the dealers of Metz; and the Marseillais, led by an aggressive bookseller called Mossy, dominated the trade in Aix.[80] The quarto also sold well in seats of intendancies and other important administrative bodies such as the estates of Languedoc, which met in Montpellier; but there are too many counter examples, like Lille and Limoges, for one to make much of this tendency. And to a certain extent, subscriptions flourished in Protestant cities: Nîmes, Montepellier, Montauban, and La Rochelle. In the case of Montauban, which ranked fifteenth in subscriptions and twenty-fifth in population, the connection between Protestantism and Encyclopedism seems to have been particularly strong: 78 of the 105 quartos sold there were subscribed through Crosilhes, a dealer who catered to a Huguenot clientele and who often ordered the works of Voltaire and Rousseau along with Protestant editions of the Bible and Psalms. Buchet and Gaude, who collected all but 3 of the 212 subscriptions in Nîmes, also dealt heavily in Protestant and Enlightenment literature; so their subscribers probably included a large proportion of Huguenots.[81] And one Huguenot subscriber, a merchant in Sedan called Bechet de Balan, indicated an affinity between his religious beliefs and his interest in the *Encyclopédie* by the way he placed his order: Je vous prie . . . de me faire passer . . . ce Dictionnaire encyclopédique dont vous m'avez parlé et relié proprement en veau; y joindre s.v.p. quelques sermons des meilleurs pour lire dans nos heures de dévotion le dimanche en famille.''[82] Of course the special appeal

---

80. The quarto also sold badly in Colmar, which had a Conseil Supérieur, and in Strasbourg, the nucleus of the book trade in Alsace. But the Alsatian trade was oriented more toward Germany than France, despite the Frenchification of the elite of the province. It may be, too, that the *Encyclopédie d'Yverdon*, which was strictly prohibited from central France, outsold the quarto in Alsace.

81. The peculiar mixture of Protestant and Enlightenment books in the orders of Buchet, Gaude, and Crosilhes is clear from dozens of their letters to the STN. On Sept. 5, 1776, for example, Buchet ordered twenty-six copies of a Protestant Bible and thirteen copies of *Le Christianisme dévoilé*. On March 30, 1779, he complained about the severity of the inspections in the local guild, ''qui ne laisse rien entrer dans nos magasins de ce qui est susceptible d'être arrêté. Comme cette Bible (n'est) pas tolérée par le fanatique [sic], qui compose cette nouvelle inquisition, on ne manquera pas de faire un procès verbal et de m'interdire pour toujours, si je n'avais la sage précaution de l'éviter.'' Nîme's population of 41,195 made it the eleventh largest city in France, and it came ninth in the ranks of the *Encyclopédie* subscriptions—a good showing in comparison with Orléans, which was a little larger in population but took only half as many *Encyclopédies*.

82. Bechet de Balan to STN, March 9, 1777.

of the *Encyclopédie* for Huguenots does not mean that it lacked attraction for Catholics. On the contrary, it sold especially well in some intensely Catholic cities, which had large endowments of ecclesiastical institutions, notably Angers, Chartres, and Auch. The population of Angers, an administrative center where the clergy was exceptionally influential,[83] came to only a third of that in Nantes, a great commercial city only a short distance down the Loire. Yet the Angevins bought nearly three times as many quartos as the Nantais.

The relative failure of the subscription campaign in Nantes —Nantes ranked sixth in population and thirty-eighth in sales of the quarto—seems extraordinary, except that it failed just as badly in other port cities. The quarto sold poorly in Le Havre, Brest, Sète, and Toulon, and did not sell at all in Lorient, Saint-Malo, Cherbourg, Dieppe, Calais, and Dunkerque. It did fairly well in Marseilles, Bordeaux, and Rouen, but not if one considers their population. Marseilles ranked third in population and sixth in subscriptions, far behind Toulouse and Besançon, which were much smaller cities. And Bordeaux and Rouen were seats of parlements and academies, in which an old-fashioned patriciate set the tone of intellectual life, often to the exclusion of the merchants. Manufacturing cities gave the *Encyclopédie* a still worse reception. With the exception of Cambrai, the quarto sold very badly in all the great textile centers of the North.

	Population	*Subscriptions*
Lille	61,467	28
Amiens	39,853	54
Reims	31,779	24
Saint-Omer	20,362	5
Valenciennes	19,016	13
Abbeville	17,660	26
Cambrai	15,608	57
Beauvais	13,183	8
Sedan	10,838	2
Saint-Quentin	10,535	16

83. On the predominance of the church in the life of Angers see John Mc-Manners, *French Ecclesiastical Society under the Ancien Régime. A Study of Angers in the Eighteenth Century* (Manchester, Eng., 1960).

Heavy industry had not developed much in France by 1780, so it may not seem surprising that few *Encyclopédies* were sold in the industrial cities of the future: 13 copies in Clermont-Ferrand and Saint-Etienne and none at all in Roubaix, Tourcoing, and Mulhouse. There was some manufacturing in a few cities with high *Encyclopédie* sales—metallurgy in Grenoble and textiles in Tours, Nîmes, and Montpellier. But the only clear case of a manufacturing and commercial center in which the sales of the quarto rose disproportionately above the size of the population was Lyons. Lyons, however, was a special case, not only because it served as the headquarters of the quarto enterprise but also, as will be seen, because Duplain tampered with the subscriptions in his home territory.

It is tempting, therefore, to advance a general hypothesis: the *Encyclopédie* did not especially appeal to merchants and manufacturers but rather to a heterogeneous public of noblemen, clerics, and a group sometimes identified as the *bourgeoisie d'Ancien Régime*—that is, notables, rentiers, officials, and professional persons, as distinct from the modern industrial bourgeoisie.[84] Thus Besançon and Lille may indeed represent opposite extremes in the literary market place: on the one hand, an old-fashioned city, encrusted with institutions of the state and church; on the other, a city ready for the leap into the nineteenth century, unencumbered by tradition.[85] It seems odd that the *Encyclopédie* sold so much better in the former than the latter—unless one sees the book as a representative product of the Old Regime rather than a prophetic work about the new. Still, speculative map-reading and slippery correlations between sales and cities do not provide material for sound conclusions. One can only use them to formulate a hypothesis—the *Encyclopédie* appealed primarily to a traditional mixed elite, rather than to the commercial and industrial bourgeoisie—which can be tested against two remaining sources of evidence: the subscription record of the Franche-Comté, where the social distribution of the quarto can be studied in detail, and the letters of the booksellers, where one can catch a few glimpses of the quarto's readers.

84. See Pierre Goubert, *L'Ancien Régime* (Paris, 1969), I, chap. 10.
85. That the two cities may be taken to epitomize the two extremes is clear from recent work in urban history, notably Claude Fohlen, *Histoire de Besançon* (Paris, 1965) and Louis Trenard, *Histoire d'une métropole. Lille. Roubaix. Tourcoing* (Toulouse, 1977).

Diffusion
Subscribers, A Case Study

During the century after its incorporation in the kingdom (1674), Besançon acquired a set of institutions that made it look like a perfect specimen of the provincial capitals of the Old Regime—a military *gouvernement,* a parlement, a *bureau des finances,* an academy, a university, and a host of judicial and fiscal offices. It acquired so many of them, in fact, that it seemed to be all superstructure. Aside from a textile manufactory, which employed twenty-eight artisans, it had no industry of any importance, and its commerce existed largely to supply the needs of the soldiers, *parlementaires,* lawyers, and royal officials who poured into the city in the wake of Louis XIV, doubling its population and transforming its appearance. Elegant town houses, in the style of Louis XV and Louis XVI, grew up along the four main streets, which ran parallel through the center of the town between the imposing, neoclassical intendancy to the south and the vast new Caserne Saint-Paul to the north. Religious edifices proliferated to such an extent that perhaps one-fourth of the land within the city walls belonged to the church—that is, to the archbishopric and the cathedral chapter, the seven richly endowed parishes (the church of Saint-Pierre had forty-one priests and sixty-eight chapels attached to it), and a dozen or so monasteries and convents. It was hardly possible to cross a street in Besançon without seeing someone wearing a robe or a sword. About one out of every forty persons belonged to the regular or secular clergy, and one in every seven served in the army. (The population had reached 32,000 by 1789, of whom about 800 were religious and 4,500 were soldiers.) The local *almanach* gives the impression that the town contained nothing but monks, priests, soldiers, magistrates, lawyers, and officeholders. It lists 73 councilors and 18 other officials in the parlement, 157 lawyers and attorneys, 37 members of the bailliage court, 17 receveurs des finances, 22 members of the Bureau des finances including 10 trésoriers de France, 16 top officials in the intendancy, 19 municipal councilors and aldermen, 15 administrators in the Direction des Fermes, and so on, through a maze of offices in the Eaux et Fôrets, the Domaines et Bois du Roi, the Régie Générale, the Poudres et Salpêtres, the Monnaie, the Juridiction consulaire, and many more. These were the men who sat in Besançon's academy, patronized its theater, joined its three masonic lodges, and sent their sons to its flourishing, ex-Jesuit Collège and its university.

But who read its books? By 1780, the vast majority of the Bisontins—95 percent of the men and 60 percent of the women —could read; and the town had a public library and a *cabinet littéraire* as well as four booksellers. Although little is known about the cultural life of the artisans, shopkeepers, and other *petites gens* who made up the bulk of the population, they probably remained intensely Catholic; for local historians stress that the Counter Reformation continued as a powerful force throughout the province and its capital until the end of the century, unchecked by Jansenism or the Enlightenment. In short, prerevolutionary Besançon seems to have been a closed and conservative little world—an outpost of Bourbon bureaucracy in a remote and backward province, the last place in which one would expect to find much of a market for the *Encyclopédie*.[86]

The market did not look promising to the town's leading booksellers, Charmet and Lépagnez cadet—at least not at first. "Ce livre avait beaucoup de succès en bien des pays, mais il ne me paraît pas qu'il prenne ici," Charmet observed. He did not even try to sell any subscriptions, and Lépagnez doubted that he could sell more than two dozen.[87] After a few weeks of sounding his customers, however, Lépagnez began to realize that he had under-estimated the demand. He asked the STN to rush as many prospectuses as possible to him in early June 1777, and for the next six months his letters serve as a measure for the spread of the *Encyclopédie* fever throughout Besançon and its hinterland. By June 10, he had sold his two dozen subscriptions and thought he might sell four. By June 20, he had made 48 sales and by June 30, 72, even though he had not dared to use the STN's prospectuses, owing to their poor paper. Attractive prospectuses on good paper were essential to his sales campaign, he explained; so he had

86. The above account is based on the *Almanach historique de Besançon et de la Franche-Comté pour l'année 1784* (Besançon, 1784) and from Fohlen, *Histoire de Besançon*. Fohlen characterizes Besançon as "une ville . . . où les idées philosophiques avaient relativement peu pénétré, l'attachement aux traditions était fortement enraciné" (p. 260).

87. Charmet to STN, May 12, 1777, and Lépagnez to STN, May 13, 1777: "27 exemplaires sont bien suffisant pour Besançon, même en me donnant tout le mouvement possible."

printed his own, hoping to collect still more subscriptions.[88] By August 22, he had sold 154; a week later he expected to pass the 200 mark, and at the end of September he had reached 260 and was still going strong. Having sold 338 by November 19, he reported that he would not handle subscriptions for any other book until "le feu de l'*Encyclopédie* sera passé.

He became stalled at that point for another year—not because the demand had been satiated, he claimed, but because of Laserre's surreptitous cutting: "Il est vrai que si l'on eût été assez adroit pour ne faire aucune suppression à la première édition, j'en aurais placé 600, et je n'aurais supporté aucuns reproches, au lieu que je n'en ai placé que 300 dont je reçois des reproches continuels. C'est un fait vrai."[89] Evidently many customers paid close attention to the text as well as the paper on which it was printed, and the attacks on the quarto had scared them off. Eventually Lépagnez and an allied bookseller from Dole called Chaboz collected 52 more subscriptions, making 390 in all and finally exhausting the market. "Ayant farci ma petite province de 390 exemplaires de votre Encyclopédie in-quarto . . . il n'est plus possible de trouver place à aucune. Vous devez être bien content,"[90] Lépagnez concluded in December 1779.

Meanwhile, he had collected 100 subscriptions for an edition of Rousseau's works published by the Société typographique de Genève; and he expected to sell 100 more, although he complained that the book trade had gone into a severe slump.[91] His competitor, Charmet, had left the *Encyclopédie* field to him but had dealt heavily in the works of d'Holbach, Helvétius, and Lamettrie, thanks to the protection of the intendant, Bourgeois de Boynes, who had agreed to burn fake copies of confiscated books in return for *des civilités palpables*— specially bound editions of the philosophes, which he kept for

---

88. The paper of the prospectuses served as a model for that of the book, and Lépagnez emphasized that his customers paid a great deal of attention to the quality of the paper when they made purchases. Lépagnez to STN, June 30 and Aug. 28, 1777.

89. Lépagnez to STN, Feb. 28, 1779.

90. Lépagnez to STN, Dec. 14, 1779.

91. Lépagnez to STN, Aug. 30, 1780: "Ne croyez pas, je vous prie, que je fais ici une grande consommation de livres. Je vous jure qu'après l'*Histoire universelle*, l'*Histoire ecclésiastique*, celle de l'Eglise gallicane, la Bible de Vance, l'*Encyclopédie* et le Rousseau, le reste me laisse dans une vacance depuis deux ans."

his own library.[92] Instead of being shut out of this tradition-bound city, the Enlightenment had poured into it and had even penetrated its most powerful and prestigious sectors.

The extent of the penetration can be measured from a subscription list Lépagnez published in 1777 when he had accumulated 253 subscriptions. As Duplain's records show that Lépagnez's final total of 390 subscriptions—which included 52 collected by Chaboz—were the only ones sold in the Franche-Comté, Lépagnez's list covers 65 percent of the quartos in his province. It is particularly valuable because it identifies almost all the subscribers by *qualité* or occupation.[93] Of the 253 subscribers listed by Lépagnez, 137 came from Besançon, the rest from smaller towns, mainly Dole, Pontarlier, Poligny, Vesoul, Arbois, Lons-le-Saunier, Gray, and Auxonne. Figures 6 and 7 illustrate their social position, according to estate and occupation.

About half of the quarto customers in Besançon and two-fifths of the others in the Franche-Comté belonged to the privileged orders. The graphs may overrepresent them slightly because a few of the military officers and councilors of the parlement may have been wrongly classified as noblemen. But the margin of error is too small to affect the proportions, and the importance of the *privilégiés* seems even greater if one considers their minority position within the population as a whole.[94] By way of contrast, the artisans, shopkeepers, day laborers, and servants who made up about three-quarters of

92. Charmet to STN, Oct. 18, 1775.
93. The list was printed and bound into volume I of the copy of the quarto in the Bibliothèque nationale, Z.2658. It has been republished by John Lough in *Essays on the Encyclopédie of Diderot and d'Alembert* (London, 1968), pp. 466–473.
94. Judging from their high rank and, in most cases, their aristocratic titles, the military subscribers did not contain any commoners, but some of the *parlementaires*, who were mostly councilors in the traditionalist if not reactionary parlement of Besançon, could have belonged to families that had not yet attained hereditary nobility. See Jean Egret, ''La révolution aristocratique en Franche-Comté et son échec (1788–1789),'' *Revue d'histoire moderne et contemporaine*, I (1954), 245–271. In his *Histoire de Besançon*, p. 210, Claude Fohlen concludes, ''Cette noblesse de robe tient le 'haut du pavé,' en l'absence d'une noblesse d'épée numériquement peu importante et d'une bourgeoisie presque inexistante''; but he does not cite any source for that statement. By compiling information from Besançon's almanacs and other sources, Daniel Roche estimated the population of adult males as follows: clergy, 9.9 percent; nobility, 2.4 percent; liberal professions, nonnoble officeholders, and administrators, 4.5 percent; bourgeois rentiers, 3.7 percent; merchants and manufacturers, 2.9 percent; and artisans, laborers, and servants, 76.1 percent. *Les lumières en province*, ''Troisième partie: Annexes et illustrations.''

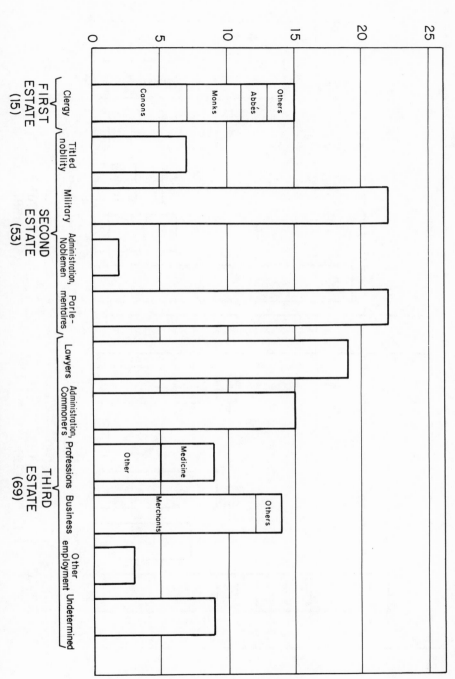

Figure 6. Subscribers to the Quarto in Besançon

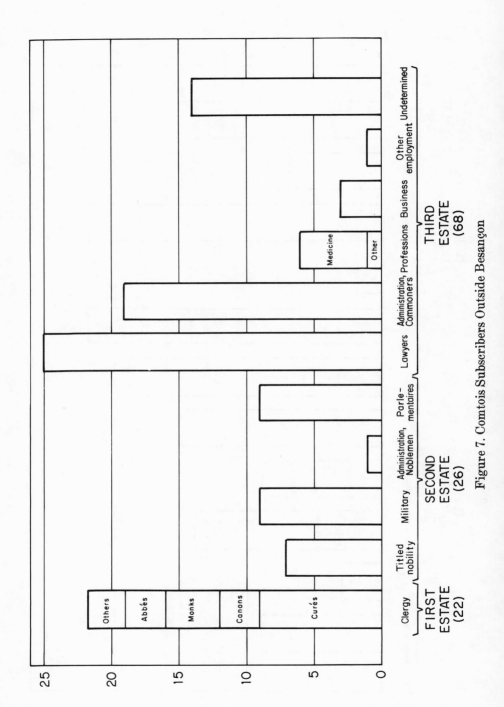

Figure 7. Comtois Subscribers Outside Besançon

Besançon's population do not appear at all among the quarto's subscribers; nor do the peasants and shopkeepers who formed the bulk of the population in the rest of the province. The *petites gens* simply could not afford the book. If it interested them, they might have consulted it in the *cabinet littéraire* or reading club organized by Lépagnez; and some of them may remain hidden in the "undetermined" category of the graphs. But the *Encyclopédie* appealed primarily to the traditional elite—the men connected with the ecclesiastical, military, and judicial institutions of Besançon. And, incidentally, the subscribers were men; women accounted for only 3 of the 253 subscriptions on Lépagnez's list.

As a garrison town, Besançon might seem likely to provide a good many subscribers from the upper ranks of the army, but one would not expect to find so many of its clergymen among the subscribers for a book that remained on the Index. There they are, however, both regular and secular, not only seven *chanoines de la Métropole* but also nine curates from the outlying towns and villages. Perhaps they were sophisticated men who could overlook the anticlericalism of the *Encyclopédie* in order to enjoy its intellectual riches, or perhaps they turned some of its sharpest remarks against their own superiors, for ideological ferment was spreading throughout the lower clergy on the eve of the Revolution, and the quarto did not attract any customers among the prelates of the Franche-Comté. Nonetheless, as the Counter Reformation had swept through the province with unusual force, it seems surprising that 19 percent of the non-Bisontin subscribers should have been clerics. One wonders what went through the heads of Blanchot, curé de Bourguignon-la-Charité, and Porcherot, curé de Joux, when they turned over the pages of their quartos.

The parlements had condemned the *Encyclopédie* as vociferously as the church, but twenty-two of Besançon's subscribers were connected with its parlement, including thirteen councilors and three presidents. The book had a great appeal for men of the law, not only in Besançon, where 14 percent of the local lawyers bought it but also in the smaller towns, where lawyers formed the largest group of subscribers. Four of Besançon's eighteen doctors subscribed, followed by various other professional persons—military engineers and *architectes,* a notary, an apothecary, and the principal of Besançon's important *collège*. Royal officials subscribed almost as heavily as clergymen and lawyers, so the quarto reached some important figures in the province's power structure, including two *lieutenants criminels*, three of the twelve *subdélégués*, and two of the four *secrétaires à l'intendance*, if not the intendant himself. It did not sell well among businessmen outside Besançon (they accounted for only 3 of the 116 subscribers), but the Bisontin subscribers included a dozen merchants, a *directeur général des Fermes*, and the only manufacturer of the city, a certain Détrey, who ran a small textile business.

Two of those businessmen, Détrey and Chazerand, served in the municipal government of 1793, and a subscriber called Rambour, who was *contrôleur des entrées de la ville*, led the rather mild Bisontin version of the Jacobin movement. Two subscribers who were canons of the Métropole also played leading parts in the Revolution, which produced a schism between the upper and the lower clergy of the province. The abbé Millot was elected as a deputy to the Estates General but resigned and then helped form the municipal government of 1790. And the abbé Seguin became constitutional bishop, a deputy to the Convention, and president of the departmental directory. The fourteen deputies to the Estates General from the Third Estate of the Franche-Comté included three subscribers (Bidault, *lieutenant-criminel* in the bailliage court of Poligny, and Blanc and Grenot, both lawyers in Besançon); and its composition suggests the importance of the legal profession in the Comtois revolutión, for it included six bailliage magistrates, seven lawyers, and one notary. Of course, other subscribers, particularly those from Besançon's reactionary parlement, probably became counterrevolutionaries and émigrés. It would be absurd to conclude that because a few future revolutionaires can be identified from the subscription list, subscribers in general favored the Revolution—just as it would be misleading to make too much of the fact that Besançon's subscribers came primarily from its notables. Where else would they have come from? But it seems significant that such a heavy proportion of the ruling elite in such a remote province should have wanted to buy the *Encyclopédie* and should have wanted to buy it so badly. The *feu de l'Encyclopédie*, as Lépagnez called it, burned brightest among the traditional leaders of provincial society.

Diffusion in France

Does the case of the Franche-Comté bear out the hypothesis that the *Encyclopédie* had little appeal for the commercial bourgeoisie? Not entirely, for even though almost all of the quartos in Besançon were bought by *privilégiés* and *bourgeois d'Ancien Régime,* a small proportion went to the city's small merchant class. To be sure, one cannot support a general interpretation with such slight statistics, and one cannot take Besançon to typify France. Particularism went so deep under the Old Regime that no two provinces shared the same culture, and "France" itself was a geographical expression, often applied to the Parisian Basin or Ile de France. But the *Encyclopédie* market transcended regional boundaries, and something of its general character can be seen from the correspondence of the men who did the marketing. Although booksellers rarely discussed their customers, merchants who dabbled in the book trade often included some observations about the reading public in their letters. Curiously, their worst customers came from their own kind—that is, from other merchants, as the following examples indicate.

Barre, a merchant in Nantes: "Les négociants ne pensent guère à la littérature."

Gosselin, a merchant in Lille: "Je ne doute point, Monsieur, que vous ne trouviez dans le reste de la France de quoi vous dédommager amplement du peu de goût qui règne dans notre ville pour la littérature. Nous commençons seulement à sortir de cette léthargie qui a enchaîné l'Europe pendant plusieurs siècles. Notre climat ou plutôt notre sol n'est point fécond en gens studieux ; et il y a cinquante ans on n'aurait pas trouvé une seule bibliothèque passable dans tout Lille."

Bechet de Balan, a merchant in Sedan : "Notre ville n'offre point d'amateurs de littérature. Les libraires n'y font rien . . . et je dirai à la honte de nos citoyens que l'esprit de vendre des draps et d'amasser est le seul qui les décore. Vous ne croiriez pas, Monsieur, qu'on dédaigne même les talents et que l'on néglige d'en donner dans l'éducation des enfants. Les arts agréables leur semblent une chose inutile."

Volland, "ancien officier de la Reine" in Bar-le-Duc: "Je ne prévois pas que vous puissiez les [the quartos] vendre ici, les ayant proposés à tout le monde ici, et personne jusqu'ici n'en est venu chercher un exemplaire. Ils sont plus avides de commerce que de lectures, et l'éducation y est absolument négligée . . . Vous ne trouverez pas le débit de vos livres ici. MM. les nobles ne sont pas riches, et les négociants aiment mieux apprendre à leurs enfants que 5 et 4 font 9 ôtez 2 reste 7 que de leur apprendre à faire le bel esprit.''[95]

These remarks might represent nothing more than social prejudice, but they were made by merchants about merchants. The letter writers did not seem to be trying to impress the directors of the STN, who were merchants also, and no other letters in the vast correspondence of the STN refer to a peculiar lack of demand for literature among any other social group. Moreover, the letters that deplored the cultural backwardness of merchants all came from towns where the sales of the quarto were disproportionately low. The STN never heard anything bad about the book-buying habits of the merchants in Lyons and Marseilles. In fact, a bookseller in Marseilles called Caldesaigues mentioned the occupation of his first thirteen subscribers to the quarto, and nine of them were merchants.

The other references to individual subscribers in the STN's correspondence all concern notables and noblemen,[96] but they are too rare and scattered to have much significance, except

95. Barre to STN, Sept. 15, 1781; Gosselin to STN, July 7, 1775; Bechet de Balan to STN, March 9, 1777; and Volland to STN, July 23, 1780. After emphasizing the lack of interest in literature among his fellow merchants in his letter of Sept. 15, 1781, Barre added, ''Cependant j'ai de mes amis qui ne dédaignent pas les lettres.'' He eventually sold sixteen sets of the quarto. His customers may have included some merchants because he remarked in a letter of Aug. 7, 1781, ''J'en ai pourvu tous mes amis.'' However, Vallet fils aîné, a bookseller in Nantes, complained to the STN in a letter of Aug. 19, 1779, that he had great difficulty in selling the *Encyclopédie;* and Pelloutier, a Nantais merchant, warned the STN in a letter of Feb. 2, 1785, not to expect any orders for books from the merchants of the Caribbean, ''contré où on s'occupe trop à gagner de l'argent pour chercher à acquérir de l'esprit.''

96. For example, the STN received the following reports of sales: on July 9, 1777, from Harlé of Saint-Quentin, a report of a sale to the comte de la Tour Du Pin Chambli; on Nov. 27, 1781, from Volland of Bar-le-Duc, a sale to a ''gentilhomme'' in Bar and to the chevalier de Jobart, capitaine reformé in Nancy; on Aug. 12, 1777, from Robert of Bar-le-Duc, sales to the chevalier de Longeau, lieutenant des maréchaux de France in Bar, to the Baron de Bouret, chevalier de Malte in Bar, to M. de Trouville, chanoine de la Madeleine in Verdun, and to Baudot, procureur du roi in Bar; on Feb. 7, 1781, from Entretien of Lunéville, a sale to an unnamed lawyer; and on March 20, 1778, from Mondon of Verdun, sales to eight persons, ''tous de la première classe,'' including an ''officier du régiment de Bresse.'' On Oct. 1, 1778, Favarger reported a sale to Baillas de Lambrède, commis des guerres at Marmande near Toulouse; in an entry dated June 9, 1780, in Brouillard B, the STN noted a sale to a lawyer from Nancy called Le Gros, and in his subscription list Duplain identified six individual subscribers: ''le curé de Cherier,'' ''Dusers, chevalier de Saint-Louis,'' ''Guiget abbé de la Croix Rousse à Lyon,'' ''le comte d'Orsey à Paris,'' ''le comte de Neuilly à Versailles,'' and ''Giraud curé à Vichy.''

in two cases, which show how the *Encyclopédie* spread through particular milieux, thanks to the intercession of well-placed individuals. The first of these intermediaries was an artillery captain in Saint-Dizier called Champmorin de Varannes. He arranged to sell four quartos to other officers and offered to set himself up as a distributor of the STN's books in military circles, where, he noted, there was great interest in the works of Rousseau. Although his proposal never came to anything, it illustrates the strong demand for Enlightenment literature among army officers, which also shows up in the letters of booksellers in garrison towns like Besançon, Metz, Verdun, and Montpellier. The second informal *Encyclopédie* salesman for the STN was a friend of Ostervald's and a councilor in the Parlement of Paris called Boisgibault, who bought a quarto himself and sold three others to colleagues in the parlement. Unlike Champmorin, he did not seek a commission but arranged the sales ''par le désir ardent que je vous connais aussi pour contribuer à la propagation des sciences et pour répandre un livre qui en est le dépôt.'' Boisgibault wrote as an admirer of Turgot and Malesherbes and also of Switzerland, where he hoped to travel in order to enjoy ''le spectacle délicieux de la liberté et de cette fierté noble qu'elle inspire à tous les hommes, qu'elle égalise.'' His role in the distribution of the *Encyclopédie* expressed an ideological commitment that may have been mainly rhetorical but seems to have spread rather widely among the younger *parlementaires,* whose opposition to the government in 1787 and 1788 proved to be crucial in precipitating the Revolution.[97]

If it seems certain that the *Encyclopédie* appealed to some elements at the top of French society, it is impossible to know exactly where the downward penetration of the book stopped. Only once did a correspondent of the STN mention price as an obstacle to its diffusion. A Protestant schoolteacher in Caen, who kept body and soul together by a small, clandestine book business—though he described himself merely as ''un simple particulier qui tient un nombre de jeunes gens Protestants en pension dont je désirerais de faire des citoyens utiles et vertueux''[98]—informed the STN that he was too poor to buy

97. Boisgibault to STN, July 11, 1781, and April 4, 1781. On the parlement's role in the prerevolution see Jean Egret, *La pré-révolution française (1787–1788)* (Paris, 1962).
98. Chaudepied de Boiviers to STN, June 21, 1777.

the quarto. Most literate Frenchmen must have been excluded from the subscription list for the same reason, but thousands of them belonged to *cabinets littéraires,* where they could read as much as they wanted for as little as one and a half livres a month. Booksellers frequently set up these reading clubs simply by subscribing to a few periodicals, using their stock as a library, and arranging quarters behind their shop as a reading room. Although it is difficult to know what went on inside them or who their members were, it seems probable that the *cabinets littéraires* became important centers of idea diffusion in the late eighteenth century. In 1777 Nicolas Gerlache, a small bookseller in Metz, derived almost half his income from his *cabinet.* It had 379 members, and its library included the quarto *Encyclopédie* as well as a good deal of illegal literature.[99] Choppin of Bar-le-Duc told the STN that he ordered the quarto specifically for his *cabinet littéraire.* It kept company in his shop with the works of Rousseau, the *Système de la nature, Le Compère Matthieu, Vie privée de Louis XV,* and other prohibited books, which Choppin also bought from the STN, for he explained that "ce sont ces sortes d'ouvrages dont j'ai le plus de débit."[100] Buchet of Nîmes, Lair of Blois, and Charmet of Besançon ordered the same material for their *cabinets littéraires;* several other booksellers who handled the quarto subscriptions probably did likewise; and their customers probably included many readers of modest means. As Mercier put it in his *Tableau de Paris,* "N'avez-vous point de bibliothèque? Pour quatre sols vous vous enfoncez dans un cabinet littéraire, et là, pendant une après-dînée entière vous lisez depuis la massive *Encyclopédie* jusqu'aux feuilles volantes."[101]

What then can one conclude about the readership of the quarto in France? The subscriptions sold well among men of the robe and of the sword—*parlementaires* and army officers in particular. They sold even better among professional men— lawyers above all and also administrative officials and clergy-

---

99. Gerlache described his *cabinet* in several letters to the STN, notably an undated letter received in May 1777.

100. Choppin to STN, April 25 and July 28, 1780.

101. L.-S. Mercier, *Tableau de Paris,* 12 vols. (Amsterdam, 1783–89), IV, 3. The profusion of *cabinets littéraires* deserves further study. For some additional references see Daniel Mornet, *Les origines intellectuelles de la Révolution française (1715–1787),* 5th ed. (Paris, 1954), pp. 310–312.

men. But they sold badly among men who lived by trade and manufacturing, at least in the north and northeast and in most port cities, if not in Lyons and Marseilles. The example of Besançon, where merchants comprised a small but significant minority of the subscribers, should make one wary of equating trade with disinterest in literature. But that equation seemed valid to several of the STN's correspondents, and several others mentioned only notables when they identified their subscribers. Despite the inconclusive character of the evidence, therefore, it seems likely that the *Encyclopédie* had a far greater appeal for the *bourgeoisie d'Ancien Régime* than for the commercial and industrial bourgeoisie. The evidence thins out when it comes to the circulation of the *Encyclopédie* among the common people. Shopkeepers and artisans might have had access to it in *cabinets littéraires* or by borrowing, and some of them might have been able to afford the octavo after it finally penetrated the French market, but they are never mentioned in any of the sources. Although one cannot exclude the possibility that the *Encyclopédie* reached a great many readers in the lower middle classes, its main appeal was to the traditional elite—the men who dominated the administrative and cultural life of the provincial capitals and small towns.

Diffusion Outside France

The quarto may have represented provincial Enlightenment in France, but it expressed enlightened cosmopolitanism everywhere else. Its sales pattern looks like an itinerary from a Grand Tour: London, Amsterdam, Brussels, Paris, Lisbon, Madrid, Naples, Venice—and beyond, to Munich, Prague, Pest, Warsaw, Moscow, Saint Petersburg, Copenhagen, and Hamburg. It reached most of the great capitals of Europe, but it did not flood them, as it did when it arrived in the provincial capitals of France. The quarto spread itself too thinly to penetrate much of the non-French public. Yet its non-French diffusion is worth studying because it shows how far the *Encyclopédie* vogue reached and whom it touched.

The quarto publishers sold 7,257 copies of their book in France and 691 in the rest of Europe—that is, they provided more than half the *Encyclopédies* that existed in France before 1789 but less than 10 percent of the rest in Europe. The continental market was supplied primarily by the first two folio editions and the octavo. The editions of Lucca and Leghorn, which together amounted to only 4,500 sets, seem to have sold mainly in Italy. And the *Encyclopédie d'Yverdon,* though it was a very different book, creamed off much of the *Encyclopédie* market in the Low Countries. As these other editions accounted for the great majority of *Encyclopédie* sales outside France, the sales record of the quarto cannot be taken as a measure of *Encyclopédie* diffusion on a continental scale. When studied in company with the dossiers of the European booksellers, however, it provides a complete picture of the quarto's distribution. And that picture covers almost every corner of the European book trade, as Figure 8 illustrates.

As might be expected, the density of quartos was thinnest in eastern Europe, and the letters from the booksellers of the east convey a sense of cultural as well as geographical remoteness. Weigand and Köpf of Pest, for example, wrote that they could not order many *Encyclopédies.* They had taken a few sets of the octavo and an *Encyclopédie d'Yverdon,* but not many of their customers could read languages other than Latin and Hungarian. The country had only begun to revive after centuries of Turkish oppression, and Enlightenment was spreading slowly.

Notre établissement n'est presque fondé que sur de livres d'assortiment, le caractère universel de la nation et d'un pays qui a été envahi pendant des siècles entiers par des guerres cruelles et sanglantes, n'étant pas encore assez près du dernier degré de culture pour produire des auteurs qui puissent garantir les libraires des frais d'impression . . . L'*Encyclopédie* de laquelle vous nous faites une exacte description n'est pas celle-ci dont la Société de Berne nous a fourni quelques tomes, et nous ne pouvons pas encore donner commissions de cet ouvrage. Cependant, nous en espérons donner dans peu de temps, l'ouvrage n'étant encore que peu connu chez nous, surtout quand la liberté de penser et d'écrire qui a été rendue à notre heureux royaume par notre auguste souverain, se sera répandue dans les individus.

In the end, Weigand and Köpf ordered only one quarto, and they were sorry they did. When it finally reached them after four months in entrepôts and barges along the Lech and the Danube it took two years to sell. And after they turned it over to their customer, "un des premiers seigneurs de notre pays," he discovered that it was missing six sheets. Although

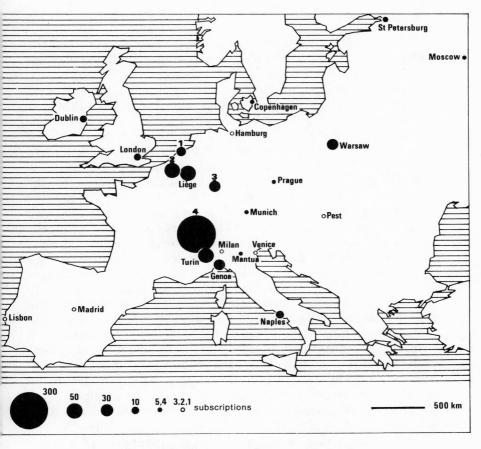

Figure 8. The Diffusion of the Quarto Outside France

This map is based on Duplain's subscription list and the STN's record of the sales that it made from its own stock—the 208 quartos that it received as its share of Panckoucke's 500 (for an explanation of that transaction see Chapter VII). Some of those sales took place after the dissolution of the quarto association in February 1780, when the agreement on maintaining a common wholesale price came to an end. At that time the STN and Panckoucke cut their prices and sold their surplus to other dealers, who in turn were able to act as wholesalers. The map does not, therefore, represent the final destination of all the quartos outside France, and it especially exaggerates the importance of four cities that mainly served as entrepôts; Neuchâtel, Geneva, Brussels, and Liège. In most other cases, however, it gives a fairly accurate indication of the quarto's diffusion throughout the Continent.

The numbers represent areas where the subscriptions were too dense to be shown city by city:
1. The Netherlands: Amsterdam, The Hague, Haarlem, Leiden, Utrecht, Maastricht.
2. The Austrian Netherlands: Brussels, Ypres.
3. The Rhineland: Frankfurt, Homburg, Mannheim, Worms.
4. Switzerland: Basel, Geneva, Lausanne, Neuchâtel, Nyons, Soleure.

the STN finally furnished replacements, they feared that the faulty merchandise had made them lose one of their most important clients.[102]

Christian Rüdiger of Moscow had an equally difficult time with five folio *Encyclopédies,* which he ordered from the STN on March 13, 1777. The STN shipped them on May 31, but they did not arrive until August 1778, owing to the onset of an early winter. They, too, were full of *défets,* which had to be replaced from Paris and did not arrive until late in 1779. But the Parisian supplier (probably Panckoucke) sent the wrong sheets. By the time the STN had straightened out this second snafu, it was well into 1781, and Rüdiger had lost two of his customers. The STN tried to console him by offering the quarto, but he took only four sets. Instead of weighty literature, his customers favored "nouveautés de la dernière fraicheur, surtout aussi dans le genre libre et gaillard, à figures indécentes."[103] He even sent specially to Leipzig in order to have this frothy stuff hauled to him overland by sleigh during winter. Normally, he received his orders by ship, via Frankfurt, Lübeck, and Saint Petersburg. With luck, he could get a letter to Neuchâtel in a month, and it could get its shipment back to him four months later. He paid in bills of exchange on Amsterdam, but his trade suffered badly from a decline in the Russian currency: the value of the ruble dropped from 40 to 36 Dutch florins and the value of the French livre tournois increased from 22 to 30 copecks during the 1780s. Despite all these difficulties, he imported huge quantities of French books. Evidently Diderot and *Thérèse philosophe* helped his customers get through the Moscow winters.

The evenings of Saint Petersburg must have been nearly as long, and many of them were spent in reading, for J.-J. Weitbrecht, the most important bookseller of the city, explained to the STN, "Nous ne vendons guère qu'en hiver, pour lequel nous nous préparons seulement l'été."[104] Weitbrecht served a court clientele that seemed to be eager for fashionable French literature, although it bought only five quartos. Charles Guillaume (or Karl Wilhelm) Muller, another bookseller of Saint Petersburg, dealt primarily in

102. Quotations from Weingand and Köpf to STN, Sept. 17, 1781, and March 6, 1784.
103. Rüdiger to STN, July 12, 1787.
104. Weitbrecht to STN, Nov. 16, 1778.

Russian books, but he ordered a great many Enlightenment works from the STN, including three quartos and some heavy doses of Voltaire and Rousseau. Muller claimed to do business with "tous les lecteurs de ce pay-ci."[105] Although it is impossible to know who they were, they probably included a good many westernized courtiers and foreigners. They needed their books before November, when the ice closed around Saint Petersburg. And they were willing to pay dearly in order to have some contact, through the written word, with the world of the philosophes, for the books' price increased enormously in the course of their long trip. In fact it cost twice as much to send a letter from Neuchâtel to Saint Petersburg (4 livres 10 sous) as to buy an ordinary book in the STN's home office (about 2 livres).

The STN found it almost equally difficult to reach Warsaw, which was closer in space but not in time because shipments normally took three months, going by land via Strasbourg, Ulm, Krems, and Cracow. In Poland, as in Russia, the audience for French books seems to have been aristocratic. Only three booksellers in Warsaw carried French books; two of them, Michel Gröll and Joseph Lex, timed their orders to coincide with the sessions of the Polish Diet, because as Lex explained, "Elle attirera les seigneurs polonais." A Strasbourgeois who had come to Warsaw in search of his fortune in 1771, Lex had learned to profit from the spread of French culture in the east. He had become "connu et bien vu de tout ce qu'il y a de mieux dans cette ville" by selling books to noblemen, while his sister sold dresses to their ladies. Thanks to a *cabinet littéraire,* which he had established with a subsidy from the king, he expected to find several customers for the quarto in this milieu:

J'ouvre le premier du mois prochain un cabinet littéraire pour lequel la plupart des seigneurs de cette ville se sont abonnés. Je profiterai des occasions de leurs visites pour trouver des souscripteurs pour votre *Encyclopédie* dont le prospectus promet de très grands avantages. Mais pour y réussir d'autant mieux, il faudrait que vous eussiez la complaisance de m'en faire passer au moins 3 exemplaires à mesure que l'ouvrage sortira de presse. J'en garderai un pour mon cabinet

---

105. Muller to STN, Aug. 17, 1778. Although the above statements about Muller's book orders sound vague, they and all similar remarks on literary demand are based on a careful reading of voluminous dossiers, and could be supported by statistics drawn up from the STN's account books.

littéraire et trouverai aisément à placer les 2 autres. Par ce moyen j'en vendrai à coup sûr beaucoup, attendu que lorsqu'un seigneur a quelque ouvrage nouveau, les autres veulent aussitôt l'avoir aussi. Mais comme celui-ci est un peu coûteux, je ne vous en demanderai, Messieurs, qu'à mesure que j'en recevrai la commission; et je vous préviens que je ne trouverai à en placer beaucoup que lorsqu'une fois j'en aurai un exemplaire sur les lieux.

By pursuing this strategy, Lex soon sold thirteen quartos— all of them, it seems, to Frenchified aristocrats. One copy went to the Russian ambassador, who required a special binding on which he stamped his coat of arms. For his part, Gröll sold eighteen copies, despite some initial pessimism: "Il n'y a rien à faire avec des grands ouvrages dans ce pays-ci." Thus both booksellers did rather well with the quarto, although they agreed that their clients preferred light literature: "que des choses amusantes et intéressantes, surtout pour les dames, et tout au plus 4 ou 6 exemplaires, peu d'ouvrages profonds et sérieux," according to Lex's formula.[106]

Wolfgang Gerle of Prague also notified the STN that his customers did not care for anything too deep or too long: "Les grands seigneurs qui entendent le français ne se soucient guère de ces sortes d'ouvrages." He ordered five quartos in addition to at least two octavos and a dozen copies of the second folio. But these *Encyclopédies* represented only a tiny proportion of the books that he provided for French readers throughout the Habsburg empire. Unlike the dealers farther to the east, he seemed well informed about the western book trade. He frequented the Leipzig book fair and cultivated several Swiss suppliers, haggling knowledgeably over terms. But his shipments from Neuchâtel often took two months to reach him, traveling via Basel, Schaffhausen, Ulm, and Nürnberg. The War of the Bavarian Succession brought his affairs to a standstill for a while in 1779 and deprived him of two quarto subscribers who were army officers. And he also had difficulties with subscribers, who did not want to pay for the four extra volumes, and with censors, who took their time about letting the quarto through customs. He seemed to operate in the middle range of the international trade, half way between suppliers in the Low Countries and Switzerland, on

106. Quotations from Lex to STN, May 13, 1780; March 31, 1779; and Jan. 19, 1778; Gröll to STN, June 16, 1781; and Lex to STN, March 31, 1779.

the one hand, and consumers in outposts like Moscow and Warsaw, on the other.[107]

The other German dealers also occupied this area, and they traded heavily in *Encyclopédies,* but not in the quarto. The quarto–octavo war had produced two distinct zones of influence in the international market: Panckoucke's consortium concentrated on France, and the sociétés typographiques of Bern and Lausanne worked primarily in *l'Allemagne,* a term that they applied to everything east and north of the Rhine up to the Slavic countries. The Bernois had marketed most of their books in this territory for years. Their director, Pfaehler, had contacts in all the major German cities, including Frankfurt am Main and Leipzig, where he regularly attended the book fairs, and Heidelberg, where his brother ran a book shop. Indeed, the Neuchâtelois thought that the octavo publishers had such a strong hold on the German market that they virtually gave up on it.[108] But when the French market began to run dry, they attempted to cross the Rhine, using Swiss agents such as Serini of Basel, Petitpierre of Basel, and Steiner of Winterthur. These dealers peddled quartos on sales trips, which usually carried them through Frankfurt and Leipzig, but they only managed to pass out prospectuses and to confuse their customers, who could not keep the editions straight. Serini himself got the fake and real versions of the quarto mixed up and offered both of them to his clients along with the octavo, which he apparently favored. The octavo always won. In fact, Serini never thought the quarto would have much of a chance in Germany. He would spread the word in Leipzig and Frankfurt, he wrote to the STN in 1779, "Mais je doute que je puisse placer de l'*Encyclopédie* quarto. L'édition octavo a été annoncée dans tous les coins en Allemagne."[109]

---

107. Gerle to STN, March 7, 1778. Gerle recounted his difficulties with the extra volumes in a letter of Aug. 26, 1780: "Les altercations que j'ai eues avec mes abonnés pour votre édition de l'*Encyclopédie* et leur mécontentement de devoir payer 4 volumes de plus que dans l'édition de Pellet [Gerle was confused on this point] sont allés si loin qu'on a voulu me rendre l'ouvrage entier ou me forcer de perdre ces 4 volumes moi-même." Gerle complained about the censors in letters of July 28 and Nov. 20, 1779, but by April 21, 1781, he was expecting his trade to improve greatly, owing to the reforms of Joseph II.

108. STN to Duplain, April 7, 1779, and STN to Panckoucke, June 24, 1779.

109. Serini to STN, March 27, 1779. See also the discouraging remarks in Serini's letters to the STN of Nov. 22 and 29, 1777: "Cette entreprise [the

When the STN attempted a direct assault on the German market, it failed completely. In the summer of 1779 Bosset spread prospectuses and sales talk everywhere on a tour that took him through Basel, Strasbourg, Rastadt, Mannheim, Frankfurt, Darmstadt, Mainz, Karlsruhe, Hanau, Kleve, Koblenz, Cologne, Bonn, and Düsseldorf. Although several booksellers promised to urge the quarto on their clients, none of them ever placed any orders, evidently because the octavo and the *Encyclopédie d'Yverdon* had satisfied the demand. For example, Mainz, an active and prosperous town, absorbed twenty octavos and not a single quarto.[110] The STN's attempts to woo German booksellers through its correspondence also brought discouraging replies. Joseph Wolff of Augsburg said that he had two unsold sets of the *Encyclopédie d'Yverdon* on his hands and he did not want to waste his money on any more editions ''après qu'il y a déjà tant milles des exemplaires dans le monde.''[111] Dealers in Frankfurt and Cologne found only two customers for the quarto, although they promoted it in the local press. The quarto's biggest success came in Mannheim, where Fontaine sold twenty-seven sets, despite reports of a decline in the demand for French books by one of his colleagues, who could not sell anything in French at the Leipzig fair and who warned the STN, ''L'on a peine de croire comme depuis l'absence de la cour la lecture française a fait place à l'allemande.''[112]

Bruere of Homburg in Hesse disputed this view. The German public wanted to read French literature, he claimed, but German booksellers did not want to provide it:

Les libraires allemands ne voient qu'avec un oeil jaloux la préférence que les gens du bon ton donnent à la littérature française sur la lit-

---

quarto] aurait été excellente, mais elle sera gâtée par deux différentes éditions. La librairie en Suisse vient [*sic*] un vrai brigandage.'' On the STN's dealings with its other agents see STN to Gerle, Dec. 13, 1779; STN to Petitpierre, March 13, 1779; and STN to Steiner, April 24, 1779.

110. Bosset to STN, July 24, 1779. Although the social and cultural history of Mainz has been thoroughly studied, little is known about the reading habits of its citizens during the eighteenth century. See F. G. Dreyfus, *Sociétés et mentalités à Mayence dans la seconde moitié du XVIIIe siècle* (Paris, 1968), pp. 495–497.

111. Wolff to STN, Sept. 27, 1779. Like Wolff, German booksellers usually wrote in legible but ungrammatical French rather than in German.

112. La Nouvelle librairie de la cour et de l'Académie of Mannheim to STN, July 4, 1787. See also Deinet of Frankfurt am Main to STN, June 30, 1779; Hollweg & Laue of Frankfurt to STN, Sept. 23, 1780; and Société typographique de Cologne to STN, Aug. 22, 1782.

térature allemande. De là ils envisagent les ouvrages français comme une branche de commerce étranger, qui nuit au débit de leurs productions nationales. Cela est si vrai que lorsque'ils voient un ouvrage français un peu avantageusement annoncé, ils se hâtent de le faire traduire en allemand, pour affaiblir le débit de l'original.[113]

Bruere even argued that the German dealers would not publicize the quarto because they wanted to favor a German translation of it. A translation did get under way in Frankfurt am Main, but it soon floundered, as Bosset learned during his sales trip of 1779:

J'ai vu . . . M. Warrentrap et Venner, qui ont fait l'entreprise de l'*Encyclopédie* en allemand dont j'ai vu le premier volume sur du papier très mince et un tas de manuscrits pour la suite, à laquelle travaillent, dit-il, une trentaine de savants. Je crois que ce sera un meilleur ouvrage que l'*Encyclopédie* française. Les Allemands sont plus profonds. Mais ce sera les Callendes Grecques, et elle ne fera pas de tort à la nôtre. Plusieurs libraires m'ont assuré qu'elle ne se continue pas.[114]

Although Bruere's remarks about booksellers sounded somewhat too conspiratorial to be believed, they did indicate the character of the potential market for the *Encyclopédie* in Germany. By "les gens du bon ton" he meant the German princelings and the members of their courts. He himself sold a quarto to the Landgrave of Hesse Homburg and to the Prince of Anhall Schaumburg, and he peddled his other books in the "cours du nord."[115]

J.-G. Virchaux of Hamburg served a similar clientele in northern Germany and Scandinavia. When he referred to his customers, he mentioned only "princes souverains" and "seigneurs suédois." He even warned the STN about the effect of its plebeian paper on such personages: "J'aimerais bien que vous prissiez de plus beau papier pour vous éditions . . . C'est en partie ce qui fait vendre; car comme je ne fournis

113. Bruere to STN, Aug. 12, 1779. Although he developed a small book business of his own, Bruere was not a bookseller but a man of letters and a protégé of the Landgrave of Hesse Homburg.
114. Bosset to STN, July 24, 1779. In a letter to Bruere of Aug. 19, 1779, the STN said that it considered the German edition to be economically unfeasible, owing to the high cost of paper in Germany as well as the cost of the translation.
115. Bruere to STN, Feb. 12, 1781. Bruere also took out at least a dozen subscriptions to the octavo *Encyclopédie*, and in a letter of Oct. 19, 1780, he said that he bought books for the libraries of the Landgrave of Hesse Cassel, the Count of Bückburg, and the Prince of Phillipsthal.

absolument que les souverains du nord et les plus grands seigneurs chalands . . . vous sentez bien, Messieurs, qu'il ne m'est pas à propos d'avoir des éditions peu riantes. Ne pourriez-vous pas tirer du papier de Périgord?'' Although these remarks may have been calculated to impress a supplier in order to get shipments on credit, Virchaux had nothing flattering to say about his customers' tastes when it came to the contents as opposed to the appearance of books. They liked superficial literature, he explained, and especially pornography—above all, if it had illustrations. Books on politics, natural history, and a few other scientific subjects did fairly well, but not those on theology and law. He therefore had little hope for the *Encyclopédie* in his territory: ''Les grands et volumineux ouvrages étant du débit le plus ingrat dans les pays du nord, nous ne pouvons pas, Messieurs, nous charger de l'*Encyclopédie* in-quarto . . . Nous avons deux exemplaires de celle d'Yverdon, qui nous sont fort à charge.'' In the end, he sold only two copies—one went to a ''très grand seigneur très riche''—and he attributed his lack of success to the American war as well as the frivolity of his public: ''Nous ferons tout notre possible pour faire connaître votre *Encyclopédie:* mais tant que la guerre et les affaires politiques fixeront les esprits, il n'y a guère d'apparence qu'on en vende beaucoup; et en général les ouvrages d'aussi longue haleine sont peu recherchés à présent.''[116]

Claude Philibert of Copenhagen indicated the *Encyclopédie* was equally unsuited for his clientele, although he sold several octavos and a dozen copies of the *Encyclopédie d'Yverdon.* He reprinted the STN's prospectus, but it brought in only two sales: ''Jusques ici je n'ai que des espérances, car on aime très peu souscrire ici et très peu ou point donner des avances, pas même un sol. Le jeu et les plaisirs vont avant tout.''[117] The STN also sold a copy in Copenhagen through Johann Heinrich Schlegel, who had interceded in an unsuccessful attempt to persuade the king and queen of Denmark to accept the

116. Quotations from Virchaux's letters to the STN of Jan. 9, 1779; Nov. 25, 1780; Dec. 19, 1778; Sept. 4, 1779; Aug. 2, 1780; and May 26, 1780.

117. Philibert to STN, June 24, 1777. Two years later, in a letter of June 29, 1779, Philibert reported, ''Ce n'est pas ma faute si je n'ai pas eu des souscriptions pour votre édition de l'*Encyclopédie.* C'est assez annoncée, mais ici tout va lentement et l'on veut voir avant que de se résoudre à acheter. Aussi il est inutile de proposer des souscriptions. Je comptais pourtant bien d'en avoir quelques unes, mais l'édition in-octavo est venue la traverser.''

dedication of the quarto and had helped the publishers of the Leghorn edition to sell thirty subscriptions. After taking soundings, however, Schlegel wrote that he could do no more: the Danish market for the *Encyclopédie* was exhausted.[118]

Reports about the demand for *Encyclopédies* in London sounded like those from other parts of northern Europe. After making the rounds of the book shops and placing notices in the *Morning Herald,* Jean-Baptiste d'Arnal warned the STN not to expect many sales in England: "Je vous réitère au reste, Messieurs, que le zèle pour la littérature se refroidit tous les jours, que les esprits anglais se dégénèrent, et que nos seigneurs aiment mieux acheter 20 billets d'opéra à un Vestris, à un Noverre, à une Allegrante que de dépenser 20 guinées en livres."[119] But D. H. Durand, a Swiss pastor in London, sold thirteen quartos, despite competition from the distributors of the Leghorn edition;[120] and more quartos probably reached England through Panckoucke, who had far better contacts with the London booksellers than did the STN. So the English market probably was not as bad as it appeared to be in d'Arnal's letters. Getting the books across the Channel did not prove to be an insurmountable problem, despite the war (France was joined by Spain in June 1779 and by The Netherlands in December 1780 in the coalition against Great Britain). The STN sent its *Encyclopédies* to London via Frederick Romberg Co. of Ostend, which used ships with double passports.[121] It also sent a dozen quartos from Ostend to Ireland, and they arrived without mishap, although they took eight months to get there, as Romberg had trouble finding a safe neutral ship. They went to the book store of Luke White in Dublin along with works by Rousseau, Voltaire, and Buffon and *La vie privée de Louis XV.* When the STN offered him Raynal's *Histoire philosophique,* White refused, explaining that he preferred an Irish edition. The French Enlightenment clearly had spread across the British Isles.

Most of it did not come from France, however, but from The Netherlands, where philosophic literature was printed and

---

118. Schlegel to STN, May 30 and July 12, 1777.
119. D'Arnal to STN, Feb. 5, 1782.
120. In a letter of Dec. 4, 1777, Durand said the Leghorn edition was preferred, even though it cost 30 guineas, because it contained a full set of plates.
121. Bosset, from Amsterdam, to STN, Sept. 3, 1779: "Ils [Romberg frères] ont obtenu un passeport des deux nations, et il n'y a pas de risque de tout." Presumably Bosset was referring to the Habsburg and British authorities.

reprinted in even greater quantities than in Switzerland. For precisely that reason, the Dutch market did not absorb many quartos. Bosset tried to crack it during his trip of 1779, but the Dutch treated him as a competitor and an enemy, and in any case they had already bought too many other *Encyclo-pédies* to have any interest in his. Rousseau's publisher in Amsterdam, Marc Michel Rey, was willing to dine with Bosset but not to do business with him. Rey accused the STN of pirating his editions of Rousseau (it was secretly preparing a new edition at that very moment), and he said that he still had not sold twenty-five sets of the second folio *Encyclopédie,* half the number for which he had subscribed almost ten years earlier. Harreveld, another of the great Amsterdam dealers, claimed that he had sold fifty sets of the other editions, mainly the *Encyclopédie d'Yverdon,* and could sell no more. Pierre Gosse Junior of The Hague, who had bought up all of Félice's *Encyclopédie,* had marketed it so aggressively in the Low Countries that the smaller booksellers would not buy the rival editions, for fear of antagonizing him—or so they told Bosset. He did not know what to believe in such hostile territory and moved on to the Austrian Netherlands, having sold only a few quartos among the highly literate and highly competitive Dutch.[122] South of the Rhine, he found another world. The best bookseller in Antwerp, which had dominated the publishing industry in northern Europe during the late sixteenth century, was a certain "Mme. Moredon, qui roule carosse en ne vendant que bréviaires."[123] The book trade hardly existed in Louvain, despite the presence of 3,000 university students. It flourished in Liège and Maestricht, important centers of pirate publishing, but they could not absorb any more quartos because Plomteux of Liège, the STN's partner in the quarto enterprise, had satiated the whole region.

The letters that the STN received from its correspondents in the Low Countries generally confirmed Bosset's reports. Gosse warned that he would defend the *Encyclopédie d'Yverdon* until he had sold the last set,[124] and that moment seemed

---

122. Bosset reported his dealings with the Dutch booksellers in letters of Aug. 30, Sept. 3, and Sept. 7, 1779.

123. Bosset to STN, Sept. 13, 1779.

124. Although Félice's *Encyclopédie* was foundering on the Dutch market, Pierre Gosse Junior wrote bravely to the STN on Sept. 5, 1777, "Quant à votre *Encyclopédie abrégée* [a backhanded reference to Laserre's cuts] *in-quarto,* vous

far away because several booksellers reported that its over-
abundance had destroyed the market for the other *Encyclopédies* and that it was selling at two-thirds and even half its
subscription price. Murray frères of The Hague did not want
to hear about the quarto because they were "embarrassés de
l'édition d'Yverdon, de l'in-folio, et même de l'in-octavo."[125]
Changuion of Amsterdam reported that advertisements for
the quarto in the local press had produced no response and
that "le public se lasse enfin de toutes ces duperies."[126] Luzac
of Leyden said he would continue to run notices in the *Gazette
de Leyde* if the STN wanted, but he doubted that they would
do any good, "vu que le public est inondé d'autres éditions
et que ce livre se vend aujourd'hui à moitié prix."[127] And
Aubertin of Rotterdam warned that the quarto would not sell
in his town: 'J'ai trouvé que les amateurs avaient ou l'édition
de Paris ou celle d'Yverdon et que dans une ville de commerce
ces ouvrages ne sont pas d'un goût général et ainsi d'un débit
fort borné."[128] Aubertin's comments echoed the reports of
disinterest in literature among merchants in France. But the
main problem in the Dutch market was excess supply not lack
of demand. "Les provinces sont tellement farcies d'*Encyclopédies* que les libraires non plus que les particuliers ne veulent
en entendre parler,"[129] Aubertin wrote. "A chaque vente de
livres où il s'en trouvent, les prix déclinent. La Société typographique de Berne m'a encore envoyé des prospectus de sa
seconde édition. Je les ai répandus sans le moindre succès."

---

n'ignorez point que je suis intéressé à l'*Encyclopédie in-quarto d'Yverdon*, et
comme elle est toujours très accueillie, que je la crois préférable à toute autre, je
ne me chargerai d'aucune avant que le peu d'exemplaires qui en restent, dont
mon cher père est seul possesseur, soient placés et que je n'en pourrai plus
acquérir."
125. Murray frères to STN, Dec. 8, 1780.
126. Changuion to STN, Dec. 24, 1781.
127. Elias Luzac & van Damme to STN, June 2, 1780.
128. Aubertin to STN, Nov. 12, 1779.
129. Aubertin to STN, Sept. 22, 1780. A few years earlier, however, the Low
Countries had still looked like good territory for *Encyclopédies*. Chatelain et fils
of Amsterdam, who had sold over 60 sets of the first edition, informed Dessaint of
Paris on June 29, 1769, that the Dutch market was ripe for another one: Archives
de Paris 5AZ 2009. And in the first circular letter about the *Encyclopédie méthodique* (Jan. 1778), Deveria of Liège claimed that the demand continued to be
strong in cities such as Ghent, where the bookseller Gimblet had sold more than
250 sets of the earlier editions. Amsterdam, Bibliotheek van de vereeniging ter
bevordering van de belangen des boekhandels, Dossier Marc Michel Rey.

In the diffusion of the *Encyclopédie,* Holland became a sort of burned-over district, the very opposite of Hungary, Poland, and Russia.

The Low Countries also served as an entrepôt of *Encyclopédies* for Portugal and Spain, although the STN used other routes as well—mainly the Trans-Alpine roads to Turin and Genoa and the river route from Lyons to Marseilles. Veuve Bertrand et fils of Lisbon found that one crate sent via Genoa took eleven months, one via Amsterdam took six months, and one via Ostend never arrived at all. Jacques Mallet of Valencia recommended that the STN use small boats out of Marseilles that could hug the coast and dart into port whenever English frigates appeared. He distrusted the neutral ships from Genoa, "car les Anglais ne respectent aucun pavillon et prennent tout."[130] Duplain and the STN insured their shipments, but the costs and delays damaged their business badly until the end of the American war. Antonio de Sancha of Madrid reported that a load of books sent via Amsterdam and Cadiz took a half year and that its handling costs came to more than its wholesale value. Other shipments took longer and cost more: their outcome could never be predicted, because the long-distance trade remained unstable until the peace settlement of 1783.

Even after normal traffic resumed on the routes to Spain and Portugal, the publishers still had to grapple with a greater problem: the Inquisition. Bertrand and another foreign bookseller in Lisbon, Jean-Baptiste Reycends, smuggled a few quartos and octavos into their shops, along with some *livres philosophiques* that were also taboo in France. But they never dealt extensively in the illegal trade, at least not with the STN, although they believed that the demand for foreign books was growing in Portugal.[131] The Spanish booksellers always spoke of the Inquisition with fear and trembling, even when they managed to negotiate successfully with it. Sancha of Madrid told the STN that he could not resist making a small order

130. Mallet to STN, Oct. 4, 1777.
131. On Feb. 8, 1780, Reycends wrote to the STN in an awkward French typical of foreign bookdealers, "Comme ici depuis la réforme de l'Université de Coïmbre ainsi que l'ouverture d'une Académie des sciences et des arts en cette ville les Portugais commencent d'avoir un peu plus de goût pour la lecture des bons livres, tant Latins que français et autres langues, c'est ce qui nous engage de procurer d'établir des correspondances dans toutes les villes principales de l'Europe, par cet moyen pour être mieux assortis."

after receiving its prospectus for the quarto, but "ça est une affaire si délicate dans ce pays-ci que 'de parler d'*Encyclopédie* par rapport à notre Inquisition qu'il m'a fallu une permission dudit tribunal pour pouvoir me souscrire à 3 exemplaires.'' When the STN tried to take advantage of this small opening by offering all the other books in its stock, Sancha replied that he could not take anything except a dozen copies of Robertson's *Histoire de l'Amérique* (a best seller everywhere, owing to the American war). The other books were not "convenables pour un pays si délicat comme le nôtre.'' ''Dans ce pays il faut avoir de grands soins sur les livres étrangers par rapport à notre Inquisition.'' In the end, he could not even steer the Robertson past the authorities: ''L'on a donné dernièrement un ordre à tous les ports de mer de ne pas les laisser entrer. Voyez si nous sommes dans un pays bien délicat.'' [132] Sancha, who dealt with suppliers from all over western Europe and had fifteen presses of his own, was probably the most important bookseller of Spain. Lesser houses, like Paul et Bertrand Caris of Cadiz and Jacques Mallet of Valencia, did not dare to take the slightest risk. Mallet thought he might be able to get the Inquisition to allow him to import a few quartos, but he never did.[133] And Caris considered the Spanish market hopeless: ''Quant aux ouvrages qui seraient du goût de ce pays, ils doivent être si épurés que la moindre proposition un peu équivoque ou philosophique serait arrêtée par le Saint Office. Voilà où en est réduit la littérature de ce pays.''[134]

Despite the influence of the Catholic church, the *Encyclopédie* trade in Italy resembled the Dutch rather than the Spanish variety. Only once in the hundreds of letters that the STN received from the Italian states did it hear that the *Encyclopédie* had encountered difficulties with the authorities, and on that occasion the alert came from territory that was

132. Sancha to STN, Feb. 26, 1778; April 2, 1778; Jan. 14, 1779; and July 12, 1779.

133. Mallet to STN, Oct. 4, 1777: ''Comme l'Inquisition en a prohibé partie, où sont tous les articles qui concernent la religion, nous ne pouvons les vendre sans permission, et dans ce cas, je vous la demanderai.'' The permission never came through.

134. Caris to STN, Jan. 14, 1774. A few years later, one of the Caris, who had retired to Avignon, tried unsuccessfully to ship some Voltaire and an *Encyclopédie d'Yverdon* to the shop in Cadiz. They were all sent back, he informed the STN in a letter of Aug. 18, 1780, ''ne pouvant être mis en usage par l'arrêt fulminant de l'Inquisition, aussi bête que méchant.''

eventually to be French—namely Savoy.[135] The Neuchâtelois even corresponded about the quarto with a liberal cardinal in Rome.[136] Of course Rome had condemned the book in no uncertain terms in 1759, but by 1777 it had also condemned the Jesuits, who had been the greatest enemy of the *Encyclopédie*, and it had not taken effective action against the two Italian editions. When the publishers of the Leghorn edition sent a traveling salesman around Italy, they chose a priest and gave him a 10 percent commission on every sale.[137] And when the Italian booksellers discussed their misgivings about the text, they never mentioned its impieties but only its inaccurate account of Italian geography and natural history. In the Italian prospectus for the quarto and the octavo, the Società letteraria e tipografica di Napoli invited all Italians to send in corrections, which it would forward to the publishers, so that their part of the world would get a decent treatment in the new editions. The Neapolitans also informed the STN that they had two men working hard on certain key articles, such as NAPLES and VESUVE, which needed to be rewritten in order to avoid alienating their customers.[138] None of these editorial projects ever came to anything, but they illustrate the attitude toward the *Encyclopédie* that existed everywhere among eighteenth-century booksellers: Diderot had made a good beginning, but his work needed a complete overhaul before its full market value could be realized.[139]

Operating from this premise the publishers of the Lucca and Leghorn editions had not hesitated to adapt Diderot's text to their own needs. The Lucchese had needed above all to appease the pope and therefore had presented their work as a

135. J. Lullin of Chambéry to STN, May 8, 1777: "Par ce qui est de la nouvelle *Encyclopédie* . . . nos censeurs ne veulent pas m'en permettre l'introduction; ils me font difficulté sur tout."
136. STN to Cardinal Valenti, July 12, 1779.
137. Gentil et Orr of Leghorn to STN, March 6, 1775.
138. Società letteraria e tipografica di Napoli to STN, Feb. 17, 1778, including a copy of the prospectus.
139. Società letteraria e tipografica di Napoli to STN, Feb. 17, 1778: "Nell' *Enciclopedia* francese ci sono moltissime cose inette, inutile e mal digerite che si vorrebbero sopprimere o riformare." The Neapolitans also recommended the improvements in the Leghorn edition to the publishers of the quarto: "Nell'edizione di Livorno ci sono molte cose che meritano di essere adattate, fra l'altro un bell'articolo di *Caserta* che fu fatto in Napoli." The Lausannois wooed the Neapolitans by promising to incorporate all their revisions in the octavo. In the end, however, the articles on Italy in the octavo merely reproduced the text of the quarto.

refutation as well as a reprint of Diderot's: they would confound the French heresies with a running commentary, to be written in the form of notes by a learned priest, Giovanni-Domenico Mansi. The notes, however, became thinner as the edition progressed because Clement XIII failed to execute his threats against the Lucca *Encyclopédie* and because Clement XIV became more absorbed in the dissolution of the Jesuits than in the campaign against their *bête noire*. By 1771 the publishers finished printing the seventeen folio volumes of the text and, it seems, they had sold out all or most of the edition, mainly in Italy.[140] The Leghorn edition appeared later and under more favorable circumstances, for Lucca was a tiny, frail republic, while Leghorn served as the main port of the powerful archduchy of Tuscany, and the Archduke Leopold, one of the few genuinely enlightened autocrats of the Continent, accepted the dedication of the book and protected the enterprise. The publishers claimed to have 600 subscribers in November 1769, when they were about to begin printing with a new font of Caslow type and six presses, which could turn out ninety sheets a month. They said that they had sold out all 1,500 copies by September 1775,[141] when they finished the text. But in July 1777, Giuseppe Aubert, the publisher of the Milanese philosophes and the main figure behind the Leghorn

---

140. Salvatore Bongi, "*L'Enciclopedia* in Lucca," *Archivio storico italiano*, 3d ser., XVIII (1873), 64–90. This article provides valuable information about Ottaviano Diodati, the main entrepreneur of the Lucca edition, and his relations with the Senate in Lucca as well as the papacy, but it has little to say about the diffusion of the Lucchese edition. In "*L'Encyclopédie* et son rayonnement en Italie," *Cahiers de l'Association internationale des études françaises*, no. 5 (July 1953), 16, Franco Venturi asserts that the edition sold out quickly and that it went especially into Italian libraries. John Lough, however, claims that the enterprise did not flourish, owing to difficulties in production and slow sales. See Lough, *Essays on the "Encyclopédie" of Diderot and d'Alembert* (London, New York, and Toronto, 1968), pp. 22–23. In fact, little is known about the Italian editions. Some information about the Leghorn edition can be gleaned from the correspondence of its principal entrepreneur, Giuseppe Aubert, which has been published by Adriana Lay: *Un editore illumista: Giuseppe Aubert nel carteggio con Beccaria e Verri* (Turin, 1973). Aubert's correspondence is the main source for Ettore Levi-Malvano, "Les éditions toscanes de l'*Encyclopédie*," *Revue de littérature comparée*, III (April–June 1923), 213–256. Levi-Malvano argues that both Italian editions were commercial successes, and he produces some information about the subscriptions to the Leghorn edition. They came to 600 by December 1770, of which at least 8 were in Rome, 20 in Parma, many more in Florence and Milan, and 20 in France. He also mentions four projects to translate the *Encyclopédie* into Italian and shows how the Grand Duke of Tuscany favored the Leghorn publishers.

141. Marc Coltellini of Leghorn to STN, Nov. 18, 1769.

edition, refused to handle the quarto on the grounds that he still needed to dispose of sixty unsold sets of his own *Encyclopédie*.[142] The quarto publishers expected to undersell the Leghorn *Encyclopédie* in Italy itself,[143] and they were undersold by the octavo group. So the entrepreneurs of Leghorn may have had some difficulties in their home market, despite their head start in the race to supply it.

Actually, Italy seems to have contained several different markets, just as it was composed of several different states, for the reports from the Italian booksellers varied enormously. After distributing fifty copies of the prospectus, Yves Gravier of Genoa expected to reap a rich harvest of quarto subscriptions; he did order thirty-two copies, although he eventually quarreled with Duplain and aligned himself with the octavo publishers.[144] The quarto did quite well in Piedmont. Reycends frères of Turin (cousins of Jean-Baptiste Reycends of Lisbon) sold thirty-nine sets and had not exhausted the demand by 1782, when they switched to the octavo.[145] Other booksellers in Turin ordered fourteen more sets, although they, too, preferred the octavo, as its lower price attracted more customers.[146] Giuseppe Rondi could not sell any quartos in Bergamo, and Lorenzo Manini only sold four in Cremona, but Lombardy proved to be good *Encyclopédie* territory. Not for the quarto, however; Ami Bonnet spread its prospectuses around Milan without success, and Giuseppe Galeazzi warned the Neuchâtelois that they had arrived too late on the Milanese market: he had already ordered thirty-nine octavos, and the region was full of *Encyclopédies*.[147] Tuscany seemed to be fuller. Despite a good deal of prodding

142. Aubert to STN, July 21, 1777. The reference to the 1,500 copies and the details on the type and presses come from a circular letter sent to the STN on March 6, 1775, by Gentil et Orr of Leghorn. The circular was an appeal for printing commissions, but as it also said that the publishers would welcome offers to buy their entire shop, it suggested that their business was not flourishing.

143. STN to Vincenzo of Bergamo, Sept. 30, 1779.

144. Gravier to STN, June 6, 1777; Aug. 2, 1777; March 6, 1779; and July 15, 1780. Gravier was an important wholesaler and retailer who continued to glean orders for the *Encyclopédie* until 1785. He supplied one copy to a customer in Corsica and slipped another into Spain.

145. Reycends frères to STN, Jan. 9, 1777.

146. Giraud et Giovine to STN, Jan. 31, 1778.

147. Galeazzi to STN, Sept. 25, 1779: ''Non mi conviene assolutamente caricarmi nè meno d'una copia dell' *Enciclopedia* in-quarto, poichè il paese nostro è già pieno, oltre di che quella di Losanna essendo di minore spesa molti si sono appigliati a quella.'' See also Galeazzi to STN, April 17, 1779.

from the publisher, Joseph Bouchard of Florence failed to sell a single quarto. "Ce pays-ci est rempli d'*Encyclopédies* des réimpressions de Lucques et de Livourne; les gens s'en tiennent là,"[148] he informed the STN. The booksellers of Venice, who had once planned to produce their own reprint of the *Encyclopédie*, did not give the quarto a good reception, either, perhaps for the same reason. However, Gaspard Storti, a Venetian bookseller who sent some frank appraisals of the local market to the STN, explained that "ceux qui lisent en français dans ce pays sont fort rares, et ils n'aiment ou ne peuvent faire de la dépense."[149] The STN did not have extensive contacts in Rome, nor did the other quarto publishers, apparently, for they failed to sell a single *Encyclopédie* there.[150] Naples promised to be an abundant market for the quarto, according to the Società letteraria e tipografica di Napoli. But after subscribing for sixteen quartos, the Società quarreled with Duplain and went over to the octavo publishers, from whom it eventually bought fifty sets, while stocking up on the works of Voltaire and Rousseau.[151] Apparently the land of Giannone, Vico, and Filangieri was as rich in enlightened readers as the north, where Beccaria and the Verri brothers had flourished. In general, and despite the unevenness of its cultural and political geography, Italy seems to have been a great market for the literature of the Enlightenment— not a priest-ridden backwater but a complex and cultivated

148. Bouchard to STN, June 26, 1781. Bouchard had assessed the market in the same way four years earlier, in a letter of May 27, 1777: "Pour ce qui est du manifeste de l'*Encyclopédie*, M. Duplain de Lyon me l'a participé [*sic*] il y a plus de deux mois. Ici on l'a à tout prix, vu l'édition de Lucques et la réimpression de Livourne, ce qui a rempli l'Italie." He made the same observation in letters of March 3, 1778; December 15, 1778; and March 2, 1779. Giuseppe Pagani e figlio of Florence responded enthusiastically to the prospectus, but they, too, failed to sell any quartos.

149. Storti to STN, Aug. 19, 1780.

150. Francesco Poggiali, a Roman bookseller, informed the STN that he was trying hard to sell the quarto, but his customers had been put off by a rumor that an Italian translation was being produced in Siena. Poggiali to STN, Dec. 29, 1779: "Vi devo per allora avvertire che nella città di Siena è stata posta sotto i torchi la traduzione italiana dell'*Enciclopedia* e questa farà un poco d'incaglio alla vostra." In fact, the Italian translation, like the German one, never came to anything.

151. As in many other cases, Duplain got the Neapolitans' orders wrong, failed to answer their letters, and finally responded with threats and insults when they protested at the treatment they received. "Il procedere di MM. Duplain et Comp. ci ha disgustati all'ultimo grado," the Società letteraria e tipografica di Napoli complained to the STN in a letter of Oct. 12, 1778.

land, which is only beginning to get its due in scholarship on the eighteenth century.[152]

The demand for *Encyclopédies* certainly extended beyond Europe, although it is impossible to estimate the number of sets that reached other continents. Genevan dealer Etienne Pestre informed the STN that he had bought two quartos for customers in Africa, and Meuron of Saint-Sulpice ordered one for the Cape of Good Hope.[153] The STN even consulted Benjamin Franklin about establishing an entrepot in America. That project never came to anything, nor did a similar plan that Thomas Jefferson proposed for marketing the *Encyclopédie méthodique*.[154] But Jefferson did a good deal to promote the diffusion of the *Encyclopédie* in the new republic. After reading an advertisement in the *Virginia Gazette* during a critical moment of the American Revolution, he bought a set of the Lucca edition for 15,068 pounds of tobacco. He intended it to be "for the use of the Public" and issued special orders for its protection against Cornwallis's army.[155] At the same time, he tried to find a copy for himself: "I am exceedingly anxious to get a copy of Le grande Encyclopedie, but am really frightened from attempting it thro' the mercantile channel, dear as it is originally and loaded as it would come with the enormous advance which they lay on under pretext of insurance out and in."[156] He apparently did not buy one until he succeeded Franklin as American minister to France. His residence in Paris became a clearing-house for French literature bound for America, and his correspondence between 1784 and 1789 provides some revealing glimpses of the French book trade. In 1786, for example, he informed Madison about the state of the *Encyclopédie* market: "I have purchased little

152. For a full account of Italy's intellectual life in the eighteenth century see Franco Venturi, *Illuministi italiani* (Milan, 1958——).

153. Pestre to STN, Dec. 20, 1777, and Meuron to STN, Jan. 2, 1783.

154. Ostervald and Bosset alluded to their negotiations with Franklin in a letter to the STN from Paris of April 14, 1780. Later, in 1783, they consulted Franklin about establishing an outlet for their books in America, but these discussions, conducted for the STN in Paris by the abbé Morellet, did not lead to anything. See Morellet to STN, March 25 and May 31, 1783, and a note from Morellet dated only "25"—evidently Feb. 25, 1783.

155. Jefferson to John Fitzgerald, Feb. 27, 1781; Jefferson to James Hunter, May 28, 1781; and Amable and Alexander Lory to Jefferson, Dec. 16, 1780, in *The Papers of Thomas Jefferson,* ed. Julian P. Boyd (Princeton, 1950——), V, 15, 311–12; VI, 25; IV, 211.

156. Jefferson to Charles-François d'Anmours, Nov. 30, 1780, ibid., IV, 168.

for you in the book way since I sent the catalogue of my former purchases . . . I can get for you the original Paris edition in folio of the Encyclopedie for 620 livres, 35 vols: a good edition in 39 vols. 4to, for 380 [livres] and a good one in 39 vols. 8vo for 280 [livres]. The new one [the *Méthodique*] will be superior in far the greater number of articles: but not in all. And the possession of the ancient one has more over the advantage of supplying present use. I have bought one for myself, but wait your orders as to you."[157] Madison chose to subscribe to the *Méthodique.* Jefferson did, too, but he also bought a set of the octavo edition and studied both texts whenever his curiosity compelled him down some new path in the arts and sciences.[158]

Reading

The booksellers' letters make it clear that the *Encyclopédie* reached every corner of the Continent and even crossed the ocean. It sold better in some places and in some editions than in others, but it sold everywhere, on the Russian tundra and the Turkish frontier as well as in all the major cities of the west. Unfortunately, however, the letters do not reveal much about the last stage in the life cycle of the book. Between the selling and reading of the *Encyclopédie* there exists a gap that cannot be bridged by publishing history, for it is impossible to know what went on in the minds of the readers. Some sets may never have been read at all, although most of them probably had many readers—members of *cabinets littéraires,* friends of the subscribers, and even domestic servants. Books seem to have been borrowed more extensively in

157. Jefferson to Madison, Feb. 8, 1786, ibid., IX, 265. Madison replied on May 12, 1786 (IX, 518): "A copy of the Old edition of the Encyclopedie is desireable for the reasons you mention, but as I should gratify my desire in this particular at the expence of something else which I can less dispense with, I must content myself with the new Edition for the present."

158. Jefferson bought three sets of the octavo for 528 livres, a bargain price. He apparently purchased two of them for friends. Jefferson to William Short, April 27, 1790, and Jefferson to Martha Jefferson Randolph, Jan. 20, 1791, ibid., XVI, 388; XVIII, 579. As an example of Jefferson's comparative reading of the texts see Jefferson to David Rittenhouse, June 30, 1790, ibid., XVI, 587. And on Jefferson's library see E. Millicent Sowerby, *Catalogue of the Library of Thomas Jefferson* (Washington, 1952–59), 5 vols., which shows that Jefferson owned Ephraim Chambers's *Cyclopaedia* as well as an octavo edition of Diderot's *Encyclopédie* and a nearly complete set of the *Encyclopédie méthodique.*

the eighteenth century than today, and reading may have been a different experience—less hurried, more reflective, altogether an absorbing activity in an age when men of property were men of leisure and other media did not compete with the book. Of course one can only speculate on the nature of reading in the Old Regime. Its inner workings still seem mysterious, despite the attempts of psychologists, efficiency experts, and professors of reading to decode it, speed it up, and expound it in lecture courses.[159] Nonetheless, there is some testimony from the eighteenth century about how people read books. For example, a German visitor in Paris observed: "Tout le monde lit à Paris . . . On lit en voiture à la promenade, au théâtre dans les entractes, au café, au bain. Dans les boutiques, femmes, enfants, ouvriers, apprentis lisent; le dimanche, les gens qui s'associent à la porte de leur maison lisent, les laquais lisent derrière les voitures, les cochers lisent sur leurs sièges, les soldats lisent au poste et les commissaires à leur station."[160]

Of course it does not follow that the *Encyclopédie* was read in this omniverous manner. Its size and alphabetical organization precluded cover-to-cover reading, and it probably was consulted in different ways and for different purposes—for information on a specific topic, for amusement, for systematic study, and for the thrill of discovering audacious asides that should have been caught by the censor. From the viewpoint of intellectual history, the main problem could be formulated once again as follows: did the readers of the *Encyclopédie* use it as a reference work, or did they turn to it for *philosophie*?

It must be admitted at the outset that they may not have turned to it at all. Samuel Girardet, a bookseller in the small

159. As an example of the psychological and sociological literature on reading see Douglas Waples, *What Reading Does to People* (Chicago, 1940). More recent work, including research in communication theory, does not seem to have advanced the subject very far, at least not to an outsider. But literary scholars have produced some suggestive hypotheses. See Stanley Fish, *Self-Consuming Artifacts* (Berkeley, 1972); Walter Ong, "The Writer's Audience Is Always a Fiction," *PMLA*, XC (1975), 9–21; and Ronald C. Rosbottom, "A Matter of Competence: The Relationship Between Reading and Novel-Making in Eighteenth-Century France," *Studies in Eighteenth-Century Culture*, VI (1977), 245–263.

160. Heinrich Friedrich von Storch, quoted in Jean Paul Belin, *Le mouvement philosophique de 1748 à 1789* (Paris, 1913), p. 370. These remarks, which have been handed on from historian to historian, should not be taken literally.

Swiss city of Le Locle, did not have a high opinion of his customers' interest in the quarto:

Je prévois que l'*Encyclopédie* de Genève sera un livre bien ennuyeux, vu que voici déjà près de cinq mois suivant la promesse que m'avez faite de son apparition, temps depuis lequel il ennuie déjà son acquéreur, qui n'en a encore vu aucune feuille. Que ne l'ennuiera-t-il pas quand il possédera tout ce fatras de livre, qui apparamment renchérira par la sagacité de son contenu sur le bel ouvrage d'Yverdon, que nos montagnards révèrent au point de le laisser tranquille et en repos sur le tabelar [*sic*], où ils l'ont posé dès son apparition.[161]

Panckoucke maintained that "l'*Encyclopédie* sera toujours le premier livre de toute bibliothèque ou cabinet,"[162] but it could have been a book to display rather than read. In fact, Panckoucke heard that some subscribers in Lyons were illiterate.[163] If they only used their quartos to impress visitors, however, their behavior suggests the importance rather than the ineffectiveness of the book, for it seems significant that an *Encyclopédie* on the shelf could convey prestige, like a fake coat of arms or an artificial *particule*. Perhaps by 1780 prestige had shifted to the Enlightenment, and a new phenomenon, intellectual snobbery, had been born.

In any case, the well-displayed *Encyclopédie* must have proclaimed its owner's progressive opinions as well as his learning, because no one in the eighteenth century could have ignored the notoriously ideological character of the book. All contemporary writing about it—from the *Discours préliminaire* to the attacks against it and the publicity in favor of it—stressed its identification with the Enlightenment. And insofar as there is any evidence about the response of the readers, it indicates that they looked for *philosophie* as well as information in the text. In an "Avis" at the beginning of volume 13, the quarto publishers printed an excerpt from a letter sent to them by a French doctor, who objected to some cuts in the article ACÉMELLA: "Je n'y ai pas trouvé les réflexions sur les propriétés des plantes qui terminent le dernier paragraphe du Dictionnaire. Cependant ces réflexions sont philosophiques et intéressantes." The editors replied that the "observations

161. Girardet to STN, Jan. 27, 1778.
162. Panckoucke to STN, Aug. 4, 1776.
163. Panckoucke to STN from Lyons, Oct. 9, 1777: "La faveur du public est sans exemple. Des gens qui ne savent pas lire ont ici souscrit."

philosophiques'' would come later, in the article PLANTES. They clearly could not afford to leave out any *philosophie*. A book-dealer in Loudun also wrote to them that his subscribers had complained about ''divers articles de théologie traités trop dans le goût Sorbonnique, sans doute pour favoriser d'autant mieux sa circulation en France, mais ces entraves à la liberté de penser ne plaisent pas à tous les lecteurs.''[164]

The attraction of the book's *philosophie* for eighteenth-century readers also seems clear from a project of some French expatriates in London, who announced that they would compile all the *morceaux philosophiques* that Félice had cut out of Diderot's text and would publish them as a supplement to the *Encyclopédie d'Yverdon*. In this way, they argued, Félice's subscribers could recuperate the most valuable part of the *Encyclopédie* without having to purchase an entire new set from the other Swiss publishers. Actually, this plan seems to have been a maneuver by Pierre Gosse Junior of The Hague to save his unsellable surplus of Yverdon *Encyclopédies;* and like so many publishers' projects, it never was executed. But it suggests why Félice's *Encyclopédie* floundered, while Diderot's went through edition after edition: ''L'éditeur ne s'est pas conformé aux diverses façons de penser des lecteurs,''[165] the announcement explained—that is, the readers wanted *philosophie,* and Félice had merely given them information.

It is difficult to believe that eighteenth-century readers did not seek information in the *Encyclopédie,* but it would be anachronistic to assume that they used it in the same way that modern readers use modern encyclopedias. Diderot and d'Alembert meant to inform and to enlighten at the same time. Their basic strategy was to argue that knowledge had to be philosophic in order to be legitimate; by seeming merely to purvey knowledge, they struck against superstition. To distinguish between the informative and the philosophic aspects of the *Encyclopédie* is to separate what the authors meant to be inseparable and misconstrue the meaning of the book for their readers. Although it is impossible to penetrate the minds

164. Malherbe of Loudun to STN, Sept. 14, 1778.
165. The quotations come from the announcement in the *Gazette de Leyde* of Aug. 10, 1779, published in the name of a ''Société typographique de Londres.'' Gosse's connection with this group is apparent from subsequent notices in the *Gazette de Leyde* of Aug. 24, 1779, and May 16, 1780. On his trip through the Low Countries, Bosset learned that the ''Société typographique de Londres'' was a front for Gosse. Bosset to STN, from Brussels, Sept. 13, 1779.

of those readers, one can enter into their libraries and occasionally catch glimpses of them, bending over the pages of the *Encyclopédie*. Here, for example, is a scene from the autobiography of Stendhal:

Mon père et mon grand-père avaient l'*Encyclopédie* in-folio de Diderot et d'Alembert, c'est ou plutôt c'était un ouvrage de sept à huit cents francs. Il faut une terrible influence pour engager un provincial à mettre un tel capital en livres, d'où je conclus, aujourd'hui, qu'il fallait qu'avant ma naissance mon.père et mon grand-père eussent été tout-à-fait du parti philosophique.

Mon père ne me voyait feuilleter l'*Encyclopédie* qu'avec chagrin. J'avais la plus entière confiance en ce livre à cause de l'éloignement de mon père et de la haine décidée qu'il inspirait aux p[rêtres] qui fréquentaient à la maison. Le grand vicaire et chanoine Rey, grande figure de papier maché, haut de cinq p[ieds] dix pouces, faisait une singulière grimace en prononçant de travers les noms de Diderot et de d'Alembert. Cette grimace me donnait une jouissance intime et profonde.[166]

Such were the sensations of a young *Encyclopédie* reader in a wealthy bourgeois family of Grenoble during the 1790s. Other readers in other settings probably had different experiences, which can never be captured and catalogued. But whatever they felt, they must have known that in their hands they were holding one of the most challenging books of their time, a book that promised to reorder the cognitive universe and that would therefore produce some gnashing of the teeth among the local priests—unless the priests themselves were subscribers.[167]

166. Stendhal, *Vie de Henry Brulard*, ed. Henri Martineau (Paris, 1949), I, 379. This reference was kindly supplied by Mr. Joseph Gies.

167. Priests in many provinces may have bought the *Encyclopédie* as frequently as they did in the Franche-Comté. According to Louis Madelin, a list of forty *Encyclopédie* subscribers in Périgord contained the names of twenty-four curés. Madelin, *La Révolution* (Paris, 1913), p. 18. And the clergy accounted for six of the forty-two *Encyclopédie* owners whom Daniel Mornet identified in his study of private libraries. Mornet, ''Les enseignements des biblothèques privées (1750–1780),'' *Revue d'histoire littéraire de la France*, XVII (1910), 465, 469. The correspondence of the STN occasionally contains casual references to priests such as the following, in a letter from Antoine Barthès de Marmorières of Oct. 12, 1784: ''Un bon prêtre m'avait prié de vous demander au dernier rabais possible votre *Encyclopédie* de Genève.''

VII

⋎⋎⋎⋎⋎⋎⋎⋎⋎⋎⋎⋎

SETTLING ACCOUNTS

In January 1780, the quarto partners met in Lyons to settle their affairs. As they had conducted their business by conspiracy from the beginning, they brought it to an end in a dramatic dénouement, which is worth following in detail, not only for what it reveals about the spirit of early modern capitalism but also as an episode in the pre-Balzacian *comédie humaine*—or a *drame bourgeois,* as Diderot might have put it.

The last act in the speculation on the quarto began in October 1778. By then Panckoucke and his partners had fought off rival *Encyclopédies,* some fake, some real, from Geneva, Avignon, Toulouse, Lyons, Lausanne, Bern, and Liège. Their own *Encyclopédie* had evolved from project to project and edition to edition, becoming more profitable and more difficult to manage at each stage. For a while in the summer of 1778 it had almost spun out of control. But the settlement of the contract dispute on October 10, 1778, made it seem possible for the quarto publishers to bring "la plus belle [entreprise] qui ait été faite en librairie"[1] to a happy ending: the Lyonnais had been bought off; the octavo group was retreating from France; the Liégeois had made peace; and Duplain had accepted Panckoucke's terms for the third quarto edition. Nothing remained to be done, apparently, except the production and distribution of the final volumes, the collection of the last payments, and the division of the profits. The Traité de Dijon prescribed procedures for winding up the affair. It even required the associates to meet every six months in order to

1. STN to Panckoucke, Aug. 20, 1778.

324

make sure the accounts were in order, although in fact they met only twice—in February 1779, when Duplain gave an interim report on the state of the speculation, and in February 1780, when the associates liquidated it. But the quarto did not end in the prescribed manner, and its ending did not fit neatly into the twelve-month period between the two meetings. As soon as the publishers signed the contract for the third edition, their plots began to thicken again. They failed to resolve their conflicts in February 1779. And their final *règlement de comptes* produced an explosion.

The Hidden Schism of 1778

In June 1778 Duplain, who often came up with bright ideas about how to sell *Encyclopédies,* suggested that he and Panckoucke each market 500 sets of the third edition on their own. The proposal had the advantage of simplifying his operation in Lyons because it would permit him to dispose of hundreds of quartos *en bloc* instead of dealing with dozens of booksellers and individual subscribers. And it appealed to Panckoucke because Duplain promised to reserve the rich Parisian market for him. Not only would Panckoucke monopolize the wholesale business in the capital, he would also be free from any difficulty in getting rid of his 500 sets because he would allot 208 of them to the STN for its 5/12 interest in his portion and 41 each to Plomteux and Regnault for their 1/12 interests. Sales were booming at this time. There seemed to be no danger that the associates would become competitors rather than collaborators, and so Panckoucke accepted Duplain's proposal in July.[2]

In early November, Duplain sent the following letter to Batilliot, the Parisian banker who specialized in publishing speculations:

Je vais vous faire une belle affaire, mon cher ami, mais c'est à condition que personne n'en saura rien, pas même Panckoucke, notre ami commun. Cherchez un ou deux colporteurs qui ayent la confiance du public, et chargez-les de recevoir des souscriptions de l'*Encyclopédie* de Pellet dont je vous envoie le prospectus. Partagez avec eux le bénéfice. Vous voyez qu'en plaçant 13, vous gagneriez plus de 1400 livres. Vous annoncerez, ou pour mieux dire vos gens annonceront, que les 4

2. STN to Panckoucke, June 7, 1778, and Panckoucke to STN, July 21, 1778.

premiers volumes sont en vente à Genève. Vous ne parlerez du tout point de moi . . . Je me charge de l'entrée à Paris. Brûlez ma lettre.

Instead of burning the letter, Batilliot turned it over to Panckoucke, and Panckoucke sent a copy to Neuchâtel, restricting his commentary to one acid remark: the letter revealed "la vilaine âme de Duplain." It also demonstrated that the struggle for the *Encyclopédie* market had produced a secret war among the quarto associates as well as an open war between them and their rival publishers. Secrecy was the first line of defense that Panckoucke urged upon the STN. Duplain must not know that they knew. While he raided their territory, they must mount a counter sales campaign, and they must store all the incriminating evidence they could find until the moment when they could use it most effectively against him. An open split now could bring down the whole speculation. But why had Duplain attempted this stab-in-the-back, and why had Batilliot revealed it to Panckoucke?[3]

Batilliot's role can be explained easily: Panckoucke had just saved him from bankruptcy. "Il ne périra pas,"[4] Panckoucke told the STN. "Mais il m'en a l'obligation." Duplain seems to have acted from the opposite kind of motivation: revenge, compounded by a desire to enrich himself at his associates' expense. He resented his defeat in the bargaining over the contract for the third edition, and he felt entitled to a larger portion of the profits. Had he not mounted the

3. Panckoucke sent the copy to the STN in a letter of Nov. 6, 1778, with the following explanation: "Vous pouvez, Messieurs, juger de la vilaine âme de Duplain par la lettre ci-jointe. Il était bien convenu qu'en nous chargeant de ces 500 il ne ferait plus aucune négociation à Paris. Vous voyez comme il tient parole . . . Il faut de votre côté écrire partout pour en placer en province. Nous n'avons pas de temps à perdre . . . Ne parlez point à Duplain de la lettre ci-dessus. Tout cela servira dans l'occasion. Une brouillerie actuelle ne servirait qu'à nous nuire."

4. Panckoucke to STN, Dec. 22, 1778. Panckoucke indicated that the rescue operation was a big affair: "Vous ne sauriez craire encore combien les affaires de Batilliot m'ont tourmenté." He first mentioned it in his letter of Nov. 6, 1778: "Batilliot vient d'éprouver une faillite énorme, mais il s'en tirera. Nous nous mettons 5 à 6, et nous le cautionnons. C'est un brave homme qu'il nous importe de conserver." Bookdealers were often trapped in bankruptcies, many of them fraudulent, during the eighteenth century, so they warned their allies when they thought a bankruptcy might be imminent. Thus on Nov. 22, 1778, only a few weeks after plotting with Batilliot against Panckoucke, Duplain warned the STN to beware of both men: "Nous vous dirons sous le plus grand secret et par attachement pour vous que les sieurs Milon de la Fosse, banquiers à Paris, manquent, que Batilliot y est pour une somme énorme. Nous craignons bien le contrecoup pour notre ami Panckoucke, et comme nous savons que vous négociez avec Batilliot, nous vous en disons deux mots."

whole operation by himself? Had he not collected the subscriptions, organized the printing, sent out the shipments, dunned the customers for payment, and in general driven himself to the point of exhaustion, while Panckoucke dreamed up projects in Paris and the STN complained about its cut in the printing? He had engineered the greatest publishing speculation of the century, singlehanded. His associates had been more of a hindrance than anything else. Yet the contracts awarded them half the profits. Very well, he would help himself to his rightful share, even if he had to use some dubious methods. Of course, Duplain never put this decision in writing. But his actions spoke for themselves, and his attitude sometimes showed through letters that he wrote in a rage, particularly during his quarrel with the STN over its printing of volume 6:

Comment, Messieurs, vous nous exposerez à perdre dans un instant et notre fortune et une riche spéculation en imprimant un volume exécrablement mal, et il ne nous sera pas permis de nous plaindre? Si nous l'avons fait en termes trop amers, c'est que réellement nous étions grevés, et qu'il faut à un coeur surchargé un épanchement. Nous travaillons jour et nuit pour la réussite de l'affaire, et il semble, Messieurs, que vous fassiez tout ce que vous pouvez pour la détruire.[5]

The associates therefore approached their first shareholders' meeting as if they were preparing for a civil war. From November onward, Panckoucke and the Neuchâtelois adopted a conspiratorial tone in their letters about Duplain while remaining amiable and businesslike in their letters to him. "Nous devons prendre nos précautions par rapport à Duplain de manière qu'il ne puisse pas même soupçonner qu'on se défie de lui," the STN wrote to Panckoucke. "Il faut cacher jusqu'à nos soupçons," Panckoucke echoed in reply.[6] A new note also came into Duplain's letters at this time. Instead of marveling at the inexhaustible selling power of the quarto, he suddenly warned his partners that the demand for it had disappeared. The flow of subscriptions had dried up, he informed the STN on November 10, 1778. Each associate should make a supreme effort to spread circular letters, publish notices, and push the book in his commercial correspondence. Duplain himself had just sent out 200 circulars; if they did

5. Duplain to STN, Feb. 9, 1778.
6. STN to Panckoucke, Dec. 15, 1778, and Panckoucke to STN, Jan. 3, 1779.

not bring in fresh subscriptions, the associates would have quartos rotting in their warehouses for years and years. Duplain's letters also stressed his difficulties in getting the subscribers to pay, while he emptied his own coffers in order to meet his printing and paper bills. On December 1, he told the STN that he was 150,000 livres in arrears. The closer he came to the meeting in which he was to submit his accounts to the associates, the worse the speculation sounded.

Panckoucke did not get upset at Duplain's lamentations. He knew that one had to allow for a good deal of ''honest graft'' in the book trade, and he felt confident that Duplain would not go as far as outright swindling: ''Duplain fera usage de toute l'activité de son âme pour augmenter les frais, mail il ne divertira pas les fonds.''[7] The STN, however, suspected that Duplain was secretly printing extra sets of the third edition in order to sell them on the sly.[8] Regnault, who did some secret surveillance for Panckoucke in Lyons, reported that the press-run of the third edition was larger than the contract stipulated. Panckoucke then began to feel alarmed: ''Nous avons affaire à un homme très fin et très avide, qui ne mandera [*sic*] pas mieux de nous surprendre.'' Panckoucke's own efforts to sell the quarto in Paris had got off to a bad start, and he still feared that Duplain might have skimmed the cream off the market by a clandestine sales campaign. So he demanded that Duplain reduce the third edition by 500 copies or at least cancel the agreement to divide the 1,000 sets. ''Il faut l'y contraindre, l'y forcer,'' he wrote anxiously to the STN. But Duplain would not budge. He insisted that they maintain both the agreement and the pressrun, even though the market had been sated.[9]

At first Panckoucke tried to protect himself by keeping back

7. Panckoucke to STN, Dec. 22, 1778.
8. STN to Panckoucke, Dec. 15, 1778. At that time, Duplain had just asked Panckoucke to provide some extra copies of the plates and was refusing to let the STN print anything in the third edition—two indications to the Neuchâtelois that the printing was larger than he claimed.
9. Panckoucke to STN, Jan. 3, 1779. Regnault's report on the pressrun, which later proved to be inaccurate, was mentioned in this letter. Panckoucke expressed his demands to Duplain in letters of Dec. 22 and 26, 1778, Bibliothèque publique et universitaire de Genève, ms. suppl. 148. Duplain explained his refusal to the STN in a letter of Jan. 21, 1779. He noted that the printing of the third edition was so advanced—it had then reached volume 16—that the association would lose a great deal by reducing it.

the plates.[10] But as the STN objected, that tactic would not deter Duplain if he were seriously trying to swindle them; it would merely make him suspect their suspicions. In fact, it made him furious. He blamed Panckoucke's failure to deliver the plates for the refusal of many subscribers to make their payments; and he told the STN that he would never again do business with Panckoucke.[11] So there was little that Duplain's partners could do, except look long and hard at his accounts in Lyons while trying to prevent him from seeing through their own disguise.

"Je crois qu'il sera bien important de ne rien laisser appercevoir à Lyon," Panckoucke advised the Neuchâtelois. "Il faut voir le compte, l'examiner de sang froid, et faire ensuite nos observations. Je voudrais que ce fût le plus calme d'entre vous qui vînt à Lyon, ou bien il faudrait que vous y vinssiez deux." He proposed that they lodge together at the Palais Royal—"une grande auberge sur le quai de la Saône où l'on est fort bien"—so that they could coordinate their maneuvers in private. The Neuchâtelois should bring all their contracts and correspondence, in case they became locked in a debate with Duplain and needed evidence to support their arguments.[12] The Swiss publishers promised to follow Panckoucke's instructions, though their unhappiness with the whole business was increasing every day. They had become embroiled in a new quarrel over their printing with Duplain, and they grumbled that Panckoucke had become so absorbed in his Parisian speculations that he had neglected the sales of the quarto as well as their anti-Duplain defenses. They did not even like his choice of hotel: the Palais Royal was too far from Duplain's shop. But they agreed to send Bertrand there on January 25. Sensing a need to draw closer to his ally in the face of the adversary, Panckoucke offered to change his reservations to the Hôtel d'Angleterre: "Si vous arrivez avant moi à Lyon, je vous prie, Messieurs, de ne rien entamer que de

10. Panckoucke to STN, Dec. 22, 1778: "On ne peut pas nous imposer sur la vente. Les planches sont notre sûreté."

11. Duplain to STN, Dec. 20, 1778: "Enfin, nous vous avouerons que c'est la dernière affaire que nous aurons avec lui." Duplain must have sent some angry letters to Panckoucke, considering Panckoucke's letter to him of Dec. 22, 1778: "Entendons-nous, concilions-nous, et ne nous détruisons pas." Bibliothèque publique et universitaire de Genève, ms. suppl. 148.

12. Panckoucke to STN, Dec. 22, 1778.

concert. Nous avons à forte partie, et soyez sûrs qu'on cherchera de toutes les manières à nous surprendre.''[13]

In early January, the STN asked to postpone the meeting for a few weeks, because Bertrand had developed a serious illness. His condition deteriorated, so in early February Ostervald and Bosset went to Lyons in his place. Soon after their departure, he sent them a reassuring letter: he could now write without suffering too badly from headaches. But soon after their return, on the morning of February 24, he died in his bed. He was forty years old and left three children and a widow, Ostervald's younger daughter, Elisabeth. Bertrand had worked hard to make a success of the STN. He had been a man of great learning and advanced ideas. Had he lived to complete his labor on the STN's augmented edition of the *Description des arts et métiers* and some of his other literary projects, he might have won a place as a minor *philosophe*. Instead, he left a void in the Ostervald family and in the STN, which was only partly filled when his widow took over some of the commercial correspondence. In one of the rare, noncommercial remarks in that daily marathon of letter-writing, Ostervald mentioned his loss to Panckoucke: ''Nous venons de perdre le professeur Bertrand, notre gendre et associé, décédé hier matin des suites d'une fièvre bilieuse dont tous les secours de l'art n'ont pu le délivrer. Il vous sera facile de vous peindre l'amertume de notre situation dans ce moment. Daignez la partager et nous continuer votre amitié.''[14] Panckoucke replied, ''Il est affreux d'être moissonné si jeune. Je sais tout ce qu'une telle perte doit vous causer d'embarras, d'amertumes.''[15] It was a brief moment when the businessmen spoke to each other only as human beings. Immediately afterward they resumed their commercial discourse, for the *Encyclopédie* was moving too fast, as a speculation, for them to pause and contemplate the constants of the human condition. It had been necessary for them to take stock of their affair in Lyons while Bertrand was dying in Neuchâtel.

13. Panckoucke to STN, Jan. 7, 1778 (a slip for 1779) in reply to the STN's letter of Dec. 29, 1778.

14. Ostervald to Panckoucke, Feb. 25, 1779. Ostervald then continued: ''Quelque fâcheux que soit toujours un événement de ce genre dans nos circonstances, nous ne laisserons que de suivre nos affaires selon le plan qui nous a dirigés jusqu'ici.''

15. Panckoucke to Ostervald, March 7, 1779.

A Preliminary *Règlement de Comptes*

Duplain, however, had tried to persuade the associates that they did not need to hold the meeting. At the end of December, he suggested that Panckoucke cancel it because there were no dividends to distribute. He had reinvested the income from the first two editions in the preparation of the third. There would be profits enough to divide once the third edition had been distributed. And until then, Panckoucke need not worry about the financial management of the enterprise, which Duplain had left to his partner in Lyons, "Messieurs Veuve d'Antoine Merlino et fils, seigneurs suzerains d'un million d'écus romains, et je suis encore leur caution vis-à-vis de vous avec trois immeubles considérables que la vente de mon fonds m'a mis dans le cas d'acquérir. Ainsi dormez bien sur les deux oreilles." Panckoucke rejected this request, but not because he doubted Duplain's solvability. "Je n'ai point d'inquiétude pour les fonds," he explained to the STN. "Mais je crois nécessaire, indispensable, quoiqu'en dise Duplain, de nous rendre à Lyon pour avoir un compte. Il ne faut pas se fier à toutes ses belles promesses." The STN agreed completely: the crucial thing was to get Duplain to open up his books and produce a "compte . . . tout dressé . . . pour l'examen et la vérification."[16]

Exactly what examination and verification entailed is difficult to say. Panckoucke said the process would take five or six days and clearly expected Duplain to give a detailed account of all income and expenditure, with supporting evidence from letters and ledgers, in case any associate challenged it. The point of the exercise, as the STN put it, was to "voir clair" and then to pass an *acte* certifying Duplain's accounts. This formal agreement would join the other contracts in the *Encyclopédie* dossier maintained by each associate. It could be used as a weapon in the final *règlement de comptes* because it committed Duplain to a certain version of the affair—the number of subscriptions received, volumes printed, livres expended and collected by a certain date—which he could not modify in the future. In short, it would help Panckoucke and the STN to protect themselves from fraud.

Given the vituperation and the conspiring that had pre-

16. Panckoucke to STN, Jan. 11, 1779, with a copy of Duplain's letter, dated Dec. 31, 1778, and STN to Panckoucke, Jan. 17, 1779.

vailed among the associates before the meeting, the *acte* that they signed at the end of it makes strange reading: it suggests that nothing but harmony had reigned in the association (see Appendix A. XV). The associates all agreed that the first two subscriptions had been filled. Duplain therefore had to account for the difference between the revenue of 6,150 quartos and the cost of their production, which had been fixed by his contracts with Panckoucke. The *acte* then took note of his claim to have applied that profit to the expenses of the third edition. It certified that the third edition was being printed at a pressrun of 2,375 (4 reams, 15 quires instead of the 4 reams, 16 quires prescribed by the contract of October 10, 1778) and that those copies would be sold "en société," except for the 1,000 sets that Panckoucke and Duplain had agreed to split. Not a very revealing document. What had really happened at the meeting?

The examination of Duplain's accounts provided an occasion for a good deal of horse trading. Each of the associates had his own projects and priorities. Duplain wanted to subordinate everything to the speedy liquidation of the quarto, a goal that was unobjectionable in itself but that entailed the division of the 1,000 sets and some painful friction with the STN. By investing all the association's income in an attempt to produce the third edition as quickly as possible, Duplain had failed to pay the STN's bills. He also had refused to give it a share in the printing because his second objective was to profit from his role as a printing contractor. The STN wanted to be paid, to receive a volume from the third edition to print, and to get the big printing job that had eluded it ever since it had formed its original partnership with Panckoucke.

Panckoucke's interest had shifted from the quarto to the *Encyclopédie méthodique,* much to the dismay of his associates. When he arrived in Lyons, he produced a contract for this new *Encyclopédie* which he had drawn up with some speculators from Liège and now wanted the quarto associates to ratify. In earlier negotiations with the Liégeois, Panckoucke had agreed to sell them the plates and entry to the French market for 205,000 livres. But then he decided to cancel the sale in order to take over the *Méthodique* himself, keeping his quarto associates as partners in the new enterprise. The associates, however, did not want to sacrifice their share of the 205,000 livres or to become entangled in a new speculation before unsnarling the old one. They already resented Panckoucke's neglect of their sales campaign for the third edition. And Duplain was insensed about his slowness in producing the plates, while the STN objected to his laxness in supervising Duplain. In short, the association seemed to be fissuring in many different directions, some of which cut across the basic cleavage between Duplain and the Panckoucke group. But none of these rifts appeared openly, and the associates patched up the most frayed areas of their enterprise by making concessions to one another.[17]

Panckoucke had most to concede, and the Swiss had most to complain about. Feeling the need to reinforce his alliance with the STN, Panckoucke concluded a separate treaty with Ostervald and Bosset on February 13. The STN acknowledged its commitment to accept the 208 quartos that fell to it from its 5/12 share in Panckoucke's 500, and it reaffirmed its commitment to pay the 92,000 livres that it owed for its original portion of Panckoucke's plates and privilege. Panckoucke, "désirant obliger Messieurs de la Société typographique," then gave them a free 5/24 share in the quarto edition of the *Table analytique,* which he was about to produce with Duplain. He and Duplain had secretly hatched this plan in their contract of September 29, 1777. The only *Table* that Panckoucke had mentioned in his letters to the Neuchâtelois was the folio edition, which he had unsuccessfully tried to sell to them in June 1777. At that time, he had explained that he had bought the original Mouchon manuscript for 30,000 livres; he would sell it for 60,000 livres; and he would guarantee that it would fetch 128,000 livres in profits. Since then, times had changed. Panckoucke had shifted his allegiance decisively to the STN, and he did not want the Neuchâtelois to feel cut out of a speculation that he had planned behind their back. But the speculation itself looked alluring as ever. The *Table* provided a useful summary and index to Diderot's vast text. And it seemed certain to sell well in the quarto format because it could be marketed in the wake of the quarto *Encyclopédie,* which had been far more successful than the folios. Panckoucke was convinced that it would be bought by most of

---

17. The negotiations concerning the *Encyclopédie méthodique* are discussed in detail in Chapter VIII. Although the quarto associates did not keep a record of their meeting in Lyons, their transactions can be pieced together from references scattered throughout the STN papers.

the quarto subscribers and even by persons who did not own *Encyclopédies*. His gift of 5/24 share in the enterprise therefore had considerable value. And he ingratiated himself still further with the Neuchâtelois by promising to intervene with Duplain so that they would get to print the quarto *Table,* a matter of six quarto volumes.[18]

Duplain, who was to manage the quarto *Table* as he had managed the quarto *Encyclopédie,* encouraged the Neuchâtelois to expect they would get the printing job. He also mollified them by paying several of their bills of exchange, shortly before the meeting began. And he gave them a volume from the third edition to print. This last concession not only kept their presses going, it also quieted their doubts about the size of the pressrun; for they would know from their own printing that the third edition really contained no more than 2,375 copies, as Duplain had testified in the *acte* passed at the meeting. Duplain and the Neuchâtelois even laid plans to collaborate in pirating the Genevan edition of the works of Rousseau. If the Neuchâtelois did not dismiss all their suspicions about "notre homme de Lyon," they certainly returned home feeling much better about him.

The issue that had upset Panckoucke's relations with Duplain concerned the printing of the plates, the division of the 1,000 sets, and the proposal to reduce the printing of the third edition. Panckoucke gave way on all of them. He went further: he granted Duplain a permanent 12/24 share in the plates and privilege of the *Encyclopédie* as a replacement for the temporary share that had been allocated for the quarto. This grant was probably worth as much as the gift to the STN of a share in the *Table* because it gave Duplain a claim on all of Panckoucke's ancillary speculations, including the *Encyclopédie méthodique.* Why had Panckoucke been so accommodating? The STN had already convinced him about the need to supply Duplain with the plates, and Duplain had shown that it would be unreasonable to reduce the pressrun of the third edition because he had almost printed half of it. Perhaps it was also too late to rescind the agreement on the 1,000 sets, since each associate had been selling his portion privately

18. Panckoucke made his original offer to the STN in a letter of June 16, 1777. He finally published the folio edition of the *Table* himself, in two volumes, printed by Stoupe in 1779. See Panckoucke's circular of July 31, 1779, and his letter to the STN of July 21, 1778.

for half a year. But Panckoucke had criticized Duplain's
stewardship so vehemently that it is difficult to see why he
should have rewarded Duplain for it—unless one takes account of the *Encyclopédie méthodique*. More than anything
else, Panckoucke wanted to get his associates to accept his
new contract with the Liégeois. By granting Duplain a permanent share in the plates and privilege, he would not only win
Duplain's support for the new enterprise but also give him a
stake in it.

Still, Panckoucke and the STN had secretly expressed
grave doubts about Duplain's integrity. Had he defended his
management so effectively as to win them over? There is no
record of the exchanges across the conference table, but those
who had negotiated with Duplain—shrewd men like Favarger
and d'Arnal—agreed that he could be extremely persuasive.
He may indeed have persuaded the associates that they were
threatened by a severe drop in demand. They knew from their
own sales campaigns that the subscription flow had virtually
stopped. So they probably saw nothing suspicious in his
gloomy report about the prospects for the third edition. Since
he quieted their suspicions about the size of its pressrun and
bound himself to produce the profits from all 6,150 sets of the
first two editions, he seemed to have little room for peculation.
And finally, they discovered a secret weapon for their defense.
Somehow, when Duplain's back was turned, they got access
to his subscription register and secretly copied it. There was
nothing suspicious about the list itself, which accounted for
7,373 subscriptions, but it could protect the associates against
fraud, in case Duplain falsified the subscription report in the
accounts that he would present after the completion of the
third edition, when they would meet again to divide profits
and liquidate the entire enterprise.[19]

The Lyons meeting of 1779 was not, therefore, the straightforward, amiable affair that it appeared to be in the *acte*
passed by the associates before they dispersed. Each man ar-

---

19. Ostervald and Bosset alluded to the secret copying of the subscription list
in a letter to Mme. Bertrand of Feb. 13, 1780, which referred to the ''registre
. . . que nous relevâmes furtivement et sans qu'il s'en doutât il y a un an.'' The
copy itself is in the STN papers, along with related documents about the Lyons
meetings of 1779 and 1780 discussed at the end of this chapter. One of the documents, a memoir by Plomteux from Feb. 1780 (see Appendix A. XIX), makes
clear that Duplain painted a very black picture of the third edition during the
meeting of Feb. 1779.

rived nursing his own grudges and desires, and each session was charged with cross-currents and suspicions. Instead of exploding, however, the tension was dissipated by a process of shuffling projects and trading off concessions. But the associates did nothing in February 1779 to purge the distrust that had alienated the Panckoucke group from Duplain and that remained repressed until the final reckoning of February 1780.

The Feud Between Duplain and the STN

As soon as the associates returned from the summit conference at Lyons, the atmosphere clouded over again. Their commercial correspondence showed that it had become harder than ever to sell quartos, and their expansive, conciliatory mood passed as the market shrank. To be sure, they had already sold an unprecedented number of *Encyclopédies*. But after February 1779, their final balance sheet began to look less sensational than they had originally expected. Minor credits and debits took on major importance. The STN and Duplain, in particular, spent the next twelve months hassling and haggling. Although the inventory of their quarrels makes sad reading, it reveals a great deal about the relationships between printer and publisher, associate and manager, during the final stages of the enterprise.

Some of the disputes concerned trivial issues. For example, Duplain refused to pay 90 livres 15 sous of the STN's printing bill because he had to have the frontispiece of volume 35 redone in Lyons. The STN had printed it as "troisième édition, à Genève" which contradicted his attempt to pass off the third edition as a new, Neuchâtel product. The Neuchâtelois claimed that they had merely remained faithful to the copy he had sent to them. So every time one house sent a statement (*compte courant*) to the other, there was a fresh dispute over an entry for 90 livres 15 sous. But usually Duplain and the STN argued over more important sums. In billing Duplain for its expenditure on paper, the STN included a charge for the *chaperon* or extra quire for every ream it had purchased. Early modern printers used the *chaperon* for pulling proofs and replacing defective sheets or sheets spoilt in the printing. Without it, they could not produce 500 acceptable sheets from every ream of paper they bought. The STN therefore felt entitled to reimbursement for its *chaperon* payments, but Duplain refused to be debited for anything more than 500 sheets per ream, making a discrepancy of 1,066 livres for 111 reams of *chaperon* in the accounts.[20]

Even more important and acrimonious was the dispute over 31 misplaced quartos. On March 14, 1778, the STN ordered 52 sets of the third edition, which were to be charged to its account as a wholesaler and which it intended to distribute among its own customers, that is, to individuals who had subscribed through the STN instead of going directly to Duplain. It heard nothing about the order until October 13, when Duplain wrote that he was sending 20 copies of volumes 1 and 2. The STN replied that it would be glad to receive those 20 but that it needed 32 more to complete its order of 52. Duplain's shipper had actually sent all 52 volumes 1–2 on October 1, but he had not informed the STN that they were en route. The subsequent shipment of 20 volumes was meant for subscribers of the first two editions who had to be supplied from the third, owing to difficulties in coordinating the shipping operation. Duplain had asked the STN to take those 20 subscriptions on its account because they concerned old STN customers in Northern Europe who lay outside the range of his own commerce. So he read the STN's request for 32 more quartos as a new order, while the STN understood his shipment of the 20 copies as a partial fulfillment of the old one. He therefore sent off 32 additional volumes 1 and 2 to Neuchâtel, bringing the total to 104 instead of 52. As soon as they received notification of this last shipment, the Neuchâtelois realized what had happened and explained the double misunderstanding to Duplain. He trebled the confusion, however, by interpreting the explanation as an attempt to escape

20. As this issue is of some importance in printing history and analytical bibliography, it seems worthwhile to cite the following passage of the ''Mémoire contre Monsieur Duplain'' that the STN submitted to the arbitrators who finally settled their dispute in Feb. 1780: ''Chacun sait que toutes les fois qu'on travaille dans une imprimerie, il se trouve toujours plusieurs feuilles de papier ou défectueuses ou que les ouvriers salissent, gâtent, et déchirent, sans parler de celles qui servent pour les épreuves. C'est la raison pour laquelle celui qui fournit le papier en ajoute toujours un certain nombre en sus de ce qu'il faut, et ce surplus se nomme *chapelet* ou *chaperon*. Ici la Société invoque en toute confiance l'usage généralement reçu dans tous les lieux où l'on fait rouler des presses et supplie Messieurs les arbitres de faire déclarer sur le cas tel libraire de cette ville qu'ils jugeront à propos d'appeler. Il n'y a de différence que dans le nombre de feuilles ainsi ajoutées pour chaperon. Quelquefois il va à une main par rame; le moins est une main pour deux rames.''

from an obligation to pay for 104 sets. The drop in demand compounded the problem because neither associate wanted to be stuck with surplus *Encyclopédies*. The STN agreed to take the first 20 of the extra sets because they had been subscribed by its own clients; and it found a subscriber for one of the other 32. But it refused to accept the final 31, which had a wholesale value of 8,526 livres. Duplain would not budge. As each new volume came off the press, he sent 31 unwanted copies to Neuchâtel, and the STN sent them back again. Each associate debited the other for their wholesale price and their transport. And each shipment touched off a new flurry of letters full of recriminations. The STN offered to submit the dispute to arbitration, but Duplain refused. He was so obdurate, in fact, that he seemed to act in bad faith, for a close reading of the correspondence shows that he was mainly to blame for letting a misunderstanding mushroom into a feud that poisoned relations between him and the Neuchâtelois until the very end of the enterprise, when it was settled in the STN's favor by a panel of arbitrators.[21]

Duplain and the STN also kept up a running battle about two minor items in their accounts for 1778. The STN had printed volume 15 of the first two editions and had shipped it to Lyons in sixty-four crates via Pion of Pontarlier on June 20. But Duplain had not received the crates by July 1, when he planned to send off volumes 15, 16, and 17, as well as a volume of plates, to the subscribers. With the passing of each day after the July 1 deadline, Duplain grew increasingly anxious about the STN's volume. The delay sent repercussions throughout the enterprise: it held up his own shipments; the lag in the shipments caused a postponement in the payments due from subscribers upon the receipt of the books; and the postponement of the payments cut badly into the income needed to produce the later volumes. By July 24, Duplain had sent three letters to Pion without receiving any reply or any indication of what had happened to the shipment from Neuchâtel. Unable to wait any longer, he sent his clerk to search for it along the route between Pontarlier and Lyons. The clerk ran into the crates at Beaufort: Pion had simply taken

21. This affair was discussed in dozens of letters and memoranda for almost two years. The crucial letters for interpreting the original misunderstanding are: STN to Duplain, March 16, 1778; Duplain to STN, Oct. 13, 1778; STN to Duplain, Oct. 17, 1778; Duplain to STN, Oct. 20, 1778; and Duplain to STN, Feb. 18, 1779.

his time about forwarding them and had neglected to notify Duplain by the customary *lettre d'avis*. His wagoner had also done a bad job of covering the crates. So when they finally arrived at the end of the month, ten sheets had been ruined by rain. In a fury, Duplain had the sheets reprinted in Lyons and charged the STN 275 livres for them as well as 124 livres for the clerk's journey. He even demanded compensation for the loss of a month's revenue, owing to the delay of the July 1 shipment—a matter of 200,000 livres at ½ percent interest per month, or 1,000 livres. And he raged at the Neuchâtelois for leaving him with his storerooms stuffed with *Encyclopédies,* which might have been seized at any moment if the local clergy had turned against him: "Nous sommes ici accablés de livres, sans cesse dans la crainte de quelque délation au clergé et dans l'huile bouillante. Vous seuls, Messieurs, causez nos peines."[22]

The Neuchâtelois replied that the fault was Pion's, not theirs. Although they had not been obligated to meet any deadline, they had sent the volume off on time. According to commercial law, the shipping agents and wagoners were responsible for delays that occurred en route. And there were established procedures for dealing with damaged merchandise. Duplain should have had a *procès-verbal* drawn up as soon as the shipment arrived, so that he could hold the wagoner liable. To foist the liability onto the STN was absurd, for the STN had ceased to be responsible for the shipment as soon as it left Neuchâtel. By the time of the Lyons meeting in February 1779, Duplain's temper had cooled, and he tacitly conceded the case to the STN by accepting an account statement it had submitted on January 6 that excluded all reference to the compensation he had demanded. But the punitory debits reappeared in the accounts that he sent to the STN later in 1779. The STN refused to accept them, and they provided material for a great many bitter remarks in the commercial correspondence between the two houses until the time of the liquidation of the quarto, when the arbitrators decided the issue in favor of the STN.[23]

22. Duplain to STN, July 28, 1778.
23. The most important of the many letters and accounts exchanged on this subject are: STN to Duplain, July 15, 1778; Duplain to STN, July 24, 1778; Duplain to STN, July 28, 1778; STN to Duplain, July 28, 1778; Duplain to STN, Aug. 7, 1778; STN to Duplain, Aug. 26, 1778; and Duplain to STN, Sept. 2, 1778.

Ostervald and Bosset returned from the conference of February 1779 with the first part of the copy for volume 19 of the third edition in their baggage. When they began to print it, they noticed that it contained the same textual errors as volume 19 of the first two editions. Yet Duplain was trying to sell the third edition, under the STN's imprint, as "supérieure à l'autre pour l'exécution, la correction etc.,"[24] and the contract for it allotted the abbé Laserre 3,000 livres for improving the text of the earlier editions. The STN concluded that "notre bon abbé" had collected his wages without doing any work at all. It felt distressed at having loaned its name to an *Encyclopédie* that "fourmille de fautes innombrables et très grossières que l'on reprochait avec tant de raison à l'in-folio et qu'un homme doué de bon sens n'aurait jamais laissé passer s'il avait lu avec attention." And it complained that Duplain did not care about the quality of the work: he had gouged his partners for the salary of "son bon ami l'homme de l'Eglise" and then had failed to supervise him—a neglect that bordered on mismanagement and that could damage the quarto's sales as well as the STN's reputation.[25]

But as explained in the previous chapter, the STN had far more serious complaints about Duplain's management. It blamed him for the "faux calcul par un animal de prote,"[26] which had committed the quarto publishers to provide the public with the full text of the original seventeen folio volumes and the four volumes of the *Supplément* in twenty-nine volumes in-quarto. That was a typographical impossibility, which Linguet had exposed with his usual verve in perhaps the most widely read journal in Europe. Duplain had had to extend the quarto to thirty-six volumes and to give away three of them free. But how could he constrain the subscribers to pay for the extra four? As they entered into "la crise des volumes excédant le nombre 29,"[27] the quarto associates trembled; they knew that they would face a great many contested bills, cancelled subscriptions, and law suits—all because of Duplain. In fact, Duplain only managed to fit the text into thirty-six volumes by giving each volume "une grosseur mon-

24. Duplain to STN, July 10, 1778.
25. STN to Plomteux, May 1, 1779. For further information on Laserre's editing see Chapter V.
26. Ibid.
27. Duplain to STN, April 17, 1779.

strueuse"—120–130 sheets (about 1,000 pages) instead of the 80–90 sheets (about 700 pages) that his associates considered appropriate for a quarto.[28] As the subscribers paid by the volume and the printers were paid by the sheet, this tactic narrowed the association's profit margin. But it fattened Duplain's profits because he received a set amount from the association for every sheet he had printed—and he had them printed at a far lower price than he was paid, except for the volumes done by the STN. He cited this exception as a reason for expecting the STN to grant a delay in the payment of its printing bills. The STN replied tartly by citing his rake-offs as grounds for expecting him to pay his bills on time.[29] And then it made a more damaging discovery about his manipulation of his expenses. In July 1779, the STN hired the assistant foreman (*second prote*) from Pellet's shop to help with its printing, which had then reached volume 19 of the third edition. The new employee, Colas, had supervised the printing of the first two editions of the same volume for Pellet. He was therefore able to explain why it contained so many more sheets in the first editions—12 sheets, or 96 pages, to be exact— than in the third, even though the text was exactly the same. Pellet had had his compositors widen the spaces between words and stretch out paragraphs until they ended on new lines ("printer's widows"), so as to use up as many sheets as possible. This tactic cost the association 744 livres, which the STN suspected Pellet of sharing with Duplain.[30]

Instead of raising all these matters with Duplain, the associates kept the most serious of them secret, in order to confound him with incriminating evidence at the final settling of accounts. But they could not contain all their discontent. New issues kept arising, especially from subscribers who complained about the disorder in Duplain's shipments and billing. For example, in mid-April 1779, Duplain and the STN exchanged account statements. Each statement contained one company's version of the debits and credits that the other had

28. STN to Plomteux, May 1, 1779.

29. Duplain to STN, Dec. 1, 1778, and STN to Duplain, Dec. 5, 1778: "D'ailleurs nous ne sommes point d'humeur de faire des avances, ayant besoin [de] nos fonds; et quant à ce que vous nous marquez sur le bénéfice que nous pouvons avoir sur l'impression, vous en avez vous, Messieurs, un bien plus considérable sur la quantité de volumes que nous savons que vous avez fait imprimer au-dessous du prix stipulé."

30. STN to Panckoucke, Feb. 25, 1779, and March 23, 1779.

accumulated over the last few months. The accounts were widely different. In a tightly argued letter that went on for seven pages, the STN disputed twelve items from Duplain's account. There were the first four volumes of the set of the third edition for the marquis de Boissac, which should have been sent to Heidegger of Zurich instead of being debited to the STN: 40 livres. There was the shipment of thirty-one copies of volumes 7 and 8 of the first edition, which Duplain had sent to Neuchâtel via Geneva instead of following its instructions to use the cheaper route through Pontarlier: a difference of 5 livres 10 sous. There was the misfiled down payment of Buchet of Nîmes, who had subscribed for his twenty-six sets through Duplain, not through the STN: a debit of 288 livres to be erased. There was the delay in Duplain's shipment to the STN of its subscribers' copies of volumes 5–8 of the third edition, which arrived long after Duplain's bill for them: 1440 livres, which should be carried over from this *compte courant* to the next. There was Duplain's supplementary payment for one of the STN's shipments—a bill that the STN had settled with Pion, that Pion's wagoner had collected a second time in Lyons, and that Duplain had inserted in the STN's account: another debit of 360 livres to be struck out. And so it went—a series of imbroglios, which were not terribly important individually but, when taken together, represented 9,151 livres and several months of accumulated bad feeling.

Duplain would not give an inch on any of them. In some cases he was right. Buchet's 288 livre down payment actually had been made to the STN, which had misfiled it and which acknowledged its error as soon as Duplain produced convincing evidence of the mistake. But most of the disputes resulted from the disorder of Duplain's operations in Lyons. The associates received a great many complaints from subscribers whose *Encyclopédies* had arrived late, or with missing or damaged sheets that Duplain refused to replace, or with too many copies of one volume and too few of another. In his haste to collect payments, Duplain sometimes wrote bills of exchange on booksellers who had not yet received their shipments. When they complained, he threatened to sue. They retaliated with countersuits or took their complaints to his associates, who then became embroiled in dozens of quarrels between Duplain and the subscribers. Each quarrel hurt the association, either by sapping its profits or by eroding the associates' confidence in Duplain's management. "Les plaintes sont très multipliées," Panckoucke lamented to the STN. "J'ai une peur horrible que tout cela ne finisse par des procès."[31]

The STN found it especially difficult to cope with Duplain because it dealt with him in three capacities: as an associate, as an *Encyclopédie* wholesaler, and as a printer. It could not keep those roles distinct. Duplain took sums that were due to the STN from its subscribers and subtracted them from its printing bills. Then he refused to pay those bills on the grounds that he had not settled his multiple quarrels with the STN over subscriptions and shipments. He refused to honor the bills of exchange that it wrote on him, forcing it to make emergency loans, at a high interest rate, to cover its own engagements. He did pay some of its bills but never in full, for he always found fault with some item. In January 1779, he paid most of what he owed for the first three volumes it had printed, but in April he refused to pay for the last two and revived disputes over items that had been settled in the January account. He never settled the April account at all—until forced to do so by arbitration after the liquidation of the quarto. The STN complained bitterly: "Généralement vous agissez à notre égard avec une rigueur étonnante. Selon vos prétentions le fruit de tout notre travail serait réduit à rien . . . Nous vous supplions de nous traiter comme les derniers de vos imprimeurs."[32] But the last year of the STN's relations with Duplain produced nothing but refusals on bills of exchange, arguments over accounts, and a growing conviction on its part that Duplain had gone from mismanagement to peculation.

Marketing Maneuvers

Duplain's quarrels with the STN symptomized a change in the character of the enterprise. The associates had always distrusted one another, but as they approached the moment when the final balance sheet would be drawn up, they conspired and maneuvered with greater intensity, hoping to make

31. Panckoucke to STN, June 1, 1779.
32. STN to Duplain, April 24, 1779.

the most from a market in its last stages of contraction. That the demand had dropped drastically had become clear in November 1778, when the STN expressed the fear that "toutes les ressources soient épuisées en France."[33] Favarger sold only three subscriptions on his tour of southern and central France in late 1778, and Panckoucke had sold only twelve in Paris by April 1779. The Parisian situation was unusual, however, because other editions had glutted the market and because Panckoucke had neglected his sales campaign. As Duplain and the STN complained, he had shifted his allegiance from the quarto to the *Encyclopédie méthodique.* Although he honored his commitment to delay the publication of the prospectus of the *Méthodique,* he let word leak out that his super-*Encyclopédie* was well under way; the STN then objected that the leaks amounted to "l'équivalent d'un prospectus."[34] Panckoucke's negligence probably hurt the sales of the quarto, but it was far less damaging than another form of insubordination among the associates: secret price-cutting.

Panckoucke and the STN had committed themselves in their contract of February 13, 1779, to maintain the subscription price of the quarto. But the division of the 1,000 sets and the decline in demand inevitably produced competition instead of cooperation among the associates. Soon after the meeting of February 1779, a rumor spread that Plomteux had offered special terms for his portion of Panckoucke's 500 copies. Duplain warned that any deviation from the terms stipulated in the prospectus would ruin the sales of the rest of the edition. Two months later Duplain heard that Regnault was selling his portion with an offer of free transport and a year's credit —an enticement that amounted to a reduction in the sales price and would undermine Duplain's attempts to collect cash from the association's subscribers. This deviant salesmanship may have contributed to the deterioration of Regnault's relations with Panckoucke, for in 1777 and 1778 Regnault had served as Panckoucke's secret agent in Lyons (his principal activity apparently was to spy on Duplain), and by June 1779 their friendship had been destroyed by a quarrel that threatened to end up in court. Panckoucke him-

33. STN to Duplain, Nov. 22, 1778. On Nov. 29, 1778, the STN warned Panckoucke that "la province est rassassiée, ce que nous avons appris par un de nos commis, qui est de retour d'un voyage qu'il a fait dans les provinces méridionales de France, où il n'en a placé qu'une couple."
34. STN to Duplain, April 10, 1779.

self eventually deviated from the prescribed terms of the subscription, although he did not actually lower the price of his quartos. And later in 1779, the STN began to tempt its customers with various selling devices: six months' credit; three free copies for every twenty-four orders instead of one for every twelve; free stitching (the STN evaluated *brochage,* which involved folding and sewing the sheets, at 4 percent of the wholesale price); and free transport as far as Lyons. By the end of the year, Plomteux reportedly had traded some of his quartos for books that he expected to market more easily, and a few quartos had sold in Lyons for 240 livres—54 livres off the original wholesale price. Although it would be an exaggeration to say that the bottom had dropped out of the market, the demand fell off seriously enough in 1779 to increase the pressure that was building up among the associates. And while the pressure rose, Duplain attempted to dispose of the rump of the third edition by some adroit maneuvering.[35]

First, he offered to trade his share in the quarto *Table* and in the *Encyclopédie méthodique* for the 208 sets that the STN had acquired as its portion of Panckoucke's 500. The difficulties of the market made the STN feel inclined to accept. But Panckoucke secretly advised it to refuse, basing his argument on Duplain's avarice: "C'est un homme avide que ce M. Duplain et qui aime l'argent avec fureur." It followed that 208 quartos were worth more than a half interest in the *Table,* which would produce 4,000 sales according to Panckoucke's estimate, and the more remote prospects of the *Méthodique.* But Panckoucke was also concerned about the need to prevent Duplain from cutting himself loose from their speculations: "De quelque manière que vous traitiez, il faut

35. The quarrels over pricing stand out in Duplain to STN, Feb. 18, 1779; Duplain to STN, April 17, 1779; and STN to Panckoucke, May 2, 1779. Regnault's role in the enterprise remains obscure, but a letter to him by the STN of June 6, 1779, shows that he and Panckoucke were expecting to take their quarrel to court. The STN's remarks on stitching came in letters to Bosset of Aug. 7, 1779, and to Noel Gilles of Aug. 15, 1779. Panckoucke offered credit in return for a slight increase in the price of the quarto in a circular dated July 31, 1779. And the price-slashing in Lyons was reported in letters from the STN to Bosset, Sept. 16, 1779, and from d'Arnal to the STN, Nov. 12, 1779. By this time, however, only a few volumes of the quarto remained to be sold, and it seems unlikely that many of them were marketed at such a low price, especially outside the Lyons area, where sales were heaviest. As mentioned in the previous chapter, Jefferson found that the quarto sold for 380 livres—that is, virtually the same amount as the original subscription price—in Paris in 1786.

que Duplain reste intéressé pour une part et l'obliger surtout à agir personnellement pour placer l'édition. Duplain brûle de se retirer, et son activité nous est nécessaire. Je vous prie de ne pas me compromettre. Duplain ne me pardonnerait de la vie de vous avoir donné un avis qui croise ses intérêts.'' This line of reasoning convinced the Neuchâtelois. The fact that Duplain had made the offer was reason enough to refuse it, they replied to Panckoucke. And it did indeed seem odd that Duplain should want to acquire more quartos at the very moment when they seemed to be least in demand.[36]

Duplain did not give up, however. In fact he tried to coerce the STN into accepting his offer. He sent the first eight volumes of the STN's 208 copies to Paris instead of Neuchâtel—a move that not only held up the STN's attempts to market them but also decreased their value by overloading them with 1,466 livres in transport costs. The Neuchâtelois had expected to sell most of their 208 sets in northern Europe. Faced with the expense of the unnecessary Lyons-Paris-Neuchâtel journey and the threat that the next twenty-eight volumes would follow the same route, they might be expected to take the easy way out and trade the books to Duplain. Duplain did not make this threat openly. He presented the Paris shipment as a retaliatory measure. The STN had offered to sell its copies with six months' credit, he explained to Panckoucke, and that derogation from the terms of the subscription would ruin his own sales. Besides, he had contracted with Panckoucke to supply 500 quartos to the Parisian market. To Paris they would go, regardless of Panckoucke's subsequent agreements with his own associates.[37]

The Neuchâtelois replied that Duplain's argument was ''fautif à tous égards.'' Not only had they honored their commitment to keep up the price (they did not begin price-cutting until August 1779), but also the agreement to divide the 1,000 sets did not require Panckoucke to receive his share in Paris. Any such provision would have been absurd and would have been rejected by Panckoucke and the STN together. The Neuchâtelois could only interpret Duplain's ac-

36. Panckoucke to STN, March 7, 1779, and STN to Panckoucke, March 14, 1779.

37. Duplain to Panckoucke, March 12, 1779, from a copy in Panckoucke's letter to the STN of March 18, 1779.

tion as an attempt to force them into complying with whatever secret strategy he had devised to wind up the affair with the greatest possible profit for himself, no matter what the cost to his associates. They therefore resolved to resist, and to keep their suspicions hidden: ''Vous nous avez conseillé très sagement de dissimuler jusqu'au bout avec lui et de ne pas marquer notre juste mécontentement. En vérité la chose devient de jour en jour plus difficile,'' they told Panckoucke. They stuck to their decision to refuse Duplain's request for the 208 copies, and the 1,466 livres in unnecessary transport costs became one more item in the long list of what they referred to as their ''griefs contre Duplain.''[38]

Meanwhile Duplain tried to persuade Panckoucke to divide up the rest of the third edition as they had done with the first 1,000 copies. About 450 or 500 sets remained to be sold, he said. He would apportion them among the associates, and this time he would ship them wherever they pleased so that each man could dispose of his copies as he liked. There was one thing to be said for this proposal: it would prevent further deterioration in the relations among the associates by abandoning the agreement against price-cutting and by leaving them free to sell the last quartos on the best terms they could find. Panckoucke felt tempted to accept it, although he had no illusions about Duplain's ''avidité sans bornes,'' as he explained to the STN. True, Duplain had sent the 208 copies to Paris in order to keep the STN out of the provincial market, and he had refused to give Regnault his copies in Lyons for the same reason. But Duplain could find even more ways to sabotage their sales; for he controlled the machinery of the enterprise in Lyons, and there was little the associates could do, at a distance of 150 leagues, to keep him in check. It was too late to recoup their losses from the rake-offs on the printing, although Panckoucke would try to protect them from any major embezzlement by holding back the last volume of plates until the accounts had been settled. But they might prevent further fraud by dividing up the last copies and selling them on their own. In that way they could put an end to the quarrels over the last marketing and bill-collecting operations and could hold the association together until the final reckoning

38. STN to Panckoucke, March 23, 1779, and March 14, 1779. See also STN to Duplain, April 24, 1779.

of the accounts—assuming that they could continue to hide their distrust of Duplain. "Au nom de l'amitié, ne me compromettez jamais," Panckoucke concluded. "Nous avons, malgré les plaintes que nous pourrions faire, à nous concilier jusqu'à ce que tous les comptes soient rendus."[39]

The STN saw two objections to Duplain's proposal: it believed that he exaggerated the number of unsold sets, which it estimated at 400 instead of 450–500; and it thought that he would have an unfair advantage in marketing them, for he could pick off the last of the subscriptions that had been flowing into the association's central office, while they would have to sell their portion on the open market where the demand was weakest.[40] A few days later, Duplain put his proposal formally to the associates in a circular letter: "Il reste de la troisième édition un nombre de 480 exemplaires, et nous voyons avec peine qu'il ne s'en vend plus. Nous croyons pour l'avantage de tous les intéressés qu'il convient de se partager ce nombre, parce que chacun ayant ses moyens, il sera plutôt épuisé."[41] The STN rejected this proposition adroitly, by telling Duplain that he had done such a good job of marketing seven or eight thousand *Encyclopédies* for the association that it was sure he could sell a mere "queue" of 480. Panckoucke then came around to the STN's opinion. He did not trust Duplain's version of the subscriptions, he confided to the Neuchâtelois. So he would reject the proposal and refuse to ship the third volume of the plates to Lyons.[42]

Panckoucke's delaying tactics infuriated Duplain, who asked the STN to intervene for the release of the plates. The Neuchâtelois agreed to do so, but at the same time they insisted that Duplain surrender the remaining volumes of their 208 sets in Lyons, where d'Arnal had rented a warehouse to receive them. This bargaining worked. Duplain delivered the books to d'Arnal in Lyons rather than to Panckoucke in Paris, and the STN persuaded Panckoucke to send off the last volume of plates, arguing that further stalling on the plates

39. Panckoucke to STN, March 18, 1779, which included a copy of Duplain's proposal from a letter by Duplain to Panckoucke dated March 12, 1779.

40. STN to Panckoucke, March 23, 1779. The STN also supported Panckoucke's resolution to hold back the third volume of plates "à titre d'épouvantail" and reassured him that it would keep their secrets: "Nous sommes de votre avis de garder des ménagements jusqu'à la clôture, et nous ne vous compromettrons jamais."

41. "Lettre commune à Messieurs les intéressés," April 1, 1779.

42. STN to Duplain, April 7, 1779, and Panckoucke to STN, April 25, 1779.

would alert Duplain to their suspicions. The associates therefore backed away from a confrontation once again in the spring of 1779. But their positions had hardened, their mistrust deepened. Duplain had only strengthened the secret collaboration between Panckoucke and the STN by maneuvering to get most of the copies of the third edition into his own hands, while paradoxically—or so it seemed—the demand had continued to decline, increasing the pressure on the associates to find some way to market the last of their *Encyclopédies*. Then, just when the pressure was becoming hardest to bear, Duplain came up with another proposal.[43]

The Perrin Affair

In July 1779, Duplain traveled to Paris and warned Panckoucke that the future of the third edition looked even blacker than it had seemed in February. The flow of subscriptions had completely dried up. It seemed certain that the associates would be stuck with 400 quartos when they met at the end of the year to liquidate the enterprise. They would have to divide up the surplus sets and market them individually. But by then they might find it impossible to sell the quarto on any terms, except as scrap paper, because the market was already sated, and it would be ruined irredeemably by the publication of the *Encyclopédie méthodique* in 1780. Fortunately, however, Duplain had found an "entrepreneur" who had caught the *Encyclopédie* fever. He had fired the imagination of this man, whom he did not mention by name, with the prospect of a speculation on the last copies of the quarto. So the associates could dump their unwanted sets on him. To be sure, the entrepreneur demanded extraordinary terms—156 livres per set, 53 percent of the normal wholesale price—but they would be lucky to get rid of their last *Encyclopédies* at any price. And he would take a huge number—all their leftovers, which amounted to 422 sets.[44]

---

43. On this brief cooling-off period in the relations among the associates see Duplain to STN, April 17, 1779; STN to Duplain, April 24, 1779; STN to Panckoucke, May 2, 1779; and Panckoucke to STN, June 1, 1779. In the last letter, Panckoucke agreed to send the plates because of the supreme importance of keeping their suspicions hidden and avoiding an open break with Duplain.

44. On the terms of this deal see the contracts of Aug. 3 and Aug. 13, 1779, in Appendix A. XVII; and on the negotiations see Panckoucke to STN, Aug. 3, 1779, and Duplain to STN, Aug. 3, 1779.

Panckoucke found this offer very attractive. He had not been able to sell more than two dozen sets of the 200 or so sets that he had retained after splitting the 1,000 with Duplain a year ago. He wanted to wind up the quarto quickly in order to devote all his attention to the *Encyclopédie méthodique.* And he felt hungry for the impeccable bills of exchange, which Duplain's entrepreneur promised to provide and which would mature in only six months. The entrepreneur seemed delightfully oblivious to the dangers of the book trade, because he merely asked that the prospectus for the *Encyclopédie méthodique* be held back until August 1780, whereas all the insiders of the business already knew that the great work would soon burst upon the market. And Duplain pushed the proposal hard, using his carrot-and-stick methods of persuasion. He said that his man would also purchase all of Panckoucke's portion of the 1,000 sets they had divided; if Panckoucke refused, Duplain would sell his own portion at half price to the entrepreneur, who would then be able to undersell the other associates. Panckoucke knew that Duplain was a tricky operator, but he wanted to turn this trickiness outward, so that the associates should not be victims of it. Duplain had someone else on his hook. Better to let him land the catch before it got away. They would never have such an opportunity again.[45]

Such was the advice that Panckoucke sent to Neuchâtel on August 3, 1779. In a confidential postscript, he also recommended that the STN attach a condition to its acceptance of Duplain's offer: it should insist on receiving the printing job for the quarto edition of the *Table analytique,* with an allotment of 1,000 livres for the editorial work to be done on each of the four volumes. Duplain sent a letter on the same day, stressing the same themes, with a confidential new twist of his own. If the Neuchâtelois wanted to refuse his proposal, they should tell him so; but they should inform Panckoucke that they had accepted it. Duplain would then supply them with their 5/24 share of the leftover quartos, which they could sell on their own terms behind Panckoucke's back. But he urged them to agree to his proposal so that they could all liquidate the quarto quickly and move on to speculations on the *Encyclopédie méthodique* and the *Table.*

The STN chose to maneuver against Duplain as Panckoucke

45. Panckoucke to STN, Aug. 3 and Aug. 13, 1779.

had recommended rather than to mislead Panckoucke as Duplain had suggested. It sent its consent to the proposal, with the proviso that Duplain let it print the *Table*. Duplain agreed, noting that the decision to sell the surplus 400 now looked wiser than ever because the demand had completely evaporated.[46] Two contracts consummated the sale (see Appendix A-XVII). The first, August 3, 1779, concerned only Panckoucke and Duplain. Panckoucke committed himself to accept the transaction that Duplain was to arrange with the entrepreneur—who was referred to only as "on"—and Duplain was to come up with the bills of exchange that "on" would supply for all the remaining quartos, Panckoucke's as well as those of the association, at 156 livres a set. The second contract, dated August 13, revealed that "on" was a "M. Perrin, commissionnaire à Lyon." Duplain certified that he had received 65,832 livres in Perrin's notes for 422 sheets of the quarto and that Perrin would produce another 24,060 livres for the 160 sets that Panckoucke was holding at his disposition in Paris.[47]

Instead of saving the associates from internecine strife, as Panckoucke had hoped, the Perrin deal brought them to the brink of civil war. It embroiled the STN and Duplain in a quarrel about the printing of the *Table analytique,* and it ultimately confirmed Panckoucke and the STN in their suspicions that Duplain was swindling them. The printing job mattered a great deal to the Neuchâtelois because they needed to keep their shop busy after completing their last volume of the *Encyclopédie* in May. By August they had had to fire half their work force, but they retained an "assortiment de bons ouvriers" in the hopes of finding a big enough commission to relieve "nos presses oisives."[48] They had been expecting since February that relief would come in the form of the *Table analytique*. They had made elaborate preparations for the work, discussing typeface, costs, pressrun, prospectuses, and other details in a series of letters with Panckoucke. And Panckoucke had led them to believe that Duplain would give them

46. Duplain to STN, Sept. 2, 1779.
47. Duplain kept the original version of the contract of Aug. 13 and sent copies to the associates. As the contract did not provide for any free thirteenth sets, it might be said to have given Perrin his quartos for about 60 percent of the normal wholesale price.
48. STN to Bosset, Aug. 30, 1779, and STN to the Société typographique de Lausanne, June 1, 1779.

the job. Duplain, however, planned to produce the *Table* as he had produced the text; that is, he wanted to profit from his role as middleman by contracting the printing to the firms that did cut-rate work for him in Lyons. He therefore felt reluctant about accepting the STN's condition for its agreement to the Perrin sale, and as soon as he had received the agreement, he began to back away from his commitment on the printing. He told the STN that it would have to do the work as cheaply as the cheapest of his Lyonnais printers. He explained quite frankly that he felt entitled to a cut in the printing payments as a reward for his success as an entrepreneur. He had already gathered 1,500 subscriptions for the *Table,* he said, and he expected to take in still more as soon as he published its prospectus. The Neuchâtelois immediately protested that this new demand violated the Perrin agreement. But Duplain would not budge. If they refused to accept his terms, he said, they could drop out of the Perrin deal and take their 5/24 share in Perrin's 422 quartos. Perrin would be delighted to return them because he was beginning to realize that they were unsellable. For the same reason, as Duplain knew very well, the Neuchâtelois could not afford to add more quartos to their own stock. Nor could they continue much longer to maintain a large shop without a large printing job. So they gave in and agreed to work on his terms. They never got the commission, however, because he kept postponing the printing; and when they complained, he stopped answering their letters. So the *Table* provided material for another dossier to be argued over at the liquidation in Lyons.[49]

While the STN grappled with Duplain over the *Table,* Panckoucke tried to come to grips with the elusive Perrin. Perrin proved to be difficult to deal with because, according to Duplain, he demanded that Panckoucke pay for the transport of his 160 quartos from Paris to Lyons. Panckoucke

49. The story of the preparations for the quarto *Table* has considerable interest, especially for typographical history, but it would require too long a detour to be recounted here. The most important of the many letters concerning it, all from 1779, are: STN to Panckoucke, Feb. 25; Panckoucke to STN, March 7; STN to Panckoucke, March 14; Panckoucke to STN, March 18; STN to Panckoucke, March 23; STN to Panckoucke, May 9; Panckoucke to STN, June 1; Panckoucke to STN, Aug. 3; STN to Bosset, Aug. 7; Duplain to STN, Sept. 2; Duplain to STN, Sept. 9; STN to Panckoucke, Sept. 19; Duplain to STN, Sept. 20; STN to Panckoucke, Sept. 30; STN to Panckoucke, Oct. 15; and Duplain to STN, Nov. 9.

refused, on the grounds that the contract of August 13 only obligated him to hold Perrin's quartos "à sa disposition" in Paris. But according to Duplain, Perrin would back out of the sale if Panckoucke did not assume the transport costs, which were certain to be very expensive. In fact, Duplain warned that Perrin was looking for an excuse to cancel the deal. But Panckoucke stood his ground, even though Perrin threatened to take him to court—according to Duplain.

"According to Duplain" seemed to be the only form through which Perrin could express himself. Panckoucke never dealt directly with Duplain's entrepreneur and did not even learn his name until Duplain described him in a letter as "un nommé M. Perrin, commissionnaire de Strasbourg, qui a une maison à Lyon, une je crois à Paris, et enfin un homme excessivement riche dont je vous réponds."[50] That sounded suspiciously vague to Panckoucke, who by now had learned to suspect everything emanating from Lyons. "Je vous serai obligé," he wrote to the STN,

de vous en faire informer par M. d'Arnal, sous main, si ce M. Perrin n'existe pas, si même ce n'est qu'un prête-nom. C'est une friponnerie de Duplain que nous ne devons pas souffrir, parce que nous n'avons consenti à la vente de ces exemplaires que parce qu'il nous a assuré qu'on n'en vendait plus du tout et que même il *nous a menacé si nous ne voulions pas consentir à cette vente qu'il était déterminé à vendre à ce prix ses exemplaires.* Priez M. d'Arnal de prendre les informations les plus secrètes. Nous ne devons pas souffrir d'être la dupe du très avide Duplain.

In a postscript, Panckoucke reminded the STN to keep to their strategy of hiding all suspicions: "Si les informations de M. d'Arnal sont conformes à mes soupçons, il faut les tenir très secrètes, laisser aller M. Duplain en avant etc."[51]

Industrial espionage is no invention of the twentieth cen-

50. Panckoucke to STN, Sept. 10, 1779, citing a letter he had recently received from Duplain.
51. Panckoucke to STN, Sept. 10, 1779. The emphasis is Panckoucke's. For some reason, perhaps because he was upset, Panckoucke wrote two letters to the STN on the same day about his suspicions. The second version read: "J'ai demandé à Duplain le nom de l'acquéreur de *l'Encyclopédie.* Il m'a nommé un certain Perrin. Je crains bien encore que ce soit un tour qu'il nous joue et qu'il ne soit lui-même l'acquéreur sous le nom de ce Perrin. Quoiqu'il puisse en être, il faut user jusqu'au bout de tous les ménagements possibles. Il faudra bien que ce Perrin se fasse connaître, sa demeure, sa qualité. Je désire que mes soupçons ne soient pas fondés."

tury, and it was not new to the speculation on the *Encyclo-pédie*. The quarto associates had spied on one another from the beginning. During its first negotiations with Panckoucke, the STN had conducted a secret investigation of him, using his neighbors as informants and Perregaux as an agent. It commissioned an underground bookdealer called Quandet de Lachenal to do another investigation in 1781. Panckoucke had received confidential reports about Duplain's operations from Regnault, and Louis Marcinhes kept the STN informed about the activities inside the Genevan printing houses. Spying seems to have existed everywhere in the eighteenth-century book trade. The word *espion* occurs frequently in the correspondence of the booksellers, just as it appears in the titles of some of their most popular books: *L'Espion turc, L'Espion anglois, L'Espion dévalisé*. It covered a wide range of activities, some of them rather innocent. Like many traveling salesmen, Favarger was spying when he interviewed workers, peeked at sales registers, and filched newly printed sheets. More serious was the practice of bribing workers to provide sheets for pirated editions. The STN warned Beaumarchais that the printers in Kehl were certain to supply sheets of his Voltaire to pirates who wanted to beat him to the market. And it also put its customers on guard against double agents in the underground book trade—men like Desauges of Paris, Poinçot of Versailles, and Mallet of Troyes, who ordered illegal books and then turned the suppliers over to the police. The French police even had agents in the STN's own shop, according to reports from Jacques-Pierre Brissot, who probably was a police spy himself. Spying had become such a common activity that it seemed natural for the STN to plant a man in Duplain's shop. In fact, it had received secret reports on the shop from d'Arnal for two years before Panckoucke asked it to investigate Perrin, and so it requested d'Arnal to undertake the investigation.[52]

The STN shared all of Panckoucke's suspicions about Perrin: "Plus nous examinons l'affaire de la vente . . . et plus nous y découvrons d'obscurités, de contradictions même,

---

52. On the STN's warning to Beaumarchais see Ostervald to Bosset, May 3, 1780; on its relations with Brissot see Robert Darnton, "The Grub-Street Style of Revolution: J.-P. Brissot, Police Spy," *Journal of Modern History*, XL (1968), 301–327.

et de procédés suspects."[53] After receiving his instructions from the STN, d'Arnal replied that he would undertake the mission, although it would not be easy:

Vous nous chargez, Messieurs, d'une commission bien difficile à exécuter. Comment sans se compromettre pouvoir questioner, interroger, surtout étant connus pour avoir des relations avec vous? Vous sentez, Messieurs, que nous sommes par cela même très suspects à ceux qui pourraient nous donner quelques éclaircissements. Cependant, nous avons engagé un ami, qui est fort lié avec M. Perrin, à nous prêter son ministère pour sonder ledit sieur. Il usera de toute l'adresse possible pour tâcher d'arracher audit sieur son secret, et s'il n'y réussit pas, personne ne pourra se flatter de réussir. Mais ledit Perrin étant chargeur de M. D xxx, est par cette raison très intéressé à le ménager. Il est fort à craindre qu'il ne se détermine pas aisément à le trahir . . . Dès que notre espion aura fait quelque découverte, nous nous hâterons de vous en instruire.[54]

D'Arnal added that Perrin ran a shipping firm, Carmaignac et Perrin, which was believed to be wealthy and which did indeed have a branch in Strasbourg. It struck him as most unlikely that a shipper would speculate in publishing, so he agreed with the hypothesis that Duplain was using his shipping agent as a straw man. But he would have to proceed with extreme caution if he wanted to acquire enough evidence to prove the case against Duplain, for Duplain knew that d'Arnal was intimately connected with the STN, and Duplain's connections with Perrin might be just as close. D'Arnal therefore kept in the background. He counseled his spy to move carefully and to wait for a casual encounter with Perrin, as a special visit might arouse suspicions.[55] They had not made contact by October 10, the last date on which "P xxx" was mentioned in the letters from d'Arnal that have survived in the papers of the STN. At that point, d'Arnal's man disappears from the documents. He may well have produced some important information for Ostervald and Bosset four months later, when they arrived in Lyons for the final settling of accounts. D'Arnal himself fought at their side throughout that last encounter with Duplain. But there is no record of his activities. So one can only say that the final phase of the

53. STN to Panckoucke, Sept. 19, 1779, in Bibliothèque publique et universitaire de Genève, ms. suppl. 148.
54. D'Arnal to STN, Sept. 24, 1779.
55. D'Arnal to STN, Sept. 28, 1779.

quarto's history involved an espionage operation directed against the manager of the enterprise by two of the associates.[56]

While d'Arnal pursued his investigation in Lyons, Panckoucke and the STN continued on their own to accumulate incriminating evidence against Duplain. The STN learned from a French bookseller that Duplain had claimed, before the Perrin sale, that the third edition was virtually sold out—an innocent remark, perhaps, but one that made the STN wonder where Perrin's 422 quartos came from.[57] Panckoucke discovered a printed flyer, which had been circulating among French booksellers and which offered the quarto at a discount in Pellet's name. He concluded that it represented an underhand sales campaign by Duplain similar to the Batilliot plot of November 1778: "Je suis persuadé que Pellet n'est que le prête-nom de Duplain, et que ce Perrin n'est qu'un homme de paille ou un prête-nom comme Pellet. Enfin, ne dites rien, et laissez-le agir. Si cette vente [that is, the Perrin sale] n'est pas réelle, il faudra bien qu'il nous tienne compte."[58] Why should Duplain continue his efforts to acquire and sell quartos, if the demand had been exhausted and the supply turned over to Perrin, the associates wondered. Duplain's surreptitious marketing seemed especially insupportable to Panckoucke because in October Panckoucke received a copy of the prospectus for the *Table analytique,* which Duplain had written and published without consulting the associates. It promoted Duplain's quarto *Table* by denigrating the folio *Table* that Panckoucke had just finished printing—an unnecessary swipe at one of Panckoucke's side speculations, which threw him into a rage.[59]

Just when the associates were finding it most difficult to contain their suspicions and their anger, Duplain sent them a proposal for selling still more quartos to Perrin with a highly suspect "Aperçu" about the present state of the sales:

56. D'Arnal probably included the reports on his investigation in the private correspondence he maintained with Bosset, his father-in-law, which has not been preserved.

57. STN to Panckoucke, Sept. 19, 1779, Bibliothèque publique et universitaire de Genève, ms. suppl. 148.

58. Panckoucke to STN, Oct. 2, 1779.

59. Panckoucke to STN, Oct. 11, 1779, and STN to Panckoucke, Oct. 15, 1779.

6589 placés à divers
500 remis à M. Panckoucke
500 à moi [Duplain]
422 vendus suivant la dernière convention à 156
livres [to Perrin]
_____
8011[60]

8011 turned out to be an important number for the quarto enterprise. It was Duplain's figure for the output of marketable *Encyclopédies*—the number of complete sets that he had put together from the sheets sent in by his printers, the number that he had marketed and for which he agreed to be held accountable. The contracts for the three editions had set pressruns totaling 8,550, and later documents indicate that 8,525 copies were actually printed. Why was there such a disparity between the number of quartos produced and the number that Duplain claimed to have sold? Duplain avoided discussion of production figures, but he indicated that his printers had supplied him with material for far more than 8,011 sets, because he emphasized the number of *défets* he had accumulated. *Défets* were sheets that had been spoiled in printing, transport, or warehouse work. Since a spoilt sheet could ruin a volume and a ruined volume could ruin a set, Duplain had his workers put aside all the spoilt sheets they found in the shipments that poured into his warehouses from the twenty or so shops that printed the thirty-six volumes of the three editions. To unpack, check, assemble, store, and ship out all those *Encyclopédies* was a complex operation. Duplain's employees worked in confusion and haste, spoiling still more sets as they proceeded. In the end, he said, they filled two warehouses with *défets*. But 100 to 130 sets, perhaps even more, could be salvaged from the *défets* by printing a few missing sheets. Duplain therefore proposed that the associates authorize him to put together a scrap edition and to sell the new quartos at 156 livres a set. He thought Perrin might be interested in buying them.[61]

60. Duplain to STN, Oct. 11, 1779.
61. Duplain to STN, Oct. 11, 1779: ''Je craignais bien que les *gachis* énormes qu'on a faits dans mes magasins, dans les envois, nécessités par la promptitude de l'opération, qui a obligé d'employer de toutes sortes de personnes, ne nous eussent privé de bien d'exemplaires, et ne nous eussent pas mis dans le cas de faire une nouvelle récolte que voici. Comme le tome 36 va finir, je vais faire ranger

Duplain did not explain why the associates should spend more money to produce more *Encyclopédies* at a time when the demand seemed to have hit bottom. Nor did he account for Perrin's willingness to buy another hundred quartos so soon after discovering—according to Duplain—that there was no market for those he had acquired by his previous purchase. Strangest of all was Duplain's claim that only 6,589 of the 8,011 copies had been sold to subscribers; for the secret subscription list that the Neuchâtelois had copied from his books in February 1779 accounted for 7,373 subscription sales. Even the figure of 8,011 complete sets seemed questionable, because Duplain had insisted that Panckoucke print the plates at the much higher pressrun of 8,600. ''J'ai reçu le calcul en gros des exemplaires envoyés par M. Duplain,'' Panckoucke wrote to the STN after receiving the ''Aperçu.'' ''Les planches sont le thermomètre de la vente. On les tire à 8,600. En supposant exact le calcul que nous avons fait à Lyon de 8,309 et en y joignant 130 ou 150 provenant des défets, nous nous rapprocherions beaucoup du nombre exact, mais nous serions bien éloignés du calcul de M. D. Tout cela ne peut se vérifier que sur les lieux.'' Only one explanation seemed to make sense of all those puzzles and paradoxes: Duplain had turned the enterprise into a gigantic confidence game.[62]

Panckoucke and the STN had pretty well adopted that view by the autumn of 1779. But they continued to deal with Duplain as if they suspected nothing. They refused his proposal for the scrap edition but maintained a polite, businesslike tone

---

les defets, et ce qui va vous surprendre, c'est que j'ai deux magasins remplis. Il est donc question de savoir si la compagnie veut suivre mon avis, qui est de refaire les feuilles qui compléteront les exemplaires; et je crois qu'avec une très modique dépense nous ferons 100, peut-être 130, peut-être plus encore d'exemplaires . . . Quant à moi, Messieurs, mon avis est qu'on travaille sur le champ aux défets, qu'on refasse les feuilles qui produiront avec un gros avantage des exemplaires complets, et qu'on vende 156 livres les exemplaires revenants nets de l'opération, si M. Perrin, qui a acheté les autres, le veut.'' The term *défets* was used loosely to cover both spoilt sheets and leftover sheets from sets that had been spoiled. Extensive spoilage created a large surplus of leftover sheets and therefore a possibility of constructing a scrap edition.

62. Panckoucke to STN, Oct. 25, 1779. Panckoucke stressed this point again in a letter to the STN of Oct. 31, 1779: ''On tire les planches de l'*Encyclopédie* à 8,600. Il faut donc qu'il y ait 8,600 de discours.'' As mentioned, the STN had believed earlier that Duplain might be cheating on the size of the editions, but his willingness to let them print a volume from the third edition had quieted their suspicions. He could have had extra copies of that volume printed, however, or he could have ordered an unusually large printing of it in the first two editions.

in their correspondence with him, while venting their anger in their letters to one another. That was a difficult role to play, especially for the Neuchâtelois, who complained to Panckoucke, "Nous observons qu'en général nous sommes constamment sur la défense et uniquement occupés à parer les bottes qu'il plaira de nous porter. Cette position n'est rien moins que la plus avantageuse. N'aurons-nous point en main de quoi agir un peu offensivement contre lui?"[63] Their desire to strike back had grown almost ungovernable in June, when they went through a final round of their financial quarrels with Duplain. Having completed the printing of their last volume, they had notified him that his debt had mounted to 35,000 livres. Duplain replied that he owed 6,000 livres, and refused to honor 29,000 livres in the STN's bills of exchange. The STN tried to collect its bills through d'Arnal, protesting that "on ne peut rien de plus cruel que leurs procédés à notre égard."[64] But d'Arnal replied that Duplain claimed "qu'enfin il n'avait rien à démêler avec vous, Messieurs, et que c'était par pure complaisance qu'il répondait aux lettres que vous lui écrivez."[65] Soon afterward Duplain stopped answering most of the STN's letters in addition to rejecting most of its bills. There was little that the STN could do but wait to retaliate at the final accounting, while fuming secretly to Panckoucke, who replied in kind.

PANCKOUCKE: Je suis bien persuadé que ce Perrin n'est qu'un homme imaginaire ou tout au moins un prête-nom. D. est avide et ne se pique d'aucune délicatesse. Il faut bien se laisser vendre les exemplaires, et puis nous lui demanderons compte, la vente à Perrin n'ayant pas eu lieu . . . Faites en sorte que Duplain ne soupçonne rien.

STN: Tout concourt à appuyer vos conjectures quant à notre associé touchant le véritable acquéreur des exemplaires vendus en dernier lieu, et nous avons lieu d'espérer que nos amis de Lyon découvriront encore quelque chose à ce sujet. Mais quoiqu'il en soit, nous ne pourrons manquer d'en savoir le vrai lors de la réddition [*sic*] des comptes . . . Nous ne sonnerons mot jusqu'à l'époque fatale, qui heureusement n'est pas éloignée.

PANCKOUCKE: Je pense toujours que ce Perrin est un homme de paille. Duplain nous joue. Je ne serai point dupe de sa cupidité.

---

63. STN to Panckoucke, May 9, 1779.
64. STN to d'Arnal, June 23, 1779. "Leur" referred to "Duplain et compagnie." Companies normally used the plural form in their commercial correspondence.
65. D'Arnal to STN, June 27, 1779.

And so the dialogue continued, as suspicion piled on suspicion, intrigue compounded intrigue, and the plots and subplots gathered together until everything culminated at the same point: the *règlement de comptes* at Lyons in February 1780.[66]

The Anatomy of a Swindle

Duplain and his associates argued over the accounts for sixteen days—until February 12, when they agreed on a general settlement. The STN and Duplain then continued to argue over their particular accounts, which were finally resolved by arbitration on February 21. Almost a month of complicated and impassioned wrangling; and to reconstruct it, only the flotsam and jetsam that it left in its wake and that Ostervald and Bosset swept up and took home in their baggage. In one of the letters that he wrote while the storm was raging. Bosset complained about having to recreate the accounts from "les chiffons de Duplain."[67] The historian must work from Bosset's own scrap paper, the *chiffons* of *chiffons*: notes taken during debates, sums scrawled in the course of strategy sessions, and memos dashed off at critical moments. In the absence of coherent documentation and in the interest of clarity, it seems best to begin with a summary of all the swindles that Duplain's partners discovered, even though the discoveries came at different stages of a long, confused struggle.[68]

The most incriminating evidence that Duplain's associates brought to Lyons was the subscription list they had secretly copied from his books a year earlier. They kept this weapon in reserve and forced Duplain to come up with a list of his own in order to justify his version of the financial situation. Then they retired to their inn, compared the two lists, and discovered a monumental swindle. Thanks to Bosset's margin notes on his copy of the secret list, one can follow the fraud, subscription by subscription, as Duplain's associates unraveled it. Every time Bosset found an entry on the first list (the secret one of 1779) that was not on the second (the falsified

66. Panckoucke to STN, Sept. 27, 1779; STN to Panckoucke, Oct. 3, 1779; and Panckoucke to STN, Oct. 15, 1779.
67. Bosset to Mme. Bertrand, Feb. 13, 1780.
68. Most of the notes can be found in two dossiers of the STN papers entitled "Procès STN contre Duplain" and "Dossier Encyclopédique."

one of 1780), he jotted the number of missing subscriptions in the left-hand margin. Thus, "4 . . . Bergeret de Bordeaux 58 souscriptions," meant that Bergeret had bought 58 subscriptions by February 1779 but Duplain attributed only 54 to him in 1780. In the right-hand margin, Bosset penciled in subscriptions reported by Duplain that had escaped their attention when they compiled their secret list in 1779. Thus, "Veuve Brun de Nantes 4 souscriptions . . . 1," meant that Duplain credited Veuve Brun with 5 subscriptions, though the associates had found only 4 in his books. Bosset added up the numbers in the right-hand margin: 137. Then he added that sum to the total on the secret list: 137 + 7,873 = 8,010—only one short of the 8,011 that comprised their entire stock according to Duplain's final report. Although he never found that last missing quarto, Bosset was able to account for every other *Encyclopédie* on Duplain's list. And by adding up the figures in the left-hand margin, he was able to show precisely how many subscriptions Duplain had hidden: 978. The differences between the two lists is represented graphically in Figure 9.

The secret list included the 500 quartos that Panckoucke had taken from the 1,000 divided with Duplain, but it did not mention Duplain's 500, nor Perrin's 422. Thus when Duplain bewailed the collapse of the subscriptions to his associates in February 1779, he knew that all three editions had been sold out, except for the 500 he had dumped on Panckoucke. He had hidden the sales in order to collect the full amount for them, while paying nothing for 500 of the quartos that he sold in the association's name and paying for 422 at half price through the phony intermediary of Perrin. The subscription flow really had abated by 1779, but enough orders continued to trickle in for Duplain to compound the swindle by a more audacious stroke. Posing as Perrin, he offered to buy back Panckoucke's unsold *Encyclopédies* (a matter of 166 sets) at half price. And after Panckoucke agreed, Duplain attempted to procure still more cut-rate quartos for "Perrin" by salvaging 200 sets from the *défets*. The associates refused because at last they realized that Perrin was a straw man. Duplain's "Aperçu" of October 1779, which later proved to be an accurate preview of his final version of the subscriptions, showed that the three editions had produced only 8,011 sets and that only 6,589 of that number had been sold directly to subscribers, the other

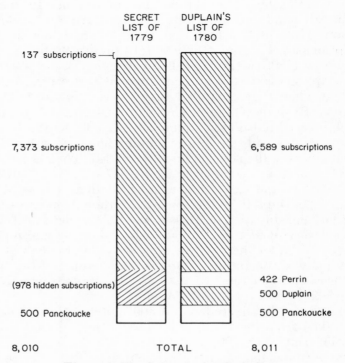

Figure 9. The Subscription Swindle

The only surviving version of the subscription lists is the copy in Bosset's handwriting in the STN papers, dossier "Procès Duplain contre STN"; the only reference to the secret copying of it is in the letter from Ostervald and Bosset to the STN, dated from Lyons on February 13, 1780: "le registre des souscriptions . . . que nous relevâmes furtivement et sans qu'il [Duplain] s'en doutât il y a un an." (For the text of this key letter see Appendix A. XIX.) Bosset's margin notes accord perfectly with a memo that he wrote entitled "Souscriptions dans le premier registre qui ne sont point dans le dernier." Both manuscripts fit the version of the swindle described in the "Premier mémoire de M. Plomteux" (see Appendix A. XIX). These three documents make it possible to reconstruct Duplain's list of 1780.

The only ambiguity concerns the dating of the 137 sales that appeared on the second list and not on the first. It would seem that those sales occurred during the twelve months after February 1779. In that case, the list of February 1779 covered 98 percent of the sales and not the full 8,011 subscriptions. But Plomteux's memoir and some other notes by Bosset indicate that the 137 sales had been made by the time the secret list was compiled. Moreover 16 of those 137 subscriptions concerned the first two editions, which certainly had been sold out by then. So Duplain was right when he complained about the decline of the subscription rate during the last twelve months of the enterprise: all three subscriptions had already been filled. It was the very success of the sales that made it possible for him to swindle his partners by pretending that the third edition was a failure.

The manuscript for the secret list is complicated by the provision for the free thirteenth copies. Its entry for Bergeret actually read "4 . . . Bergeret de Bordeaux 58 souscriptions pour 54." meaning Bergeret had taken out 54 subscriptions and therefore had earned four free sets for the four dozen subscribed. The entry for Bergeret on Duplain's list must have read "Bergeret de Bordeaux 54 souscriptions pour 50." Duplain frequently trimmed off a few copies from large orders like Bergeret's. But more often he eliminated entire orders, the most flagrant case being Gaston of Toulouse, who ordered "130 . . . 130 pour 120 souscriptions," none of which appeared in the list of 1780.

The comparison of the two lists raises one final question: why did Duplain hide 978 subscriptions from his associates when he acquired only 922 cut-rate and free *Encyclopédies* from them? The answer seems to be that he would have more than covered the difference of 56 sets if he had received the additional 166 from Panckoucke. He could also fall back on the free thirteenths that he was hiding. And he probably produced more than the 8,011 sets that he declared, or at least planned to produce a scrap edition.

1,422 having gone to Panckoucke (500), Duplain (500), and Perrin (422). By merely counting up the 7,373 subscriptions on their secret list, the associates could measure the extent of the swindle. By this time, Duplain's attempt to wring a few more livres from the Perrin fraud seemed almost comical. He threatened that Perrin would sue, if Panckoucke did not pay for the transport of the 166 sets that were to go back to Lyons from Paris. Panckoucke was quite ready to meet an *homme de paille* in court; but feigning a desire to avoid confrontation, he agreed to cancel the sale of the 166 sets and continued to hide his knowledge of Perrin's identity in order to ambush Duplain at the meeting of February 1780.[69]

The ambush could not succeed unless the anti-Duplain forces could prove the validity of their secret list. D'Arnal's

69. On Nov. 14, 1779, Panckoucke wrote to the STN: "Il [Duplain] vient de me mander que désirant m'éviter un procès, il a engagé Perrin à me rétrocéder les 160 exemplaires de mes *Encyclopédies*, qu'il voulait à Paris à 156 livres sans port. J'ai accepté cette rétrocession, qui me confirme de plus en plus que tout cela n'est qu'un jeu et que Perrin n'est qu'un prête-nom. Je suis bien sûr qu'il n'y a eu aucune procédure et qu'il n'a pu y en avoir." Nine days later, Duplain warned Panckoucke, "Vous avez tort, mon bon ami, de ne pas croire au procès que vous faisait à juste titre le sieur Perrin." On Nov. 27, Panckoucke replied solemnly, and without revealing his knowledge about Perrin, that he stood by his version of the transport costs but would take back the disputed sets. On the same day he sent a copy of both letters to the STN, which had experienced enough of Duplain's bullying and bluffing to appreciate the irony of the situation.

spy may have got Perrin to talk, but d'Arnal probably gave Ostervald and Bosset an oral report on the results of his investigation, and there is no record of it. Once Duplain submitted his own subscription list, however, his associates had a way to verify the fraud. They wrote letters to several of the booksellers, whose subscriptions had been falsified according to a comparison of the two lists. The letters simply asked how many quartos each bookseller had purchased, and the answers confirmed Bosset's calculations. So the Panckoucke group possessed irrefutable proof that Duplain had cheated them of subscriptions that could be valued as high as 287,532 livres.[70]

Once they got Duplain's list of 1780 into their hands, Panckoucke and his partners had an opportunity to examine one of its most interesting entries: "Audambron de Salasy & Jossinet 535 souscriptions pour 494." That meant that the Lyons firm of Audambron et Jossinet had subscribed for 494 quartos at the usual wholesale price for booksellers of 294 livres plus one free set for every twelve subscriptions, which brought their total up to 535. The same firm appeared on the secret list of 1779 with "535 pour 535," that is, as having subscribed for all 535 quartos without the benefit of the baker's dozens. The secret list did not mention the price of the subscriptions, but because they concerned local sales, there was good reason to suspect that they had occurred at the retail price rather than at the wholesale *prix de libraire*. Audambron and Jossinet were not *libraires* but businessmen, like Perrin. And like Perrin, they had every appearance of operating as a false front to hide sales that Duplain had secretly made at the ex-

---

In retrospect, it seems odd that Panckoucke should have accepted the Perrin deal in the first place. But the situation still looked confused in the summer of 1779— and at that time Panckoucke believed that Duplain had produced about 8,400 complete quartos, not 8,011—so it probably appeared credible that the Perrin sale would rid them of their last sets at a moment when the market had been sated.

70. On the letter-writing campaign see Ostervald and Bosset to the STN from Paris, March 10, 1780: "Il a fallu écrire plusieurs lettres de Lyon pour nous assurer par des lettres des souscripteurs de la fausseté des registres." As an example of the replies they received see Ranson of La Rochelle to STN, Feb. 19, 1780. The cost of the swindle for the Panckoucke group can be estimated in different ways. A total of 287,532 livres represents the value of the 987 sets at their wholesale price without deducting anything for the free thirteenths, even though Duplain probably sold several baker's dozens. The Panckoucke group's half interest in the enterprise entitled it to half that sum, minus "Perrin's" payment of 65,832 livres, making 77,934 livres. But Panckoucke and his supporters actually developed a different argument in demanding compensation, as shown below.

pense of the association. The Panckoucke group therefore investigated Audambron and Jossinet. They discovered that the firm had not taken out any subscriptions at all but merely had operated as Duplain's local sales agency, having agreed to work for him at a commission of 15 sous per volume or 7½ percent of the wholesale price instead of the 25 percent profit received by genuine retailers. Therefore Duplain had cheated on the sales in Lyons just as he had attempted to cheat on the Parisian sales through the intermediary of Batilliot.[71] He had sold 535 quartos at 384 livres each (205,440 livres) instead of 494 quartos at 294 livres each (145,236 livres), thereby swindling the association of 60,204 livres.

As Duplain had twice used straw men to hide secret sales, it also seemed possible that he had sold more than the 8,011 sets mentioned in his accounts. His associates did not even know how many quartos he had printed, although the sum of the pressruns probably was 8,525. In that case, Duplain had 514 incomplete (*défectueux*) sets—a number that seemed excessive (6 percent of the output). Had Duplain sold some of those 514 quartos on the sly? His offer to salvage 200 of them for another half-price sale to Perrin sounded suspicious, especially as the 200 sets reappeared in the credit column of his accounts for 30,000 livres. They would have fetched 54,390 at the wholesale price (deducting for free thirteenths). So Duplain might have swindled his partners for another 54,390 livres. They believed he had. In their calculations of his fraud they included that sum to cover "le reste de l'édition, qui est sûrement vendu." But without a genuine subscription list from 1780, they could not prove their case; and they had to reduce their charge to criminal negligence, as they explained in a memorandum entitled "Griefs contre M. Duplain," which attacked his general management of the enterprise. Having in all likelihood cheated the association on the salvageable *défets,* Duplain sold the rest of them, a huge mountain of scrap paper, to Jossinet for 20,000 livres. And, to the discomfiture of the STN, he also sold the speculation on the quarto edition of the *Table analytique* to his associate Amable Le Roy for 50,000 livres. All of these maneuvers and manipu-

---

71. The STN's papers do not contain a full report on this investigation, but the STN made the accusation emphatic—"nous [le] savons et sommes en état de le prouver"—in a deposition that it submitted to the arbitrators of its case against Duplain. "Dernières observations" and "Mémoire contre M. Duplain" in the dossier "Procès STN contre Duplain."

lations lay behind Duplain's version of the association's income, which came to 1,851,588 livres in all.[72]

Duplain's version of the expenses seemed equally suspicious. The contracts entitled him to make some money from his role as middleman in the printing because they set fixed prices for every sheet, regardless of the actual cost. But he seems to have gone beyond the bounds of customary profit-taking. He told Favarger that he cleared 1,500 livres from the printing of every volume of the first two editions; he might have made more, because the STN averaged 5,612 livres in profits from the four volumes it printed for those editions at the price stipulated in the contracts. His profit margin was even greater in the case of the third edition. So his rake-offs on the printing probably totaled about 75,000 livres.[73]

There was nothing the associates could do to force Duplain to disgorge that sum, but they had evidence that he had deliberately multiplied the number of sheets in order to increase his profits from the rake-offs at their expense. As explained above, the STN knew from its assistant foreman, who had worked on the *Encyclopédie* in Pellet's shop, that Pellet had used fraudulent techniques of spacing and paragraphing—

72. As mentioned, the associates suspected Duplain of printing extra copies at several points before the final meeting. The preliminary *règlement* of Feb. 10, 1779, stated that 6,150 sets of the first two editions had been printed, as was required by the first two contracts. But it said that the third edition was being printed at 2,375 copies, although the third contract set the pressrun at 2,400. The commercial correspondence of the associates contained several contradictory references about the size of the third edition during the next few months. For example, on Feb. 18, 1779, Duplain instructed the STN to print volume 19 of the third edition at 4 reams 17 quires (2,425 copies) "pour fournir aux imperfections des deux éditions premières." And in a letter to Panckoucke of March 23, 1779, the STN calculated the "montant de l'édition" at 2,360. Panckoucke's final accounts on the plates, which he called "le thermomètre de l'affaire," showed that they were printed at 8,600, as he had claimed, but they may have included more spoilt sheets than did the printing of the text. All things considered, 8,525 is probably an accurate estimate of the gross output of quartos. The critical issue was the number of sets that were ruined. Bosset asserted that Duplain had sold 54,390 livres' worth of those *défectueux* sets in a memo that he wrote for the associates, "Tableau de ce qui devrait nous revenir de l'entreprise."

73. Without going into the long calculations behind this estimate, one should note two considerations. Owing to the printing done by the STN, Duplain profited from the difference between the contractual and the actual printing costs in the case of thirty-two volumes from the 6,150 copies of the first two editions and thirty-five volumes from the 2,375 copies of the third edition. Duplain had to allow some profit to the printers who worked for him, but he forced them to accept very hard terms, particularly in the case of the third edition. Favarger's remark occurred in a letter to the STN of July 15, 1778: "Il me dit qu'il y a environ 1500 livres à gagner pour lui de faire imprimer ici plutot que chez nous."

tricks that might seem trivial, but that expanded volume 19 by ninety-six unnecessary pages, worth 744 livres. Since Pellet worked hand in glove with Duplain, it seemed probable that they had collaborated on this swindle and that Duplain had made similar arrangements with his other printers. The STN denounced this "connivance punissable" in its "Griefs contre M. Duplain." It held Duplain responsible for making the volumes far too fat—up to 136 sheets per volume instead of the 110–115 that it considered as a maximum for the quarto format and the 90 that had originally been planned, "ce qui faisait son compte comme imprimeur mais non point celui de l'entreprise."

Having padded the volumes, Duplain went on to pad his expense account. On January 28 he submitted a "Compte général du coût de chaque volume," a strange document, because he took the volume rather than the edition as the unit in summarizing his expenses. Thus he charged the association 37,214 livres for printing volume 1 in all three editions, 33,590 livres for printing volume 2, and so on, making 1,361,385 livres in all. He received a fixed rate for every sheet, but the number of sheets per volume varied. So by lumping the editions together, he could slip a few fictitious sheets into the charges for each volume without arousing suspicion. Unknown to him, however, the associates had been looking for slips for more than a year. They took his "Compte général" back to their inn, procured a copy of the first, second, and third "editions," and started counting sheets. One can imagine them surrounded by stacks of quartos, thumbing through seventy-two volumes sheet by sheet, and calling off sums to Bosset, who kept a tally on one scrap of paper, scribbled calculations on another, and recorded his conclusions in one simple arithmetic expression:

> 1,361,385 [Duplain's reported costs, in livres]
> 1,234,296 dépense réelle
> _____
> 127,089 trop porté[74]

In addition to the printing costs, Duplain had lined up a great many other items in the debit column. He put down

---

74. The notes, in Bosset's hand, are on loose sheets accompanying the STN's memorandum, "Dernières observations à l'article des 13me et 2 livres 10 sous par exemplaire" in the dossier "Procès STN contre Duplain."

3,000 livres for the use of Pellet's name on the title page, 27,000 livres for the ransom paid to Barret and Grabit, 33,150 livres to Laserre for preparing the copy for the first two editions, and another 3,000 livres for his work on the third edition. The STN found this last debit particularly objectionable because it believed that Laserre had not done any new work at all on the copy for the third edition. But 3,000 livres was a trifle compared with Duplain's expense allowance, which had caused such strife during the contract negotiations for the third edition. Panckoucke had agreed to increase the sum fixed for Duplain's expenses to 16,000 livres. But Duplain wrote into the contract a clause acknowledging his claim to have a "dépense . . . infiniment plus considérable" and indemnifying him with an "insurance" provision. He was permitted to charge the association for transporting the third edition from Lyons to Geneva, even though he planned to keep it in Lyons. The charge was construed as a fee for assuming the responsibility for any losses that might come from a raid on his shop by the Lyonnais authorities. The real danger, however, was not that the authorities would confiscate the books but that Duplain would inflate his fee. The contract did not set a fixed price for his insurance service, although the associates could have calculated the price easily enough, using the normal transport cost per quintal as a multiplier for the estimated tonnage of the volumes. "N'avoir pas pris la plume pour chiffrer nous coûtera bien cher," the Neuchâtelois wrote to their home office on February 10, 1780. Duplain demanded 104,000 livres for the fictitious transport costs, and they did not see how they could parry that blow. They also objected that he was demanding "sacrifices horribles, tels que de lui allouer 60,000 livres pour menus frais." He had larded his expenses shamelessly, they lamented, and he fought every attempt to trim the fat off them.[75]

Finally, the associates wanted compensation for Duplain's general mismanagement of their affairs. Each of them had been pummeled with complaints from customers whose

75. Duplain reported his expenses in a "Premier compte" submitted on Jan. 28; they can also be known from notes that Bosset took on that occasion, although there are some discrepancies between those notes and some other documents. The STN protested bitterly against the 3,000 livres given to Laserre for his work on the third edition in its "Griefs contre M. Duplain." It is not clear why Duplain should have allotted 33,150 livres to Laserre for the first two editions instead of the 30,600 livres required by the contracts.

quartos had arrived late, or with missing volumes, or with damaged sheets, or with excess packing and transport charges. In his haste to get the *Encyclopédies* distributed and paid for, Duplain had produced chaos in his warehouses and shipments. He refused to replace damaged sheets and even to answer requests for replacements. He wrote bills of exchange on subscribers faster than the subscribers could pay for them. And when they requested a delay in the payments, he threatened to sue. Meanwhile, he had exposed the association to suits from subscribers who had contracted to pay for twenty-nine volumes and found themselves charged with thirty-three, owing to his miscalculation about the number of volumes required for reducing the folio text and *Suppléments* into the quarto format. Every time he bungled, he played into the hands of booksellers who were looking for a pretext to refuse their payments; for some of his subscribers were as unscrupulous as he, and others were incapable of paying because they went bankrupt. So when Duplain made his financial report to the associates, he accompanied his accounts with an appalling "Note des débiteurs que M. Duplain reconnais insolvables ou chicaneurs": another 128,600 livres to deduct from their profits.[76]

Each of the associates arrived in Lyons with his own set of "griefs" against Duplain. Panckoucke wanted a reckoning for the 166 quartos that Duplain had contracted to buy while masquerading as Perrin. The STN brought an enormous dossier of claims for its unpaid printing bills, which would have to be adjudicated in the course of the general settlement. And Plomteux came bearing a grudge, for he thought that Duplain was holding on to profits that should have been distributed a half year earlier.[77] It was Duplain's hold on the profits that worried the associates. He had paid all the expenses, received all the income, and controlled all the ramifications of a financial affair so large and complex that it gave him ample room to exercise all of his talent at embezzlement. The total value of his swindles is difficult to estimate. The associates could prove that he had robbed them of 171,684 livres at the very least. But that sum does not do justice to his

76. The STN summarized its complaints about Duplain's management in its "Griefs contre M. Duplain."

77. Plomteux to STN, Aug. 16, 1779: "J'ai payé exactement ma mise et j'aurais cru que parvenu à placer 7,500 exemplaires, il y aurait dû avoir une dividende des bénéfices que doit avoir rendus une affaire aussi brillante."

efforts, which probably cost them twice as much.[78] Still, 171,684 livres was an enormous amount of money in the eighteenth century. The printers of the STN normally received about 12 livres a week—relatively high wages, almost as high, in fact, as those of skilled laborers in Paris. Duplain's peculation was the equivalent of what six or seven of them would make in a lifetime. And it almost certainly amounted to a great deal more, for Duplain covered his tracks by scrambling his accounts. He followed a general strategy of deflating credits and inflating debits so that his balance sheet would show the smallest possible profit to be divided among the associates at the *règlement des comptes*. When they arrived in Lyons, they knew that they would have to take the opposite line, showing how great the actual income had been, how small the expenses really were, and how much of the profits Duplain had pocketed.

The Final Confrontation in Lyons

The full extent of Duplain's embezzlement only became apparent to his associates in the course of their long and difficult debates. But they had uncovered his principal swindles before their arrival, and they came in the expectation of finding more. Having gathered together the threads of his intrigues for more than a year, they hoped to find enough corroborative evidence from his accounts to overwhelm him at the final reckoning.

They made the usual preparations concerning stagecoaches, hotel rooms, and strategy sessions. Panckoucke, who liked to travel in style, rejected the STN's suggestion that they book rooms at the Hôtel d'Angleterre, where they had stayed during the last meeting. He preferred the Palais-Royal, a large inn with rooms overlooking the Saône. They finally

---

78. Any attempt to estimate Duplain's peculation should take account of the fact that he was swindling an association in which he formally had a half interest and Panckoucke the other half. Thus the actual cost to the associates was half the total value of the swindles. Calculating by halves, the associates could demand reimbursement of the following sums:

> 77,934 livres for the hidden 978 subscriptions
> 30,102   for Audambron and Jossinet
> <u>63,648</u>   for the fictitious sheets
> 171,684 livres

They had hard proof of all these swindles, and they had strong circumstantial evidence that Duplain's expense-padding and such had cost them another 157,000 livres, approximately.

compromised on Le Parc, where d'Arnal reserved a three-room suite according to Panckoucke's specifications: "Je désirerais que cet apartement fût bien éclairé et donnât sur la rue. Nous étions bien tristement à l'Hôtel d'Angleterre." Plomteux shared the suite with Panckoucke and traveled with him from Paris. Ostervald and Bosset arranged to arrive from Neuchâtel on the same day. In order to coordinate maneuvers with their allies, they took two rooms in the same inn. And in order to have *pièces justificatives* at hand, they came fully armed with contracts and correspondence. After setting up headquarters on January 26 or 27, the anti-Duplain forces were ready for battle on the 28th.[79]

Meanwhile Duplain prepared his accounts. The STN tried to hurry him, but with no success. "Vous devez bien imaginer que ce n'est pas l'ouvrage de trois minutes,"[80] he protested to Panckoucke. The Neuchâtelois had wanted to hold the meeting in November because they worried about leaving the income from the subscriptions in Duplain's hands and because they needed to collect their share of it quickly, in order to pay off some heavy debts in December. But Duplain would not cooperate. He refused to honor a final series of bills of exchange that the STN wrote on him, forcing it to come up with 30,000 livres to save its account with d'Arnal. And he insisted that he could not put his books in order and collect the payments for the shipments of the last volumes until February. "Je crois avoir fait faire l'impossible en vendant 8,000 *Encyclopédies,* en les imprimant, et rendant des comptes en 18 à 21 mois,"[81] he told Panckoucke. "Vos Suisses sont des gens affamés. Je souhaite qu'ils soient rassassiés, mais j'en doute bien fort. Il y a, je suis sûr, encore 400,000 livres en l'air." In the last letter that he wrote to the STN, dated January 16, 1780, he again emphasized the difficulties of liquidating so huge an

79. On the preparations see Panckoucke to STN, Nov. 22, 1779, and Jan. 6, 1780, and d'Arnal to STN, Jan. 11, 1780. The STN had originally expected the meeting to be in July or Aug. 1779 and had urged Plomteux to attend it in a letter of May 1, which stressed the danger of being duped by Duplain and the importance for Plomteux to "nous fortifier de votre appui. Croyez-nous, la chose en vaut la peine." Plomteux was on business in Paris when he got word of the final date of the meeting. "Je sens tout le besoin que nous avons de nous prêter mutuellement nos secours," he wrote to the STN from Paris on Dec. 11, 1779.

80. Duplain to Panckoucke, Nov. 23, 1779, copy in Panckoucke to STN, Nov. 27, 1779.

81. Duplain to Panckoucke, Dec. 27, 1779, copy in Panckoucke to STN, Jan. 2, 1780.

enterprise so quickly, and he warned that there would be some hard bargaining in Lyons:

J'ai plus d'occupations que je n'en puis suivre. Vous pouvez arriver à la fin de ce mois. J'espère être à même en ce temps de rendre à M. Panckoucke un compte à peu près juste. Il y a encore des sommes énormes en arrière, et vous en serez persuadés sur le compte que vous en rendra M. Panckoucke. Il n'a pas fallu être manchot pour vendre, imprimer, et réaliser en trois ans huit mille *Encyclopédies* et être en état de rendre un compte. Je vous répète, Messieurs, ce que j'ai écrit à M. Panckoucke, que mon compte sera entièrement conforme au traité, que comme je ne lui demanderai pas une obole de plus que ce que m'alloue le traité, je ne céderai pas un denier de mes droits.

In a formal sense, as Duplain's letter indicated, the accounting only involved the two signatories of the Traité de Dijon. But each of them had ceded portions of his half interest to his own associates: Duplain to Merlino de Giverdy, Amable and Thomas Le Roy, and perhaps some other Lyonnais; Panckoucke to the STN, Plomteux, and Regnault. The settling of the accounts might therefore be considered as a general stockholders' meeting. But it was really a confrontation between two camps.[82]

The battle began on January 28, 1780. "Nous avons déjà eu quelques scènes rudes au sujet de nos comptes avec M. Duplain,"[83] Ostervald and Bosset informed their home office on the following day. "Tels que les combats de coqs en Angleterre, Panckoucke et Duplain se sont donnés de forts assauts." Unfortunately, they did not send any other blow-by-blow descriptions to Mme. Bertrand, who was minding the shop in Neuchâtel and was wrought with "inquiétude sur la crise où vous êtes."[84] But their notes and memos make their general

82. The full roster of shareholders varied from time to time and cannot be known completely. Rey had dropped out of the Association at the end of 1777, and Suard apparently had ceded his 1/24th interest back to Panckoucke. In a letter to the STN of Dec. 2, 1779, Duplain listed "MM. Plomteux, Regnault, Grabit, Bougy, qui, je crois, sont tous ses [Panckoucke's] co-associés." Bougy's name does not appear in any other documents. It is hard to see how Grabit could have bought an interest in the quarto after having threatened to pirate it with Barret. And Regnault, who had quarreled with both Duplain and Panckoucke, seems not to have attended the conference. In any case, Panckoucke's principal partner was the STN, which actually owned a larger share in the quarto than he did. In a letter to the home office of Feb. 13, 1780, Ostervald and Bosset described the anti-Duplain group as "nous quatre . . . appuyés de d'Arnal comme troupe auxiliaire"—that is, Panckoucke, Plomteux, and the two Neuchâtelois.
83. Ostervald and Bosset to Mme. Bertrand, Jan. 29, 1780.
84. Mme. Bertrand to Ostervald and Bosset, Feb. 5, 1780.

strategy clear. They meant to keep their suspicions secret until Duplain had committed himself to a fraudulent report on his stewardship. They knew from Duplain's "Aperçu" of October, from the secret subscription list, and probably from their spy's reports that Duplain had swindled them on a grand scale. But he had not yet taken the final, fatal step: the submission of his accounts. Once he produced his balance sheet, the associates could hold him legally responsible for every fraud they found. They could make their own calculations of debits and credits and their own demand for the profits to be shared. If Duplain stuck by his version, they could force him to justify it with evidence about expenditures and subscriptions. Then they could strike back with counterevidence from their well-stocked arsenal in Le Parc. And if they forced him to surrender, they could make him accept their terms for a settlement. They were businessmen, not law enforcement officers. They wanted to rescue their profits, not to put Duplain in prison. But to succeed, they had to play their part astutely, to produce their incriminating material at the most effective moments, and to lure Duplain into further self-incrimination, so that in the end he faced a choice of paying compensation or going to jail.

January 28, 1780, was therefore a momentous date in the history of the *Encyclopédie*. Duplain submitted a balance sheet, which showed the following totals:

income	1,851,588	livres
expenditure	1,718,260	
profit	133,328	livres

Not a glorious finale for an enterprise that Panckoucke had expected to be ten times as profitable. Duplain also produced the list of fraudulent and bankrupt debtors, which made things look even blacker. And finally, he tried to show how bad the printing costs had been by his "Compte général du coût de chaque volume."

As explained above, Bosset took the printing expenses, counted the number of sheets in the editions, and found that Duplain had padded the printing account with fictitious sheets worth 127,089 livres. That discovery put the anti-Duplain group in an excellent position to counterattack. It showed that

Duplain had undervalued the profits by almost 100 percent, merely by manipulating the debit side of his accounts. And it gave the associates a chance to make the most of his fraud on the credits by exposing the true identity of Perrin. A surprise attack on both sides of his balance sheet might overwhelm him. But they would have to prepare the ground carefully. So they withdrew into Le Parc and discussed tactics from January 29 until January 30 or 31.

First the Panckoucke associates needed to agree about a common counterproposition on the profits: how much could they demand from Duplain in order to recover everything he had pilfered? Panckoucke and Bosset drew up drafts of the accounts as they ought to read after being adjusted for Duplain's swindles. As a basis for their calculations, they took the costs set by the contracts, the real number of sheets printed per volume (124 on the average), and the total number of *Encyclopédies* (8,011) for which Duplain had just made himself accountable. Bosset, who was more a financier than a littérateur, produced the more complete version (see Appendix A. XIX). He allotted only 20,000 livres for the "insurance" or fictitious transport costs. He demanded a rebate of 48,828 livres for the Perrin fraud. And he came up with a total profit that was 350,000 livres higher than Duplain's:

income	1,946,300	livres
expenditure	1,516,082	
profit	430,218	livres

Bosset then added another 50,000 livres to cover the sale of the quarto *Table* and 54,390 livres for the 200 extra sets that he believed Duplain had sold at the wholesale price. He also noted that the associates could claim 67,620 more livres, if they could demonstrate that Duplain had sold his 500 sets in the association's name.

To prove that last charge, however, they would have to get a subscription list from Duplain. Duplain had of course refused to put such a weapon in their hands, but they could argue that they were entitled to have an itemized *inventaire* of the subscriptions in order to decide between his version of the accounts and theirs. If he refused, they would insist that he come up with a half million livres in profits instead of the

133,328 that he had offered. If he accepted, they would get a crucial piece of evidence for the case they were constructing against him, and they would be able to compare his list with their own. It was therefore important to conceal their secret list while revealing their knowledge of Perrin and to hold fast to the demand for 500,000 livres until Duplain gave ground. "Voilà, ce nous semble, les premières propositions d'accommodement que nous pouvons lui faire et dont notre avis n'est point de nous départir, à moins qu'en nous donnant l'inventaire général que nous demandons il nous mette dans le cas de nous relâcher," Bosset concluded.[85]

Unfortunately, there is no account of the session at which Panckoucke tore the mask off Perrin. The next document in the series on the meeting is a letter of February 6, in which Ostervald and Bosset told Mme. Bertrand that they had just been through a week of fierce argument: "Ce que l'on fait un jour peut être détruit le lendemain . . . Nous nous en occupons le jour et la nuit, et il le faut bien quand on a à faire à gens de cette sorte. Mais s'il plaît à Dieu et à notre bon droit, nous en sortirons et peut-être plutôt que ne le pense Duplain, grâce à sa friponnerie avérée." On the same day, Duplain at last stepped into the trap that had been so carefully prepared for him, by producing his subscription list. According to a "Relevé des registres des souscriptions" accompanying the list, he could account for only 6,589 subscriptions, aside from the 1,000 he had divided with Panckoucke and the 422 he had sold to Perrin. After subtracting for gift copies and free thirteenths, he asserted that only 6,074 of those sets had produced any revenue for the association. By adding that revenue to the Perrin payments, he came up with 1,851,588 livres, the figure he had originally given as the total income of the enterprise. But now, after nine days of debate, the associates could prove that figure was false.

They returned to Le Parc, compared Duplain's list with

85. "Tableau de ce qui devrait nous revenir de l'entreprise," a memo on tactics that Bosset wrote as a sequel to his "Produit net de l'entreprise tel qu'il doit être réellement." Panckoucke submitted a similar memorandum to Bosset and Ostervald, which they labeled "Aperçu de l'*Encyclopédie* fait à Lyon par M. Panckoucke" (Jan. 30, 1779, in "Dossier Encyclopédique," STN papers). It is less trenchant than Bosset's but is interesting in two respects: it gave an even lower figure for total expenditure than Bosset did (1,314,493 livres), and it estimated the size of the three editions at 8,450 copies. Bosset calculated that the Perrin swindle came to 48,828 livres by subtracting the Perrin payment and an allotment for the free thirteenths from the wholesale price of the 422 sets.

the secret list they had kept in reserve, and found the 978 missing *Encyclopédies*. Then they began their letter-writing campaign to acquire further evidence of the fraud. They evidently put d'Arnal and his spy on the tracks of the Audambron and Jossinet swindle, which leaped to their eyes during the comparison of the two lists. And they set Plomteux to work on a Mémoire, which they could threaten to publish if Duplain would not settle on their terms.

Once the Panckoucke associates had prepared their final salvo, Duplain's defeat was inevitable. They apparently hit him on February 11 with every piece of incriminating evidence they could find. Still Duplain held out. On the morning of the 12th, Bosset raided his shop with a police officer, a bailiff, and an attorney, who confiscated his books. At that point he admitted the 48,000 livre Perrin swindle, but he would confess no more. Then the associates threatened to ruin his name by exposing him in court and by publishing Plomteux's Mémoire, a crushing indictment for fraud, malice, and "insatiable cupidité" (see Appendix A. XIX). They even applied pressure through his family and friends. And finally, on the afternoon of the 12th, Duplain capitulated. He agreed to pay his associates 200,000 livres if they would sweep his swindling under the rug, where it has remained until this day.[86]

Dénouement

On February 13, Ostervald and Bosset sent the happy news to Mme. Bertrand: "Nous nous empressons, Madame, de vous communiquer la fin de notre combat avec Duplain, qui heureusement est terminé sans sang répandu." They considered themselves lucky to have got 200,000 livres from Duplain because he had fought until the end to bring them down to 128,000 livres, arguing quite rightly that they would all suffer heavy losses from his attempts to collect the last payments and from his entanglement in subscribers' lawsuits. In fact, 200,000 livres was probably a fair settlement. It was almost as much as the associates would have received, if Duplain had accepted their original, rather exaggerated version of the profits; and it coincided with a later version, which they set at

---

86. On these last maneuvers see the letter of Ostervald and Bosset to Mme. Bertrand of Feb. 13, 1780.

400,000 livres.[87] Of course they had had to take extreme measures in order to wring so much money out of *le roué*, as they took to calling him. They had resorted to blackmail, both in the Plomteux Mémoire and by verbal threats to "le perdre de réputation tant ici qu'à Paris."[88] And they also had made concessions on ancillary issues: the settlement with the publishers of the octavo *Encyclopédie*, the quarto *Table analytique*, and the *défets*. All of these matters were resolved in a contract signed by Panckoucke and Duplain on February 12, which liquidated the partnership they had formed three years earlier by the Traité de Dijon.

The contract made an appropriate ending to the quarto enterprise because it was a legalized lie. Having torn the façade off Duplain's embezzlements, Panckoucke now reconstructed it. He congratulated Duplain on "l'exacte vérité" of the accounts, and he singled out Duplain's report on the Perrin sale for praise. Not only did Panckoucke testify to the authenticity of the sale, he also explained Duplain's willingness to reimburse the associates for it as "un effet de la générosité de ses procédés" (see Appendix A. XVIII). That formula, which probably raised some laughs in Panckoucke's quarters, really meant Duplain had bought off his blackmailers.

The contract stated that Duplain had paid Panckoucke 176,000 livres in notes that would become due in three installments, ending in August 1782. The remaining 24,000 livres came from the octavo publishers. As explained above, they had been lobbying in Lyons to bring their war with the quarto association to an end. They had given Duplain their notes for 24,000 livres, on condition that he would get Panckoucke to open the French market to them. Panckoucke agreed, took the notes, traded them in for octavos, and sold the octavos at a discount, thereby ruining the market that he had abandoned.

The *Encyclopédie méthodique* also figured in the contract, because Panckoucke acknowledged that Duplain retained the 12/48ths interest in it that had been granted him. As Duplain later sold that interest to his old straw man, Jossinet, for 12,000 livres, the *Méthodique*, like the octavo, helped to cushion

---

87. "Griefs contre M. Duplain": "Par le tableau détaillé qui a été fait, il conste que le sieur Duplain a eu de bénéfice net avant aucun partage et dont il tient l'argent plus de 400,000 livres." The 50 percent interest of the Panckoucke associates would have entitled them to half that sum.

88. Ostervald and Bosset to Mme. Bertrand, Feb. 13, 1780.

the blow he received at the liquidation of the quarto. Duplain also found solace in the arrangement for the quarto edition of the *Table*. He had signed over that subsidiary speculation to Barret—having somehow patched up the quarrel over Barret's pirating—for 50,000 livres. By the contract of February 12, Panckoucke surrendered his share in the *Table,* thereby depriving the STN of its own portion and of its hopes to do the printing. According to Bosset's notes, Duplain had also sold his two warehouses' worth of *défets* to Jossinet for 20,000 livres, but they eventually came into the possession of Amable Le Roy, who took over the management of the enterprise during the last stages of its liquidation. Finally, Duplain profited from the sales of the surplus quartos, the 200 or more sets leftover after the distribution of the 8,011. He himself had evaluated them at 30,000 livres. And judging from his record, he may have hidden a great many more assets and embezzlements. So despite the victory of the Panckoucke group, Duplain emerged from his speculation on the quarto as a wealthy man.[89]

He could not consider his fortune safely made, however, until he settled with the STN. Ostervald and Bosset had arrived in Lyons with so many complicated "griefs" against Duplain, that they were reconciled to the inevitability of a lawsuit. But how could they sue him over his handling of a book that had been condemned by the Parlement of Paris, the French clergy, the king, and the pope? It was one thing for the French authorities to tolerate the distribution of the *Encyclopédie,* another for them to legitimize its existence in the courtroom. Fortunately, this problem had been foreseen in the

89. These transactions are clear from the final contract and the accompanying documents (see Appendix A. XVIII). The *défets*, however, gave rise to some complicated quarrels among Le Roy, the STN, Panckoucke, Duplain, d'Arnal, and Revol. Suffice it to say that the former associates continued to argue over them, with lawsuits and appeals for arbitration, for another two years. A great many subscribers never received replacements for the spoilt and missing sheets of their sets, and consequently some of them refused to make their last payments. Le Roy found his stock of *défets* insufficient to supply them. And Panckoucke concluded, "Duplain nous a trompé en falsifiant la clause de l'acte [the final contract of Feb. 12, 1780]. Le Roy ne s'est engagé qu'à fournir ce que les magasins produiraient." Panckoucke to STN, Jan. 22, 1782. But ultimately he counseled resignation: "Il est certain qu'il [Duplain] a livré ses magasins de défets et qu'on y a fait pour nous les recherches auxquelles il s'était obligé par ses actes. Il n'était pas obligé à autre chose. Il n'est pas dit dans la dernière transaction passée à Lyon qu'il sera obligé de fournir les défets lorsqu'ils manqueront. Il n'est pas parlé de réimpression . . . Ces exemplaires imparfaits ne sont pas absolument sans valeur." Panckoucke to STN, Oct. 7, 1782.

contract for the third edition, which obligated the associates to submit all disagreements to arbitration. Duplain reaffirmed this obligation in a statement that he signed on February 14. Four days later, both sides accepted an agreement on procedure. Each party was to name two arbitrators; and if the committee of four failed to agree on a decision, it would choose a single *surarbitre,* who would make the final ruling. Each party would submit his own version of the STN-Duplain account, supported with *pièces justificatives* 'and a rebuttal of the opponent's account statement. The arbitrators would then accept rebuttals of the rebuttals and would hand down a decision. The process was cheap and efficient, the antithesis of the official judiciary system, and it reveals an important feature of the semilegal and clandestine book trade: the system could not function on the principle of honor among thieves. Bookdealers cheated one another so flagrantly that they had developed their own paralegal institutions to keep themselves in check; they could not do business otherwise. The institutional response had answered the social need, beyond the pale of the law.[90]

Exactly how much Duplain owed the STN in February 1780 does not show through clearly among the welter of conflicting claims. Duplain did not dispute the printing and paper charges, which were fixed by the contracts and which made up the bulk of the STN's bills. So those charges were set aside and the differences narrowed to the disputes that had accumulated over the last two years. That procedure left plenty of room for disagreement, because according to the STN's account, Duplain owed it 23,531 livres 18 sous, and according to Duplain's the STN was 17,619 livres, 18 sous, and 3 deniers in his debt. Duplain arrived at this result by legerdemain with his debits and credits, especially the 8,526 livre debt for the disputed thirty-one sets. But his arguments did not stand up against the evidence that the STN produced from their commercial correspondence, and the arbitrators began by ruling

90. This account of the settlement between Duplain and the STN is based on the dossier ''Procès STN contre Duplain,'' which contains the contractual agreements on the arbitration, dated Feb. 14 and 18, 1780; a half-dozen memorandums and account statements that the STN submitted to the arbitrators; and the ''Sentence arbitrale'' of Feb. 21, 1780. Commercial cases, which were settled by *juges consulaires,* did not suffer as badly from costs and delays as the cases brought before the bailliages and parlements, but bookdealers often preferred the still more efficient paralegal system.

that he would have to take back all the extra sets he had sent to Neuchâtel and pay for their transport.

Aside from that dispute and some other issues that were dropped, the two sides remained separated by about 8,000 livres. They disagreed on everything that had given rise to their quarrels over the last two years, from the STN's demand for reimbursement of its *chaperon* (1,066 livres) to Duplain's claim for the travel expenses of his clerk (124 livres). The STN also demanded compensation for two new items. First, it charged Duplain for all of d'Arnal's expenses: 424 livres for brokerage fees, *protêts,* and interest on the emergency loans that had resulted from Duplain's refusal to honor its bills of exchange. Secondly, it built a clever but somewhat facetious argument around the Audambron and Jossinet swindle. Duplain had really taken a 25 percent bookseller's commission on the 535 quartos that he had fraudulently debited to Audambron and Jossinet at the wholesale price. Therefore the STN claimed that it was entitled to a 25 percent rebate on the subscriptions that it had gathered, even though it had originally collected them in the name of the association and not as a wholesaler. It would have been reasonable to consider this issue settled by Duplain's indemnity and the general liquidation of February 12, but Ostervald and Bosset wanted to squeeze every sou they could get out of Duplain. So they used his confession as grounds for inserting another 4,740 livres in his debits. Then they summarized all their arguments in an impressive tableau: on the left, six gigantic "Erreurs à notre débit dans le compte de Monsieur Joseph Duplain"; on the right, four equally large "Omissions à notre crédit." They accompanied this sheet with a seven-page "Mémoire contre Monsieur Duplain," and some other supporting documents. Having stifled their anger for so many months, the Neuchâtelois at last had a chance to vent every resentment, to demand justice for every injury, and to expose their associate as a *roué.*

How well Duplain defended himself is difficult to say, because his rebuttals have not survived. But he argued from a weak position, having already been forced to confess his mismanagement of the enterprise as a whole. He evidently tried to counterattack by arguing that his misdeeds were no worse than those of the Neuchâtelois: they had undermined his efforts to collect from the subscribers by secretly encouraging their bookseller friends to refuse payment, and they had tried to ruin his side-speculation on the *Table analytique* by secretly planning to pirate it. Bosset denied these charges in a deposition of February 14, and Duplain apparently failed to make them stick, although the second one had come close to the truth.

The four arbitrators, all distinguished lawyers and businessmen, handed down a unanimous, fifteen-page "Sentence arbitrale" on February 21. As they were settling disputes over specific sums of money, rather than determining guilt or innocence, they did not pronounce on Duplain's morality; but they showed what they thought of him by granting almost all the STN's demands. They required Duplain to pay 56,600 livres, only 2,400 short of the maximum requested by Ostervald and Bosset. The Neuchâtelois wrote home triumphantly that they had got more than they expected. And, at last, they had closed their quarto accounts: "Nous devons bénir Dieu de nous en être tirés comme cela."[91]

Epilogue

Ever since it began business in 1769, the STN had hoped to strike it rich by speculating on the *Encyclopédie*. When its opportunity came in 1776, it committed a great deal of its capital to Panckoucke's original enterprise, the folio reprint plan, hoping to profit by the huge printing job as much as by its half share in the publishing partnership. But as Panckoucke continued to maneuver, scrapping some projects and piecing together others, the STN watched its share shrink: from 1/2 of

91. Ostervald and Bosset to Mme. Bertrand, Feb. 28, 1780, and also Feb. 22, 1780. In their "Sentence," the arbitrators identified themselves as "Christophe de la Rochette, avocat, ancien échevin; Joseph-Marie Rousset, ancien échevin; Claude Odile Joseph Baroud, avocat en parlement, conseiller du roi, notaire à Lyon; Jean-Baptiste Brun, négociant à Lyon." They summarized the arguments on each side of each of the disputes but did not make an item-by-item judgment. The 56,600 livres that they awarded to the STN in a lump sum evidently included remnants of the printing bills that Duplain had not paid. The arrangements for this last payment were complicated by the STN's repugnance at accepting Duplain's notes and by its need for ready cash. By a complex series of countersignatures and operations on the Bourse, the STN converted a late-maturing note of Duplain's into a liquid asset of 50,657 livres, which it deposited in its beleaguered account with d'Arnal. For further details see Brouillard C, entry for Feb. 29, 1780.

the reprint, to 5/12 of the *refonte,* and 5/24 of the quarto—not to mention 5/48 of the *Encyclopédie méthodique.* Even more distressing, the printing commissions kept slipping through the fingers of the Neuchâtelois. They doubled the size of their shop in order to begin work immediately on the reprint in 1776. But they had to postpone that job while Panckoucke organized the *refonte,* which in turn they had to shelve when he went in with Duplain on the quarto. Duplain allowed them to print only five volumes of his gigantic undertaking, but they consoled themselves with the expectation that they would get to do the *refonte* later. Panckoucke finally destroyed that illusion in June 1778 by an agreement with some Liégeois, who developed the first scheme for an *Encyclopédie méthodique.* He extorted 105,000 livres from them in return for abandoning the *refonte* and opening up the French market. But he reversed his policy once again a half year later in a second agreement with the Liégeois. This time he abandoned the money and took over the *Méthodique*—a neat trick, but one that left the STN without any compensation for the loss of its printing commission. Panckoucke then tried to pacify the Neuchâtelois with other projects: a plan for a supplementary edition of the plates to the *Encyclopédie,* a speculation on the works of Rousseau, and the printing job for the quarto edition of the *Table analytique.* Each of these also evaporated, even the last, which went to Duplain at the final meeting in Lyons, just when the associates exposed his perfidy. In the end, therefore, the Neuchâtelois concluded that they had been made the stooges of the *Encyclopédie* adventures. As small-town Swiss, they had been outwitted and outmaneuvered by the sharpest businessmen in France; and they had learned their lesson: "Les libraires de France n'ont ni foi ni loi, ne sachant pas même distinguer ce qui est honnête d'avec ce qui ne l'est pas."[92]

Actually, the STN had not done badly in its *Encyclopédie* speculations. Its 5/12 share of Duplain's 200,000 livre settlement came to 83,666 livres. That sum plus the return from the sale of Panckoucke's 6,000 folio volumes, which had been confiscated in 1770, pretty well covered the 92,000 livres that the STN had contracted to pay to Panckoucke for its share in his rights, privileges, and plates. Having recouped its original

92. Ostervald and Bosset to STN, Feb. 15, 1780. The transactions with the Liégeois are discussed in Chapter VIII.

investment, the STN still disposed of two sources of profit: its 5/48 share of the *Encyclopédie méthodique* and the 208 quartos it had acquired from the division of Panckoucke's 500 sets. According to the most optimistic calculations of the Neuchâtelois, their interest in the *Méthodique* might someday be worth 30,208 livres. But in 1781 when they had run short of capital, they sold their shares to Plomteux for 8,000 livres. The 208 quartos were a more solid asset, despite the decline in demand. The STN evaluated them at 250 livres apiece or 52,000 livres in all, and it eventually did sell them off. Its total profit therefore came to about 60,000 livres on an investment of 92,000 livres—a return of 65 percent spread over four years, or twice what the STN would have earned had it invested its money in *rentes viagères*. The STN also did handsomely on its printing for Duplain, although its ultimate profit is impossible to estimate, owing to his chicanery. When they had tabulated all their credits and debits, however, the Neuchâtelois felt bitterly disappointed by their experience with the *Encyclopédie*. They had put their money on the most successful publishing venture of the century, and their partners had creamed off most of the profits, leaving them about half of what they thought they should have earned.[93]

Ostervald and Bosset therefore left the Lyons settlement in February 1780 with an unsatisfied appetite for profit and revenge. Their next stop was Paris, where they pursued some of the projects that had eluded them in their dealings with Panckoucke. At first they concentrated on a plan to pirate the quarto *Table analytique*, a work that could bring in 50,000 livres, they calculated, owing to the demand created by the success of the quarto *Encyclopédie*.[94] The associates had ceded the rights to the *Table* to Duplain, who in turn had sold them to the Lyonnais pirates Amable Le Roy and "ce roué de Barret."[95] By underselling Le Roy and Barret with their own pirated edition, the Neuchâtelois hoped to "enfin rendre à toutes ces honnêtes gens les tours qu'ils nous ont faits."[96] But

93. On these calculations see Ostervald and Bosset to STN, Feb. 13, 1780.

94. Ostervald and Bosset to STN, Feb. 28, 1780.

95. Ostervald and Bosset to STN, Feb. 15, 1780.

96. Mme. Bertrand to Ostervald and Bosset, Feb. 27, 1780. In Nov. 1779 the Neuchâtelois had considered pirating one of Barret's works, or at least blackmailing him with the threat of doing so, and Panckoucke had encouraged them in order to "forcer ce corsaire à nous rendre l'argent qu'il nous a extorqué. C'est un homme d'une insigne mauvaise foi." Panckoucke to STN, Nov. 6, 1779.

they would have to keep their counterpiracy secret because it not only violated their agreement with Duplain but also contradicted a formal statement they had made to him, which denied any involvement in speculations on a rival *Table.*[97] Ostervald and Bosset therefore instructed their home office to send confidential notices to certain booksellers on the quarto subscription list, warning them against subscribing to the Le Roy-Barret *Table,* as a cheaper edition was in press. Neuchâtel complied with letters to some of the STN's most trustworthy customers, including Lépagnez of Besançon, who had collected 338 subscriptions to the quarto *Encyclopédie.* Unfortunately for the STN, Lépagnez had fallen behind in his payments to Duplain; and in order to receive clemency, he informed Duplain of the plot. Duplain then dashed off a fierce letter to Panckoucke, who exchanged some hard words about the affair with Ostervald and Bosset in Paris. The Neuchâtelois tried to cover up their piracy by claiming that they had made the offer to Lépagnez before the agreement with Duplain. But that transparent lie failed to meet Panckoucke's main objection, namely that the STN's treachery could provide Duplain with a pretext for refusing to pay the 200,000 livre settlement. So despite their advanced preparations (they had even ordered a new font of type for the book), the Neuchâtelois had to cancel the project and accept another humiliating defeat.[98]

The *Table* fiasco marked a turning point in the relations between the STN and Panckoucke; soon afterward, each of the former allies began to treat the other as an enemy. Having failed to pirate the *Table,* the Neuchâtelois laid plans with the Sociétés typographiques of Bern and Lausanne to produce a counterfeit edition of Panckoucke's twenty-three volume abridgement of Prévost's *Histoire générale des voyages.*[99] At

97. The text of the statement is in STN to Ostervald and Bosset, Feb. 14, 1780.

98. Ostervald and Bosset to STN, March 15, 1780. On March 31, 1780, Ostervald and Bosset informed their home office that Duplain had just arrived in Paris and ''se plaint amèrement de nous, et il peut avoir raison.'' A year later, Barret and Le Roy quarreled so badly over their common speculation on the *Table* that they took their differences to court. Barret won the case and then offered to sabotage the sales campaign of Le Roy, whom he characterized as ''successeur de M. Duplain et son digne élève,'' by a secret arrangement with the STN. But the Neuchâtelois had been too badly burned in their dealings with all three Lyonnais to accept the offer. See Barret to STN, June 17, 1781.

99. STN to Ostervald and Bosset, undated letter, probably from April 1780, and Ostervald and Bosset to STN, April 14, 1780.

the same time, Panckoucke undercut his agreement with Bern and Lausanne by slashing the prices on the octavo *Encyclopédies* they had given him, thus ruining the market for which they had paid so dearly. He also refused to give the STN a share in the printing of the *Encyclopédie méthodique* and so snatched away the last of the printing jobs that he had dangled before them as compensation for canceling the original reprint plan. At this point, the STN began its attempts to get rid of its interest in the *Méthodique,* "afin de n'avoir rien à démêler dans la suite avec un homme peu digne de notre confiance."[100] After selling its 5/48 share to Plomteux, it cut its last ties with Panckoucke, whom it now considered as someone "qui n'est bon ni à rôtir ni à bouillir," and it concentrated on the search for a "belle occasion de prendre quelque revanche"[101] for the five-year partnership in which it had played the dupe.

The search led directly to Panckoucke's pet project, the *Encyclopédie méthodique.* The last and the largest of the eighteenth-century *Encyclopédies* had two potential advantages over Diderot's text: it could correct the errors and omissions that had been the despair of Diderot himself, and it could be methodical—that is, instead of following the arbitrary order of the alphabet, it could present a systematic summary of human knowledge, organized by subject and packaged in a series of thematic dictionaries. This plan seemed so superior to Panckoucke that he expected his new *Encyclopédie* to drive the old ones off the market, as thousands of readers would want to scrap their antiquated models, whether folio, quarto, or octavo, for the latest version, which would be recognized everywhere as the only "véritable *Encyclopédie.*"[102] It was a grand plan, but it contained a flaw that the Swiss soon detected. A pirate could easily extract all the original material from the new text, rearrange it in alphabetical order, and publish it as a *Supplément,* in all three formats, to all the previous editions. By purchasing a few supplementary volumes, thousands of *Encyclopédie* owners all over Europe could avoid buying an expensive new work from Panckoucke. And by some simple counterfeiting, the pirate could reap what Panckoucke had sown.

100. STN to Bosset, May 16, 1780. See also Ostervald to Bosset, May 14, 1780.
101. Bosset to STN, June 2, 1780, and Ostervald to Bosset, June 8, 1780.
102. Panckoucke to STN, June 1, 1779.

Curiously, the Swiss discussed this plot with the abbé Morellet, an intimate of Panckoucke's circle who had originally planned to work on Suard's *refonte* of 1776. Ostervald had come to know Morellet quite well while scouting for manuscripts in Paris and asked his advice about pirating Panckoucke: should the STN reprint some of the constituent dictionaries of the *Méthodique,* or should it attack the whole work at once by producing the alphabetical *Supplément?* Morellet replied that Panckoucke would defend himself against the first kind of aggression by publishing sections of all the dictionaries simultaneously, so that the entire *Encyclopédie* would be finished at once and the pirates could not pick off the individual dictionaries one by one. As to the second plan, Morellet had to admit that it could work, once the *Méthodique* was completed, but he thought that it went beyond the bounds of conventional trade warfare: "Vous feriez un grand tort à l'enterprise de Panckoucke véritablement immense et capable d'entraîner sa ruine, si elle venait à échouer par l'exécution de votre projet . . . Il y a quelque inhumanité à lui faire un tort si grave."[103] The STN also consulted one of its Parisian agents, an indigent bookdealer called Monory, who found the plan more terrifying than reprehensible; for, like other small fry in the book trade, he trembled at the thought of Panckoucke's power: "Quant au *Supplément* que vous proposez de l'*Encyclopédie* . . . soyez assuré qu'il fera tout ce qui dépendra de lui pour en empêcher le cours; et il pourra beaucoup contre, à ce que je crois, par plusieurs raisons que vous pouvez soupçonner."[104]

Having sounded the terrain in Paris, the STN prepared for the attack in collaboration with its two allies, the Sociétés typographiques of Bern and Lausanne, who harbored even greater grudges against Panckoucke. In December 1783 the three confederates met in Yverdon to concert strategy and to draft a prospectus for the *Supplément.* By January 1784 they had printed the prospectus and opened a subscription, which they announced in the *Gazette de Berne.*

Les sociétés typographiques de Berne et Neuchâtel et M.J.P. Heubach et Compagnie de Lausanne vont travailler de concert à compléter les

103. Morellet to STN, May 31, 1783. Morellet also objected that the STN's attack would hurt Panckoucke's authors, notably Marmontel, a close friend of his who was considering publishing with the STN.
104. Monory to STN, Dec. 25, 1783.

éditions de l'*Encyclopédie* par ordre alphabétique, à peu de frais pour les acquéreurs, et les rendre équivalentes par un *Supplément* bien entendu à l'*Encyclopédie* par ordre des matières qui s'imprime à Paris. On trouve chez chacune de ces trois maisons le *Prospectus* de ce *Supplément*, qu'elles proposent par souscription, avec les détails de ce plan et des conditions auxquelles on peut se le procurer dans les trois formats, folio, quarto, et octavo. On n'imprimera que le nombre pour lequel on aura souscrit.[105]

Then they sent their first ransom note to Panckoucke. They began by reminding him of his foul play during the quarto-octavo war. But they bore no grudge, they said, with feigned restraint; for they knew that business was business: "En poursuivant une entreprise utile pour vous, vous avez nui à ces sociétés [Lausanne and Bern]; en poursuivant une entreprise qui peut leur être utile, il serait possible qu'elles vous nuisent. Tel est l'ordre des choses dans le monde, que le bien de l'un ne peut se faire sans un peu de mal pour quelqu'autre." Next they explained how their "enterprise" would cut the ground from under Panckoucke's cherished *Encyclopédie méthodique*—not that they meant to destroy his market as he had destroyed theirs: they merely wanted to bring the *Méthodique* within the range of the poorer run of customers. They would leave the rich to him. Should he feel that they were taking the lion's share of the demand, they would be happy to negotiate an agreement with him. Perhaps they should combine forces: he could handle sales in France while they worked the rest of Europe. Or, should he prefer to run the entire operation, they might be persuaded to give up their plan—if the price were right. The letter dripped with irony and false bonhomie, and it must have given some satisfaction to the men in Bern, Lausanne, and Neuchâtel, who had been on the receiving end of most of the low blows exchanged across the French-Swiss border.[106]

105. *Gazette de Berne*, Dec. 24, 1783. By this time Heubach had reorganized the Société typographique de Lausanne as Jean-Pierre Heubach et Compagnie. He continued to employ Bérenger and to maintain the affiliation with the other two Swiss firms.

106. On Jan. 10, 1784, Heubach sent the STN an undated copy of this note, which was written by Bérenger. Heubach explained that Bérenger had sent the proposal to Panckoucke "mercredi dernier" with the approval of Bern but had not had time to submit it to the STN. All three societies had apparently agreed on it in principle a month earlier. The STN wanted to reinforce this maneuver by pirating another of Panckoucke's books, but the others objected that too much aggression might make Panckoucke unwilling to negotiate. Société typographique de Berne and J.-P. Heubach to STN, Jan. 17, 1784.

In his reply, Panckoucke tried to play for time. He could not consider the proposal, he said, until he had completed the *Encyclopédie méthodique*.[107] In March 1784, Heubach received a copy of the first volumes of the *Encyclopédie méthodique* and wrote that they would do nicely for the *Supplément,* whose prospects looked excellent: "Avez-vous de bonnes nouvelles pour le *Supplément?*" he asked the STN.[108] "Nous en recevons de plusieurs endroits des nouvelles très satisfaisantes, et nous espérons de pouvoir réaliser cette combinaison dans le courant de l'été." Three months later, the project looked better than ever: the flow of subscriptions for the pirated edition evidently remained strong, for Heubach continued to report "des lettres très encourageantes" from his commercial correspondence; and Pierre-Joseph Buc'hoz of the Académie des Sciences had agreed to prepare all the material concerning natural history.[109] But at that point, all references to the *Supplément* disappear from the papers at Neuchâtel. Like many publishers' projects, it never was realized—not, it seems, through any slackening of the zeal for piracy among the Swiss, but because Panckoucke did not produce enough of his new *Encyclopédie* for them to plunder.

In the end, therefore, and despite its claims to do business with *rondeur helvétique,* the STN proved to be as cutthroat as its associates. It even betrayed its partners when their backs were turned. It plotted with Bern and Lausanne to counterfeit Plomteux's edition of Raynal's *Histoire philosophique et politique des établissements et du commerce des Européens dans les deux Indes,* and it violated a commitment to its two Swiss confederates by scheming secretly to produce an edition of Rousseau with the Société typographique de Genève.[110] The Neuchâtelois had learned to play according to the rules of a very rough game, that is, to dupe or be duped; and they had lost whatever illusions they had had when they entered upon their *Encyclopédie* speculations. "Il ne faut pas promettre plus de beurre que de pain, ne croire que ce que l'on voit, et

107. Heubach informed the STN of Panckoucke's reply in a letter of Feb. 10, 1784. This is the last reference in the Neuchâtel papers to Panckoucke, who by then had broken relations with the STN.

108. Heubach to STN, March 1, 1784.

109. Heubach to STN, June 7, 1784.

110. STN to Ostervald and Bosset, Feb. 14, 1780, and Ostervald and Bosset to STN, Feb. 8, 1780.

ne compter que sur ce que l'on tient avec les quatres doigts et le pouce,"[111] Ostervald and Bosset concluded after their final confrontation with Duplain.

Duplain himself played the dirtiest game of all. Eighteenth-century booksellers accepted piracy and secret combinations as necessary evils, but they drew the line at fraud and swindling. So even by the lax, unwritten code of his trade, Duplain stood condemned as the villain of the *Encyclopédie* venture. To Panckoucke he was "ce vilain homme," to the STN simply "le roué."[112] When other booksellers mentioned him, they evoked the same picture of unmitigated rapacity. Jacques Revol claimed that Duplain had swindled him for 4,000 livres and had done the same to Duplain's own cousin, Pierre Joseph Duplain of Paris. Pierre Joseph, an under-the-cloak book-dealer and literary agent, had nothing good to say about his cousin in Lyons. And Revol recommended himself as a smuggler to the STN on the grounds that his way of doing business bore no resemblance to Duplain's. He had known Duplain intimately since childhood, he explained, and felt nothing but distrust for him.[113] Duplain himself acknowledged his reputation for unscrupulousness, but he attributed it to the machinations of his enemies, who had foisted upon him "les surnoms de *pirate* de *corsaire,* de *forban* que l'on prodigue jusqu'à la satiété dans les libelles platement injurieux."[114] Opinion in the book trade seems to have been unanimous: he represented literary buccaneering at its worst.

Was there nothing more to this man than an insatiable appetite for profit? The question has a certain fascination, both for economic history and the history of the human soul. But it is difficult to answer, because the contemporary picture of Duplain may be a caricature, and his personality does not

111. Ostervald and Bosset to STN, Feb. 20, 1780.

112. Panckoucke to STN, Nov. 10, 1780, and Ostervald and Bosset to STN, Feb. 20, 1780. See also Panckoucke to STN, Nov. 6, 1778, on Duplain's *vilaine âme* and similar remarks by Plomteux in a letter to the STN of March 22, 1781.

113. Revol to STN, June 24 and May 8, 1780, and Pierre Joseph Duplain to STN, May 29, 1782. Note also Revol's comment in a letter of Aug. 13, 1780: "M. Duplain avait fait mettre en prison le sieur Gauthier [a bookseller from Bourg-en-Bresse]. Nous ignorons les circonstances."

114. *Mémoire à consulter et consultation pour le sieur Joseph Duplain, libraire à Lyon* (Lyons, 1777), p. 5.

show clearly through his correspondence. His letters have a brusque style. They come quickly to the point, in a rushed, imperative manner, as if Duplain were a general issuing orders from a battlefield. He had to coordinate so many attacks on so many fronts that he easily adopted an embattled tone. In the autumn of 1778, for example, when he was bargaining with Panckoucke over the contract for the third edition, issuing its first volumes, collecting payments for the previous editions, and plotting the Perrin swindle, Duplain wrote to the STN as follows: "Nous sommes écrasés par les non rentrées, par la provision des papiers pour l'hiver, par la troisième édition dont les deux premiers volumes sont en vente, et il ne nous est pas possible de faire face encore à vos demandes." "Nous vous prions à lettre reçue de nous faire envoi du tome 24. Nous assemblons et collationnons 21, 22, 23. Nous joindrons à cet envoi tous vos défets. Nous attendons que M. Panckoucke ait fini le traité nouveau à signer pour vous envoyer un nouveau volume, et il attend, dit-il votre ratification. Cela ne nous regarde pas. Tout ce que nous pouvons dire à M. Panckoucke, c'est que nos frais sont immenses . . . Nous sommes assaillis de protêts. Toulouse a en arrière 10,000, mais nous conduirons la barque au port."[115] Duplain dashed off a dozen directives like this every day, manipulating battalions of papermakers, printers, and financiers. His campaign covered France, western Switzerland, and part of the Low Countries, and it had an epic quality. Duplain meant to make the greatest possible profit from the greatest publishing venture of the era.

The single-mindedness with which he pursued this prize also reveals something of Duplain's nature. He was a gambler. He recognized that the quarto *Encyclopédie* was the chance of a lifetime, and then he staked everything he owned on its success. He sold his shop, his stock of books, his house, and his furniture and moved into a furnished room, in order to concentrate exclusively on the great affair. And once he had committed himself to this supreme speculation, he conducted it with a brutality that alienated even those booksellers whom he failed to swindle. But he did not care: he had risked everything; he could not turn back; and in the end he made a fortune. For even after settling with his associates for 200,000 livres, he

115. Duplain to STN, Sept. 15 and Oct. 9, 1778.

was a very rich man. It is impossible to calculate his profits, but he made enough to retire from the rake-offs on the printing alone.[116]

Duplain meant not merely to retire but to live *noblement;* and once he made his fortune, he began to spend it. First, he acquired a wife, a Lyonnais beauty whom he married in March 1777, when the first edition of his *Encyclopédie* had been oversubscribed and the partnership with Panckoucke had eliminated the risk of its confiscation.[117] Next, he took up pleasure trips to Paris, where he and his bride lived in offensive luxury, according to Panckoucke. They went about with a magnificent equipage and even communicated a new air of hautiness to their old servants.[118] Then Duplain began to shift his wealth from Lyons, where he had retained three buildings after the sale of his business, to a rural estate, where he planned to take up the life of a country gentleman.[119] And finally, for 115,000 livres, he bought the office of maître d'hôtel du Roi—that is, nobility. He served the king in Versailles, took to signing his

116. Judging from information supplied by Favarger in a letter to the STN of July 15, 1778, Duplain probably made about 100,000 livres on the printing. And according to F.-Y. Besnard, "on se retire volontiers des affaires, quand on a amassé de 3000 à 4000 livres de rente," that is, a capital of about 80,000 livres. *Souvenirs d'un nonagénaire,* quoted in Henri Sée, *La France économique et sociale au XVIIIe siècle* (Paris, 1933), p. 162. Duplain's wholesale and retail book business, inherited from his father, seems to have been quite large by Lyonnais standards. He sold it to Amable Le Roy for an undisclosed sum. Although the last stages of Duplain's career are obscure, their outline can be pieced together from letters of Le Roy, Revol, Panckoucke, and Bosset and Ostervald.

117. On March 10, 1777, Panckoucke slipped some congratulatory remarks in a business letter to Duplain: "Vous vous mariez, mon cher ami; je vous en fait mon compliment. La future est jeune, jolie, aimable; je vous les redouble. Le mariage est le véritable état du bonheur, quand on sait bien s'y gouverner. Je vous prie de présenter mes hommages respectueux à la demoiselle qui sera votre dame lorsque vous recevrez cette lettre. Parlons de nos affaires." Bibliothèque publique et universitaire de Genève, ms. suppl. 148.

118. On July 10, 1779, Panckoucke told the STN that Duplain and his wife were in Paris on one of their jaunts: "Tout le monde est mécontent de M. Duplain; son luxe révolte bien des gens." The anonymous author of the *Lettre d'un libraire de Lyon à un libraire de Paris* (March 1, 1779) noted (p. 9), "On parle des richesses des libraires de Paris, mais y en a-t-il deux qui ayent équipage comme Duplain?" And in speaking of Comtois, one of Duplain's servants, Revol remarked, "Les valets se ressentent de l'impudence du maître." Revol to STN, Nov. 25, 1780.

119. In a letter to the STN of Nov. 18, 1780, Revol described Duplain as "attendant son départ pour la province, où il compte se fixer." And the *Lettre d'un libraire de Lyon* made a passing comment (p. 9) about "le sieur Duplain, qui, il y a quelques années, avait à peine 40,000 livres et qui actuellement est assez riche pour penser à acheter une terre."

letters "Duplain de St. Albine," and probably spent his remaining years in *petits soupers* and *châteaux*.[120]

One event marred Duplain's triumph. On September 19, 1780, Revol informed the STN, "Madame Duplain est morte depuis trois semaines. Jugez quel chagrin pour M. le Maître d'Hôtel du Roi. Il semble que c'est un châtiment du ciel pour le punir de son avidité et de sa soif de l'or aux dépens des uns et des autres." Three years later, however, he acquired another woman. "Le sieur Duplain vient de se marier à une fille de dix-sept ans,"[121] L. S. Mercier reported to the STN from Lyons. "Comme il est fort riche, ce mariage fait bruit."

Is there another moral to this story? To the social historian, it reads like a Balzacian drama: the tale of a bourgeois entrepreneur who clawed his way to the top and then consumed his fortune conspicuously, in aristocratic abandon. In a way it is the story of French capitalism—of limited expansion and investment in status instead of production. And its supreme irony is that the vehicle for Duplain's rise into France's archaic hierarchy, only a few years away from destruction, was Diderot's *Encyclopédie*.

There was also a Balzacian flavor to Panckoucke's life. In fact, a strong dose of Panckoucke went into the confection of Dauriat, the publishing tycoon and unofficial "ministre de la littérature" in Balzac's *Illusions perdues:* "Je fais des spéculations en littérature: je publie quarante volumes à dix mille exemplaires, comme font Panckoucke et les Beaudouin. Ma puissance et les articles que j'obtiens poussent une affaire de cent mille écus au lieu de pousser un volume de deux mille

---

120. On Duplain's ennoblement see Panckoucke to STN, July 10 and June 1, 1779: "Il vient d'acheter une charge de maître d'hôtel chez le roi de 115,000 livres. Il a passé ici 8 à 10 jours, mais je ne lui ai vu que quelques heures. Sa réception a absorbé tous ses instants . . . Nous n'avons point eu le bonheur de posséder sa jolie femme." Duplain actually may have acquired his nobility by the prior purchase of the ennobling office of secrétaire du roi for 80,000 livres. Such at least was the contention of the *Lettre d'un libraire de Lyon*, p. 9, which described him as a rich arriviste "qui est actuellement à Paris pour se faire pourvoir d'une charge de maître d'hôtel chez le roi ou chez la reine, charge qui passe cent mille livres de finance; qui, pour avoir l'espèce de noblesse requise pour cette charge, achète une charge de secrétaire du roi de quatre-vingt mille livres." In a letter to Panckoucke of June 24, 1779, the STN also indicated that Duplain had purchased two offices: "C'est un homme dangereux, qui manquerait de bonne foi . . . ou . . . serait gêné dans ses paiements par ceux qu'il aura été obligé de faire pour sa savonette à vilain et sa charge de maître d'hôtel."

121. L. S. Mercier to STN, Sept. 5, 1783.

francs.'"[122] Unlike Duplain, however, Panckoucke did not invest in books in order to get out of the trade. He never stopped running after bigger and better speculations. He was a fortune hunter, but he seems to have been moved by the love of the chase. And thanks to his correspondence, one can follow him as he pursued his passion, dashing off in *turgotines* to form a partnership with Duplain in Dijon, to buy out Rey in Amsterdam, to beard Cramer in Geneva, and to settle with Plomteux in Liège. Panckoucke was at his best in those feverish moments when he dismantled one speculation in order to build another, on a grander scale—when he suddenly scrapped the Neuchâtel reprint to construct the magnificent *refonte,* for example, or when he abruptly dropped his plan to make war on the Liégeois and took over their project, transforming it into his own version of the *Encyclopédie méthodique.* His supreme period of activity came in 1778, when he nearly cornered the market for the works of Buffon, Voltaire, and Rousseau, while directing his monopoly of the *Encyclopédie.* Panckoucke felt sympathy for the ideas in those books. He developed close ties with the philosophes and wrote some philosophic works himself.[123] But ultimately he seems to have been inspired by something akin to the spirit of the robber barons in the nineteenth century. He speculated because speculation had become an end in itself, a way of life, for him. Of course, he meant to make money, and he drove a hard bargain,

122. *Illusions perdues* (Paris, 1961), p. 304. See also p. 309: "Dauriat est un drôle qui vend pour quinze ou seize cent mille francs de livres par an, il est comme le ministre de la littérature . . . Son avidité, tout aussi grande que celle de Barbet, s'exerce sur des masses. Dauriat a des formes, il est généreux, mais il est vain; quant à son esprit, ça se compose de tout ce qu'il entend dire autour de lui; sa boutique est un lieu très excellent à fréquenter." Although this description concerns a publisher of the Restoration, it applies quite well to Panckoucke and his shop, as they were described in D.-J. Garat, *Mémoires historiques sur la vie de M. Suard, sur ses écrits, et sur le XVIIIe siècle* (Paris, 1820), I, 271–275. The Panckoucke mentioned in the text was probably Charles Louis, the son of the *Encyclopédie* publisher, but Balzac might well have had in mind the senior Panckoucke, who was a legendary figure in the publishing and literary circles frequented by Balzac before he wrote *Illusions perdues* in 1836. The authors and booksellers in that novel have a strong resemblance to their forebears of the eighteenth century.

123. Panckoucke wrote translations of the classics, articles on chemistry and biology, and general essays, which he published in the *Encyclopédie méthodique.* For a list of his publications and his own account of his contacts with the philosophes see his *Lettre de M. Panckoucke à Messieurs le président et électeurs de 1791* (Paris, Sept. 9, 1791). Garat claimed, with some exaggeration, that Panckoucke maintained excellent relations with the philosophes and could have become a philosophe himself. Garat, *Mémoires,* I, 270–273.

as Beaumarchais (who was no easy customer himself) discovered in the course of their negotiations over the Voltaire manuscripts: "Je ne puis vous dire rien d'agréable sur le compte de M. Panckoucke: ses procédés envers moi sont durs jusqu'à la malhonnêteté. M. Panckoucke est *Belge,* et dix fois *Belge.*"[124] But Panckoucke had none of the narrower kind of cunning that characterized Duplain. Plomteux, who knew him intimately, described Panckoucke as "trop occupé et trop distrait";[125] and Duplain called him a "visionnaire"—a term that was meant as a pejorative but that does justice to the ambition of the greatest impresario of the Enlightenment.

Owing to this visionary temperament, Panckoucke, too, suffered from *illusions perdues.* His grandest illusion turned out to be the *Encyclopédie méthodique.* This project for a super-encyclopedia became the ruling passion of his life, his greatest speculation, and the "belle entreprise" by which he planned to crown his career. It also exemplifies the final phase of Encyclopedism and the point at which Enlightenment passed into Revolution. A great deal, therefore, can be learned by disentangling its history from the story of Panckoucke's other speculations.

124. Beaumarchais to Jacques-Joseph-Marie Decroix, a financier from Lille who took part in the negotiations, Aug. 16, 1780, Bodleian Library, Oxford, ms. Fr. d.31 (the emphasis is Beaumarchais's). Note also the opinion of Diderot as reported by Ostervald in a letter to Bosset of June 4, 1780: "En un mot, je n'aspire qu'à nous mettre hors d'intérêt avec cet homme-là [Panckoucke], après quoi nous verrons. Harlé [Ostervald's son-in-law] vous en aura peut-être parlé et vous aura dit comme à moi que Diderot l'avait assuré que c'était un homme de mauvaise foi, offrant d'en fournir la preuve." Gabriel Cramer complained bitterly about Panckoucke's management of the Geneva folio edition and referred to him as "cet homme bas et injuste." Cramer to Louis Necker de Germagny, May 25, 1777 in Theodore Besterman, ed., *Voltaire's Correspondence* (Geneva, 1965), XCVI, 189.

125. Plomteux to STN, Aug. 16, 1779.

VIII

YYYYYYYYYYYY

THE ULTIMATE *ENCYCLOPÉDIE*

While fighting pirates, guiding the quarto alliance, and ruling his publishing empire, Panckoucke undertook a project that was to dwarf all his other enterprises. He decided to create the supreme *Encyclopédie*—a book that would make Diderot's work pale into insignificance, that would encompass all of human knowledge, and that would make Panckoucke a millionaire many times over, for he planned to stake his fortune on it and to build it into the biggest enterprise in the history of publishing. This book, the *Encyclopédie méthodique,* now sits unread and forgotten on remote shelves of research libraries. It has not aroused the appetite of a single, thesis-hungry graduate student. Yet it deserves to be rescued from oblivion because it represents the ultimate in Encyclopedism. It became the grandest gamble in the competition for the *Encyclopédie* market of the Old Regime. To put it together, Panckoucke mobilized the finest talent of his time. And to prevent it from disintegrating, he labored with the last of the Encyclopedists throughout the French Revolution.

The Origins of the *Encyclopédie méthodique*

Diderot himself might be considered the father, or at least the grandfather, of this project, for it was Diderot's memoir of 1768 that precipitated Panckoucke's *Encyclopédie* speculations, sending him down a path that led from an early plan for a revised edition to the Geneva folio, the proposed Neuchâtel reprint, Suard's *refonte,* and Duplain's quarto. Thus to Panckoucke, the quarto was really a detour from a long-term plan to publish a revised and expanded version of Diderot's text, which would be written by a whole team of second-generation philosophes under the direction of Suard, Condorcet, and d'Alembert. He continued to orient his speculations toward this plan and to pay Suard for working on it until the summer of 1778.

Suard later claimed that he had worked long and hard. He rented a "petit apartement" adjoining his lodgings in 1777 for 300 livres a year to serve as a "bureau de travail" and installed a small library in it, which included three folio *Encyclopédies,* two *Suppléments,* one *Encyclopédie d'Yverdon,* one *Description des arts et métiers,* one set of the abbé Rozier's *Journal de Physique,* and one "Encyclopédie anglaise," presumably the *Cyclopaedia* of Chambers. He evidently attacked the folio *Encyclopédies* with scissors and paste, cutting out the errors and grafting on additions, which he extracted from the other reference works. He incorporated notes that he had prepared for his *Dictionnaire sur la langue française* and new articles that he received from his stable of writers. For help in assembling and blending together this mass of material, he hired a copyist for 800 livres a year and a "commis intelligent" for 1,200 livres. The enterprise must have produced a heavy traffic of philosophes in Suard's apartment, as well as a great deal of activity in salons, cafés, and academies; for Suard had connections everywhere in the Republic of Letters and d'Alembert and Condorcet had promised to help him recruit contributors. Suard later told the STN that he had produced several boxes full of notes and drafts and had received 6,300 livres in salary from Panckoucke. "Pendant plus de 18 mois, je n'ai été occupé et n'ai pu m'occuper que du projet de cette rédaction, [dont] l'exécution n'a été suspendu qu'après bien des lenteurs et des incertitudes."[1]

In June 1778 Panckoucke suddenly brought Suard's operation to a halt. It had built up a great deal of momentum, but

---

1. Suard to STN, Jan. 11, 1779. Suard described his work as follows: "Je me suis occupé plus d'un an à comparer l'*Encyclopédie* originale avec le *Supplément,* avec celle d'Yverdon, et avec d'autres dictionnnaires; j'ai fait faire des tables de plusieurs ouvrages; j'ai recueilli un grand nombre de notes, qui devraient servir à corriger ou augmenter différents articles de la nouvelle *Encyclopédie* à mesure que la rédaction se faisait etc. J'ai plusieurs cartons remplis de mon travail; mais ce travail ne pouvant avoir par lui-même aucun ensemble, ne peut être présenté sous une forme régulière." The reference works he used are mentioned in Panckoucke to STN, Jan. 17, 1779.

it was headed for a collision with a faster moving *Encyclopédie,* which had emerged unexpectedly from Liège in January, when a Liégeois bookseller named Deveria issued a prospectus under an Amsterdam address and began collecting subscriptions for an *Encyclopédie méthodique.* Deveria proposed not merely to correct and enlarge Diderot's text, as Suard was doing, but also to reorganize it "par ordre des matières." Instead of arranging the material in alphabetical order, he would group it by subject and publish a series of small encyclopedias on law, medicine, natural history, and so forth. Thanks to this "methodical" organization, his readers would not have to piece together cross references and comb through dozens of heavy tomes in order to get a coherent view of a subject.

Deveria's method had philosophical as well as practical advantages over Diderot's. Diderot had cut the universe of knowledge into separate segments and had strung them out according to the arbitrary order of the alphabet—an arrangement that "ne peut plaire qu'au demi-savant ou à l'ignorant." But Deveria would bring out the underlying rational connections between different parts of knowledge. He would succeed where Diderot had failed, because he would realize the true spirit of d'Alembert's Preliminary Discourse.

Sans cet ordre précieux, qui peut seul intéresser le génie philosophique, ce recueil, morcelé dans un dictionnaire, n'est pas plus une encyclopédie qu'un amas de pierres taillées et numérotées n'est un palais: c'est à l'arrangement seul des différentes parties d'un bâtiment qu'on reconnaît le génie de l'architecte, comme on reconnaît au Discours préliminaire, non à la disposition alphabétique de l'*Encyclopédie,* le grand homme qui embrasse toute les connaissances, qui en voit l'ordre, les rapports, les liaisons.

Thus Deveria did not reject the philosophical strain in the original *Encyclopédie;* he extended it. And he did not claim any originality for his "methodical" view of knowledge, but attributed it to the forefathers of the Enlightenment. "Quant à l'ordre des matières, les libraires ne s'attribuent point la gloire de cette idée; ils la doivent à Bacon, à la voix publique, et au jugement de plusieurs savants qui ont bien voulu les éclairer."[2]

2. *Prospectus d'une édition complète de l'Encyclopédie, rangée par ordre des matières et dans laquelle on a fondu tous les Suppléments et corrigé les fautes des éditions précédentes* (Amsterdam, 1778), dossier Marc Michel Rey, Bibliotheek

At the same time, Deveria tried to win subscribers by offering good terms. Anyone who signed up on time would become a kind of stockholder in the enterprise and would receive a rebate on the subscription which would be pegged to the success of the sales. The subscriber should be prepared to pay 756 livres, Deveria estimated, but could expect to get back 420 livres if, as seemed likely, 20,000 sets were sold. Deveria planned to produce two complete editions, a folio in 36 volumes, of which 12 would be plates, and an octavo in 144 volumes with the same set of plates. After sending a down payment of 21 livres, the subscribers could count on receiving a volume every six weeks and the whole set within five years. They would pay by installments, as the volumes arrived; and they would get a bargain, for the complete set would go on sale for 1,000 livres after the subscription closed. Deveria also stressed the excellence of the type and paper he would use. And to enlist the help of booksellers in the marketing, he sent a circular letter to the major European dealers, promising them a reduction of 108 livres on every set they sold and a free set for every dozen sales.[3]

Finally Deveria's prospectus concluded with a frontal attack on the main rival edition, Panckoucke's quarto. Not only did the quarto publishers perpetuate the "vice abécédaire," it argued, but they were attempting to swindle the public, exactly as Linguet had revealed. Either they would keep their edition down to 32 volumes by amputating more than half of Diderot's text, or they would stretch it out to 99 volumes and extort more than 1,000 livres from the hapless subscribers. For his part, Deveria guaranteed his customers against any such fraud by promising solemnly to keep his *Encyclopédie* to 36 volumes and to provide the extra volumes free, should it somehow grow beyond that limit. He expected the quarto publishers to try to denigrate his plan, but he was confident of its superiority and ready to do battle. ''Permis néanmoins aux libraires de Genève, Neuchâtel, et autres villes où cet

---

van de vereeniging ter bevordering van de belangen des boekhandels, Amsterdam, quotations from pp. 1, 2, 3.

3. Although the prospectus was anonymous, the circular accompanying it, dated Jan. 1778, was signed by Deveria, who identified himself as ''Libraire & Compagnie, rue de Ravet, derrière le Palais, à Liège.'' The circular is also in the dossier Marc Michel Rey, ibid.

ouvrage a été imprimé par ordre alphabétique, de décrier ou faire décrier cette édition : le public en jugera."[4]

Panckoucke could hardly ignore such an assault, but he had no obvious means of combatting it. Although he might be able to ride out the new version of Linguet's denunciation, he could not easily disparage the plan for an *Encyclopédie méthodique*. For Deveria's project took the concept of a revised edition one step further than Suard's *refonte*. It also contained cleverer provisions for marketing, and it clearly was going to reach the market first. Panckoucke therefore felt seriously threatened and reacted as he had done in the face of the initial threat from Duplain: first he counterattacked, then he negotiated.

The counterattack took the form of a *Mémoire,* which the quarto publishers circulated in order to destroy their rival by defamation. The *Mémoire* presented Deveria as an impostor, a charlatan, a jumped-up clerk turned "escroc littéraire" who wanted to dupe the public with a phoney subscription and probably would disappear as soon as he had collected the down payments. He had been a clerk in the Parisian firm of Veuve Babuty only two years ago and lacked the resources and reputation to produce such an enormous project, the *Mémoire* revealed. If he commanded any credit in Liège, he would not have pretended to issue his prospectus from Amsterdam. "On ne peut voir sans pitié ce ton de charlatanerie qui règne dans le prospectus," the quarto publishers remarked scornfully, taking special care to heap scorn on Deveria's plan to make stockholders of his subscribers. It was a gimmick, meant to snare cash from the gullible; for Deveria could never sell enough *Encyclopédies* to pay a rebate. The market would not bear it. In talking of 20,000 sales he overestimated as egregiously as he underestimated in setting his price at 756 livres. The book was certain to cost at least 1,200 livres, if he ever produced it; and that was impossible, because his plan was unfeasible.

This last point, however, was a sticky one for Panckoucke and his partners. They did not want to argue against a revised edition per se because they were secretly preparing one (they needed to keep their *refonte* confidential until they had sold

4. *Prospectus*, p. 4.

out their quarto), and they could not find a good argument against organizing the material by subject instead of alphabetically. They therefore stressed the personal chicanery and incapacity of Deveria. If his project could be done at all, they claimed, it would require the consent of the original publishers and the collaboration of the most distinguished savants in Europe—in short, the combination Panckoucke had put together for his *refonte*. The *Mémoire* was intended to save the market for Panckoucke's revised *Encyclopédie* by scaring the public away from Deveria's. The savagery of its slander demonstrated the seriousness of the threat from Liège.[5]

But slander did not provide an effective defense, especially against an entire consortium; and Deveria had the backing of three Liégeois speculators, Lefebvre, Desforges, and Desfontaines. Their combined wealth and influence must have been impressive because Panckoucke entered into negotiations with them under the advice of his friend Plomteux, who was a minor partner in the quarto association and a major bookdealer in Liège. On June 22, 1778, the negotiations resulted in a treaty. Panckoucke gave up the attempt to destroy the Liégeois *Encyclopédie* and instead gave it his support—for 105,000 livres. After paying this sum, the Liégeois were to receive the use of Panckoucke's plates and the right to market their book in France without opposition from him and his protectors. All of Panckoucke's partners and all of Deveria's accepted the treaty and thereby saved themselves from a commercial war.[6]

It took some persuading, however, to get the acceptance of the Neuchâtelois. On July 7, 1778, Panckoucke sent them a copy of the contract and an enthusiastic report on the new enterprise. The Liégeois had already collected 900 subscriptions and would begin printing at 2,000, he explained. They were all set to launch their *Encyclopédie,* having withdrawn their earlier prospectus and printed a new one. The greatest problem in the negotiations had been to get them to postpone

5. *Mémoire pour les éditeurs de l'Encyclopédie de Genève en 32 vol. format in-4° portant réfutation d'un Prospectus de l'Encyclopédie en 36 vol. in-folio & 144 vol. in-8°, projetté par un libraire de Liège, sous le nom d'Amsterdam,* STN papers, ms. 1233.

6. Attempts to find a copy of this *traité* in Geneva, Paris, Lyons, Amsterdam, and Liège have all failed. But its main terms, like that of its successor, a contract of Jan. 2, 1779, can be inferred from Panckoucke's letters and other documents in Neuchâtel. Merlino de Ghiverdy signed the contract for Duplain, and the STN accepted it in a formal statement sent to Panckoucke on July 28, 1778.

publication because their *Encyclopédie* would hurt the sales of the quarto, which was just going into its third edition. Fortunately Panckoucke had persuaded them to hold back their prospectus until the end of the year; and so he concluded, "Cette entreprise ne nuira jamais à celle de Lyon . . . Voilà ce que j'ai bien considéré en traitant."[7] The STN answered that it could see how the settlement protected the quarto, but what about the *refonte?* Panckoucke seemed to be abandoning a promising enterprise, which they had nurtured for two years, in favor of a difficult and risky speculation. Although at first they had resisted Suard's project, the Neuchâtelois now outdid Panckoucke in their attachment to it. But their concern was for their printing business, not Suard. They had doubled the size of their shop in order to print *Encyclopédies;* yet Duplain had only let them work on a few volumes, had quarreled at every possible opportunity, and had refused to pay their bills. They expected the *refonte* to compensate them for their disappointment with the quarto. Panckoucke had promised to let them produce Suard's entire text. In fact the contract of January 3, 1777, committed Panckoucke to keep their presses busy with a quarto as well as a folio edition. His contract with the Liégeois gave them nothing. And so they clung to "la grande affaire," as they called the *refonte,* when he held the Liégeois settlement out for their approval.[8]

In answer to the STN's objections, Panckoucke pronounced the *refonte* dead. "J'ai fait avec les Liégeois tout ce que les circonstances ont exigé. Si j'eusse différé plus longtemps,

7. Panckoucke to STN, July 7, 1778.
8. The STN had expressed its position as early as Feb. 22, 1778, when it wrote to Panckoucke about a rumor that the Liégeois had formed an *Encyclopédie* speculation and then had abandoned it. "Nous sommes fort aises d'apprendre que les projets des Liégeois soient abandonnés. Ce sera toujours un obstacle de moins à surmonter. Quant à ce qui concerne notre refonte, nous pensons en général comme vous, Monsieur, et notre idée n'a jamais été de publier actuellement le prospectus de la grande affaire, puisque ce serait de vouloir renoncer de gaiété de coeur et à pure perte au bénéfice de la troisième souscription de l'in-quarto, que nous venons de proposer au public bénévole. Mais nous estimons, comme nous le disions dans notre dernière, qu'il serait nécessaire de prendre dès à présent et entre nous nos mesures et tenir tout prêt pour que dès l'expiration de l'année courante nous soyons en état d'annoncer notre refonte." The STN then expressed its negative reaction to the Liégeois settlement in a letter to Panckoucke of July 14, 1778: "Nous voyons avec peine que sur un plan aussi difficile et dont le succès vous paraît incertain, nous ayons renoncé à notre idée de refonte, qui nous a toujours paru une très bonne chose, même apres l'édition de Lyon. N'y aurait-il pas eu moyen de vendre comme à Duplain la participation au privilège pour un certain nombre d'années?"

Paris et les provinces allaient être inondés de leurs nouveaux prospectus. Leur édition étant une édition refondue et par traité méthodique, nous ne pouvions plus penser même dans l'avenir à une refonte de notre part.'' Plomteux and Ghiverdy, Duplain's agent, had supported his decision, Panckoucke added. Moreover, the Liégeois would not get the use of the plates—which were a crucial asset in all Panckoucke's negotiations—until they had paid the 105,000 livres. Better to hold out for that sum than to wage a trade war for Suard's enterprise, which looked ''douteuse'' in comparison with that of the Liégeois, who now felt so confident of their success that they had decided to increase their printing to 3,000. If they failed, Panckoucke and the STN could resume work on the *refonte.* If they succeeded, the quarto group would be 100,000 livres richer and meanwhile would have drained all the profits from Duplain's *Encyclopédie,* which was already so profitable as to eclipse all other speculations: ''L'affaire de Lyon couvre tous nos frais, avec un bénéfice immense; c'est elle seule que nous devons avoir en vue.'' So the STN should not regret the demise of Suard's revision: ''Sans doute qu'en traitant avec les Liégeois nous perdons le droit de nous arranger avec tout autre, si tant est que leur édition réussisse et qu'ils nous fassent le payement des deux premiers milles. Mais de bonne foi, après cette édition refondue et nos huit mille de Lyon, qui pourra jamais penser à aucune *Encyclopédie?*''

To help the Neuchâtelois over their disappointment about the printing, Panckoucke soothed them with some kind words, ''S'il se présente quelque bonne spéculation, je m'adresserai volontiers à vous. Nous sommes liés pour la vie, à ce que j'espère, et vous me trouverez disposé à vous obliger dans tous les temps.'' And to give substance to his sentiment, he dangled two speculations before them. He described the first as a project ''concernant la queue de notre *Encyclopédie* lyonnaise,'' which he would reveal after his idea had ripened. The other was an edition of Rousseau's works. Rousseau had just died, leaving his Neuchâtel protector, Du Peyrou, as executor of his will. Word had leaked that his papers contained some valuable manuscripts, including some *mémoires* (the *Confessions*), and publishers all over western Europe were scrambling to get their hands on them. Panckoucke, of course, had inside information and was already plotting another multithousand livre speculation, which he mentioned, typically, in a breathtaking postscript: "M. Du Peyrou doit avoir les manuscripts, ou au moins une partie, de J. Jacques Rousseau. Il faudrait que vous en traitassiez avec la veuve; et de tout ce qu'elle a, j'entrerai pour moitié. Mais je ne veux pas paraître. L'un de vous pourrait faire un voyage ici au sujet. Vous pourrez donner jusqu'à trente mille livres, le tiers comptant, les deux autres tiers par acte remboursable en deux payements. Je vous en ferai faire une grande affaire et vous viendrez au secours de cette malheureuse. Il faut surtout avoir ces mémoires. Occupez-vous de cela. Parlez-en avec M. Du Peyrou et ne perdez pas de temps."[9]

The Climactic Moment in Enlightenment Publishing

This letter pacified the Neuchâtelois. They declared themselves willing to accept the Liégeois contract, to "renoncer à toute idée Encyclopédique," and to accept Panckoucke's compensatory offers. They were eager to hear more about the corollary speculation on the quarto, and they snapped up his suggestion about the Rousseau.[10] The area around Neuchâtel was Rousseau territory, just as Ferney was a Voltaire site. The two philosophes had spent so much time in the vicinity of Swiss publishing houses that they gave new life to an old industry, and nothing stimulated the trade more than their deaths; for each of them left behind manuscripts that were spectacular in themselves and that at last made it possible for publishers to produce complete versions of their works. Moreover they had seemed to incarnate two aspects of the Enlightenment, and they died within two months of each other in 1778. So their deaths touched off terrific intrigues among publishers who speculated on Enlightenment. Panckoucke, who led the field in intriguing, traveled to Switzerland in August, hoping to stake a claim to both caches of manuscripts.

Panckoucke's Swiss trip and the next eight months of his commercial operations proved to be a turning point in his life and in avant-garde publishing. For a while he seemed to be on the verge of monopolizing the production side of the Enlightenment. His trip took him through Ferney, where he took possession of Voltaire's papers; Neuchâtel, where he plotted with the STN for the acquisition of Rousseau's manuscripts;

9. Panckoucke to STN, July 21, 1778.
10. STN to Panckoucke, July 28, 1778.

Montbard, where he discussed his editions of Buffon's works with Buffon; and back to Paris, where he reinforced his monopoly of the *Encyclopédie* by taking over the Liégeois project. Those speculations required such heavy capital investment that even an "Atlas of the book trade" could not carry them all. Panckoucke therefore had to choose among them. First he decided to sell off his inventory in order to build up his capital. Then he engaged in some dizzying rounds of buying and selling. By March 1779, he had prepared to part with half his Buffon, had dropped out of the bidding on the Rousseau, had sold his Voltaire, and had bought out most of the other gamblers on the *Encyclopédie*. Although the full story of his operations during this crucial period belongs to his own biography rather than to that of the *Encyclopédie*, it is important, in order to understand the transition from the quarto to the *Méthodique*, to follow Panckoucke as he discarded one speculation after another and eventually put most of his money on the last and largest *Encyclopédie* of the century.

One can imagine how Panckoucke's conversations in Neuchâtel soared over the new range of speculations that had suddenly changed the scene of the publishing world, but the only evidence of the topics he covered is a note that the Neuchâtelois sent to him soon after his departure. The note accompanied a contract that he and the STN had just signed for an edition of a *suite des planches*. This was the speculation on the *queue* of the quarto that Panckoucke had mentioned in his earlier letters. Calculating that a large number of the 8,000 quarto subscribers would want a complete set of illustrations instead of only three volumes (the folio editions contained eleven volumes), Panckoucke and the STN agreed to produce a supplementary edition of them. They needed merely to have their original copper plates retouched or re-engraved and then to run off the number of copies that their subscribers demanded—an inexpensive, unrisky, and profitable operation. Of course it would not be feasible until the completion of the quarto. At that time the Traité de Dijon would have expired, ending Duplain's claim to a share in the plates. So Panckoucke and the STN hid their agreement from him, hoping to cut him out of the profits. The Neuchâtelois did not know that a year earlier Panckoucke had made a secret deal with Duplain for a quarto edition of the *Table analytique*. The quarto *Encyclopédie* was such a huge, complicated affair that it created several opportunities for ancillary speculations around the Paris-Neuchâtel-Lyons triangle. But by August 1778, Panckoucke and the STN had drawn together in a common distrust of Duplain and a common desire to produce the works of Voltaire and Rousseau: hence the intimate tone of the STN's note of August 25. Alluding to Panckoucke's earlier plans to establish a residence in Neuchâtel, it expressed regret that he had not been able to spend the entire summer in "votre seconde patrie." The Neuchâtelois informed him that, as agreed, they were dispatching a "lettre ostensible" to be used in the struggle with Duplain over the contract for the third edition of the quarto, and they sounded hopeful that "vous pourrez nous dire quelque chose positif sur les oeuvres de V., continuant toujours à nous occuper de celles de Jean-Jacques."[11]

In Ferney, Panckoucke had acquired an enormous collection of manuscripts from Madame Denis, Voltaire's niece and heir. He had had an inside track in the race for this prize because he had visited Ferney before and had persuaded Voltaire to prepare his papers for an enormous revised edition of his works.[12] The Rousseau manuscripts were another matter. After the rumor had spread that they contained his *Confessions,* a sure best seller, a whole pack of publishers set off after them. By August 1778, a newly formed Société typographique de Genève, which had enlisted Paul Moultou and Gabriel Cramer, seemed to have taken the lead, but it had to fight off Regnault of Lyons, Boubers of Brussels, the Société typographique of Lausanne, and other formidable competitors. The main hope of the Neuchâtelois derived from their connections with Du Peyrou, who gave a sympathetic hearing to their propositions at the end of August and revealed that

11. STN to Panckoucke, Aug. 25, 1778. The speculation on the plates never took place, and Panckoucke's letters ceased to mention it because it was superseded by his arrangement with the Liégeois of Jan. 2, 1779, and by Duplain's acquisition of a full share in the original plates.

12. Although a great deal remains to be said about the history of the Kehl Voltaire, the main outline of the story is clear from three articles by George B. Watts: ''Panckoucke, Beaumarchais, and Voltaire's First Complete Edition,'' *Tennessee Studies in Literature,* IV (1959), 91–97; ''Catherine II, Charles-Joseph Panckoucke, and the Kehl Edition of Voltaire's *Oeuvres,*'' *Modern Language Quarterly,* XVII (1956), 59–62; and ''Voltaire, Christin, and Panckoucke,'' *The French Review,* XXXII (1958), 138–143. See also Brian N. Morton, ''Beaumarchais et le prospectus de l'édition de Kehl,'' *Studies on Voltaire and the Eighteenth Century,* LXXXI (1971), 133–149 and Giles Barber, ''The Financial History of the Kehl Voltaire,'' *The Age of the Enlightenment: Studies Presented to Theodore Besterman* (London, 1967), pp. 152–170.

he was expecting a visit from Rousseau's other protector, the Marquis de Girardin, sometime in the autumn. For a while, it looked as though they could capture the treasure in their own back yard. They prepared to pay 10,000 livres and to work out a joint Voltaire-Rousseau speculation with Panckoucke: "Nous nous occupons fortement des oeuvres de Jean-Jacques . . . Si nous réussissons à terminer cette affaire, nous serons charmés, Monsieur, de vous proposer un arrangement qui resserre encore les liaisons que nous avons le plaisir de soutenir avec vous. Pour cet effet, nous vous proposons d'entrer pour une part dans votre entreprise du Voltaire. Il nous convient beaucoup mieux de la faire avec vous que d'entendre aux propositions qu'on nous fait d'ailleurs pour la contrefaire."[13]

In September, however, Girardin postponed his trip to Neuchâtel and the STN faltered in its efforts to win over a still more important personnage, Rousseau's widow, Thérèse Levasseur, who had the final say on the disposition of the manuscripts and who had not acquired any love for the Neuchâtelois during the years she had spent in their territory. Still, the STN hoped to bring her around by negotiating through Girardin and by increasing its offer to 24,000 livres with a lifetime annuity of 1,200 livres. Meanwhile, they tried to keep Panckoucke on their side: "Voltaire est une grande affaire, et vous êtes seul à même de le faire; mais cela ne veut pas dire que vous le fassiez seul; et comme nous croyons que vous serez bien aise de trouver quelques aides, nous vous offrons nos services. Il sera toujours glorieux pour vous de donner au public les portefeuilles des deux plus grands hommes du siècle, car le Rousseau sera aussi une grande entreprise. Nous espérons de pouvoir vous en dire des nouvelles avant peu. La partie des Mémoires qui existe est très considérable. Il y aura beaucoup de pièces neuves et très intéressantes. On compte sur huit volumes in-quarto. Vous serez contrefait pour Voltaire et nous pour Rousseau. La chose est plus difficile quand nous serons réunis."[14]

This prospect sounded tempting enough for Panckoucke to encourage the STN's pursuit of the Rousseau while he concentrated on the Voltaire. But he maintained an independent

13. STN to Panckoucke, Nov. 1, 1778.
14. STN to Panckoucke, Nov. 13, 1778. See also STN to Panckoucke, Sept. 15 and Nov. 1, 1778, and STN to Perregaux, Aug. 8, 1778.

policy, and to increase his independence, he built up his capital
by selling off his stock. He had intended for some time to get
out of the wholesale book trade in order to concentrate on
journalism and a few choice speculations in book publishing.
But the special circumstances of 1778 precipitated this shift
in policy, which he later described in the language of the
Bourse. "Il me semble que vous imprimez trop de livres,"[15]
he wrote to the STN a year after his big sale. "A votre place
je ne voudrais pas de magasin. Cent mille livres qui circulent
valent mieux que cent mille écus en magasin de librairie. J'ai
eu cette manie, en mars 1778. Mon fonds de livres était un
objet de quatorze cent mille livres. J'ai pensé en être la victime. Vous sentez bien que ce fonds ne rapportait pas l'intérêt
de l'argent. Je voudrais n'avoir pas pour cent mille francs de
livres, et je désirerais que mon portefeuille fût mieux garni."

Panckoucke sounded more like a financier than a publisher,
because by 1778 he had become a specialist in the financial side
of publishing and so many manuscripts came on the market at
once that he spent the year dashing from speculation to speculation. The pace became frantic in November and December,
when he attempted to liquidate an inventory worth 1,400,000
livres while chasing after the works of Voltaire and Rousseau
and directing two speculations on the *Encyclopédie* as well as a
journalistic empire. His letters from this period have a breath-
less quality. Their handwriting is neglected, their sentences
short, their phrasing rushed. On November 6, 1778, for example, he wrote to the STN, "Nous n'avons plus de temps à
perdre. L'affaire des Liégeois est toujours en suspens . . .
Tout cela exige bien des courses, des travaux, des écritures
. . . Je suis extrêmement occupé dans ce moment-ci. Je travaille à faire ma vente. Je ne garderai que les journaux, le
Buffon, le Voltaire, et mes intérêts dans l'*Encyclopédie*. J'ai
déjà vendu six gros articles. Ma vente à la chambre aura lieu
dans 15 jours . . . Il n'y a rien à faire du Rousseau sans les
Mémoires. Je puis seul faire le Voltaire. C'est une si grande
affaire que je me suis déterminé, pour m'en occuper et m'y
livrer entièrement, à vendre presque tout mon fonds. Je ne
crains pas beaucoup la contrefaçon. Savez-vous que j'ai de

15. Panckoucke to STN, Nov. 22, 1779. Panckoucke's recent acquisition of the
Mercure also determined his decision: "Cette opération me porte à réaliser le
plan que j'ai toujours eu de vendre mon fonds, hors l'*Histoire naturelle.*"
Panckoucke to STN, July 7, 1778.

quoi donner 20 volumes nouveaux et que l'auteur m'a laissé
31 volumes corrigés de sa main?"[16]

Panckoucke did not sell his entire stock of books, and those
he sold fetched only 350,000 livres in notes with a five-year
maturity. He still needed or wanted capital at the beginning
of 1779. So he invited bids for his Buffon and his Voltaire.
The Buffon apparently remained on his hands for another
year, and he only parted with half of it; but the Voltaire immediately produced a fresh round of bartering.[17]

Panckoucke had learned that each of his principal partners
in the quarto *Encyclopédie* was plotting to pirate his edition
of Voltaire, so he secretly and separately offered to sell it to
each of them. The Neuchâtelois had told him frankly that they
were considering an offer to join some pirates but that they
preferred to speculate with him on both Voltaire and Rousseau. Guessing, correctly, that the sociétés typographiques
of Lausanne and Bern were behind this remark, Panckoucke
suggested that the Swiss Confederation buy him out. "Je
n'ai point pris de résolution au sujet de Voltaire. C'est une
très riche mine d'or dont l'exploitation est difficile. Je préférerais de vendre mon manuscrit. Vous devriez vous occuper
de cet objet et vous associer avec Berne, Lausanne, et vous
gagneriez plusieurs millions. Toute l'Europe attend une nouvelle édition. Je vous donnerai les moyens de ne pas être
contrefaits. Le manuscrit me revient à cent mille livres. Je
veux doubler mon argent et 500 exemplaires. Vous n'avez pas
d'idée de ce que j'ai acquis. J'ai tout réunis, et il y a telle
correspondance qui m'a coûté deux mille écus. Enfin, je m'obli-
gerai à donner 20 volumes nouveaux."[18] At the same time he
sounded Duplain, and after some preliminary bargaining he
sent almost the same offer to Lyons. "Je vous ai dit mon

---

16. Panckoucke's remark about the *Mémoires* was a response to a rumor,
which the STN had reported, that the manuscript of the *Confessions* had been
burned.

17. Panckoucke to STN, Dec. 22, 1778: "Les embarras de ma vente m'ont
pris aussi beaucoup de temps. J'ai encore 5 à 6 articles à placer, et j'y travaille.
J'espère à la fin de janvier n'avoir que mes journaux et le Buffon. Encore je
cherche un intéressé pour ce dernier." How and when the Buffon sale occurred is
unclear, but in his *lettre circulaire* of Feb. 28, 1781, Panckoucke reminded his
customers that "Le Buffon ne m'appartient qu'à moitié." The reference to the
350,000 livres is in Panckoucke to Duplain, Dec. 26, 1778, Bibliothèque publique
et universitaire de Genève, ms. suppl. 148. Although it is impossible to know the
state of Panckoucke's finances at this time, a rumor was circulating that he was
"à la veille de faillir." *Mémoires secrets*, entry for May 20, 1779.

18. Panckoucke to STN, Nov. 6, 1778.

dernier mot sur le prix de la vente. Je ne donnerai point mon manuscrit à moins de trois cent mille livres. Il me coûte, en espèces, en argent comptant, cent mille livres . . . Comme vous n'êtes pas accoûtumé à payer de copie, ce prix vous paraît excessif; mais n'ai-je pas moi acheté le droit de faire l'*Encyclopédie* 312 mille livres, après qu'on en eût passé 4,000, à Paris; et ce droit, ne l'ai-je pas revendu à peu près ce même prix; et les Liégeois en troisième lieu, ne nous l'ont-ils pas encore rendu? Il y a plus à espérer du Voltaire que de l'*Encyclopédie*. Je sais que toute l'Europe attend une nouvelle édition.''[19]

This double-dealing never came to anything, because in March 1779 Panckoucke sold his Voltaire for 300,000 livres to a third party, a man who was at least his equal in speculations and intrigues: Beaumarchais. This sale opened the way to the great Kehl edition of Voltaire. It also provided a finale to a round of heavy betting, which illustrates how much was at stake in the speculations on Enlightenment and how great the market was for the works of the philosophes, according to the calculations of the speculators.[20]

Meanwhile, what had become of the Rousseau? Panckoucke had left the negotiating to the Neuchâtelois, who had become embroiled with Girardin and the Genevans. In January 1779 they lost in the bidding to Thérèse Levasseur and also failed to force their way into the Genevan consortium, which won it. Next, they plotted to pirate the Genevan edition with Duplain,

19. Panckoucke to Duplain, Dec. 26, 1778, Bibliothèque publique et universitaire de Genève, ms. suppl. 148. Although Panckoucke asked for 300,000 livres from Duplain and for 200,000 livres from the STN, the offers were roughly the same because he also requested the STN to give him 500 free copies. His letter to Duplain went on to make an alternate proposal for a half interest in the Voltaire with a complicated provision for refunding the notes in his portfolio. He did not underestimate the value of his ''trésor'': ''Vous parlez d'un tirage de 4,000. Vous n'êtes pas de bonne foi, mon ami. Vous tirerez 12, 15, 20 mille et vous n'en doutez pas. Je suis bien sûr que quand le prospectus aura paru, et il est tout prêt, vous en serez convaincu. J'y ai fait l'histoire du manuscrit. Vous même m'assurez dans vos précédentes que nous aurions chacun 4 à 5 cent mille livres de bénéfice, mais moi je vous en assure qu'il y a le double et le triple à espérer et que je ne me fais point illusion. Comme cette entreprise est la seule grande affaire qui reste en librairie et que je l'envisage comme un moyen de me tirer du commerce avec une fortune proportionnée avec [les] grands moyens que j'ai employés dans cet état, je suis bien décidé à ne m'en désaisir que lorsqu'on m'en fera un très grand avantage.''

20. For Panckoucke's version of his deal with Beaumarchais see *Lettre de M. Panckoucke à Messieurs le président et électeurs de 1791* (Paris, Sept. 9, 1791), pp. 17–18.

who was to lobby for a monopoly of the French market. But this plan also collapsed, and the STN fell back on a series of similar conspiracies, first with Regnault, then with Heubach, then with Boubers. In the end they had to accept defeat and a reprimand from Panckoucke: "Vous êtes bien la cause, Messieurs, que j'ai manqué cette grande et superbe entreprise. J'en aurais donné jusqu'à 40,000 livres. L'associé de Genève a dîné aujourd'hui avec moi. Je sais qui leur a vendu le manuscrit. Ce n'était pas là le cas de marchander. J'ai un véritable regret d'avoir laisser échapper cette occasion. L'associé en a placé plus de dix mille dans sa tournée."[21]

But Panckoucke had only himself to blame. He could have pursued the Rousseau. Instead, he let it pass; he abandoned the Voltaire; he dropped half the Buffon—all in order to devote himself to the *Encyclopédie méthodique,* which became the supreme speculation of his career.

The Liégeois Settlement

Perhaps it was a tender spot in Panckoucke's tough-minded calculations that made him favor the *Encyclopédie* over the other works of the philosophes. It had been his first great project, and he meant it to be his last, he told the STN in July 1779: "Tout le plan de l'*Encyclopédie méthodique* est tracé. J'ai déjà passé trois traités; les censeurs sont nommés, et je compte finir ma carrière par une belle entreprise."[22] But how was it that he came to speak of the *Méthodique* as if it belonged to him and to choose it over all the other options that tempted him at the end of 1778?

Whether or not sentiment determined it, Panckoucke's choice was precipitated by a crisis in Liège. Deveria and his associates turned on one another, entangling themselves in law suits, and seemed incapable of extricating their *Encyclopédie* from the courts before the January 1779 deadline for the publication of its new prospectus. In early November 1778,

21. Panckoucke to STN, June 1, 1779. In its reply, dated June 6, 1779, the STN tried to justify itself as follows: "Nous n'avons agi que d'après ce que vous nous aviez conseillé d'en traîter avec M. Du Peyrou, qui s'est laissé mener par M. de Girardin, et celui-ci par M. de Moultou de Genève, qui étant en possession d'une partie des manuscrits a voulu avoir part à l'entreprise et la voir exécutée sous ses yeux." See also STN to Girardin, Jan. 24, 1779, and STN to Duplain, Feb. 20, 1779.
22. Panckoucke to STN, July 10, 1779.

Panckoucke announced their bad fortune as if it were good news: "Je profiterai de leur division pour faire retarder leurs entreprises."[23] Duplain had reported that the third quarto subscription was filling very slowly; so every delay in the *Méthodique* helped the quarto; and of course if the Liégeois failed completely, Panckoucke and the Neuchâtelois could print and sell their own revision without sharing its profits. But the more Panckoucke considered the Liégeois project, the better it looked, and by the end of December he decided to take it over instead of undermining it.

Although the takeover must have occupied most of the "courses, travaux, écritures" referred to in Panckoucke's letters, he hardly mentioned the great event at all. "J'oubliais la nouvelle la plus importante,"[24] he added, almost as an afterthought, at the end of one of his hurried dispatches to Neuchâtel. "Je me suis enfin arrangé depuis deux jours avec les Liégeois. Je me mets à la tête de l'entreprise. Tout ce que j'ai eu en vue dans cette longue opération, c'est de retarder l'entreprise de six mois, car quoiqu'ils n'eussent pas réussi dans la manière dont ils l'ont conçu, ils nous auraient fait un mal effrayable." On the same day Panckoucke sent an even briefer announcement to Lyons; and soon afterward he urged Duplain to cut the pressrun of the third edition of the quarto by 500, even though several volumes had already been printed, because the quarto associates now controlled the *Méthodique*.[25]

This request, which Duplain refused, suggested an odd shift in Panckoucke's policy. At first he had advocated collaboration with the Liégeois because of the need to protect the quarto; now he wanted to cut back on the quarto because of the collaboration with the Liégeois. The references to the Liégeois settlement in his subsequent letters were brief and inconsistent: sometimes he presented it as a defensive maneuver, sometimes as a good thing in itself. It was hard to know what he was up to. But by sifting through the commercial correspondence of the next few years, it is possible, in retro-

23. Panckoucke to STN, Nov. 6, 1778.
24. Panckoucke to STN, Dec. 22, 1778. He and the Liégeois did not sign the contract for this new arrangement until Jan. 2, 1779.
25. Panckoucke to Duplain, Dec. 22 and 26, 1778, Bibliothèque publique et universitaire de Genève, ms. suppl. 148. In the letter of Dec. 26 Panckoucke stated, "Cette affaire entraîne nécessairement, pour nos intérêts communs, une diminution dans le tirage de la troisième édition." On Dec. 24, 1778, he wrote to the STN: "L'arrangement avec les Liégeois doit nous faire changer de plan."

spect, to trace the main line in the speculating that issued in the *Méthodique*.

Panckoucke explained his policy on the *Méthodique* at the meeting of the quarto association in February 1779. The associates must have considered his explanation satisfactory because they agreed to accept his new contract with the Liégeois, dated January 2, 1779, which superseded the original contract of June 22, 1778. Two and a half years later, however, the whole thing seemed mysterious to the STN. By then the quarto had been liquidated, leaving a legacy of suspicion and resentment, and the Neuchâtelois were negotiating with Plomteux, their former partner in the quarto, for the sale of the share in the *Méthodique* that was allotted to them by the second Liégeois contract. Neither that contract nor its predecessor survives in their papers, but from the correspondence that led up to the Plomteux sale it is possible to get some idea of how Panckoucke dismantled the Liégeois project and built his own *Encyclopédie méthodique*.

In order to put a price on their share, the Neuchâtelois needed to know what it represented. They therefore sent their copy of the contract of January 2, 1779, to their Paris agent, Quandet de Lachenal, and asked him to investigate Panckoucke's operation. "Ledit sieur [Panckoucke] chargé de ménager nos intérêts comme ayant part à la propriété des cuivres et de traiter en notre nom avec une prétendue société formée pour l'entreprise dont il s'agit, se trouve aujourd'hui avoir travailler pour son profit individuel, ayant réussi, je ne sais comment, à faire évanouir cette prétendue société, ce qui comme vous le sentez n'est pas absolument conforme aux lois."[26] In short, the Neuchâtelois felt swindled: "Tout ce tripotage a trop l'air d'une duperie pour ne pas nous donner quelques inquiétudes . . . Nous connaissons d'ailleurs l'homme."[27] After studying the contract. Quandet took the same view: "Cette compagnie, la forme du traité fait avec elle, tout dans cette affaire me paraît fort obscur. Ce qui me le paraît moins, c'est que le sieur Panckoucke a seul la clef de tout ce beau tripotage . . . La compagnie, le traité sont, ou je serais bien trompé, de la poudre jetée aux yeux. C'est un manteau dans lequel on a voulu envelopper des

26. STN to Quandet de Lachenal, April 3, 1781.
27. STN to Quandet de Lachenal, April 1, 1781.

vues peu loyales dont on s'est servi pour dérober à des yeux trop clairvoyants la marche tortueuse de l'opération et pour faire retomber par la suite sur une compagnie imaginaire ou simple prête-nom l'inexécution partielle ou absolue du premier traité.''[28] A good example of the conspiratorial views that prevailed among eighteenth-century bookdealers. But had Panckoucke really trapped his partners in a plot?

The affair is too murky and the evidence too sparse for one to declare Panckoucke guilty or innocent, but it seems important to take account of the judgment of Plomteux, who was an old hand in the book trade, who knew Panckoucke very well, and who traveled to Paris in the spring of 1781 in order to look into the *Encyclopédie méthodique* for himself. He did not send a full report to the STN, which had revealed its suspicions to him, but he pronounced the enterprise legitimate and healthy, and he repeated his earlier offer to buy the STN's share in it. Plomteux's verdict satisfied the Neuchâtelois. They called off Quandet's investigation; and, reassured as to the value of the speculation, they held on to their share a little longer, although they eventually sold out to Plomteux.[29]

From the commercial gossip of 1781 one could conclude either that Panckoucke had swindled his partners in 1778–1779 or that he had cut them in on a profitable enterprise. But despite this fundamental ambiguity, one can sift through the rumors for a few nuggets of hard information about the contracts of June 22, 1778, and January 2, 1779. By the first contract, Panckoucke sold the Liégeois the use of the plates and the protective covering of the privilege. By the second, he took over their *Encyclopédie*. The first was tantamount to a lease; the second involved a complete reorganization and refinancing of Deveria's operation. As far as Panckoucke's partners were concerned, the first contract represented an asset of 105,000 livres and protected their quarto both by delaying the publication of the *Méthodique* and by stipulating that the plates could not be used until the money was paid. The second contract canceled the 105,000 livre debt and allotted the plates and privilege to a new association, which Panckoucke put together for the production of his own kind of *Encyclopédie méthodique*—one that would combine Suard's

28. Quandet de Lachenal to STN, April 9, 1781.
29. STN to Plomteux, April 3, 1781, and Plomteux to STN, April 14, 1781.

organization of philosophes with Deveria's conception of sub-*Encyclopédies* organized according to subject matter. As compensation for sacrificing their claims to the 105,000 livres and for sharing their rights to the plates and privilege, Panckoucke's quarto associates received shares in his new association.

This was the aspect of the settlement that later seemed so suspicious to the Neuchâtelois. Their suspicions were also aroused by clause six of the second contract, which specified that the shareholders of the new Association could not draw on the profits of the new *Encyclopédie* until it was sold out—a provision that seemed to leave their assets in Panckoucke's hands indefinitely and to have the opposite effect of the protective clause about the plates in the first contract, which he had earlier singled out for praise. They felt especially dubious about the Association itself because Deveria vanished from it, leaving Panckoucke in his place; and Panckoucke gained complete control of it, both by acquiring some of the Liégeois' shares and by writing the contract in such a way as to put himself in charge of its administration. Since he dominated each of the contracting parties, he appeared to be selling something to himself; and he could be counted on to have arranged the transaction so that it did not hurt his interests. ''Ne vous a-t-il pas paru d'abord assez singulier,'' the STN asked Plomteux in 1781 ''que M. Panckoucke, chargé de ménager vos intérêts et les nôtres en traitant avec une société étrangère, ait trouvé moyen d'écarter tous ceux qui la composent et soit devenu en se mettant à leur place notre seule partie adversaire en quelque sorte, avec qui nous sommes appellés aujourd'hui à discuter ces mêmes entreprises?''[30]

Plomteux did not deny that the Liégeois settlement benefited Panckoucke, but he clearly believed that it had advantages for the quarto partners, too. Although he did not explain his reasoning, it probably went as follows. By the end of 1778 the Liégeois were in disarray: not only could they not

30. STN to Plomteux, April 3, 1781. The STN was blunter in a letter to Quandet de Lachenal of April 1, 1781, which explained that Panckoucke had first sold the plates ''à lui et à certains quidames à nous inconnus''; then had finagled the postponement of their payment; and finally, ''tout cela fait, le sieur Panckoucke, qui par parenthèse a été prêtre et martin dans cette affaire, comme je l'expliquerai quelque jour, ayant exclus ses prétendus associés, se mit en devoir d'exécuter son dessin et traiter avec divers savants pour lui servir de collaborateurs.''

come up with the 105,000 livres for the plates, they also seemed incapable of producing their *Encyclopédie.* Yet it was a good project. It already had attracted 900 subscribers and some wealthy investors. In January-February 1779, the quarto associates had an opportunity to take it over and merge it with Suard's operation, which could use some new capital. They had already got good value from their plates. Instead of writing off the 105,000 livres, they could graft one healthy enterprise on another and make greater profits from a greater *Encyclopédie.*

One can imagine Panckoucke developing such arguments before his associates in Lyons, but the ultimate interpretation of his two contracts must remain guesswork, and the same is true of the merger that followed them. It seems likely that the Liégeois and quarto groups each took a half interest in the new Association. In that case, a share in the quarto would have entailed a half share in the *Méthodique;* and the STN, which had possessed 5/12 of Panckoucke's initial investment in the plates and privileges (it had ceded 1/12 of its original half interest to Suard) and 5/24 of the quarto, came to own 5/48 of the *Méthodique.* Eighteenth-century publishers often divided their speculations into small portions, which could be traded, played, or cashed in. After the contract of January 2, 1779, gave them a fresh hand, the publishers of the new *Encyclopédie méthodique* began a new round of gambling. Sometime that year Plomteux bought Deveria's share of 1/5 and later acquired 1/10 from one of the three other Liégeois. In addition to these 3/10, he had 3/48 by virtue of his holdings in the quarto. And in June 1781, he bought the STN's 5/48 for 8,000 livres, although, as he put it, "c'est en tremblant que je fais de nouvelles dépenses pour un objet qui n'a été pour moi jusqu'à présent qu'une source de pertes, de chagrin, et d'inquiétude."[31]

31. Plomteux to STN, June 26, 1781, and May 1, 1781: "J'ai acquis, ainsi que je vous en ai fait part à Lyon 4/20, qui m'ont été cédé par M. Deveria à qui nous avons transmis nos droits. Un autre intéressé dans l'entreprise m'a depuis cédé deux autres vingtièmes tellement que mon intérêt dans l'*Encyclopédie méthodique* est actuellement de 6/20, indépendamment des 3/48 qui me restaient au même titre que vous pour la part des cuivres." A collection of contracts between the publishers and authors of the *Méthodique* in the Kenneth Spencer Research Library, University of Kansas, ms. 99 shows that Deveria countersigned contracts, along with Panckoucke and Plomteux, until July 1782, when his name disappears from the documents. So presumably he remained associated with the enterprise for a while after selling his one-fifth interest to Plomteux.

Such rhetoric was part of the game. Three months earlier, Plomteux had offered 5,000 livres, also "en tremblant."[32] The STN had originally valued its share at 21,875 livres, and the bartering had gone on for almost a year, interspersed with lamentations by the STN about being forced to put so low a price on so valuable an asset and by agonizing from Plomteux about "les risques de ces entreprises" and his "triste expérience" with them in the past.[33] Actually, 8,000 livres was probably a fair price for a share of about one-tenth in the speculation, but Plomteux had reason to tremble in paying it because it brought his total holdings to almost half of the entire enterprise. What happened to the other half is hard to say. Duplain sold his 12/48 for 12,000 livres to Jossinet of Lyons, who later sold them for the same price to someone else, possibly Panckoucke. Judging from allusions in the STN's letters, Panckoucke may have bought shares from the other Liégeois, too. He already possessed an interest of at least 3/48 as a result of his holdings in the quarto, and he spoke of the speculation "dont je me suis rendu maître" as if he had a controlling interest in it.[34]

Panckoucke's Conception of the Supreme *Encyclopédie*

Panckoucke was interested in something more than control. The *Encyclopédie méthodique* seized his imagination and aroused his passion for *les grandes affaires,* as he called them, in addition to his appetite for profits. At first, to be sure, he had expressed nothing but scorn for Deveria's project, which, he said, would not even result in a genuine *Encyclopédie:* "Dans le fonds, cette *Encyclopédie* n'en est pas une."[35] Even

32. Plomteux to STN, March 22, 1781. The negotiations actually began in Aug. 1780.

33. Plomteux to STN, Jan. 17, 1781, and STN to Plomteux, May 31, 1781: "Vous comprendrez aisément, Monsieur, que [nous] n'avons pu qu'éprouver beaucoup de répugnance à renoncer ainsi à des droits qui auraient dû nous rapporter une somme sans comparaison plus forte et à vendre à vil prix ce qu'on nous a fait payer si cher."

34. Panckoucke to Duplain, Dec. 26, 1778, Bibliothèque publique et universitaire de Genève, ms. suppl. 148. As mentioned, Rey sold back his 1/24 interest in the quarto; Regnault probably did the same because his name dropped out of the correspondence about the quarto in 1777. If, as seems likely, Panckoucke took up their shares, his interest in the quarto would have gone up to 5/24.

35. Panckoucke to STN, July 7, 1778.

after taking over the Liégeois speculation, he described his involvement negatively, as a maneuver to defend the quarto.[36] But his attitude changed as he became more and more immersed in the enterprise: "On travaille à l'*Encyclopédie Liégeoise*, et je ne serais pas étonné que cela réussit," he wrote to the STN in January 1779.[37] Although he still sounded cautious about its prospects after the February meeting in Lyons, his optimism rapidly rose.[38] By April, he seemed completely engrossed in organizing his new *Encyclopédie*—and he clearly considered it "his": "Je suis seul à la tête de cette grande entreprise . . . Je compte . . . faire incessament toutes les démarches nécessaires pour cette grande entreprise. J'ai eu aujourd'hui une conversation avec deux savants, qui me font espérer qu'elle réussira. Je suis accablé de besognes et d'affaires difficiles. Le temps est fort court dans ce pays-ci.'"[39] He stamped his personality on the work by incorporating the Suard and the Deveria projects in a plan that dwarfed both of them. "J'ai bien retourné cette affaire, et j'espère qu'on en sera content," he wrote in June.[40] "Ce sera là la véritable *Encyclopédie*. Bien des gens de lettres goûtent mon plan, qui ne ressemble point au plan Liégeois." What had once seemed a harebrained scheme of an insignificant clerk now promised to become the only *Encyclopédie* worthy of the name—"un superbe ouvrage et la vrai *Encyclopédie*"[41]—a book that would make the work of Diderot and d'Alembert look small by comparison.

Panckoucke made the shortcomings of the original *Encyclopédie* the central theme of his campaign to promote the *Méthodique*. He opened his first prospectus with a quotation from Voltaire's *Questions sur l'Encyclopédie,* which damned Diderot's work with faint praise as a "*succès, malgré ses défauts*" (Panckoucke's italics). "M. de Voltaire désirait ardemment

36. Panckoucke to STN, Jan. 3, 1779: "L'acte des Liégeois a été signé hier, quoique tout fût convenu il y a plus de dix jours. J'ai fait insérer que les prospectus ne paraîtraient qu'en juin avec les premiers volumes. Nous sommes sauvés de ce côté-là."
37. Panckoucke to STN, Jan. 17, 1779.
38. Panckoucke to STN, March 7, 1779: "Comme les bénéfices de l'édition Liégeoise ne sont pas prochains, je ne voudrais pas que vous missiez actuellement à cela une bien grande valeur."
39. Panckoucke to STN, April 25, 1779.
40. Panckoucke to STN, June 1, 1779.
41. Panckoucke to STN, June 15, 1779.

une nouvelle édition de l'*Encyclopédie,* où les fautes de la première fussent corrigées," the prospectus explained. "C'était pour cette nouvelle édition qu'il avait fait ses *Questions sur l'Encyclopédie.*"[42] If Voltaire's endorsement were not persuasive enough—and who would not be impressed by a pronouncement of the great man, whose prestige was then at a peak?—those who hesitated to subscribe could consider the advice of Diderot himself, whose memoir about the faults of his *Encyclopédie* could be read as propaganda for Panckoucke's. Having originally commissioned the memoir to promote his plan for a revised edition in 1768, Panckoucke made it a major element in his sales campaign of 1781. The prospectus of the *Méthodique* quoted Diderot's criticisms at length and showed how Panckoucke's *Encyclopédie* would meet them, point by point. Panckoucke went further in some of his other promotional material. The individual prospectuses of the dictionaries that were to make up the overall *Encyclopédie méthodique* stressed the inadequacies in Diderot's coverage of every subject that Panckoucke's authors treated. Even Diderot's disciple Jacques-André Naigeon, who prepared the dictionary on philosophy, revealed that the master had been dissatisfied with his own articles on philosophical subjects and had planned to improve them in a later edition.[43] And Panckoucke wrote that the last ten of Diderot's seventeen volumes of text were little more than a scissors-and-paste job, hurriedly put together by the Chevalier de Jaucourt and a team of copyists, who lifted their material from standard reference works.[44]

Panckoucke did not disparage everything about the original

42. Panckoucke published a slightly abridged version of the original "Grand prospectus" for the *Méthodique* in the *Mercure de France,* Dec. 8, 1781. He also issued it in pamphlet form and reprinted it at the beginning of the first volume of the dictionary *Beaux-Arts* in the *Encyclopédie méthodique.*

43. "Il regrettait de n'avoir pas donné à cette partie de l'histoire des progrès de l'esprit humain une attention et des soins qui répondissent à l'importance de l'objet; et il se proposait d'y suppléer dans une seconde édition. Son plan était vaste et bien conçu." "Tableau et apperçu du nombre de volumes de discours & de planches que doit avoir l'*Encyclopédie* par ordre de matières" in the dictionary *Mathématiques,* III, 16 of the *Encyclopédie méthodique.*

44. "Représentations du Sieur Panckoucke, Entrepreneur de l'*Encyclopédie méthodique*" in *Encyclopédie méthodique, Mathématiques,* III, iv: "M. le chevalier de Jaucourt, possesseur d'une bibliothèque assez considérable, s'environna d'une douzaine de copistes et de secrétaires, auxquels il faisait transcrire les différents articles des livres qu'il leur indiquait. Les dix derniers volumes, si l'on en excepte les articles d'arts mécaniques, de philosophie et de mathématiques, ont presque été en entier composés de cette manière."

Encyclopédie.[45] On the contrary, he claimed to esteem it more than Diderot did, for he wanted to persuade the public that the *Encyclopédie méthodique* would incorporate all the good elements of its predecessor while eliminating the bad. But he objected to the organizing conception of the original *Encyclopédie*, not merely to the inadequacy of the information it provided. D'Alembert's *Discours préliminaire* had correctly rooted that conception in the new approach to knowledge and nature that began with Francis Bacon, Panckoucke argued; but the main text belied that insight, for it followed the order of the alphabet, instead of organizing the material systematically—that is, according to the way man had accumulated knowledge and constructed sciences. Diderot and d'Alembert had adopted the alphabetical order as a matter of convenience, a procedure that made some sense in the beginning, when they had not realized the enormity of their undertaking and had expected to produce only ten volumes of text. But no reader could get a coherent picture of any science by roaming around in the seventeen volumes that Diderot finally produced.

Panckoucke would remedy this defect by organizing his *Encyclopédie* into twenty-six subencyclopedias, each of which would cover a branch of knowledge, ranging from mathematics and physics to commerce and *arts et métiers.* He had borrowed this idea from Deveria, though he did not acknowledge the source, but he modified the Liégeois plan in an important respect. Where Deveria had proposed to regroup the articles of the original *Encyclopédie* into separate treatises, Panckoucke retained the alphabetical order in each of his subencyclopedias, or *dictionnaires,* as he called them. In this way, they would not look like pale imitations of existing treatises, and they would retain their usefulness as reference works. Moreover, they could be read as treatises, because each author would begin his dictionary with a *table d'analyse,* laying out the principal concepts of his science and the order in which the articles should be read by someone seeking a systematic account of it. For example, a reader who wanted an introduction to physics would take Monge's dictionary off the shelf and would begin with the article "Mouvement," moving on to "Vitesse," "Puissance," "Force," and so on in the order

45. See especially Panckoucke's somewhat sanctimonious praise of the original *Encyclopédie* in a passage of the prospectus that followed Diderot's criticism of it. *Mercure,* Dec. 8, 1781, p. 53.

prescribed by Monge. Another reader might want to know what air was. Lavoisier's experiments had made this a fascinating question for a great many educated Frenchmen, who realized that the ancient, four-element theory was collapsing but could not understand the revolution in chemistry that had undermined the old view of the world. Should the confused reader reach for the dictionary of chemistry, the dictionary of physics, or the dictionary of medicine? Panckoucke explained that he should go directly to the twenty-seventh part of the *Encyclopédie méthodique,* a *Vocabulaire universel,* which not only would serve as an index to the entire work but also would be a dictionary of dictionaries—a supreme repertory of every idea and every word in the French language, each defined and classified according to its place in the structure of knowledge. The *Vocabulaire universel* would show that "air" appeared as a decomposable substance in dictionary VI (chemistry) and as an active element in dictionary II (physics).

Panckoucke evidently did not worry about the possibility that the same thing could be treated in contradictory ways in different dictionaries or that contradictory views among scientists and philosophers might not be smoothed over by the careful ordering of words and distribution of work. The link between words and things apparently struck him as self-evident; and he did not agonize over the epistemological basis of the enterprise.[46] His approach was taxonomic and organizational: if he could make sure that every word went where it belonged in each dictionary and that each dictionary fit into its proper place within the entire *Encyclopédie,* he would produce "une bibliothèque complète et universelle de toutes les connaissances humaines."[47] This ambition fired Panckoucke's imagination. He gloried in the idea of constructing "un des plus beaux monuments que les hommes, dans aucun temps, aient jamais élevé à la gloire des lettres, des sciences, et des arts."[48] The glory would be his own, for he intended to plan and assemble the super *Encyclopédie,* not merely to publish it. He

46. After explaining the plan to introduce each dictionary with a "table d'analyse," the prospectus concluded, "Par ce moyen, le lecteur voit, pour ainsi dire, d'un seul coup-d'oeil le tableau de chaque science, et la liaison de tous les mots qui y ont rapport, ou plutot de toutes les idées qui en sont les éléments." Dictionary *Beaux-Arts,* I, vii.

47. *Mercure,* Dec. 8, 1781, p. 150.

48. *Beaux-Arts,* I, v.

even proposed to write some articles himself, including a history of Encyclopedism, which would explain the evolution of the editions.[49] According to a letter from a subscriber, which Panckoucke published in his promotional material, the book would "immortalize" him.[50] Whether or not Panckoucke let his hopes soar so high in his secret thoughts, his correspondence shows that he identified himself increasingly with the venture, as it became greater and greater in size, until finally it eclipsed all his previous speculations. He had become so swept up in it by 1780, that he probably believed his own sales talk, including all the superlatives of the prospectus, which promised "le recueil le plus riche, le plus vaste, le plus intéressant, les plus exact, le plus complet et le mieux suivi qu'on puisse désirer."[51]

Panckoucke went about this task in a spirit that was both managerial and philosophical, as befitted an entrepreneur of Enlightenment. He divided the cognitive universe into its constituent parts—at first twenty branches of knowledge, twenty-six by the time he issued the prospectus, and eventually more than fifty—each of which was to be covered by a dictionary. Then he hired several persons to cut apart two sets of the original *Encyclopédie* and the *Supplément* and to file each article under one of the twenty-six rubrics. Not only was this work long and laborious (it took about a year), but it required excellent editorial judgment; for Panckoucke's rubrics did not coincide with the labels attached to the articles in Diderot's text. The cutters and cataloguers had to know their way around the borders of all the fields of knowledge and had to decide where to dispose of articles that fell in disputed territory. In fact, they could not operate at all without a polymath to supervise them. That person was probably Suard. Although Panckoucke did not mention Suard in the prospectus, where he described the way he constructed the

49. Ibid., p. xlviii. Panckoucke's projected history may have been derived from d'Alembert's, which was to be published in the Suard *refonte* of 1776. Panckoucke certainly presented himself as the mastermind of the *Méthodique* and not merely its publisher. See, for example, his Avis of October 1787, reprinted in the dictionary *Beaux-Arts*, I, lxxxi: "Nous n'avons point fait ici simplement les fonctions de libraire, nous avons fait le plan de l'*Encyclopédie* actuelle, la distribution de toutes les parties, composé la première, la deuxième et la dernière division du grand prospectus qui a été publié."

50. *Mathématiques*, III, xviii–xix.

51. *Beaux-Arts*, I, iv.

Méthodique, he had engaged Suard to mount just such a clipping and sorting operation for the *refonte* that they had planned to publish with the STN in 1776. As explained above, Suard's work had reached an advanced stage before the quarto speculation forced him to abandon it. He probably took it up again or at least turned over his material to Panckoucke, when Panckoucke shifted his interest from the quarto to the *Méthodique.*

Next, Panckoucke began to put together a team of Encyclopedists. He made the rounds of the academies and salons of Paris, signing up the best man he could find in every subject to be covered. This recruiting campaign must have involved some fast footwork and hard bargaining, for Panckoucke concluded a separate contract with the author or authors of each dictionary; but the prospectus indicated that he proceeded with Baconian orderliness. Having assembled his experts, it explained, he got them to agree on the boundaries of their sciences—no easy matter in cases where fields overlapped, as in mathematics, physics, and chemistry, or where the subject itself had not yet developed into an autonomous discipline, as in the case of economics, which still lay latent within *économie politique.* Panckoucke then presented his authors with the clippings from the original *Encyclopédie* and the *Supplément,* which obviated the need for each of them to wade through all twenty-one volumes of the original text. After close study of the clippings, the authors could assess the previous coverage of their subjects, deciding what gaps needed to be filled by further research and what primary concepts ought to receive the most thorough treatment. They were to draw up a list of key words, each of which would merit an essay, as distinct from secondary terms, which could be dispatched with dictionary-type definitions. Then they would arrange the key words in their order of importance, forming the *tables d'analyse* to guide readers who wanted to study the dictionaries as treatises. They would write a preliminary discourse, explaining the history of their subject and the main tendencies in the literature on it. And then they would write their texts. Panckoucke's clippers and filers also would put together a "table des mots communs et équivoques,"[52] which each author would consult in order to avoid repeating or contradicting the work of his colleagues. In the end, therefore, all the parts would fit

52. Ibid., p. vi.

together in a harmonious whole, forming the unbroken chain of all knowledge. Throughout Panckoucke's project ran an almost Linnaean preoccupation with naming and cataloguing and a Condorcet-like conviction that knowledge was progressive, coherent, and reducible to the dimensions of a single *summa,* twice the size of Diderot's original compendium.

Panckoucke as an Editor

Harmonious as it was in Panckoucke's conception, the *Encyclopédie méthodique* became, in the end, a monstrosity. It grew out of proportion because Panckoucke in his role as a kind of managing editor failed to keep it under control. That role remains obscure, but one can catch glimpses of it from the dossiers of one of Panckoucke's authors, Auguste-Denis Fougeroux de Bondaroy, and Fougeroux's collaborators on the dictionaries about botanical subjects.

Although he has now been forgotten, Fougeroux de Bondaroy was once as imposing as his name sounds. A man of independent wealth, he devoted himself to botany and followed his uncle, the eminent botanist Henri-Louis Duhamel du Monceau, into the Academy of Sciences. He wrote essays on crafts for the academy's *Description des arts et métiers* and on a variety of other subjects, including physiology and archeology. But he became best known as an authority on trees. It was owing to his expertise in this field that Panckoucke sought him out for the *Méthodique.* When he had first laid plans for his dictionary of botany, Panckoucke went right to the great naturalist Jean-Baptiste-Pierre Antoine de Lamarck. But as Lamarck's work on the book grew in size and complexity, Panckoucke decided to develop a separate dictionary for readers with special interests in growing plants. He therefore commissioned the abbé Alexandre-Henri Tessier, a specialist on grains from the Academy of Sciences, and André Thouin, an expert on gardening from the Academy and the Jardin du Roi, to produce a dictionary of agriculture. Agriculture also became unmanageably large, however, so Panckoucke turned to Fougeroux for a work on forestry and trees.[53]

53. This account derives from Panckoucke's ''Tableau et apperçu'' on his work in *Mathématiques,* III, 7–8, which probably makes the mushrooming of the botanical sciences in the *Méthodique* seem more rational than it was in reality. For background on Fougeroux see the article about him in *Biographie universelle,*

A contract signed on February 16, 1781 set the terms for this undertaking. Fougeroux promised to supply articles on "toute la partie de l'agriculture concernant les bois, les futaies, les semis, plantations et l'aménagement des forêts," copying or modifying the material in Diderot's *Encyclopédie* as he saw fit. In return, Panckoucke made precise commitments about payment: "Et moi, Charles Panckoucke, je promets et m'engage de payer à Monsieur de Fougeroux la somme de vingt-quatre livres par chaque feuille in-quarto dudit dictionnaire, caractère de cicéro, semblable à celui avec lequel on a imprimé le *Dictionnaire de physique* de M. Brisson, soit que ces articles soient entièrement nouveaux, soit qu'ils soient copiés en totalité ou en partie dudit dictionnaire encyclopédique. Ledit payement se fera moitié comptant à mesure de l'impression de chaque feuille et l'autre moitié à la fin de l'ouvrage, en mes billets à quatre, huit et douze mois. Je donnerai en outre à M. Fougeroux un exemplaire complet de cette nouvelle Encyclopédie méthodique." These were standard terms, judging from the dozen other contracts for the *Méthodique* that have survived. Panckoucke always paid by the piece—usually twenty-four livres for copy that produced a quarto sheet in pica, proofreading included. In several cases he also set deadlines, but he soon discovered that they were easier to write into a contract than to enforce.[54]

Having signed up with Panckoucke, Fougeroux began to receive instructions and circular letters, which kept the authors informed about the state of the enterprise. The first circular arrived in November 1782, just after Panckoucke had completed the printing of his first installment, and it

---

ed. J. F. and L. G. Michaud (Paris, 1811–1852), XIV, 496–497. Most of the following discussion is based on the Fougeroux-Panckoucke dossier in Case Wing Z 311.P188, Newberry Library. Related documents can be found in various collections in Paris, Amsterdam, Oxford, and Lawrence, Kansas—an indication of how widely the papers of the *Méthodique* are scattered.

54. Fougeroux's contract is in Schenking Diederichs, fol. 362, Universiteits-Bibliotheek of Amsterdam. It included a deadline clause, which was later annulled, according to a margin note in Panckoucke's hand. There are thirteen similar contracts in the Kenneth Spencer Research Library, University of Kansas, ms. 99, including Panckoucke's contract with Tessier, dated Feb. 15, 1781, which contains almost exactly the same wording as the contract with Fougeroux. In a few cases, however, Panckoucke deviated from the standard terms. He paid from 30 to 48 livres per sheet for dictionaries involving many collaborators. And in the contract for the dictionary of finance, signed with Digeon, directeur des fermes, on June 23, 1780, he specified that the subject should be treated "d'une manière philosophique."

showed that his enthusiasm was running high. "Je suis en état, Messieurs, de pousser cette entreprise aussi vigoureusement que vous le désirez. Et si l'on veut me seconder, je réponds qu'elle sera finie en trois ans au lieu de cinq. Il y aura l'année prochaine 18 mille rames de papier en magasin. J'ai traité avec 18 imprimeurs, qui, chacun, se sont obligés à une fonte neuve, et ces fontes sont déjà livrées en partie. Il y a actuellement, compris le volume de planches, dix ouvrages différents sous presse."[55] Panckoucke did not restrict himself to the managerial side of the publication in his subsequent letters. In July 1783 he sent Fougeroux some "observations impartiales" about forestry. A year later, he provided a "petite brochure" to be used "dans la partie des bois."[56] And he also procured information from some of his influential friends: "M. de Malesherbes a fait venir des greffes de poirier de Turgovie. Les meilleurs espèces de poirier pour le cidre sont:

le carisi) du côté de St. Germain."[57]
le gromenil)

While relaying material to Fougeroux, Panckoucke arranged for a certain Teller d'Acosta to draft some of the articles. Fougeroux needed reinforcement, for he had fallen far behind schedule and whenever Panckoucke pleaded for copy he replied with complaints about the pressure of family responsibilities and the problem of coordinating his work with that of the other authors, particularly Lamarck.[58] Meanwhile, however, Teller ran into difficulties of his own. He delegated some of the work to a still more obscure assistant, but this system of subcontracting failed to speed up the flow of copy. "Le temps que vous me donnez, Monsieur, est trop court pour vous promettre un travail bien fait pour la fin de l'année,"[59] Teller protested to Panckoucke. "Il me faudrait au moins deux ans pour rassembler tout ce qui est nécessaire et pour rédiger." Panckoucke therefore dropped the auxiliaries and urged Fougeroux to press ahead on his own: "Vous avez cela tout

55. Circular letter, in manuscript, "à MM. les auteurs de l'*Encyclopédie méthodique*," Nov. 18, 1782, Case Wing Z 311.P188, Newberry Library.

56. Panckoucke to Fougeroux, July 26, 1783, ibid.

57. Panckoucke to Fougeroux, June 27, 1784, ibid.

58. Fougeroux to Panckoucke, Jan. 30, 1783, Schenking Diederichs, fol. 363, Universiteits-Bibliotheek, Amsterdam.

59. Teller to Panckoucke, July 1, 1783, Case Wing Z 311.P188, Newberry Library.

fait dans vos papiers et dans ceux de Monsieur votre oncle [Duhamel du Monceau]. Le retard me ferait infiniment de tort. Le public se plaint qu'on lui donne toujours la même chose, et je ne puis établir la confiance qu'en lui montrant qu'on travaille sur toutes les parties.''[60] The editor wanted original work, and he wanted it in a hurry. But he could not browbeat a distinguished, fifty-year-old academician. Although he referred discreetly to payment,[61] he probably did not expect money to serve as an inducement for a man of Fougeroux's wealth. Panckoucke had to cajole copy out of his botanist, to press him gently with billets, and to sollicit interviews, always maintaining a respectful tone: ''Panckoucke présente tous ses respects à Monsieur de Fougeroux . . . Il le prie en grâce de lui donner un quart d'heure pour conférer avec lui.''[62]

After four years of coaxing, Panckoucke still had no dictionary of trees and forests, so he attempted a rather deferential ultimatum:

> Vous ne m'écrivez point la lettre que vous avez eu la bonté de me promettre et qui soit pour moi une nouvelle assurance que vous acheverez ce dictionnaire des bois dans le temps prescrit par mes engagements avec le public, c'est à dire pour 1788. Si vous aviez changé de résolution, ce serait un malheur pour moi, pour le public, pour l'ouvrage. Mais enfin en me le mandant sur le champ, je chercherai quelqu'un qui puisse vous remplacer, et alors cette partie serait fondue avec celle de M. l'abbé Tessier, car je ne consentirai à faire de cette partie un dictionnaire séparé qu'autant qu'elle sera votre ouvrage et parce que vous pouvez seul, Monsieur, par vos connaissances dans ce genre, lui donner toute la perfection dont elle est susceptible.[63]

When it became clear, two years later, that this tactic had failed, Panckoucke found another collaborator for Fougeroux. He contracted with the Genevan botanist Jean Senebier for a series of essays on plant physiology, which would follow key words in the dictionary: *Air, Bois, Boutons, Feuilles, Fleurs, Fruits, Ecorce, Lumière, Racine, Sève,* and *Végétation.*

60. Panckoucke to Fougeroux, July 1, 1783, ibid. See the similar remarks in Panckoucke to Fougeroux, July 10, 1783.
61. Panckoucke to Fougeroux, whom he addressed in the third person, July 3, 1783: ''Je mettrai aussi en banque chaque mois sur sa copie des bois, s'il le désire.'' Ibid.
62. Panckoucke to Fougeroux, July 3, 1783, ibid.
63. Panckoucke to Fougeroux, Feb. 12, 1786, ibid.

Though a man of great learning (he had translated Spallanzani and had written original works on bibliography and history as well as biology), Senebier was a modest pastor and librarian. He was also six years younger than Panckoucke, in contrast with Fougeroux, who was four years older and a member of the Académie des sciences. Therefore, Panckoucke could adopt a more peremptory tone and provide more detailed guidance in his instructions:

Je suis pressé d'imprimer. On se plaint que des dictionnaires ne sont pas commencés . . . J'ai beau dire qu'en associant d'autres personnes à mon travail que des parties seront infiniment mieux traitées. Le libraire n'est pas satisfait. Tâchons donc, Monsieur, je vous en supplie, justemment et sans que l'ouvrage en souffre, de le contenter, et voici je pense les moyens d'y parvenir.

Dans un dictionnaire il est possible d'un mot de renvoyer à beaucoup d'autres et de former par une réunion un raisonnement qui satisfasse le lecteur. Comme nous commençons, nous avons toute liberté du choix d'une lettre ou d'une autre, et je préférerais quantité de pareils renvois, qui donnent le loisir de perfectionner le travail.

Je pense encore, Monsieur, que la physique végétale est encore si peu éclaircie qu'il serait trop présomptueux de prendre un parti et de se décider pour l'un ou l'autre des sentiments. Comme vous le dites très bien, il faut s'en tenir au rapport des opinions . . . Si vous connaissez quelques faits qui y apportassent des doutes, je pense qu'il faudrait ne pas les omettre, mais sans soutenir chacune de ces opinions de manière à ce que le lecteur imagine qu'on la regarde comme irrévocablement établie . . .

Au mot air, le mieux serait, je crois, de le considérer en ce qu'il est nécessaire à la végétation; et après avoir rapporté les sentiments qu'on a soutenus sur la manière dont il s'introduit dans les plantes, d'annoncer des doutes sur cette partie, si vous en avez. Il semble à la vérité, qu'il s'introduise dans l'intérieur des plantes par plusieurs moyens, qu'il s'y décompose; et c'est pour donner l'intelligence de ceci et de son propre sentiment qu'il sera nécessaire d'indiquer aux lecteurs les découvertes qu'on a faites sur la décomposition de l'air, sans entrer dans les détails qui appartiennent aux dictionnaires de physique et de chimie. Il sera donc très bon d'établir, d'après vos observations, comment il n'entre dans les plantes que de l'air fixe dissout dans l'eau que la lumière élabore et fait sortir sous la forme d'air pur . . .

Je ne crois pas, Monsieur, dans cette partie que nous traitons devoir parler de la nouvelle doctrine substituée au phlogistique. La question est trop embrouillée jusqu'ici, et je pense qu'on en est encore à disputer sur les mots, et il faut s'en tenir, comme vous le dites, à donner des explications sur les faits de la végétation, en employant les deux façons de s'annoncer et cette discrétion vient naturellement au mot étiolement.[64]

This letter shows that at least on some occasions Panckoucke acted as the Diderot as well as the Le Breton of the *Encyclopédie méthodique*. Not only did he shuffle assignments and indicate how an author should attack a subject, he also leaped into the midst of the most disputed territory in the scientific discussions of his time. He virtually told Senebier what Senebier would doubt and what Senebier would observe. Panckoucke's botanist had recently completed a series of experiments that showed how light affected vegetable nutrition and put him in the line of scientists like Marcello Malpighi, Charles Bonnet, Joseph Priestley, and Jan Ingenhousz, who developed the chemical theory of plant physiology. Nonetheless, Panckoucke did not hesitate to lecture Senebier about fire and air. In fact, he warned him against pronouncing phlogiston dead, even though other contributors to the *Encyclopédie méthodique* had already buried it. Of course, it would be extravagant to expect that a businessman like Panckoucke could assimilate the notion of the oxygen cycle that Fourcroy and Guyton de Morveau were developing with Lavoisier at that very time. Panckoucke retreated to a safe, Baconian stance on theory and tried to leave room in his book for competing scientific systems. What seems remarkable is that he should discourse about *air fixe* and *air pur* as if he had met them in the laboratory, not that he should use the vocabulary of his age; and that he should tell a leading botanist how to write about *physique végétale,* not that he should hand out assignments. As an editor, Panckoucke did not believe in laissez-faire.

His interventions, however, did not step up the production of the writers. In November 1788, Panckoucke issued another circular letter to them:

Messieurs,

Je suis accablé de plaintes des souscripteurs de l'*Encyclopédie;* les choses en sont même aujourd'hui à un point que je ne dois plus vous cacher ce qui se passe.

La lenteur de la publication de quelques parties dont il n'a encore rien paru, le ralentissement de plusieurs autres, dont on n'a publié que

64. Panckoucke to Senebier, Nov. 10, 1788, ibid.

quelques volumes, l'incertitude du temps où l'ouvrage sera fini, ont rendu ma position extrêmement critique . . .

Le sort de l'*Encyclopédie,* Messieurs, dépend entièrement de vous. J'ai des engagements, je désire de les remplir, j'y sacrifierai ma vie et ma fortune; mais je ne puis rien sans vous, Messieurs. Il n'est plus question de faire de vaines promesses, c'est un engagement positif et solennel de votre part que je réclame.

Panckoucke noted that some of the authors had produced nothing whatsoever in six years, despite the deadlines stipulated in their contracts. Fougeroux had promised and prevaricated for precisely that length of time, and Panckoucke reminded him in a personal postscript that he had committed himself "de la manière la plus solennelle de mettre sous presse cette année. Il est nécessaire qu'il y ait un demi-volume cette année; M. de Lamarck est assez avancé pour cela."[65]

That is the last letter in Fougeroux's dossier. No dictionary of trees and forests appeared in 1788, nor in 1789, nor in the remaining years of the eighteenth century. It was not published until 1821, and its author was not Fougeroux but the botanist-Girondist Louis-Augustin-Guillaume Bosc. Feeling that he owed an explanation to those subscribers who had survived four decades of revolution and war, Bosc revealed in the preface that Panckoucke had not wanted to produce a separate dictionary of trees: he had engaged Fougeroux to write for the dictionary of agriculture, but Fougeroux had fallen so far behind in the writing that the agricultural segment of the *Encyclopédie* could not be issued, unless Panckoucke cut the articles on trees out of it and left them to be covered in a subsequent dictionary.

Far from having followed Panckoucke's original, well-wrought plan, the *Encyclopédie*'s segmentation into dictionaries had proceeded in a haphazard fashion. The book had assumed its awkward size and shape because planning had had to give way to human foibles of the sort epitomized by Fougeroux: "M. Fougeroux de Bondaroy était âgé, était infirme, et par caractère remettait toujours au lendemain ce qu'il avait projeté de faire la veille," Bosc explained. "Aussi, quand les premières feuilles fournies par mes collaborateurs Tessier et Thouin furent prêtes à être livrées à l'impression, n'avait-il pas encore écrit une ligne." After missing his first deadline, Fougeroux had promised to make the second. But "malgré

65. Circular dated Nov. 1788, ibid.

cette promesse, la première partie de l'ouvrage de M. Fougeroux de Bondaroy ne parut pas en 1787, car son état physique et moral s'aggravait de jour en jour. Il ne put fournir, aux pressantes sollicitations de M. Panckoucke, que deux ou trois feuilles, dont la moitié n'était pas de lui; enfin la mort vint le frapper en 1789. Alors la Révolution éclatait; alors le commerce de la librairie s'anéantissait.''[66]

The Authors of the *Méthodique*

Neither Fougeroux nor anyone else can be taken to typify Panckoucke's Encyclopedists. They were a diverse lot—doctors, lawyers, professors, government officials, and littérateurs—but they had one thing in common: eminence. Although much of the gilt has now worn off their names, they stood out as the top intellectuals in France during the 1780s. At that time intellectual distinction tended to be measured by membership in the learned societies of Paris, which were at the peak of their influence on French cultural life. Of the seventy-three principal contributors to the *Méthodique* in 1789, fifteen belonged to the Académie des sciences, seven to the Académie française, seven to one of the other Parisian academies, eighteen to the Société royale de médecine, and eight to the Société royale d'agriculture. The density of academicians was twice as great among Panckoucke's contributors as it had been among Diderot's.[67] Panckoucke's group also had accumulated so many additional memberships in provincial and foreign academies that they did not list their full academic pedigrees under their names in the *Méthodique*. The two authors of the *Dictionnaire de physique,* for example, appeared on its title page simply as members of the Académie des sciences. But one of them, Joseph-Jérôme Lefrançois de La Lande, belonged to learned societies in Berlin, London, St. Petersburg, Bo-

66. Avertissement in *Dictionnaire de la culture des arbres et de l'aménagement des forêts,* pp. v–vi. See also Panckoucke to [Senebier], Oct. 22, 1789, Archives de Paris, 8 AZ 278.

67. Of the 141 authors in Diderot's group identified by John Lough in *The Contributors to the ''Encyclopedie''* (London, 1973), there were 14 members of the Académie des sciences, 12 members of the Académie française, and 5 members of the other Parisian academies. Allowing for men who belonged to more than one academy, that comes to about the same number of academicians as in Panckoucke's group, which was half as large.

logna, Stockholm, Göttingen, Rome, Florence, Cortona, Mantua, and Haarlem; and the other, abbé Charles Bossut, held memberships in St. Petersburg, Bologna, Utrecht, and Turin. Panckoucke made the predominance of academicians among his authors a selling-point for the book: "Cent auteurs de la capitale en sont actuellement occupés, et la plupart sont ou de l'Académie française, ou de celle des sciences, ou des inscriptions," he boasted, with some exaggeration, in his sales literature of 1789.[68] He did not settle for rank-and-file immortals, either, but signed up the leading members of the learned societies: the marquis de Condorcet, perpetual (that is, permanent) secretary of the Académie des sciences; Jean-François Marmontel, perpetual secretary of the Académie française; Félix Vicq d'Azyr, perpetual secretary of the Société royale de médecine; Pierre-Marie-Auguste Broussonet, perpetual secretary of the Société royale d'agriculture; and Antoine Louis, perpetual secretary of the Académie royale de chirurgie.

The academies provided the main institutional base for the Encyclopedists and also opened their way to special niches in the complex superstructure of the Old Regime. Four of the authors of the biological dictionaries—Louis-Jean-Marie Daubenton, Jean-Baptiste-Pierre-Antoine de Lamarck, Antoine-François de Fourcroy, and André Thouin—held appointments in the Jardin du roi as well as the Académie des sciences. All but Lamarck also belonged to the Société royale d'agriculture. And while holding those positions Daubenton taught in the Collège royal, and Fourcroy served as a royal censor. It was from the interweaving of roles like these that scientific careers were made in the eighteenth century and that scientists were supported. The top-ranked *pensionnaires* of the Académie des sciences received pensions of 2,000 livres and in return served the state. By granting patents (*privilèges*) and discrediting quacks, for example, they became the official guardians of the

---

68. Panckoucke, *Abrégé des représentations et du mémoire sur l'Encyclopédie qui doit paraître le 14 ou le 21 mars* (1789), Case Wing Z 311.P188. Newberry Library. The information in this and the following paragraphs comes from the *Encyclopédie méthodique* itself, from the *Almanachs royaux* of the 1780s, and from several biographies and biographical dictionaries. Of the latter, Michaud's *Biographie universelle*, has yet to be replaced, though it still cannot be relied on, and Charles C. Gillispie, ed., *The Dictionary of Scientific Biography* (New York, 1970–76), 14 vols. is excellent for the more famous figures.

line that ostensibly separated legitimate from illegitimate scientific activity.[69]

Support and service took the form of teaching for several Encyclopedists—not in the traditional faculties of the universities but in technical schools, oriented toward the requirements of the state. The abbé Bossut and Gaspard Monge taught geometry at the Ecole du génie at Mézières. Jean-Pierre-François Duhamel taught metallurgy at the Ecole des mines. Edme Mentelle and Louis-Félix Guinement de Kéralio taught geography and tactics at the Ecole militaire. And Honoré-Sébastian Vial du Clairbois and Blondeau taught engineering, shipbuilding, and mathematics in the Ecole de la marine. These men discussed the same subjects in the classroom that they expounded in the *Méthodique:* their writing grew directly out of their professional activity. This was a new tendency in scholarly work, which usually had been done by amateurs in the past, and it left its mark on the technical dictionaries of the *Méthodique.* They were written by technical officials. Jean-Marie Roland de la Platière, an inspecteur des manufactures, produced most of the vast dictionary of *Arts et métiers.* Nicolas Desmarets, also an inspecteur des manufactures, did *Géographie physique.* De Surgy, a former premier commis des finances, wrote *Finances.* And Gaspard-Clair-François-Marie Riche de Prony, an inspecteur des ponts et des chaussées, wrote *Ponts et chaussées.* The main exception to this rule was the dictionary of commerce by the abbé Nicolas Baudeau, a well-known Physiocrat, but a man who had only a theoretical knowledge of his subject. He could have learned a great deal from Panckoucke.

Panckoucke assembled a distinguished group of doctors to write on medicine and lawyers to write on law. He stressed that the *Dictionnaire de médecine* came from "vingt médecins, presque tous de la Société royale de médecine."[70] Its principal author, Vicq d'Azyr, had made the Society into a great center for public health and, with the help of Turgot, had aligned medicine with service to the state. As doctor to the Comte d'Artois and later to the queen—Marie-Antoinette reportedly

69. See Roger Hahn, *The Anatomy of a Scientific Institution: The Paris Academy of Sciences, 1666–1903* (Berkeley, 1971), especially chap. 3.

70. Panckoucke, "Sur le retard que l'*Encyclopédie* a éprouvé de la part de plusieurs auteurs," in the dictionary *Histoire,* I, 8. Actually, Panckoucke listed only eighteen authors of the *Dictionnaire de médecine* in 1789, and only eleven of them were members of the Société royale de médecine.

called him *mon philosophe*—he became something of a courtier and empire-builder.[71] His principal collaborators—Dehorne, Charles-Louis-François Andry, Nicolas Chambon de Montaux, Jean Colombier, François Doublet, Dieudonné Jeanroi, and Jean Verdier—were familiar with the ailments of the aristocracy. When Andry looked back on a career of treating well-born patients, he reflected, "J'ai gentilhommisé la médecine."[72] Panckoucke's lawyers also had distinguished practices. Most of them—men like Pierre-Paul-Nicolas Henrion de Pansey, Jacques-Vincent Delacroix, and abbé Antoine-René-Constance Bertolio—were well-known figures of the Paris bar, although at least two—Jean-Philippe Garran de Coulon and Jacques Peuchet—seem to have been garret types, who were hungry for their assignments. Perhaps the former left the spadework to the latter. An avocat au parlement called Le Rasle coordinated the writing, with help from Antoine J. Boucher d'Argis, an eminent conseiller au Châtelet, who had contributed more than 4,500 articles to Diderot's *Encyclopédie.*

The men of letters who wrote for the *Méthodique* came from the salons and academies of the capital. Their paths all led to the Académie française and crossed, en route, at the Tuesday evenings of the baron d'Holbach, rue Royale Saint-Roch, at the dinners of Mme. de Marchais in the Pavillion de Flore, in the offices of the *Mercure* and the *Gazette de France,* and in the antechambers of Versailles. They ruled the republic of letters, but their world seemed to shrink during the last years of the Old Regime as the great figures of the Enlightenment aged and died. Marmontel and Nicolas Beauzée of the Académie française produced the only dictionary in the *Méthodique* devoted to literature proper. Suard and his fellow journalist-academician, the abbé François Arnaud, originally planned to produce the dictionary of beaux-arts; but Arnaud dropped out and Suard switched to the dictionary of music, which he eventually relinquished to another of his salon companions, Pierre-Louis Ginguené. Beaux-arts went to one of Diderot's old collaborators, Claude-Henri Watelet, another academician, and then, after Watelet died in 1786, to a Diderot protégé, Pierre-Charles Lévesque of the Académie des inscriptions et

71. See the article on him by P. Huard and M. J. Imbault-Huard in *The Dictionary of Scientific Biography,* XIV, 14–17.
72. *Biographie universelle,* I, 687. For the specialities of each contributor see the identifications given at the beginning of the dictionary *Médecine,* I.

belles-lettres. Antoine-Chrysostome Quatremère de Quincy, who had already established himself as an authority on archeology and classical architecture, undertook the dictionary of architecture. Gabriel-Henri Gaillard, of the Académie française and the Académie des inscriptions et belles-lettres, wrote the dictionary of history. And Pierre-Louis Lacretelle, a lawyer and littérateur from the *Mercure,* produced the dictionary of metaphysics, logic, ethics, and education. These latter-day philosophes seemed to come from the immediate circle of Panckoucke and his brother-in-law Suard. Jean-François La Harpe, who moved in and out of that circle, remarked in his waspish way, that an ''esprit de parti'' had dictated Panckoucke's choice of Encyclopedists: ''Le libraire Panckoucke, qui est à la tête de l'entreprise, a choisi tous ceux que lui a désignés M. Suard, son beau-frère; et c'est ainsi que toutes les entreprises littéraires seront conduites, quand il y aura un libraire à la tête.''[73] In fact, the philosophes whom Suard had assembled for Panckoucke's *refonte* in 1776 probably did provide the nucleus of Panckoucke's team for the *Méthodique.* At the very least, they all belonged to the same world.[74]

Elevated as it was, this world had room for some lowly littérateurs, who probably turned out a good deal of the text of the *Méthodique.* A certain amount of subcontracting probably took place, as Panckoucke's dealings with Fougeroux de Bondaroy suggest. And when Panckoucke's luminaries failed to produce copy, he assigned the work to lesser lights. Nicolas-Etienne Framery, a hack writer who often did odd jobs for Panckoucke, put his shoulder to the wheel with Ginguené af-

73. La Harpe, *Correspondance littéraire* (Paris, 1801), III, 302. La Harpe, who had fallen out with Suard during the quarrel over the music of Gluck and Piccini, objected that ''des parties très importantes sont confiées à des hommes très médiocres,'' notably Condorcet, Naigeon, and the abbe Baudeau. Ibid.
74. For information on this literary world see the memoirs of Morellet, Marmontel, and Garat. Of the eleven authors mentioned in the contract for Suard's *refonte,* six appeared in Panckoucke's original prospectus for the *Méthodique:* Suard, Arnaud, Marmontel, Louis, Condorcet, and d'Alembert. Panckoucke explained that ''la santé et les diverses occupations de M. d'Alembert ne lui permettent pas de partager notre travail; mais du moins il a promis de nous remettre différentes additions qu'il a faites, il y a longtemps, à plusieurs de ses articles de mathématiques et qu'il avait destinées aux futures éditions de l'*Encyclopédie;* par ce moyen, il aura part encore à l'édition du dictionnaire que nous annonçons.'' *Beaux-Arts,* I, viii. D'Alembert died in 1783, before he could contribute anything. A seventh member of the Suard group, Antoine-Léonard Thomas, died in 1785, after four or five years of nursing his ill health in the provinces. The four others, La Harpe, Morellet, Saint-Lambert, and Petit did not join Panckoucke's second term.

ter Suard dropped the dictionary of music. When Panckoucke needed an article on a craft or trade for the *Dictionnaire des arts et métiers,* he put Jacques Lacombe to work. Lacombe had qualified for the law, and had acquired a mastership in the booksellers' guild of Paris, before joining Panckoucke's stable of writers. Although he retained his guild membership, he seems to have been a satellite in Panckoucke's book business, and he produced the more frivolous of the *Méthodique* dictionaries: *Chasses et pêches, Art aratoire, Amusements des sciences, Encyclopédiana,* and *Jeux mathématiques.* These were scissors-and-paste jobs, which Panckoucke assigned to Lacombe in order to squeeze more money out of the subscribers by producing more dictionaries. He also planned to sell them as separate works, hoping to tap the market for light literature. (The *ana* were mainly jokes and the *jeux mathématiques* puzzles.) Soon after the outbreak of the Revolution, Panckoucke set another of his Encyclopedists, Jacques Peuchet, to compile a *Dictionnaire de l'Assemblée Nationale.* Panckoucke thought the market could take five volumes; but Peuchet, an unemployed lawyer turned journalist and revolutionary bureaucrat, produced only one—it was actually volume 2 and did not get beyond the letter *a*—before Panckoucke scrapped the project.

The light supplementary dictionaries did not suit the heavy, learned tone that prevailed throughout the *Encyclopédie méthodique,* and the men who put them together differed from the other Encyclopedists in an important respect: they wrote for a living. The great majority of Panckoucke's authors did not. They received pensions from their academies, salaries from their professorships, stipends from the honorific appointments, and subsidies from their journals.[75] A few of the Encyclopedists enjoyed independent incomes, drawn from the traditional sources of wealth in the Old Regime: land, *rentes,* and offices. The only one of them to operate at the level of high finance was Watelet, the art critic, who was a receiver general in the state's baroque system of taxation. Watelet may have been even wealthier than Panckoucke himself, although

75. This last source of income was more important than has generally been realized. The government, which granted privileges for the journals, attached pensions to their revenues and then awarded the pensions to prominent men of letters. Panckoucke said that his journals subsidized more than a hundred *pensionnaires.* ''Mémoire pour M. Panckoucke relatif aux journaux dont il est propriétaire,'' in the dictionary *Histoire,* I, 29.

he had fallen into the red by a million livres when he died.[76] Louis-Bernard Guyton de Morveau, who produced the dictionary of chemistry, must have derived a handsome income from his office of avocat général in the Parlement of Dijon before he moved to Paris. Boucher d'Argis lived at a similar level as a conseiller in the Châtelet; and both of them, like most officeholders, probably received *rentes* from their estates. Fougeroux de Bondaroy and Quatremère de Quincy were the only others to live like old-fashioned *savants*, relying on their private wealth for the leisure to pursue scholarship. The rest of Panckoucke's group received pay for their scholarly or artistic work—even if it came through sinecures, for France had not yet entered the era of the mass reading public and university-based research.

Most of the pay came from the crown. In fact, dependence on the state stands out as the dominant element in the economic and occupational background of the Encyclopedists. The great majority of them held royal appointments of some sort, and several served the royal family directly—Nicolas-Sylvestre Bergier as confesseur du roi, François Robert as géographe du roi, Marmontel as historiographe du roi, Vicq d'Azyr as premier médecin de la reine, Dehorne as premier médecin de Mme. la Comtesse d'Artois, Desmeunier as secrétaire ordinaire de Monsieur (the future Louis XVIII, then comte de Provence), Beauzée as secrétaire-interprète de Monseigneur Comte d'Artois (the future Charles X), Mentelle as historiographe de Monseigneur Comte d'Artois, and Gaillard as secrétaire ordinaire de Monseigneur Duc d'Orléans. Twelve of Panckoucke's seventy-three authors were censeurs royaux. Such men recognized power and patronage as basic facts of life. Indeed, Garat later described Suard as "un intermédiaire et un ambassadeur entre le gouvernement et la littérature."[77] The government did not attempt to coopt and corrupt talent but to reward it; and in staffing his publication with government men, Panckoucke merely recruited those who had received most recognition. At least half his authors derived their basic income from the state, and almost all of them received official honors and honoraria. Far from suffering as "alien-

76. John F. Bosher, *French Finances 1770–1795: From Business to Bureaucracy* (Cambridge, Eng., 1970), p. 106.

77. Dominique-Joseph Garat, *Mémoires historiques sur la vie de M. Suard, sur ses écrits, et sur le XVIIIe siècle* (Paris, 1820), II, xxii; see also II, 90.

ated intellectuals,'' they served the state and gloried in the benefits it showered upon them.

Two Generations of Encyclopedists

What distinguished Panckoucke's Encyclopedists from Diderot's? The question raises the possibility of comparing two key groups of intellectuals: the men who expressed the Enlightenment at mid-century, when it burst upon the public scene, and their successors at the end of the Old Regime, when Enlightenment passed into revolution. Such comparisons are extremely difficult to make, however. Like all forms of ''prosopography,'' they can be more misleading than revealing,[78] so it is necessary to advance some caveats before plunging into the analysis. The Encyclopedists were too diverse a group, in the 1780s as in the 1750s, to represent an entire generation of intellectuals. The two *Encyclopédies* contain too many different ideas, values, theories, ideologies, protests, and expressions of sentiment to be reduced to simple formulas. And it would be the height of reductionism to explain their content by the social background of their authors. Nonetheless, Diderot's *Encyclopédie* was recognized in his time, and has been studied ever since, as the supreme expression of the Enlightenment. The *Encyclopédie méthodique,* which has hardly been studied at all,[79] was a self-conscious attempt to extend the work of Diderot. As explained in the prospectuses and preliminary discourses, each book was built on the idea that the world was not composed of booming, buzzing confusion but of forces that could be perceived, reduced to basic principles, and ordered coherently by the human mind. This rational order could be reproduced in a single book—that was the audacious message of d'Alembert's *Discours préliminaire,* of the tree of knowledge accompanying it, and even of the subtitle of the work: ''dictionnaire raisonné des sciences, des arts et des métiers.'' The method of the *Méthodique* derived from the same faith in reason, and the authors of both books shared that faith, however much they may have disagreed about other

78. See Lawrence Stone, ''Prosopography,'' *Daedalus* (winter 1971), pp. 46–79.

79. The only important study is George B. Watts, ''The *Encyclopédie méthodique,*'' *Publications of the Modern Language Association of America,* LXXIII (1958), 348–366. Although it is based on only a few printed sources, it bears the mark of Watts's solid understanding of eighteenth-century publishing.

issues. Given those two enormous works, produced by two huge teams of writers, and inspired by the same vision, one does not face a comparison of apples and oranges but of comparable intellectual products. And in studying the producers, one is not attempting to obliterate thought with sociology but to understand the worlds out of which the *Encyclopédies* emerged.

Thanks to the labor of several scholars, notably Jacques Proust, John Lough, Frank Kafker, and Richard Schwab, 160 of Diderot's collaborators have been identified. Using similar sources, one can piece together biographies for all but ten of the seventy-three men that Panckoucke listed as "auteurs de l'*Encyclopédie* actuelle" in 1789. Panckoucke gave the names and occupations of those ten (six were doctors), but he added that twenty-five more persons, who preferred to remain anonymous, had contributed *articles considérables,* making a total of about 100 Encyclopedists when the Revolution broke out. The number could be increased by compiling the signatures after all of the articles; for not only did the principal authors delegate work to assistants, they scattered during the Revolution and were replaced, in part, by new men, who continued to labor—unless they gave way to still more replacements—until 1832. But a complete inventory of all the articles and all the authors would only confuse the picture of Panckoucke's authors in the 1780s. He had assembled his full crew, raised the structure, and completed part of the work in all of the sections of his "monument," as he called it, by 1789. The seventy-three contributors whom he listed at that time did the basic work on the book and can be taken to represent the second generation of the Encyclopedists in contrast to the 160 men of Diderot's generation. The contrast involves problems concerning the representativeness of the data, the adequacy of the statistical base, the validity of the classification scheme, and the danger of giving a false impression of precision by reducing a fluctuating population of cranky philosophes to graphs and charts. But with caution and some suspension of disbelief, one can skirt around those problems and arrive at some reasonable conclusions.[80]

A comparison of the age distribution in the two groups shows how distinctly Panckoucke's Encyclopedists constituted

80. Panckoucke's identifications of his seventy-three authors appear in the dictionary *Mathématiques,* III, xxviii. For further information see Appendix D.

a new generation. Not only did Panckoucke publish his first volume thirty-one years after Diderot's but also his authors were relatively young. Almost half of the sixty-three contributors whose age can be determined were in their twenties and thirties in 1782, when the first volume appeared; and only sixteen were more than fifty years old. Their average age was forty-one, just old enough to have made their mark when Panckoucke recruited them and to be at the height of their careers when the Revolution broke out. Panckoucke, who turned forty-six at the end of 1782, belonged to the revolutionary generation of Encyclopedists himself, in contrast to Diderot and d'Alembert, who died in 1784 and 1783, at the ages of seventy and sixty-five, respectively. The first Encyclopedists belonged to the France of Louis XV. Had they all survived to 1782, their average age would have been sixty-six.[81]

Panckoucke did everything he could to present his *Encyclopédie* as a continuation of theirs, but only eight of his authors had contributed to the original text, and only five had written for the *Supplément*. Most of the second-generation Encyclopedists probably got to know the original work by reading it in their youth, after it had already become a cause célèbre and even something of a classic. When Panckoucke sent them the clippings from Diderot's text, they reacted as if they had received fragments from a bygone era. For example, Vicq d'Azyr, who had been three years old when the *Encyclopédie* first appeared and seventeen when Diderot published the last volume of text, found the remarks on anatomy in the old book to be unintelligible: "Après les avoir rassemblées, ce qui a exigé un grand travail de ma part, j'ai vu que ce recueil ne serait d'aucune utilité." He could make sense of his subject only by putting the pieces together in a new way and by seeing them in a new light. He then produced a dictionary of comparative anatomy, rather than an old-fashioned description of the organs of humans. Moreover, all of Diderot's articles on medicine struck him as outdated: "Ceux qui compareront

81. The average age of Diderot's contributors has been computed from the vital statistics of 118 of them that appear in Lough, *The Contributors to the "Encyclopédie."* Of course these averages provide only a crude measure of the time gap separating the two groups. The very idea of a "generation" remains ambiguous, although it has served for a long time as an organizing concept in French literary history. See Clifton Cherpack, "The Literary Periodization of Eighteenth-Century France," *Publications of the Modern Language Association of America*, LXXXIV (1969), 321–328.

notre travail . . . avec celui de nos prédécesseurs verront que ce dernier nous a très-peu servi et que cet ouvrage peut être regardé comme nouveau. La nosologie, l'hygiène, la médecine vétérinaire, la médecine légale, la jurisprudence de la médecine, et la biographie médicale ou n'existent point ou sont absolument tronquées dans l'ancienne *Encyclopédie.*'' Several of the other young Encyclopedists expressed similar reactions to the work of their predecessors—not only in fields like physics and chemistry but also in subjects outside the sciences. Gaillard, who had been twenty-five in 1751, complained that the first *Encyclopédie* hardly mentioned history; and Robert, who had been fourteen, found its treatment of geography appalling: ''La géographie de l'*Encyclopédie* in-folio est défectueuse à tous égards: c'est un tissu d'erreurs, de méprises, et d'inexactitudes de toute espèce.'' Even before the Revolution transformed their world, the last of the Encyclopedists felt that they belonged to a new intellectual era.[82]

Unlike their readers, who were scattered all over the kingdom, the Encyclopedists came primarily from northern and eastern France. Only nine of the sixty-eight contributors to the *Méthodique* whose geographical origins can be ascertained were born south of the line from Rennes to Lyons. Most of their birthplaces lay along an arch, which curved from Picardy through Champagne, Lorraine, and Burgundy to the Lyonnais. Eighteen were born in Paris, only one in Montpellier, and none at all in the great cities of the Midi: Marseilles, Bordeaux, and Toulouse. Southern cities provided more contributors to the first *Encyclopédie*, especially in lower Languedoc. Montpellier, a great center for medical science, supplied Diderot with four authors (and at least eleven of his contributors had studied there), while Lorraine, which produced seven, was somewhat less fertile in Encyclopedists than it proved to be for Panckoucke. The most barren land for both *Encyclopédies* lay in a triangle formed by Brest, Bayonne, and Lyons, and in both cases the France to the north of the Loire dominated the south. Only the first *Encyclopédie* contained a significant proportion of foreign contributors: sixteen, of whom seven came from Geneva. Panckoucke may have employed more

82. For these reactions see the letters and other material from his authors that Panckoucke included in his ''Tableau et apperçu.'' Of course one should make allowances for the fact that Panckoucke wanted to make his *Encyclopédie* look like a vast improvement over Diderot's. The remarks by Vicq d'Azyr come on pp. 2–3 and those by Robert on p. 12.

than his single Genevan botanist, but he did not list any foreigners on his roster of 1789, perhaps because he wanted to emphasize the national character of the enterprise.[83]

It is much harder to place the Encyclopedists socially than to locate them geographically. Any author could belong to several different groups, depending on where one classifies him amidst the fluctuating and overlapping social categories of the Old Regime: estate, status, class, occupation, and wealth. But the socio-occupational categories used by Jacques Proust and Daniel Roche do justice to the complexity of intellectual life in the eighteenth century; and when applied to Panckoucke's seventy-three authors, they make it possible to draw some comparisons between the two generations of Encyclopedists.[84]

Men from the first two estates made a significant contribution to both books, in view of their small proportion in French society: they made up about 2 percent of the population, 29 percent of the first Encyclopedists, and 20 percent of the second. Their role seems less impressive if one considers that they belonged at the top of an educated elite in a largely illiterate country, but it deserves to be recognized because many scholars treat the *Encyclopédie* as a product of the bourgeoisie.[85]

---

83. Of course one cannot create cultural geography out of sixty-three birthplaces, but this pattern confirms other cartographical studies. See, for example, Roger Chartier, Marie-Madeleine Compère, and Dominique Julia, *L'Education en France du XVIe au XVIIIe siècle* (Paris, 1976), especially chap. 3. It does not, however, conform to the pattern of *Encyclopédie* consumption described in Chapter VI. On the southern Encyclopedists see Jacques Proust, *L'Encyclopédisme dans le Bas-Languedoc au XVIIIe siècle* (Montpellier, 1968). Although nothing resembling nationalism appears in Panckoucke's writing, he did stress the national importance of the *Méthodique* in 1789: ''C'est un monument national, qui sert de modèle aux étrangers.'' Avis to the thirty-first installment, April 1789, in the dictionary *Manufactures*, III, xiv. See also ''Représentations du sieur Panckoucke, entrepreneur de l'*Encyclopédie méthodique*, à Messieurs les souscripteurs de cet ouvrage'' in *Mathématiques*, III, xiii.

84. On this classificatory scheme see Jacques Proust, *Diderot et l'Encyclopédie* (Paris, 1967), chap. 1 and Daniel Roche, ''Milieux académiques provinciaux et société des lumières'' and ''Encyclopédistes et académiciens'' in François Furet, ed., *Livre et société* (Paris, 1965–70), I, 93–184; II, 73–92.

85. See especially Albert Soboul, *Textes choisis de l'Encyclopédie* (Paris, 1952) and the orthodox Marxist interpretations of Jean Luc, I. K. Luppol, and V. P. Volguine, which are discussed critically in the most important work on this subject: Proust, *Diderot et l'Encyclopédie*, 11–13. Proust himself adopts a Marxist interpretation of the *Encyclopédie* and the Encyclopedists, although he acknowledges the difficulty of getting a grip on that shifty and elusive phenomenon, the eighteenth-century bourgeois.

Figure 10. Two Generations of Encyclopedists, Geographical Origins

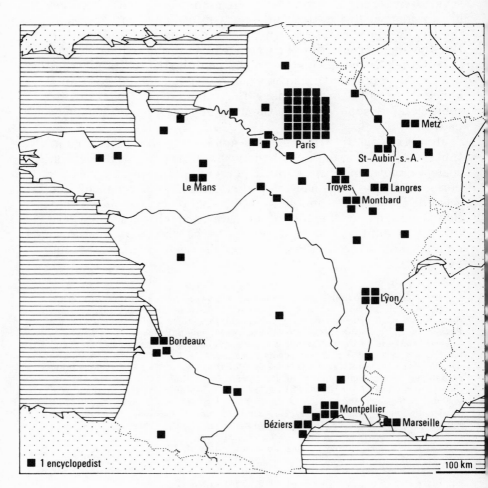

Birthplaces of Diderot's Encyclopedists

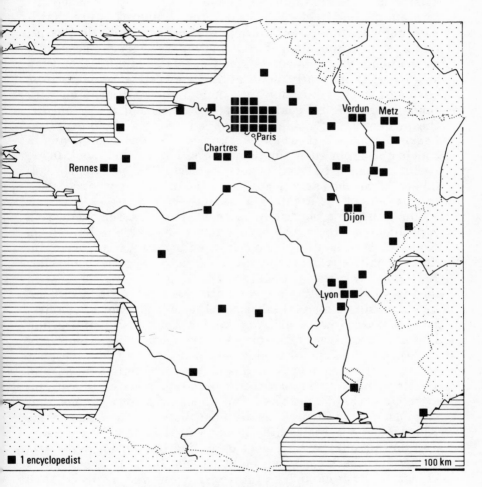

Birthplaces of Panckoucke's Encyclopedists

443

Neither the clergymen nor the noblemen of the *Méthodique* typified their order, however. All but one cleric, Bergier, were *abbés*. They had taken religious vows but lived secular lives— Bossut as a professor, Bertolio as a lawyer, Baudeau as a journalist, and so on. The noblemen did not include any great aristocrats. Pierre-Louis Ginguené was an impoverished littérateur; Condorcet, Fourcroy, and Lamarck were hardworking scientists; Jean-Gérard de Lacuée, comte de Cessac, and François-René-Jean de Pommereul were army officers. All of the noble Encyclopedists stood out as savants rather than seigneurs. Apparently none of them owned a *seigneurie* or derived income from anything that might be associated with feudalism in 1789. From the first to the second *Encyclopédie,* the proportion of noblemen dropped by half. A sign of *embourgeoisement?* It would be abusive to draw so grand a conclusion from such small numbers, but neither book made room for the privileged orders as such. The *privilégiés* collaborated on the same terms as all the other Encyclopedists— as members of the Republic of Letters.

The bourgeois contributors seemed as far removed from capitalism as their noble colleagues were from feudalism. None of them had any connection with manufacturing or industry, and only one was involved in trade: the bookseller Lacombe, who was probably more of a hack writer than a businessman. The second generation Encyclopedists did not include a single manual laborer. At least 6 percent of the first Encyclopedists were artisans, and a good many of the unidentified 11 percent probably came from workshops, too. They were not workers in the modern sense of the word, however, but master artisans—men who made watches, flutes, silks, and jewelry. Diderot did justice to the importance of this preindustrial world of work; and Panckoucke did not renounce that emphasis, even though he did not list artisans among his Encyclopedists. Instead, he incorporated Diderot's articles— and those from the even more extensive *Descriptions des arts et métiers*—in the massive, eight-volume *Dictionnaire des arts et métiers* by Roland. The future Girondist felt as dissatisfied as the rest of his colleagues with the work of Diderot: ''Vous savez combien toutes les parties que j'ai traitées, et que j'ai encore à traiter, l'ont été mal dans la première *Encyclopédie,''* he wrote to Panckoucke. ''Les savants négligeaient les arts, les

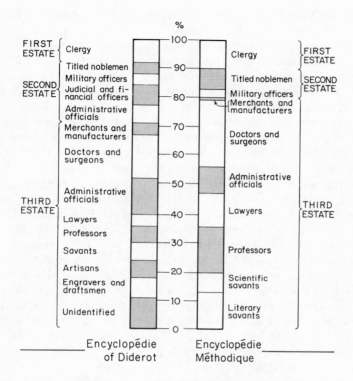

%

FIRST ESTATE { Clergy

Titled noblemen — 90 —

SECOND ESTATE { Military officers
Judicial and fi-
nancial officers — 80 —

Administrative officials

Merchants and manufacturers — 70 —

Doctors and surgeons — 60 —

— 50 —

THIRD ESTATE { Administrative officials

Lawyers — 40 —

Professors

Savants — 30 —

Artisans — 20 —

Engravers and draftsmen

Unidentified — 10 —

— 0 —

Clergy — FIRST ESTATE

Titled noblemen — SECOND ESTATE

Military officers
Merchants and manufacturers

Doctors and surgeons

Administrative officials

Lawyers — THIRD ESTATE

Professors

Scientific savants

Literary savants

Encyclopédie of Diderot   Encyclopédie Méthodique

Figure 11. Two Generations of Encyclopedists, Social Position

artistes négligeaient les lettres; ceux-ci ne pouvaient rendre ce qu'ils savaient, ceux-là ne savaient rendre ce qu'ils ignoraient; il en est résulté, dans les principes et dans les faits, un galimatias inintelligible à tout le monde." He claimed to have outdone all his predecessors, including Diderot, in making contact with the workers: "Je me suis rendu habitant des ateliers; je me suis fait ouvrier."[86] It would be wrong, therefore, to conclude from Figure 11 that the second *Encyclopédie* involved fewer workers than the first, especially as Panckoucke did not list any draftsmen or engravers among his contributors. Actually he did acknowledge the importance of the artists as well as the *artistes* (skilled workers) who stood

86. "Tableau et apperçu," pp. 48–49.

behind the book, but neither he nor Diderot considered them important enough to be named.[87]

The most significant difference between the two sets of Encyclopedists concerned the categories of professional groups: doctors, lawyers, professors, and "savants." Panckoucke used almost three times as many lawyers, proportionately, as Diderot and half again as many doctors. The increase of doctors (including one surgeon, Louis) may have been less pronounced than it looks on the bar graphs because eighteen of Panckoucke's doctors worked on only one of his dictionaries, *Médecine*. But *Médecine* eventually ran to thirteen volumes, dwarfing all the other dictionaries except *Botanique*, which also contained thirteen volumes. So the doctors produced a large part of the second *Encyclopédie*. The proportion of professors also swelled between the 1750s and 1780s. As already explained, they did technical teaching and in some respects resembled the technological officials, who made important contributions to each of the *Encyclopédies*. Both groups shaded off into the "savants," a term that was applied to men who would be called scholars and scientists today. Panckoucke's savants composed 20 percent of his contributors as opposed to 6 percent of Diderot's. The two groups differed in two other respects: a great many of the earlier savants wrote about subjects that had no connection with the way they earned their living, as in the case of Watelet; and several wrote about diverse subjects that had no connection with one another, as in the case of the chevalier de Jaucourt and Diderot himself. These amateurs and generalists disappeared almost entirely in the *Méthodique*, where the botanists wrote about botany and made their living from it, and the littérateurs wrote about and lived from literature. The doctors, lawyers, professors, and savants made up 70 percent of Panckoucke's Encyclopedists. They overlapped and intermingled

87. At one point in 1791 Panckoucke did pay tribute to "MM. Fossier et Deseve, dessinateurs très habiles" and to his head engraver: "M. Benard, chef-graveur, à qui l'*Encyclopédie* doit les plus grandes obligations, qui en a suivi les travaux avec un zèle éclairé et une constance infatigable, a seul dirigé toute la gravure, ayant sous lui soixante graveurs, qui l'ont secondé dans ce travail très long, très difficile par l'immensité des détails qu'il embrasse." Panckoucke, "Sur le Tableau encyclopédique et méthodique des trois règnes de la nature" in the dictionary *Histoire*, V, 14. Of course if room were made on the bar graph of the authors of the *Méthodique* for those sixty-three persons, it would look entirely different. It is possible, too, that Diderot's engravers represented a larger proportion of his collaborators than the 7 percent noted on the first bar graph.

and belonged to the same world—a world in which knowledge was being divided into fields dominated by a few outstanding experts. An enclosure movement was taking over French culture, led by the men of the *Méthodique*—that is, by professionals, whose predominance in the second *Encyclopédie* shows how far professionalization had advanced in the second half of the eighteenth century.[88]

From Voltairianism to Professionalism

What Panckoucke gained in expertise he lost in *philosophie.* The missing element of the *Méthodique* was Diderot, d'Alembert, Montesquieu, Voltaire, Rousseau, Turgot, Quesnay, and d'Holbach—the great figures of the first *Encyclopédie,* who were either dead or too old to contribute to another massive work. The most important authors in Panckoucke's group— Monge, Lalande, Fourcroy, Guyton de Morveau, and Lamarck —tended to be scientists in the modern sense of the word rather than philosophes in the style of Voltaire. The distinction may seem arbitrary, for the philosophes shaded off into the scientist in many writers, notably Voltaire himself, and ''science'' retained its connotation of general ''connaissance'' throughout the 1780s.[89] Furthermore, Naigeon and Concorcet, who made major contributions to the second *Encyclopédie,* can be considered the intellectual heirs of Diderot and d'Alembert, who created the first. But Panckoucke's book celebrated the coming of age of a new generation of intellectuals, and it differed significantly in tone from the *Encyclopédie* of the philosophes.

Panckoucke set that tone himself, both as an author and an organizer of the work. His own articles in the *Méthodique* concerned elevated subjects—''Discours sur le beau, le juste et la liberté,'' ''Discours sur l'existence de Dieu,'' ''Discours sur le plaisir et la douleur''—but they suffered from intel-

---

88. Professionalism is used here to indicate specialization in a field of study from which one derives one's support. It is not used in the more rigorous manner adopted by sociologists. See Talcott Parsons, ''Professions'' in David L. Sills, ed., *International Encyclopedia of the Social Sciences* (New York, 1968), XII, 536–547.

89. See the first definition of ''science'' in the standard prerevolutionary edition of the *Dictionnaire de l'Académie française* (Paris, 1778), II, 484: ''Connaissance qu'on a de quelque chose. 'Je sais cela de science certaine. Cela passe ma science.' ''

lectual flatness. Panckoucke discoursed easily on the existence of God without communicating any sense of adventure or risk and without hazarding a word that could offend an orthodox Catholic. At times he sounded deistic: "Toute la nature atteste un Dieu"; at times pietistic: "Devant le vrai chrétien, toutes les conceptions humaines ne sont que faiblesse et misère pour celui dont l'esprit n'a pour but que le trône de l'éternel: c'est de la vertu l'effort le plus sublime que cette rénonciation au monde, cet abandon de soi-même, surtout dans les personnes d'un haut rang."[90] His religious ideas contained an admixture of elements derived from his Jansenist parents and philosophe friends, mainly Buffon.[91] A great many men of his generation struggled to forge a coherent view of the world from such incompatible sources, but Panckoucke apparently arrived at a workable philosophy without much strain. Instead of plunging into murky areas or wrestling with logical problems, he approached his subjects in a belletristic manner. The beautiful, he argued, is a fixed standard, representing the ultimate in man's esthetic development. Its expression has varied, but only as a result of the different degrees of civilization in the past, the Greece of Pericles and the France of Louis XIV being high points. Beauty will be attained in the future in proportion to man's capacity for cultivation. And Panckoucke's discovery of its nature will resolve all the other problems that have tormented philosophers for centuries. "Ces pensées sur le beau étant vraies," he concluded, "Les disputes éternelles sur ce mot sont terminées, et toutes les questions même qui tourmentent depuis tant de siècles les philosophes sur le juste, l'injuste, la vertu, l'honnête, l'utile, le décent me paraissent résolues, en admettant les mêmes principes."[92]

The subject of liberty inspired Panckoucke to reflect on the need for mastery of the passions, not on any social or political questions. Most men are brutes, he observed, especially in the lower classes. Manual laborers resemble savages, who in turn

90. "Discours sur l'existence de Dieu" in the dictionary *Logique, métaphysique et morale*, I, 358 and "Discours sur le plaisir et la douleur," ibid., II, 50.

91. On Buffon see "Discours sur l'existence de Dieu," in *Logique, métaphysique et morale*, I, 358; and on Panckoucke's Jansenist background see his *Lettre de M. Panckoucke à Messieurs le président et électeurs de 1791*, p. 25.

92. "Discours sur le beau, le juste, et la liberté" in *Logique et métaphysique* I, 238.

are hardly better than animals, owing to their lack of education and the harshness of their climates. Panckoucke took his text from William Robertson's *Histoire de l'Amérique,* not from Rousseau. Some savages are so brutish, he explained, that one would have to mangle their bodies and pull out their nails to make them feel as much pain as that suffered by a well-bred European at the slightest scratch. Pleasure and pain increase in proportion to sensibility and civilization; and freedom grows with refinement, for liberty is the antithesis of animality. In extolling noble savagery and attacking the arts and sciences, Rousseau had got things backwards. ''L'homme devient d'autant plus libre qu'il a l'esprit plus cultivé . . . qu'il fait un grand usage de sa raison et de ses lumières: de sorte que l'on peut dire qu'il y a d'autant plus de cette liberté dont nous parlons, que la société où l'on vit est plus perfectionnée et que les arts et les sciences y sont plus florissants.'' Here was a rationale for the *Encyclopédie méthodique,* though Panckoucke did not say so explicitly. In helping perfect the arts and sciences, his book would advance the cause of all mankind. Of course the people who bought it might be expected to appreciate that message better than the people who scavenged the rags for its paper and transformed the rags into reams and pulled the reams through the presses and hauled the printed sheets over mountains, valleys, rivers, and plains everywhere from Paris to Moscow. But Panckoucke did not pretend to make room for men of toil and sweat in his vision of the good life. He addressed his Enlightenment to the elite—to those capable of beauty, goodness, and happiness; for, as he concluded, ''la naissance, le rang, la fortune, le talent, l'esprit, le génie, la vertu sont donc les grandes sources du bonheur.''[93]

The polite Enlightenment advocated in Panckoucke's three articles did not establish any party line to which the other 100,000 articles adhered. The *Méthodique* was too vast to be contained within any ideology; and insofar as Panckoucke made his influence felt, it was as an organizer rather than an essayist. Although each author shaped his own text (sometimes, as in Senebier's case, with a good deal of intervention from the editor), Panckoucke apportioned the assignments and molded the finished dictionaries into a single *Encyclo-*

---

93. Ibid., p. 239 and ''Discours sur le plaisir et la douleur,'' p. 45. For Panckoucke's comparison of savages and ''nos forts de la douane, nos portefaix,'' see ibid., pp. 40–41.

pédie. The proportions of the *Méthodique* as a whole suggest Panckoucke's sense of the intellectual topography of his time. His overall scheme kept changing, but according to the last and longest version of it, which he expounded in 1791, it had the following structure (the dictionaries are ordered according to size):

Histoire naturelle	9 vols.	Grammaire, littérature	3	,,
Médecine	8 ,,	Finances	3	,,
Jurisprudence	8 ,,	Commerce	3	,,
Arts et métiers	8 ,,	Marine	3	,,
Botanique	5 ,,	Manufactures	3	,,
Antiquités, mythologie	5 ,,	Physique	2	vols.
Histoire	5 ,,	Anatomie	2	vols.
Métaphysique, logique		Chirurgie	2	,,
morale, éducation	4 vols.	Bois, forêts	2	,,
Economie politique et		Police, municipalité	2	,,
diplomatique	4 vols.	Beaux-arts	2	,,
Art militaire	4 ,,	Musique	2	,,
Architecture	4 ,,	Minéraux	1	,,
Assemblée Nationale	4 ,,	Géographie physique	1	,,
Mathématiques	3 ,,	Artillerie	1	,,
Chimie, métallurgie,		Ponts et chaussées	1	,,
pharmacie	3 ,,	Vénerie, chasses, pêches	1	,,
Agriculture	3 ,,	Encyclopédiana	1	,,
Géographie et histoire		Amusements mathé-		
anciennes	3 ,,	matiques et		
Géographie moderne	3 ,,	physiques	1	,,
Théologie	3 ,,	Arts academiques	½	vol.
Philosophie	3 vols.			

Of course size does not serve as a measure of intrinsic importance, especially as some dictionaries grew beyond the bounds that Panckoucke set, but Panckoucke designed his *Encyclopédie* with great care and invested great sums in the execution of the design. So the overall proportions of the *Méthodique* provide a general indication of what he judged most worthy of emphasis—and perhaps even of what subjects would be most central to the interests of an educated public.

Panckoucke evidently expected his customers to have a strong interest in the life sciences. He lavished more of his resources on natural history and botany, which were written by two of his finest Encyclopedists (Daubenton and Lamarck), than on any other subject. Next came the medical sciences and the dictionary of *Arts et métiers* with its offshoot *Manufac-*

tures. By contrast, chemistry, the science undergoing the most spectacular development at that time, received only three volumes and physics two. Thus the *Méthodique* bears out Daniel Mornet's contention that eighteenth-century readers developed a keen interest in the sciences that seemed to bring them close to nature—not the nature of abstract, mathematical forces but of field trips, rock collections, and *cabinets d'histoire naturelle.*[94] Panckoucke's *Encyclopédie* also illustrates the eighteenth century concern with putting science to work to improve farming, manufacturing, shipping, and transportation. The liberal arts or humanities received about as much emphasis as the utilitarian or applied sciences. Although Panckoucke allotted only three volumes to literature and grammar, he gave a great deal of space to history, philosophy, fine arts, and the classics. In general, the sciences outweighed the other subjects in the *Méthodique:* they took up about half of it, as opposed to the humanities (25 percent) and what today would be called the social sciences (13 percent), the remainder being devoted to heterogenous subjects, like the *arts académiques* (dueling, dancing, riding, and swimming).

Again, counting cannot convey content. One can do justice to the intellectual substance of those 125½ volumes only by reading and analyzing them, a task that lies beyond the limits of this book. But just as one can get a sense of the intellectual life of a university by examining its faculty and its course offerings, so one can develop some idea of the culture conveyed in the *Méthodique* by studying its authors and its distribution of subjects. Browsing through the *Méthodique* is like strolling through a univrsity: first one passes the small but elegant mathematics building; Art and Music stand to the left, History and Literature to the right; the natural sciences dominate a vast quadrangle near the gymnasium and swimimng pool; and beyond them loom the law and medical schools. One has entered the modern world, in which subjects belong to departments and certified experts rule over carefully demarcated territories. Some areas of study had not assumed their modern shape by 1791. Chemistry, for example, had not yet shaken off metallurgy and pharmacy; and economics lay inchoate in three different dictionaries: *Economie politique et diploma-*

---

94. Daniel Mornet, *Les sciences de la nature en France au XVIIIe siècle* (Paris, 1911). See also, Jacques Roger, *Les sciences de la vie dans la pensée française du XVIIIe siècle* (Paris, 1963).

tique, Finances, and *Commerce.* But taken as a whole, Panc-koucke's *Encyclopédie* illustrates the emergence of the modern notion of autonomous disciplines. It was encyclopedic in a neutral way: it covered everything and put everything in its place. In that respect it had more in common with today's encyclopedias—and certainly with nineteenth-century works like the *Encyclopédie moderne* (Didot frères, Paris, 1846–1848) and the *Grand dictionnaire universel du XIXe siècle* (Larousse, 1866–1876)—than with the *Encyclopédie* of Diderot and d'Alembert. The earlier work was also universal in its coverage, but instead of slicing knowledge into segments, it treated all intellectual activities as parts of an organic whole, symbolized by the tree of knowledge.

Diderot and d'Alembert labeled the central trunk of that tree "philosophy." It grew out of the faculty of "reason," and one of its remotest outcroppings was a small branch called "revealed theology," which sprouted next to "knowledge of good and evil spirits: divination, black magic." Philosophy and theology occupied separate, three-volume structures on Panckoucke's conceptual campus. Panckoucke put an expert in charge of each: philosophy went to Naigeon, Diderot's atheistic disciple, and theology to Bergier, the king's confessor. Far from becoming entangled in debate, the two men concentrated on putting their own houses in order by cleaning up the clutter from the first *Encyclopédie.* Although he paid tribute to Diderot, Naigeon condemned the philosophical articles in the earlier work for amateurism. His own dictionary provided a thorough survey of the major schools of philosophy. It was a solid, somewhat pedantic reference book, which apparently had no difficulty in getting past the royal censor.[95] Bergier had earned a reputation as a leading opponent of the philosophes; and in undertaking his dictionary, he emphasized his determination to expunge the heresies that had made Diderot's work effective as an instrument of Enlightenment: "Dans plusieurs autres [articles] on étale les objections des hérétiques, et l'on supprime les réponses des théologiens catholiques . . . De ces divers défauts il en résulte un plus grand, c'est que la doctrine de l'*Encyclopédie*

95. See Naigeon's prospectus in *Beaux-Arts,* I, xxxv; his letter to Panckoucke of Feb. 16, 1788, in *Mathématiques,* III, 15–23; and the dictionary of philosophy itself, which became more adventuresome in the two volumes published after 1789.

est un tissu de contradictions.'' And like all the second-generation Encyclopedists, he objected to the lack of professionalism
in his predecessors. ''Les articles faits par des théologiens,
surtout par M. Mallet, sont en général assez bien; les autres,
composés par des littérateurs mal instruits ou infidèles, ont
été servilement copiés d'après les controversistes Protestants
ou Sociniens.''[96] Diderot had reveled in contradictions and had
enjoyed the points where subjects ran into each other. Instead
of treating philosophy as an autonomous subject, he had
breathed it into everything he did. It ran through his entire
Encyclopédie, from A to Z, as an informing spirit. He had
expressed it with wit, irreverence, and passion, interjecting it
between the lines, stocking it in unexpected corners, and slipping it into audacious cross-references like ''ANTHROPOPHAGES
. . . Voyez EUCHARISTIE, COMMUNION, AUTEL, etc.'' Diderot's
mordancy disappeared in the vast stretches of *Wissenschaft*
that Panckoucke laid out for the later Encyclopedists.

In reinforcing the tendency for knowledge to become com-
partmentalized and science professionalized, Panckoucke did
not turn his back on the Enlightenment. The intellectual movement associated with the philosophes included a commitment
to decipher the secrets of nature by scientific investigation
as well as determination to crush *l'infâme*. The *Méthodique* extended the former aspect of the Enlightenment and eliminated
the latter. In this way it became completely acceptable to the
regime, for the authorities had never objected to Diderot's
attempt to survey all the arts and sciences, only to his use of
that survey as a cover for unorthodox *philosophie*. Panckoucke
stripped off that cover and rearranged the sciences in a way
that would not offend anyone in power. Once isolated in
separate dictionaries, subjects like philosophy could be kept
within official limits. Every volume was censored; in fact
twelve of the seventy-three authors were censors themselves.
Not only did the book appear ''avec approbation et privilège
du roi,'' but also by special concession Panckoucke received
an extra-long, forty-year privilege. He said that he would
never have been able to accomplish so much, ''si l'administration n'avait pas eu la bonté de nous seconder dans les différentes demandes que nous lui avons faites.''[97] And he dedicated a dictionary to almost every minister of Louis XVI.

96. *Beaux-Arts*, I, xxxiv.
97. *Mathématiques*, III, vii.

Far from threatening the established order, the *Encyclopédie méthodique* appeared with the royal stamp of approval, virtually as an official publication.

Launching the Biggest Book of the Century

The official support of the enterprise certainly suited Panckoucke's interest as a businessman, and Panckoucke became more and more interested in the *Méthodique* as time went on. He kept tinkering with it, rearranging its parts and expanding its scope. The original blueprint of 1779 looked small in comparison with the gigantic scheme of 1789–1791, which has just been described. One can follow the evolution of the project during those ten years through Panckoucke's correspondence with the STN. Although his letters do not contain enough information to reconstruct the full business history of the book, they show how Panckoucke launched the speculation and how it grew, as he continually enlarged his plans and postponed the date for their realization.

Panckoucke originally planned to publish the first five volumes of the *Méthodique* in April 1779, but when April arrived, he decided to delay them and the publication of the prospectus until December. Having accumulated only a few manuscripts by mid-June, he revised that decision and put off publication for another year.[98] By mid-July he had signed contracts with three authors, had lined up his censors, and had completed plans for all twenty of the dictionaries that he then expected would constitute the entire *Encyclopédie*.[99] In August he thought that he could get the first volumes out by July 1780. By October he had signed contracts for two more dictionaries and had become so absorbed in the *Méthodique* that he neglected the quarto, according to complaints by the STN.[100] This conflict of interest disappeared with the liquidation of the quarto in February 1780, but the next months produced additional complications and delays. "Cette entreprise est hérissée de toutes sortes de difficultés," Panckoucke wrote in September. "Je compte cependant en publier le prospectus en 1781, mais il ne paraîtra rien avant deux ans. Je ne veux pas jouer ma fortune au hasard. Les souscriptions me décide-

98. Panckoucke to STN, June 15, 1779.
99. Panckoucke to STN, July 10, 1779.
100. Panckoucke to STN, Oct. 15, 1779.

ront.''[101] By the end of 1780, he had signed the contracts for all twenty dictionaries, but he had stopped talking euphorically about the "belle entreprise" that would cap his career. He now felt torn between hope and despair: "J'ai déjà passé vingt actes. C'est un énorme projet, qui, à quelques fois, m'élève l'âme et dans d'autres moments l'effraie.''[102]

Panckoucke's optimism turned downward because of the difficulties in coordinating and financing the simultaneous production of twenty treatises. Although he rarely mentioned his work in his letters to Neuchâtel, he did not disguise the fact that it had got off to a bad start: "C'est un projet de dure conception et exécution. Les deux volumes de physique que j'ai publiés ne réussissent pas trop et ne peuvent convenir à ce plan. L'auteur (c'est M. Brisson) n'a pas bien saisi ce qu'il fallait faire.''[103] The failure of Brisson's dictionary threatened to ruin the whole enterprise because it was the first to be finished. Panckoucke had expected to launch his subscription campaign in the wake of its success, and he counted on the income from the subscriptions to finance the production of the next nineteen dictionaries. He therefore withdrew Brisson's work from the *Encyclopédie* and marketed it as a separate book, while commissioning Monge to produce a new dictionary of physics and postponing the subscription once again.[104]

All these extra expenses and delays produced recriminations from the other publishers who had invested in the *Méthodique,* but Panckoucke's partners seem to have been easy to handle in comparison with his authors, who could not even agree on the phrasing of the general prospectus. It still had not appeared in April 1781, when Plomteux reported to the STN on the state of affairs in Paris: "Vingt auteurs différents presque jamais d'accord sur leurs opérations, trouvent chaque jour quelques changements à faire. J'ai dû convenir avec M. Panckoucke, et vous conviendrez vous-même, que la chose du monde la plus difficile est de concilier vingt gens de lettres tenant tous très fortement à leur opinion et très peu disposés à la sacrifier aux intérêts du libraire. Ce n'est plus M. Brisson qui est chargé

101. Panckoucke to STN, Sept. 28, 1780.
102. Panckoucke to STN, Nov. 10, 1780.
103. Panckoucke to STN, March 31, 1781.
104. The STN papers contain a *lettre circulaire* by Panckoucke dated Feb. 28, 1781, in which he offered Brisson's *Dictionnaire général, universel & raisonné de physique* to the booksellers of Europe, without indicating that it had ever had any connection with the *Encyclopédie méthodique.*

de la physique. Il a fallu employer un autre auteur à cette partie essentielle. Vous sentez que ces changements doivent en apporter dans le plan et que ce serait une chose interminable pour tout autre que M. Panckoucke, à qui il ne faut en vérité pas en vouloir. Son courage a besoin d'être soutenu, et c'est ce que je chercherai à faire pour nos intérêts communs.''[105] By May 1781, Panckoucke had settled with the printers and paper suppliers but still had not reached a final decision about the format of the work and its pressrun. He decided to publish the prospectus and assess the public's reaction to it before taking the last, irreversible steps in the production process. He now referred to the *Encyclopédie méthodique* darkly as the book that would send him to his grave: "Les encyclopédies ont empoisonné ma vie et me conduiront au tombeau. J'y ai sans doute gagné, mais les soins, les soucis, les inquiétudes, les travaux ont été énormes.''[106]

The prospectus finally appeared in December 1781. It showed how much the enterprise had grown in size and expense since Panckoucke had taken it over from the Liégeois in 1779. Panckoucke now announced twenty-six dictionaries instead of twenty, in addition to the *Vocabulaire universel* and the seven volumes of plates. The book would appear in two formats: a quarto, which would have three columns to the page and would run to forty-two volumes of text, and an octavo in two columns and eighty-four volumes. The seven quarto volumes of plates would be the same for both formats and so would the price: 672 livres. The subscription would close on July 1, 1782. After that, the book would cost 798 livres,

105. Plomteux to STN, April 14, 1781.

106. Panckoucke to STN, May 8, 1781. By this time Panckoucke had called upon his associates to furnish additional capital. Plomteux tried to clarify the situation, which had become badly clouded over, in a letter to the STN from Paris on May 1, 1781: "M. Panckoucke s'est réservé la gestion de toute l'affaire, a contracté ici avec les imprimeurs et les papetiers et fournira chaque année un compte de toute l'entreprise. Ma mise sera de 5,000 livres par mois, à commencer en octobre ou novembre prochain, et continuera sur le même pied jusqu'à ce que les souscriptions puissent fournir à la dépense. C'est donc du succès de l'entreprise que dépend le plus ou moins de fonds qu'il faudra avancer et qui ne peut être actuellement fixé. On n'est plus certain du nombre du tirage, ni du format. Le goût du public qu'on pressentira par le prospectus déterminera l'un et l'autre. Vous sentez, Messieurs, et vous êtes convenus vous-mêmes à Lyon, qu'il n'était pas possible d'entreprendre une affaire de cette importance sans faire quelques tentatives pour reconnaître les moyens de la faire réussir. C'est ce qu' avait [fait] le projet du Dictionnaire de physique qu'il a fallu sacrifier. Voilà tout ce que je sais.''

and Panckoucke promised solemnly to maintain that retail price, no matter what should happen to him or his business, in order to protect the investment of the subscribers. They would get a bargain, he insisted: more than half again as much text as in the original *Encyclopédie* at half the price. They would pay for the volumes as he issued them, in installments, at 12 livres per quarto volume of text and 24 livres per volume of plates. In this way, the subscribers would spread a series of relatively small payments, usually 24 livres each, over a period of five years, easing the strain on their finances—and on Panckoucke's, too, although he did not mention this consideration, for he needed a steady flow of capital to pay for the printing operation. He promised to complete the job by July 1, 1787—that is, to produce a mammoth edition of forty-two-volume quarto and eighty-four-volume octavo *Encyclopédies,* written, mostly from scratch, by the most distinguished savants in France and printed, on the best paper and with the finest type, by the leading printers of Paris, in only five years.

Panckoucke did not say how much he would pay his writers, but if his twenty-six contracts resembled his earlier agreements with Fougeroux and others, he must have planned to spend at least 200,000 livres on copy.[107] That sum looked small in comparison with the projected production costs. Of course Panckoucke could not estimate the printing charges accurately until he knew how many subscriptions he had sold. But he could make some rough guesses. Suppose he produced 4,000 sets of the quarto and he paid the printers and paper suppliers at the same rate as that stipulated in his contracts with Duplain: his production costs for the quarto edition alone would come to 562,800 livres. In fact, printing expenses were much higher in Paris than in the provinces or Switzerland—30–40 percent higher, Panckoucke later claimed.[108] And the octavo *Méthodique,* which required new typesetting, could have brought Panckoucke's total expenses into the range of two million livres. That was the estimate that he himself produced in the prospectus: "C'est un objet de dépense de près de deux millions," and it was a staggering sum, even for someone who

---

107. Later, in his "Représentations du sieur Panckoucke, entrepreneur de l'*Encyclopédie méthodique,* à Messieurs les souscripteurs de cet ouvrage" (1789), in *Mathématiques,* III, viii, Panckoucke wrote, "La seule copie de cet ouvrage nous revient à plus de six cent mille livres, quoique, dans nos premiers calculs, nous n'eussions pas cru qu'elle dût nous en coûter même deux cent."

108. Ibid., p. xiii.

had recently closed the accounts of Duplain's *Encyclopédie,* where the production costs had come to a million and a half. Speculation on such a scale did not occur often in the eighteenth century. So in the prospectus, Panckoucke seemed to be taking a deep breath before plunging into icy, unknown waters: "Une édition complète de l'*Encyclopédie,* par ordre de matières, nous a paru si effrayante au premier coup-d'oeil, que, quelque habitude que nous ayons des grandes entreprises en librairie, ce n'est qu'après y avoir très mûrement pensé et avoir considéré la possibilité de son exécution sous toutes les faces, que nous nous y sommes engagés et que nous avons résolu de l'entreprendre."[109]

The speculation on the *Encyclopédie méthodique* became so intertwined with the other *Encyclopédie* enterprises that its history cannot be understood separately from theirs. It was both a predecessor and a successor of the quarto, for it descended from Diderot's memoir of 1768 and Suard's *refonte* of 1776, yet it did not take shape until Panckoucke took over Deveria's project of 1778. While splicing together speculations on the *Encyclopédie,* Panckoucke also wove other huge enterprises into his grand strategy of publishing. For a few months in 1778 and 1779, he almost monopolized the works of Voltaire, Rousseau, and Buffon as well as the *Encyclopédie.* Although it is not possible to follow all of these speculations to their conclusion, it is important to appreciate their convergence; for they represent the final flowering of the Enlightenment under the Old Regime. The production of *Lumières* may have gone through a more critical period around 1750, when many of the great books first burst into print. But seen as a diffusion process, the Enlightenment went through an equally important phase in the late 1770s, when speculators scrambled furiously to produce massive editions of the works of the philosophes for a kind of "mass" audience —not the illiterate masses, of course, but ordinary readers scattered throughout western Europe. Although it has never been noticed, this second harvest of Enlightenment publishing deserves close study because it represents the high point in the spread of philosophic works before the Revolution. The *Encyclopédie* played a crucial part in this diffusion process, both in the 1750s and in the 1770s, both in its folio format and

109. *Mercure,* Dec. 8, 1781, p. 248.

in-quarto, both as a collection of heresies and as a compendium of sciences. The scientific element predominated in the ultimate *Encyclopédie* of the Enlightenment, but Panckoucke had hardly published half of it before 1789, when the fate of his final speculation became bound up with the fate of France in a revolution that transformed the organization of culture as well as society and politics.

I X

YYYYYYYYYYYY

ENCYCLOPEDISM,
CAPITALISM, AND REVOLUTION

The final stage in the history of the *Encyclopédie* during the eighteenth century looks like an entanglement of "isms": Encyclopedism, capitalism, Jacobinism, and related tendencies such as professionalism and *étatisme*. By following the *Encyclopédie méthodique* through the stormy last years of the century, it should be possible to see how those abstract phenomena operated in concrete situations and how the history of the book fit into the larger context of history in general.

Panckoucke's Folly

By December 1781, when he issued the first prospectus for the *Méthodique* and opened his subscription campaign, Panckoucke realized that he might be standing on the brink of a two-million livre disaster. Nonetheless, he resolved to push ahead. The prospectus committed him to producing a forty-two-volume, three-column quarto edition and an eighty-four-volume octavo edition of the new text by July 1, 1787. It gave the public seven months, from December until July 1782, to purchase the book at the subscription price of 672 livres—a spectacular bargain, it proclaimed, considering what the subscribers would get for their money: the greatest encyclopedia of all time, a compendium of everything known to man, the most useful work ever written, and a book that was a library in itself.

Panckoucke's sales talk might sound suspicious to anyone who knew the inside story of his previous *Encyclopédies*.

A cynic could even read the prospectus as an implicit confession about the tricks and frauds involved in the marketing of the quarto, for Panckoucke kept insisting on what he would *not* do to sell the *Méthodique*. He would not cut the price after the subscription was closed. He would not extend the subscription beyond its deadline under any pretext whatsoever. He would not publish any of the dictionaries separately or produce any subsequent version of the book, which could make the present model fall in value. He would not permit any deviation from the high standard of printing and paper exemplified by the sample pages he distributed with the prospectuses. He would not allow any flagging in the production schedule : the dictionaries would come out regularly, in twenty-three installments, the first few volumes in July 1782, the last without fail, five years later. And he would not attempt to squeeze more money out of the subscribers, who were to pay eleven livres for each volume of text they received, by producing more than the prescribed number of volumes. Panckoucke considered it so important to assuage the public's fear of this last ruse, which had given rise to more complaints than any other of Duplain's devices for mulcting the subscribers of the quarto, that he promised emphatically to keep the sets down to forty-two quarto and eighty-four octavo volumes, or, if he exceeded that limit by more than three volumes of text and one of plates, to give away the extra volumes free of charge. Every promise to the new subscribers corresponded to a low blow inflicted on the old ones, and Panckoucke was to break almost all of his promises in the course of his management of the *Méthodique*—not because he meant to swindle his customers but because the speculation spun out of his control.[1]

By March 1782 it became clear that the enterprise had got off to a disastrously bad start : only a tiny number of subscriptions had trickled in—less than 30, in the case of the

---

1. That Panckoucke keyed his sales campaign to the public's fear of being duped by the same techniques that he and Duplain had used in marketing the quarto seems evident from a close reading of the prospectus. See especially the passage on p. 151 of the version in the *Mercure* of Dec. 8, 1781, which begins : ''Quoique nous ayons toujours rempli avec la plus scrupuleuse exactitude les engagements que nous avons pris avec le public, cependant, comme il est aujourd'hui plus en garde que jamais contre toute espèce de souscription, nous nous croyons obligés, dans une entreprise de cette importance, de lui donner toutes les assurances qui peuvent établir une confiance réciproque, soit pour l'exécution de l'ouvrage conformément au *Prospectus,* soit pour le temps de la livraison des volumes, soit pour leur nombre, soit enfin pour fixer irrévocablement le prix.''

octavo edition—yet Panckoucke had counted on 5,000 subscriptions and needed about 4,000 to cover his costs. "Nous étions à cette époque au désespoir de nous y être engagés," he later explained.[2] "Nous la [the *Encyclopédie méthodique*] regardions comme absolument désespérée. Cependant nous avions fait des achats de papier considérables, presque tous les actes étaient passés avec les auteurs, et toute notre fortune compromise." He faced a terrible choice: he would either have to reorganize the whole speculation in some way that would win over the public, or he would have to abandon it, with a loss of hundreds of thousands of livres and a great deal of pride. Despite the enormity of his investment in the general preparations and the first volumes, which were due to appear in July, Panckoucke might have saved his fortune had he declared himself defeated. But he was a gambler. He decided to raise the stakes and to remain in the betting for another round.

On March 16, the *Mercure* announced that Panckoucke had scrapped the octavo edition and would publish the quarto with pages made up of two columns rather than three, as in his original plan. The two-column format would be vastly more pleasing to the eye, but it would necessitate a reduction in the size of the paper and an expansion in the number of volumes. Panckoucke now promised to fit his text into fifty-three volumes of "papier carré fin de Limoges" instead of forty-two volumes of "grand raisin." The type, petit romain cast especially for the book by Fournier of Paris, would remain the same—and so would the price. But anyone who wanted to purchase this typographically superior *Encyclopédie* would have to move fast because Panckoucke now felt constrained to close the subscription on April 30. He would open a second subscription on May 1, but it would cost 751 livres. Even at that price the *Méthodique* was a bargain, he assured the public; anyone who failed to take advantage of his offer by April 1783, when the second subscription was to close irrevocably, would have to buy the book at a retail price of

2. Panckoucke, "Nouveaux éclaircissements," *Mathématiques*, III, xx. Panckoucke made clear that it was not merely the octavo subscription that had failed, because he stressed that by March 1782 he had accumulated only "un petit nombre de souscripteurs de ces deux formats." He later said that the two subscriptions together had produced only 400–500 subscribers. *Manufactures, arts et métiers*, seconde partie, III, vliv. See also *Beaux-Arts*, I, lx: "Nous observerons que le public n'a voulu d'aucun de ces formats . . . et qu'au mois de mars 1782 nous étions au désespoir de nous être engagés dans cette grande entreprise et que nous la regardions comme absolument impossible."

888 livres. Of course those who had already subscribed could request the return of their down payment (36 livres), if they did not want to accept Panckoucke's new terms. But he made his terms seem more attractive than ever, for he had originally set the retail price at 798 livres. He wanted to stampede the public into subscribing by playing on its bargain-hunting instinct. That strategy could not succeed if word got out that the enterprise was on the verge of collapse. So Panckoucke described his original subscription campaign as a smashing success: he had only reorganized it in order to satisfy the public's desire for an improved format, he explained. He was forging ahead with the printing; two volumes were already in press and there would have been twelve, if the founders had been able to supply the type fast enough. The first volume of plates had nearly been completed, and a sample copy could be inspected in his shop at the Hôtel de Thou by anyone who doubted his ability to execute his plan as he had promised.[3]

This maneuver, which Panckoucke later described as a desperate "dernier effort pour ramener le public,"[4] apparently worked wonders. Panckoucke later claimed that it brought in 5,000 subscriptions, and according to a statement he issued in 1788, 4,042 subscribers signed up at the 672-livre rate before May 1, 1782.[5] On that date Panckoucke published a second general prospectus, which repeated the propositions of the first, except for some clauses about the new format and price and a few unobtrusive remarks about the size of the new quarto edition. Instead of offering to give away any volumes that should exceed the limit, Panckoucke now indicated that the text might run to three or four volumes more than the fifty-three that he expected to produce. In that case, the subscribers would pay for the extra volumes at the subscription rate of eleven livres each; and if, against all expectations, he should have to produce still more, they would pay only six

3. *Mercure*, March 16, 1782.
4. Panckoucke, "Eclaircissements relatifs à un premier titre d'une souscription à 672 [livres]," *Mathématiques*, III, lx.
5. "Lettre de M. Panckoucke, en date de novembre 1788, écrite aux auteurs de l'*Encyclopédie*," *Mathématiques*, III, xiv. This figure included an undetermined number of persons who had originally subscribed to Deveria's *Encyclopédie*. Panckoucke gave them the choice of accepting his terms or trying to get their down payments back from Deveria, whose role in the enterprise received only the scantiest recognition in the original prospectus. See *Mercure*, Dec. 8, 1781, p. 152. For Panckoucke's remark about 5,000 subscriptions see *Manufactures*, III, xliv. In fact, the total seems to have been 4,850.

livres per volume.[6] The 751-livre subscription did badly. Only about 808 customers signed up before May 31, 1786, when he closed it—three years later than he had said he would.[7] But it was remarkable that he could sell almost 5,000 subscriptions after the market had already absorbed 25,000 sets of the original *Encyclopédie;* and the installments paid by the subscribers seemed certain to provide enough capital for him to continue production and to cream off a handsome profit in the end—if nothing went wrong.

Between 1782 and 1786, things seemed to go right. Panckoucke had found a formula for taking advantage of the demand for a revised *Encyclopédie.* The demand certainly existed, despite the proliferation of the earlier editions, for the STN received several letters from French booksellers who considered the *Méthodique* superior and more sellable than its predecessors. In March 1780, soon after the first rumors about an "encyclopédie par ordre de matières" had spread through the booksellers' grapevine, Machuel of Rouen reported, "Il y a bien des mécontents pour l'édition in-quarto, et la nouvelle qui va paraître est préférée." Carez of Toul made the same observation a year later; and so did Gaches of Montauban, who regretted subscribing to the quarto, "vu qu'on nous propose une nouvelle édition, qui se fait à Paris dans une méthode qui selon mes faibles lumières rendra cet ouvrage plus utile et plus parfait." Judging from these letters, Panckoucke had a sense of what would sell: the public wanted an improved and reorganized *Encyclopédie,* not just pure Diderot. Back in 1778, Lair of Blois had informed the STN that he regretted the demise of the original Liégeois *Méthodique*—which Panckoucke was then secretly appropriating for himself—because it "aurait été admirable, si elle eût été exécutée suivant le plan qu'ils m'ont envoyé: 36 volumes in-folio ou 144 in-octavo avec plus de 3,000 planches: abandonner l'ordre alphabétique des mots pour suivre celle des matières était la vraie, l'utile, et l'intéressante *Encyclopédie,* qu'on ne peut donner trop tôt." So for all its extravagance, Panc-

6. Panckoucke reprinted the text of the second prospectus in *Beaux-Arts,* I, li–lviii.

7. Panckoucke did not reveal the size of the second subscription, but in his "Lettre . . . aux auteurs" of 1788 he said that the first subscription included 4,042 subscribers. In his notice "Sur les prétendus bénéfices actuels de cet ouvrage," printed in *Mathématiques,* III, xvi, he said that they "forment plus des cinq-sixièmes de la totalité."

koucke's plan did correspond to the continuing demand for Encyclopedism among eighteenth-century French readers.[8]

Having got a firm grip on the French market, Panckoucke reached out toward the rest of Europe. In 1783 Jacques Thévin, bookseller to the court of Spain, came to Paris to negotiate the purchase of 300 sets and the rights to a Spanish edition with plates to be supplied by Panckoucke. Upon his return to Madrid, he and another prominent dealer, Antonio de Sancha, issued a Spanish translation of the prospectus and reportedly filled a subscription, thanks to support from the government and the Inquisitor General himself. Panckoucke eventually sold 330 French *Encyclopédies* in Spain, and the Spaniards produced eleven volumes of their translation—an indication of how much the Spain of Charles III differed from that of Charles II and perhaps also of how much Panckoucke had watered down Diderot's text. In the summer of 1788, however, the Spanish Inquisition confiscated all the *Encyclopédies* in the warehouse of Panckoucke's Spanish agent, who fled to France, proclaiming himself a victim of fanaticism. Although Panckoucke believed he had the backing of a special board of censors named by the Council of Castile, he never recovered the 330 subscriptions; and the Spanish authorities cut short the publication of the translation.

An Italian translation got under way in Florence, reportedly with a 60,000-ducat subsidy from Leopold, the enlightened Grand Duke of Tuscany. Panckoucke angled for a subsidy of his own from Catherine II of Russia, and he boasted that the Turks thought so highly of his book that they, too, planned to translate it. Although none of these projects came to anything, they indicate how well the enterprise was succeeding. It even penetrated the American market, thanks to the help of Thomas Jefferson, who doctored its article on the United States and recruited several subscribers, including Franklin, Madison, and Monroe.

The success of the *Méthodique* also became apparent from the number of counterfeit editions it inspired. Panckoucke noted, with a touch of pride, that it was being pirated in five places by 1789: Padua, Venice, Milan, Nice, and Liège. He might have added the Swiss edition, which, as explained above, typified the way the pirates operated. The Liégeois project

8. Machuel to STN, March 31, 1780; Carez to STN, Dec. 17, 1781; Gaches to STN, Jan. 3, 1782; and Lair to STN, Nov. 11, 1778.

also provides an equally good example of *Encyclopédie* privateering. In a printed circular letter to the major booksellers of Europe, a "Société typographique de Liège," backed by a group of anonymous "négociants-entrepreneurs," offered to supply counterfeit *Méthodiques* for 677 livres a set—74 livres less than the cost of Panckoucke's second subscription. They promised that "l'édition que l'on propose, faite scrupuleusement sur celle de Paris, n'en différera que par l'infériorité du prix et la supériorité de l'exécution." The Liégeois also proposed a deluxe edition of 200 copies on extra-large paper. And they tried to edge into Panckoucke's market by winning over the middlemen. They offered bookdealers a discount of 150 livres for each copy and a free thirteenth copy for every twelve subscribed, making a magnificent profit of 2,477 livres for the dealer who could sell thirteen sets at the retail price. Attractive as they seemed, these schemes for counterfeit *Encyclopédies* all collapsed. By expanding and Italianizing Panckoucke's work, the Paduans kept their enterprise going until 1799, and the Liégeois completed an eight-volume edition of the dictionary of theology in 1792. But no publisher found it possible to reproduce the whole set of dictionaries, not because of any slackening in the general zeal for piracy but because the *Méthodique* became so enormous that it threatened to exceed the capacity of Panckoucke himself.[9]

9. For information on the various *Encyclopédie* projects outside France see: Thiriot to STN, April 8, 1783; *Mercure*, March 28, 1789; "Représentations du sieur Panckoucke, entrepreneur de l'*Encyclopédie méthodique*, à Messieurs les souscripteurs de cet ouvrage" in *Mathématiques*, III, xiii; "Lettre de M. Panckoucke à Messieurs les souscripteurs de l'*Encyclopédie*" in *Histoire*, V, 1; Panckoucke's Avis on the thirtieth and forty-eighth installments in *Manufactures*, III, xi, xlii; *Correspondance littéraire, philosophique et critique par Grimm, Diderot, Raynal, Meister, etc.*, ed. Maurice Tourneux (Paris, 1880), XIII, 135; and *Mémoires secrets*, entries for Aug. 25, 1783, and Jan. 12, 1787. The quotation comes from a printed circular, sent to the STN with a covering letter by J. J. Tutot and C. J. Renoz of Liège, on Dec. 16, 1783. Tutot and Renoz were the leaders of the Société typographique de Liège, which seems to have been created primarily in order to counterfeit the *Méthodique*. On Jefferson's involvement with J. N. Démeunier's article on the United States in the *Méthodique* see *The Papers of Thomas Jefferson*, ed. Julian P. Boyd (Princeton, 1950——), X, 3–11. Jefferson suggested that Panckoucke establish an agent in Philadelphia for collecting subscriptions. Jefferson to David S. Franks, March or April 1783, ibid., VI, 258. But as he noted in a letter to Francis Hopkinson of Jan. 26, 1786 (ibid., IX, 224), he found it difficult to get good service from Panckoucke, even in Paris: "I have sent several times to M. Panckoucke for the three livraisons of your *Encyclopédie* not yet delivered. The last answer this evening, after sending ten times in the course of the day, is that he will send me both yours and Doctr. Franklin's tomorrow morning."

The *Méthodique* suffered from gigantism. With every installment it seemed to get bigger, farther away from completion, more unmanageable, and less marketable. One cannot tell how this disease overcame the book because the manuscript sources thin out after 1781, making it impossible to reconstruct the inside story of its publishing history. But Panckoucke issued so many prospectuses, memoirs, Avis, circular letters, fliers, and other advertising material that one can follow the main lines of its evolution.

The authors proved to be the hardest element to manage in the enterprise. Their copy was usually too lengthy and too late, and it frequently failed to arrive at all. Brisson's botched dictionary of physics had ruined Panckoucke's plan to begin publication in 1781; so the prospectus set July 1782 or August "au plus tard" as the publication date of the first installment, and it promised that at least six volumes would be out by the end of the year.

On August 10, however, Panckoucke notified the subscribers through the *Mercure* that the first volume, now recast as a dictionary of jurisprudence, would not appear until October because its author, the abbé Rémy, had died, leaving his papers impounded and in disarray. Rémy's main associate had died the previous year, and their deaths were not the only ones to plague the *Méthodique:* Court de Gébelin died in 1784 before he had produced a single word for his volume on *antiquités,* which put that dictionary back three years, according to Panckoucke; Guenau de Montbeillard died in 1785, leaving the dictionary on insects in a shambles; and Watelet died in 1787, forcing Panckoucke to chase the manuscript for the dictionary of beaux-arts through the courts and to the home of his heir before it could be rescued, reworked, and published a year later.[10]

10. Panckoucke originally issued his Avis as broadsides and as notices in the *Mercure,* but he later reprinted them in certain volumes of the *Méthodique.* Several of the broadsides may be consulted in Case Wing Z 45.18, ser. 7, Newberry Library. A comparison of all three versions of the texts shows that Panckoucke did not alter them, so for the sake of convenience they may be cited as they appear in the *Méthodique.* The information above, for example, comes from the Avis on the eighteenth and the twenty-second installments reprinted in *Beaux-Arts,* I, lxxv and lxxxi and also from the "Tableau et apperçu," which Panckoucke published at the beginning of *Mathématiques,* III. On Panckoucke's difficulties with the Watelet manuscript see his letter to Watelet's heir, the comte d'Angiviller, March 15, 1787, ms. Fr. c.31, Bodleian Library, Oxford: "Je viens de faire mettre à votre adresse un exemplaire de la traduction du Lucrèce, 2 volumes in-quarto, et du Mengs, aussi deux volumes in-quarto, que j'ai fait faire

Other authors simply dropped out of the enterprise, as in the case of Fougeroux de Bondaroy. The abbé Baudeau withdrew from the dictionary of commerce, which went to Guillaume Grivel, a lawyer; and Digeon had to be replaced on the dictionary of finance by Rousselot de Surgy, a retired premier commis des finances.[11] Vicq d'Azyr remained in charge of the huge dictionary of medicine, but he contracted the writing out to so many colleagues that he lost control of it. Most of his collaborators failed to provide their copy on time, and those who made their deadlines suffered from the negligence of those who did not; for Panckoucke would not pay until the entire manuscript was ready for printing. This situation produced some discontent among men like Fourcroy, who complained that his work went unrewarded,[12] and ultimately Panckoucke had to rebuke Vicq for negligence:

Tant que vous compterez sur vos associés, vous n'irez point. Vous ne ferez pas un volume en 4 ans. Je vous avais donné le projet d'une

pour l'ouvrage de M. Watelet, qui est sous presse depuis trois ans. Je suis en avance sur cette édition de plus de quarante mille livres. Je vous supplie, Monsieur le comte, de donner vos ordres, afin que les manuscrits soient délivrés sans délai à M. Dussault.''

11. See the contracts of Grivel, Digeon, and Rousselot de Surgy in Kenneth Spencer Research Library, University of Kansas, ms. 99.

12. Fourcroy worked on the dictionary of medicine before taking over the dictionary of chemistry in May 1791. By that time he depended on the payments from Panckoucke and protested when they did not arrive: ''Vous savez que j'ai travaillé sans relâche à l'*Encyclopédie* depuis 1787. Je me trouve avoir fait actuellement la matière de plus d'un volume, et cependant je n'ai pas encore reçu 400 livres, parce qu'il n'y a pas en effet le 10ème de mon travail imprimé. J'ai cru que d'après nos conventions de 1789 on irait beaucoup plus vite qu'on n'avait été jusques là, et je ne me suis pas plaint. Vous avez été témoin de l'ardeur que j'ai mise à défendre les intérêts de l'*Encyclopédie* et du zèle que j'ai tâché d'allumer parmi mes confrères. Quoique j'aie en partie réussi, puisqu'on imprime actuellement trois demis volumes à la fois, cela va si lentement relativement à ce que j'ai fait en mon particulier que les avances de mon temps et de mon travail me deviennent onéreuses et que je ne puis pas continuer avec la même activité sans être en partie indemnisé . . . L'état de mes affaires exige que je reçoive une partie du prix de mon travail à mesure qu'il est terminé. Je vous prie donc de me faire savoir si je puis vous demander 150 livres par mois d'ici à un an ou dix-huit mois. J'ai fait au moins pour 4,000 livres de copie remise à M. Faure et déposée chez M. Vicq. Je travaille toujours à la suite, et j'en suis à la lettre D, tandis qu'on imprime tout doucement la lettre A, qui n'est pas encore finie . . . Il faut bien que je sois sûr d'une certaine somme par an pour que je puisse me livrer à ce travail avec la même ardeur.'' Panckoucke scribbled on the top of this letter, ''Je prie Monsieur Vicq d'Azyr de prendre lecture de cette lettre et lui réitère ma prière pour me donner jour et heure pour régler nos affaires.'' Fourcroy to Panckoucke, Oct. 31, 1790, Bibliothèque historique de la ville de Paris, ms. 815.

lettre circulaire que vous aviez d'abord approuvé et que vous ne m'avez pas renvoyé. Je n'irai voir ni M. Goulin, ni M. Caille, ni qui que ce soit, que vous, Monsieur, avec qui je serais charmé de correspondre. Comment pouvez-vous espérer, Monsieur, d'avoir de la copie à temps de tous vos coopérateurs, lorsque je n'ai jamais pu concilier les auteurs de la géographie, qui n'étaient que deux et qu'il a fallu séparer? Si vous comptez sur l'exactitude de vos coopérateurs, ils désoleront votre vie; vous désolerez le public, les souscripteurs; et vous m'empêcherez de remplir mes engagements. Il faut faire votre livre avec vos lumières ou avec les livres qui existent, en ayant à vos ordres deux à trois secrétaires intelligents.[13]

The deaths and broken deadlines forced Panckoucke to revise his production schedule over and over again, making the date for the completion of the book seem increasingly remote. He did not keep his promise to publish six volumes by the end of 1782. In fact, he did not even publish the first installment until November of that year; and when it came out, it disappointed a great many subscribers because it contained only the first volume of jurisprudence, half of the first volume of *arts et métiers,* and half of the first volume of natural history. Half-volumes could hardly be read, much less bound; but Panckoucke used them in most of his installments. In January 1783 he issued the second installment: the first half-volume of volume 1 of literature, and the first half-volume of volume 1 of geography. In April 1783 he issued the third: the first volume of plates, the first volume of commerce, and the second half-volume of volume 1 of *arts et métiers.* In August 1783 he issued the fourth: the second half-volume of volume 1 of geography, the second half-volume of volume 2 of jurisprudence, the second half-volume of volume 1 of literature, and the first half-volume of volume 2 of *arts et métiers.* He continued in this fashion until the very end. It suited him because he could hurry his authors into providing at least part of their copy and could publish many parts at the same time, thereby advancing all of his dictionaries along a common front, instead of stringing them out one behind the other. Not only would publication by disparate half-volumes squeeze capital more quickly from the subscribers (Panckoucke sold two half-volumes for the price of one whole volume), but more important it would foil the pirates; for it would be difficult to counterfeit and market half-volumes that came out at widely

13. Panckoucke to Vicq d'Azyr, undated, Archives de Paris, 8AZ 278.

spaced intervals. The only objections to Panckoucke's strategy came from the subscribers, who had to keep pile after pile of loose sheets sorted and stocked pending the day when they could be assembled, bound, and shelved as a coherent set of volumes.[14]

That day seemed farther and farther away, as Panckoucke got deeper and deeper into his dictionaries. Although he continued to put out a few half-volumes every two or three months, he clearly had fallen behind schedule by May 1786, when he had issued only thirty-one volumes of text. It seemed impossible for him to produce the twenty-two remaining volumes by his final deadline, July 1787.[15] And what was more distressing, he could tell from the ground he had covered that a much longer road lay ahead: he could not get all of his *Encyclopédie* into fifty-three volumes of text. He even faced an overrun of plates. He had produced five of the promised seven volumes of plates and had not yet exhausted *arts et métiers* nor begun other subjects, notably beaux-arts and natural history, which required a great many illustrations. What had gone wrong? The fault was Diderot's, Panckoucke explained in a note to his subscribers of May 1786. In circumnavigating their subjects, the new Encyclopedists had discovered far more deficiencies in the work of their predecessors than they had anticipated. They felt obligated to compensate for the shortcomings of the original *Encyclopédie* by extending the scope of their own. ''C'est la grande imperfection de cette première *Encyclopédie,* reconnue et avouée par M. Diderot lui-même, qui a nécessité une augmentation de volumes,'' Panckoucke concluded. He had already made a great deal of this argument in his original advertising, but he re-emphasized it at this point in order to present the escalation of his *Encyclopédie* as a boon rather than an additional expense to the subscribers. It would be not twice but three times as long as

14. The dates and contents of the first twenty-six installments can be ascertained by compiling information from Panckoucke's Avis and his notice entitled ''Epoques où ont paru les vingt-six premières livraisons,'' all of which he reprinted in *Beaux-Arts,* I, lxii–xcvii. That the subscribers objected to Panckoucke's publishing strategy is clear from his attempts to answer their complaints in his Avis, notably the Avis on the fourth installment, ibid., lxiv.

15. The prospectus of May 1, 1782, committed Panckoucke to complete production ''dans cinq ans, à compter du premier juillet de cette année.'' In a subsequent clause, however, it said that the subscribers would receive the complete set by Dec. 1787 ''au plus tard.'' *Beaux-Arts,* I, lviii.

Diderot's, he wrote triumphantly; it would have not 30,000 but 100,000 additional articles; it would contain not twenty-six but thirty definitive dictionary-treatises . . . and so it would include not three or four extra volumes, as he had anticipated earlier, but twenty. The additional cost of 6 livres per volume was of little importance, considering that the subscribers would receive the greatest book of the century, a book that would be more valuable than a 20,000-volume library.[16]

Panckoucke's rhetoric inflated with his price; and it did not persuade all the subscribers, for the Avis accompanying the next installments took on a beleaguered tone. They dwelt increasingly on the subscribers' criticism and complaints and presented Panckoucke as a kind of martyr, staggering under the weight of his responsibilities: "Le public s'est engagé dans cette grande entreprise par la confiance qu'il a eue en nous et que nous croyons avoir méritée par vingt-cinq années de travaux, qui souvent lui ont été utiles. Qu'il nous la continue cette confiance, qu'il seconde notre zèle et notre courage; nous en avons besoin pour soutenir le poids de cette énorme entreprise."[17] By December 1786, it was clear that the "Atlas de la librairie" had stumbled into another crisis. He did not produce any volumes for the next six months. When he resumed publication in May 1787, he had only reached volume 38 of the text. But although he had fallen hopelessly behind in production, he resolved to forge ahead—and in fact to expand from thirty to thirty-six dictionaries. Heraldry now emerged as a separate dictionary from within history, artillery from within war, music and architecture from within beaux-arts. New subjects, like *bois et forêts* and *arts académiques* (riding, dueling, dancing, swimming), had demanded dictionaries of their own. And the old dictionaries had continued to grow: finance from one to three volumes, literature from two to three, botany from two to five, jurisprudence from three to eight, *arts et métiers* from four to ten. Panckoucke's stable of authors had also expanded: he had eighteen men working on medicine alone, and a basic staff of seventy-three. He pushed them as hard as he could, he explained, prodding them with notes about "les craintes et les alarmes des souscripteurs."[18] Relations must have been strained, because the writers could

---

16. Avis of May 1786, *Beaux-Arts*, I, lxxvi.
17. Avis of May 1787, *Beaux-Arts*, I, lxxix.
18. Ibid.

not produce copy fast enough to satisfy their employer. "Notre *Encyclopédie* va mal," Panckoucke had confided to a friend in 1783.[19] "Les auteurs ne travaillent point et cela me désole." Publicly he attributed all delays to deficiencies in the original *Encyclopédie,* "où tout est un véritable chaos." "Croit-on qu'on fasse un bon livre dans un temps déterminé, comme une pièce d'étoffe?" he asked his subscribers in his Avis of May 1787. He did not remind them that in 1782 he had established a putting-out system for his authors by a series of contracts, which had specified the number of volumes to be written and the time allotted for the writing. But the subscribers had only to consult the prospectus to learn that Panckoucke had arranged for all copy to be in by July 1785 so that he could finish the printing by July 1787.[20]

In October 1787 Panckoucke made another attempt to save the speculation. He realized that even if he persuaded the subscribers to swallow their anger over the breaking of the deadline, he would have enormous difficulty in getting them to pay for the overrun in production. By that time he had published forty-two of the seventy-three projected volumes of text, and the subscribers had paid 644 livres apiece. The next installment would take them up to or over the 672 livres that most of them had contracted to pay. How could he prevent them from stopping their payments and demanding the surplus volumes free, as he had originally promised? Instead of meeting that problem head-on, he tried to skirt it by a tactic that he called a *combinaison.* In a special appeal, he offered to provide the subscribers with two new publications, an encyclopedic atlas and a series of plates on natural history. Although this supplement was optional, no *Encyclopédie* would be complete without it, he insisted; the subscribers could receive it, at a special reduced rate, along with their regular installments—provided they kept up with their payments. Most of the subscribers accepted; so, as Panckoucke later remarked, the maneuver "saved" the *Méthodique* by preventing an interruption in the flow of capital.[21]

19. Panckoucke to comte de Lacepède, Aug. 24, 1783, in Roger Hahn, "Sur les débuts de la carrière scientifique de Lacepède," *Revue d'histoire des sciences,* XXVII (1974), 352.

20. Panckoucke described the terms of the original contracts in his prospectus of May 1782, which he reprinted in *Beaux-Arts,* I, lviii.

21. "Réponse de M. Panckoucke à M. le comte d'Hulst" in *Mathématiques,* III, xix. Panckoucke provided a somewhat confusing explanation of his proposal

In fact, the supplementary art work merely postponed Panckoucke's reckoning with the subscribers for less than a year. The crisis came to a head again in November 1788. Panckoucke was then preparing the thirtieth installment, which would contain the fifty-third volume of text—that is, the final volume, according to the prospectus. Far from having reached the end of his *Encyclopédie*, however, he realized that he had come only half way. He now estimated the overrun at forty-six to forty-eight volumes rather than twenty, which would make the text about a hundred volumes in all. He was in a "position extrêmement critique," he explained in a circular letter to his authors. Subscribers were overwhelming him with complaints. Some threatened to sue if he did not return their money. The Spanish contingent had canceled its 330 orders. Five hundred others had simply stopped claiming their installments—a kind of subscribers' strike, which cut his income drastically. Instead of relying on the installment payments to finance production, he had to draw on his own capital. He had gone 150,000 livres into the red and was running a deficit of 60,000 livres a year. Expenses were rising, losses accumulating everywhere. "Mille événements que je n'ai pu ni prévoir ni calculer m'ont convaincu, qu'après m'être chargé de la plus grande et de la plus pénible tâche dont aucun libraire se soit jamais avisé, il pouvait ne me rester que le désespoir de l'avoir entreprise."

Still, he would make a last attempt to rescue their *Encyclopédie*. He asked each author to speed up the flow of copy. Those who could not finish their texts within three or four years should resign and find replacements. Those who could rally behind him should prepare a statement on the extent of their work and the time it would be finished. Panckoucke

---

in his Avis of Oct. 1787 in *Beaux-Arts*, I, lxxxviii–xcii. Essentially, he offered to credit the payments made by the 672-livre subscribers to their accounts, until an eventual settling, when they would be paid off in surplus volumes of the text. But he made it appear as if the atlas and plates would be a bonus given to them in order to fill the 79-livre gap between the cost of their subscriptions and those of the 751-livre subscribers. Actually, as he came close to admitting, he was merely attempting to pacify them, "de sorte que les payements courants continueront d'avoir lieu." Ibid., p. xci. He also slipped into the Avis a claim that volumes 5 and 6 of the plates contained 169 plates more than he had promised to supply, which entitled him to debit the subscribers accordingly.

would assemble the statements into a *Tableau,* showing exactly what remained to be done, dictionary by dictionary, so that he could win back the subscribers who doubted that he would ever complete the book. He would issue the *Tableau* with another special appeal to the subscribers, asking for an extension of three or four years and forty-six or forty-eight volumes. He stood ready to sacrifice "ma vie et ma fortune" for the supreme *Encyclopédie;* and with the authors' support, he still felt confident that it could be saved.[22]

Panckoucke published his appeal in March 1789. He warned the subscribers that they had to confront a question of life or death for their *Encyclopédie.* Threatened by lawsuits, deluged by complaints, overwhelmed by costs, he would have to abandon the enterprise unless they accepted some radical measures to save it. First, he asked them to pay a "supplément de souscription" of 36 livres, which he would credit against the final installment. Then he proposed to reopen the subscription and to furnish the new subscribers with the unclaimed volumes of the 500 old subscribers who had failed to pay for their installments and who would be struck off his list unless they caught up with their payments within two months. And finally, he announced that he would issue the dictionaries as separate publications, to be financed by still more subscriptions. In this way, he could attract customers with special interests who did not want to buy the entire set of dictionaries, and he could supply them from his overstocked inventory.[23]

Unfortunately, each of those steps violated the terms of the prospectus and exposed Panckoucke to lawsuits from the subscribers. In order to head off attacks on that front, he turned once again to his protectors in Versailles. An unsigned memorandum by a subordinate of the Garde des sceaux advised the head of the judicial system to provide Panckoucke with legal shelter: "Le sieur Panckoucke méritant la protection de l'ad-

---

22. "Lettre de M. Panckoucke en date de novembre 1788," *Mathématiques,* III, xiii–xv. In this letter, Panckoucke set the loss of the Spanish subscriptions at 300. But in his "Lettre de M. Panckoucke à Messieurs les souscripteurs" (1791) in *Histoire,* V, 1, he wrote: "L'Inquisition s'y est emparée de mes magasins. J'y ai perdu 330 souscriptions."

23. Avis of March 28, 1789, in *Manufactures,* III, x–xi; Panckoucke, *Abrégé des représentations et du mémoire sur l'Encyclopédie* (Paris, 1789) in Case Wing Z 45.18, ser. 7, Newberry Library; and *Mercure,* March 7 and 30, 1789.

ministration par son courage et par la manière dont cet important ouvrage a été exécuté, je proposerais à Monseigneur de permettre la continuation de la souscription et d'autoriser le sieur Panckoucke à demander un supplément aux souscripteurs.'' An *arrêt du conseil* followed in February, sanctioning all of Panckoucke's maneuvers. And in March Panckoucke published a letter from the Directeur de la librairie, which attested that he had done nothing illegal in the eyes of the officials who administered the law. It was a classic case of deploying protections and invoking privilege (or ''private law''), just two months before the opening of the Estates General.[24]

Aside from persuading his subscribers to put up with this latest *combinaison*, Panckoucke had to win their support for a much more unpalatable policy in the spring of 1789: he asked them to accept his decision to expand the text to 124 volumes in 51 dictionaries rather than 100 volumes in 36 dictionaries, as he had indicated in his last estimate, and to extend the deadline until 1791. He did not disguise the fact that he had originally promised to compress the text into 53 volumes and to publish it by 1788. But he insisted that anyone who read the *Tableau* accompanying his appeal would realize that his authors could not do justice to their subjects without writing far more than could have been foreseen in 1782. The *Tableau* substantiated this argument by a fifty-page survey of all the ground to be covered in the *Méthodique*. Each Encyclopedist explained why he needed more space and time— and in doing so provided some revealing information about the way he worked.

Guyton de Morveau described the difficulty of writing articles on chemistry at a time when his science was being revolutionized: ''Je ne sais comment font ceux qui fournissent tous les ans un volume; pour moi, je ne sors pas de chez moi, je ne fais pas autre chose, je suis tout entier à cette besogne, au point de négliger même mes affaires domestiques, et je n'avance point. Quand j'arrive à un article, je trouve dans mes recueils trois ou quatre fois sa longueur de notes, de matériaux, de parties toutes rédigées, et il me faut des semaines,

24. The memorandum is in the Archives Nationales, V¹549, fol. 334; the *arrêt* at fol. 357. This box contains several reports on the *Méthodique* that show how effectively Panckoucke mobilized his protectors in early 1789.

des mois pour l'amener au point que je désire.''[25] Similarly, Monge explained that he had to delay the dictionary of physics until he could see his way clear to the essence of the obscure *fluides élastiques* that had just emerged from outdated notions of air, fire, and water.[26] Vicq d'Azyr found the medical sciences in a shambles and felt compelled to piece together his own system of comparative anatomy, a much longer task than he had anticipated. Lamarck used a similar argument to justify the expansion of the dictionary of botany. No one, not even Linnaeus, had ever attempted such a systematic account of the entire world of plants, he asserted. It would be a great feat to finish by 1791 and to keep the work down to five volumes instead of the two that had been envisaged earlier (it eventually ran to thirteen). ''Si nous exécutons exactement notre plan,'' Panckoucke concluded, ''on doit y trouver tout ce que les hommes ont conçu, imaginé, créé depuis que l'art d'écrire est inventé. Il ne doit point y avoir un seul mot, un seul objet des connaissances humaines sur lesquels on ne doive trouver des détails satisfaisantes.''[27]

Panckoucke probably believed this propaganda, but it had a purpose: to persuade the subscribers to pay for almost three times as many volumes as they had contracted to buy. Many of them had receipts signed by Panckoucke, acknowledging their down payment of 36 livres on ''672 livres prix d'un exemplaire complet.''[28] They protested, citing the terms of Panckoucke's first prospectus, which promised to supply all extra volumes free, ''afin que les souscripteurs soient bien assurés qu'on ne veut profiter, en aucune manière, de cette indétermination pour augmenter à volonté, comme cela est quelquefois arrivé, le nombre des volumes.''[29] That clause must have been intended to disassociate the *Méthodique* from the bad reputation of the quarto, because nothing had infuriated Duplain's subscribers more than the extra costs inflicted

25. ''Tableau et apperçu,'' *Mathématiques*, III, 5. Panckoucke added that the dictionary of chemistry would run to three or four volumes instead of two, as originally planned: ''La chimie ayant entièrement changé depuis quelques années, il n'était pas possible d'offrir au public une simple refonte d'un ouvrage imprimé il y a plus de vingt ans.''

26. Ibid., p. 2.

27. Avis of Oct. 1787 in *Beaux-Arts*, I, xci and ''Représentations du sieur Panckoucke, entrepreneur de l'*Encyclopédie méthodique*, à Messieurs les souscripteurs de cet ouvrage,'' *Mathématiques*, III, xiii.

28. See the model receipt in the *Mercure* of Dec. 8, 1781, p. 153.

29. Ibid., 155.

on them when he stretched the text from twenty-nine to thirty-six volumes. In fact, Panckoucke tried to promote the *Méthodique* by contrasting it with the quarto, which he condemned for shoddiness, as if he had never had anything to do with it.[30] But soon after launching the *Méthodique,* he showed that he had no more scruples than Duplain about manipulating subscribers and maneuvering around obligations. In May 1782, he tried to cancel his commitment to limit the size of the new *Encyclopédie* by inserting a clause in his second prospectus stating that although he planned to publish the text in fifty-three volumes, it could run to fifty-seven—and "si, contre toute attente et pour la perfection de l'ouvrage, nous étions nécessités à un plus grand nombre de volumes de discours, les souscripteurs ne payeront ces derniers volumes que 6 livres, au lieu de 11 livres.''[31] He also struck the compromising phrase ''prix d'un exemplaire complet'' out of his next batch of receipts. And as he had given the first subscribers the choice of withdrawing their orders or accepting the terms of the second subscription, he considered himself immune from prosecution.

Nonetheless, a great many subscribers felt swindled. They had received and paid for fifty-three volumes; and when each additional installment appeared, they bombarded Panckoucke with ''cent lettres de plaintes'' or simply refused to claim their books.[32] A bookseller from Dijon called Mailly whipped up their indignation in a pamphlet that accused Panckoucke of profit-gouging. Each volume cost him only 4 livres to produce, Mailly argued, so Panckoucke made 7 livres on every extra volume that he issued. To prevent themselves from being fleeced, the subscribers should insist on receiving the rest of their sets free—and even then, Panckoucke would clear a 33 percent profit. Panckoucke replied that it cost him 6 livres to produce each volume, 7 counting his losses from unforeseen circumstances like the cancellation of the Spanish sales. Most subscriptions were sold through bookdealers, who took 2 livres in commission for every volume, leaving Panckoucke with less than 2 of the 11 livres that the subscribers paid for each vol-

---

30. *Abrégé des représentations,* 19.

31. Prospectus of May 1, 1782, in *Beaux-Arts,* I, lviii. Panckoucke also published the original prospectus in this volume, but he omitted the crucial clause about the extra volumes.

32. Panckoucke, ''Représentations,'' *Mathématiques,* III, xi.

ume. He was going to sell the extra volumes at only 6 livres each, so he would be the main one to lose by the expansion of the *Encyclopédie*.[33]

Other subscribers, however, thought he was robbing them in other ways. Several of them protested that the volumes contained far less than the "environ cent feuilles" promised in the prospectus. The difference between a quarto volume of 95 sheets (760 pages) and a quarto volume of 100 sheets (800 pages might seem trivial, and it certainly was difficult to perceive, given Panckoucke's practice of issuing half-volumes; but it could cut production costs by huge sums, considering that Panckoucke planned to publish 5,000 sets of 124 volumes. After some tortured reasoning, he conceded that the volumes contained an average of about 95 sheets. At that rate, he could have saved at least 100,440 livres, at the subscribers' expense.[34] So he might well have been trying to stay out of the red by surreptitiously thinning the volumes. It seems hard to believe that he would expand his *Encyclopédie*, if he expected to lose money on each of the sixty-six additional volumes that he planned to issue. But without his account books and commercial correspondence, one can only guess at his strategy, and it seems reasonable to believe his claims that he had his back to the wall. "Cet ouvrage a beaucoup d'ennemis, nous ne l'ignorons pas," Panckoucke wrote in his appeal to the subscribers. He spoke darkly of hostile pamphleteering, calumnies and "libelles." What worried him most, he confessed, were lawsuits. He felt haunted by the case of Luneau de Boisjermain against the publishers of the first *Encyclopédie:* "Un procès, dans notre position, nous paraîtrait une si suprême injustice, et le sort de deux de nos confrères, qui sont morts de chagrin et de désespoir des difficultés qu'ils ont éprouvées, a tellement frappé notre esprit que, si l'on prétend les renou-

33. Panckoucke, "Sur les prétendus bénéfices actuels de cet ouvrage," *Mathématiques*, III, xv–xvii.

34. At a pressrun of 5,000, Panckoucke could have saved 6,200 reams, worth at least 62,000 livres, by cutting each volume by 5 sheets. And as his labor costs would have come to at least 62 livres per *feuille d'impression,* he would have saved another 38,440 livres on the printing. These estimates are based on the production costs of the quarto *Encyclopédie* and therefore are conservative. In the *Mercure* of March 6, 1790, Panckoucke said that his printing costs came to 40 livres 5 sous for the first thousand impressions—far more than the costs of the quarto (30 livres for the first thousand impressions), as one would expect for work done in Paris.

veler à notre égard, nous regardons l'*Encyclopédie* comme détruite et anéantie.''[35]

Essentially, Panckoucke was pleading for the subscribers to accept the doubling and delaying of the *Méthodique,* to continue their payments, and to refrain from tying him up in court while he launched his new subscription schemes: "Il faut qu'on nous laisse toute notre liberté, si l'on veut que l'ouvrage s'achève promptement. Nous avons à peine assez de toutes nos forces pour suivre tous les mouvements, tous les rapports de cette grande machine; pour vaincre les obstacles; pour solliciter, presser les gens de lettres, les imprimeurs, les graveurs; pour répondre enfin aux souscripteurs.''[36] Panckoucke's talk of being harried into his grave by lawsuits may have been an attempt to win sympathy and to dramatize his position, but his position did indeed look critical and he did indeed have to fend off attacks in the courts.

In early 1789 a merchant in Nancy called Pichancourt sued the bookseller Bonthoux, who had sold him a subscription to the *Méthodique.* Pichancourt demanded that he be given the complete encyclopedia, on time and at the price set in the prospectus, or his money back. Arguing that Panckoucke should assume the responsibility for all damages caused by his failure to honor his commitments, Bonthoux got the *présidial* court of Nancy to summon him as a co-defendant. But Panckoucke refused to appear on the grounds that similar cases were being prepared in several other cities and that he could not defend himself everywhere at once. Instead, he fell back for the last time on his protectors in the government. He petitioned the king's council to save his encyclopedia—a work of immeasurable importance to the kingdom, "qui a la sanction de Sa Majesté"—and he won an *arrêt du conseil* that forbade the provincial courts to hear any cases concerning the *Méthodique,* whether they involved local booksellers or Panckoucke himself. All suits were to go directly to the Châtelet

---

35. ''Sur les prétendus bénéfices,'' *Mathématiques,* III, xvi. Panckoucke dwelt on the same theme in his ''Représentations,'' ibid., xii: ''Nous ne pourrions même supporter l'idée d'un procès dans l'avenir. Il est nécessaire que nous ne vivions pas dans la crainte de voir un jour renouveler les attaques que le défaut de prévoyance a suscitées aux premiers entrepreneurs et qui ont fait mourir de chagrin et de douleur les sieurs Briasson père et fils et abrégé les jours du sieur Lebreton.''

36. Ibid., p. x.

in Paris, where Panckoucke could exonerate himself once and for all, and anyone who sought justice elsewhere would be liable to the nullification of his plea and a thousand livre fine.[37]

That edict apparently put an end to Panckoucke's legal difficulties, but he could not get the king to decree that all subscribers pick up and pay for their installments. By March 1789, 500 of the 4,850 subscribers had stopped their payments. Yet Panckoucke needed the income from 4,000 subscriptions merely to cover costs, or so he claimed in his appeal, which warned the subscribers that attrition would kill the *Méthodique* if they did not stand by it.[38]

But the subscribers who had originally signed up for forty-two quarto volumes might well fear that the book would run to more than 200 and would not be finished in their lifetime. The *Méthodique* was growing like a monstrous weed, swarming across shelf after shelf and sending out tendrils that could be extended indefinitely, as long as the arts and sciences themselves continued to grow. The financial stakes of the enterprise were as staggering as its scale. Each subscriber was now being drawn into an expenditure of at least 1,422 livres.[39] If Panckoucke held on to only 4,000 subscriptions, he would collect 5,688,000 livres. So he was not exaggerating when he wrote that his encyclopedia was "de toutes les entreprises, la plus grande qu'on ait jamais exécutée dans la librairie."[40] It made Diderot's *Encyclopédie* look small, just as Diderot's seventeen-volume text had dwarfed its own predecessor, the two-volume *Cyclopaedia* of Ephraim Chambers. Panckoucke boasted that his "édifice . . . ne ressemble pas plus à l'ancienne (*Encyclopédie*) que le palais du Louvre à une chau-

37. *Arrêt du conseil* of Sept. 23, 1789, Archives Nationales, V⁶1145. See also the anonymous memorandum, presumably by a subordinate of the Garde des Sceaux who strongly favored Panckoucke, dated Sept. 9, 1789 in V¹553.

38. "Lettre de M. Panckoucke à Messieurs les souscripteurs," *Histoire*, V, 3.

39. Panckoucke, *Abrégé des représentations*, p. 14.

40. Panckoucke, *Représentations*, p. i. Of course one should make allowances for Panckoucke's hyperbolic style. There were other gigantic books in the seventeenth and eighteenth centuries, notably the sixty-four volume *Grosses vollständiges Universal-Lexicon aller Wissenschaften und Künste* (Halle and Leipzig, 1732–50), but it seems doubtful that any of them reached a pressrun of 5,000. Similarly, the unusually large editions probably occurred for the most part in the printing of relatively short works such as the *Gentleman's Magazine*, which Charles Ackers of London put out at about 10,000. See D. F. McKenzie and J. C. Ross, *A Ledger of Charles Ackers, Printer of the London Magazine* (London, 1968), pp. 12–18.

mière ou Saint-Pierre de Rome à une chapelle.''[41] But more breathtaking than its size were the risks it entailed. A sensible man would have cut his losses by trimming the speculation. A cautious man would have abandoned it. Panckoucke more than doubled its scale, and he made this gamble, the greatest in his life, just as France was erupting in revolution.

From Encyclopedism to Jacobinism

Panckoucke realized in retrospect that his timing could not have been worse: ''Nous touchions alors à un événement à jamais mémorable . . . La Revolution, qui n'a point tardé à éclater, qui a renversé tant d'états, de fortunes, détruit les plus brillantes espérances, m'a attaqué dans tous les sens.''[42] Actually, the Revolution did not openly attack Panckoucke, but it damaged his *Encyclopédie* in three ways: it drove off many of the subscribers, scattered several of the authors, and ruined most of the printers. Yet it did not destroy the book. The story of how the *Méthodique* and its authors and publisher weathered the Revolution suggests some of the complexities in the transitions from Encyclopedism to Jacobinism.

The convocation of the Estates General produced an enthusiastic response from Panckoucke. As a member of the electoral assembly for the Third Estate of Paris, he helped draft the Parisian cahier and took a strong stand on the basic demand of the ''Patriots'': the Third Estate should insist on voting by head rather than by order, he argued, even if it had to constitute itself separately as the nation.[43] Far from anticipating that the new system of government would hurt his *Encyclopédie,* he announced on April 27, 1789, that he had taken measures to expand and speed up production. Thanks to new arrangements with twenty paper manufacturers, forty engravers, and twenty-five of the thirty-six legally established printers in Paris, he soon would be able to put out two to four volumes every month. As it normally took about a year for one shop to print one of his huge quarto volumes, Panckoucke clearly meant to produce *Encyclopédies* at an extraordinary rate and on a staggering scale. He was attempting to tie up

41. *Mercure,* Dec. 15, 1792.
42. ''Lettre de M. Panckoucke à Messieurs les souscripteurs,'' *Histoire,* V, 2.
43. Panckoucke, *Observations sur l'article important de la votation par ordre ou par tête,* as cited by Panckoucke in *Mercure,* Nov. 21, 1789, pp. 81–82.

most of the (legal) printing capacity of the capital in a supreme effort to double the size of the biggest venture in the history of French publishing just a week before the opening of the Estates General and two and a half months before the fall of the Bastille.[44]

When the Bastille fell, it brought down the ancient system for producing and policing the printed word in France. In place of the thirty-six privileged printing shops, presses sprang up everywhere in Paris; and instead of issuing fine volumes, they turned out political tracts and newspapers. Two hundred and fifty newspapers burst into print during the last six months of 1789, ignoring the old censors, the old booksellers' guild, and the old privileges of journals like Panckoucke's *Mercure* and his *Gazette de France*.[45] To write and print so many newssheets required not merely the destruction of the old restraints on the press but the transformation of working conditions in the publishing industry. By August, Panckoucke realized that the revolution in the printing shops would damage his *Encyclopédie,* because the printers had shelved his copy in order to satisfy the demand for news about the National Assembly.[46] By November, he sounded as though he were fighting for his life:

Je suis peut-être, et je dois le dire aujourd'hui, le citoyen sur qui pèse le plus violemment la révolution; car il n'y a pas de mois où mes dépenses n'excèdent mes recettes de plus de 25,000 livres; mais j'ai cru devoir redoubler d'efforts, de vues, de moyens de crédit, de combinaisons, pour soutenir une machine énorme, à laquelle le sort de plus de six cent personnes est lié, persuadé qu'il était impossible, vu les

44. Avis of April 27, 1789, in *Manufactures,* III, xv–xvi. It is impossible to know how many presses existed in Paris in 1789 and how many of them were working on the *Méthodique,* but Panckoucke's printing job would have dominated the industry if the Revolution had not broken out. The typesetting on the *Méthodique* took a long time because the quarto volumes were thick (about 760 pages), double-columned, and set in small type (*petit romain*). As Panckoucke's edition contained 5,000 copies and ordinary pressruns usually came to 1,000–1,500, the printing of each volume probably took at least three times as long as that of most books—and Panckoucke was planning to produce 124 volumes, more or less at the same time. For his statement that it took more than a year to print one volume see *Beaux-Arts,* I, lxvi.

45. See Eugène Hatin, *Histoire politique et littéraire de la presse en France* (Paris, 1859), chaps. 2–8 and Claude Bellanger, Jacques Godechot, Pierre Guiral, and Fernand Terrou, *Histoire générale de la presse française* (Paris, 1969), I, 405–486.

46. Avis of Aug. 31, 1789, *Manufactures,* III, xviii.

ressources immenses de ce grand empire, que les choses ne reprissent pas leur cours ordinaire.[47]

In early 1790, Panckoucke lamented a general "désertion des ouvriers," who had left their old masters in order to put out the new journals, working for higher wages, by day and by night, in new establishments all over Paris. The old shops had either gone over to journalism or gone out of business. One of his principal producers, who used to print ten to twelve of the large, double-columned sheets of the *Méthodique* every week, had suspended operations, and Panckoucke 'found it impossible to stop the depletion of his own resources, although he sounded as though he took his losses like a patriot: "Personne ne souffre plus que nous de la révolution; mais il faut savoir souffrir pour la patrie."[48] At the end of February, he abandoned his speculation on separate editions of the dictionaries. He had not collected 20,000 subscriptions, as he had hoped, but 162—not enough to cover the costs of publishing the prospectuses. And instead of attracting new subscribers, the *Encyclopédie* as a whole continued to lose old ones: for the desertion rate increased as a result of the economic difficulties and the emigration. Noting delicately that some of his subscribers were now *absents du royaume*, Panckoucke extended the deadline that he had set for his customers to pick up their unclaimed installments without forfeiting their subscriptions —and he continued to extend it for the next few years, until it ceased to have much meaning. What these developments really meant, Panckoucke acknowledged, was that publishing of the kind that had flourished under the Old Regime had become unviable in the Revolution. "Le commerce de librairie a, pour ainsi dire, été anéanti, les principales maisons de la capitale obligées de suspendre leurs paiements," he wrote in 1791. "Nombre d'auteurs ont été détournés de leurs travaux par des fonctions publiques; toutes les imprimeries ne furent bientôt plus occupées que de brochures, pamphlets et surtout

---

47. *Mercure*, Oct. 24, 1789.
48. Avis of Feb. 8, 1790, *Manufactures*, III, xxii. In the few private letters of Panckoucke that survive from this period, he sounded less patriotic but no less beleaguered financially. On Oct. 22, 1789, he wrote to Senebier, "Nous sommes ici dans les alarmes continuelles. Le bien est encore incertain et le mal est affreux. Paye qui veut. Les tribunaux sont sans action. Il m'en coûte cent mille écus de ma fortune depuis dix mois pour soutenir l'*Encylopédie*. Si j'avais le malheur de suspendre, l'ouvrage serait détruit et je perdrais l'espérance de recouvrer mes dépenses." Archives de Paris, 8AZ 278.

de journaux de toute espèce, dont le nombre, dans la seule capitale, se monte à plus de cent. J'ai vu le moment où l'*Ency-clopédie* allait être abandonnée."[49]

Panckoucke only survived because he had shifted from book publishing to journalism before 1789—with the notable exception of the *Méthodique*—and because he swam with the tide of revolutionary journalism after July 14. In November 1789, he founded the *Moniteur universel,* which became the most important journal for parliamentary news in the Revolution. In June 1790, he launched *Le Gazzetin,* a less influential and more radical paper, which served as an antidote to his semiofficial *Gazette de France.* But the *Méthodique* continued to be his supreme speculation—the "grande entreprise où j'ai mis toute ma fortune," as he called it[50]—and he had to find some way to save it in 1790. As in his earlier crises, he cast about for a *combinaison.* He found that the demand for artists and engravers had not kept up with the demand for typographers, so he compensated for the decline in the output of the volumes of text by increasing the production of the plates. By this maneuver, he continued to put out installments and to bring in capital; most important, he prevented a massive hemorrhage of subscriptions, for he was certain that the subscribers would withdraw in droves if he interrupted production.[51] Panckoucke also tried to win over subscribers at this time by producing his gimmicky dictionaries: the *Encyclopédiana,* the anthology of tricks and riddles, the work on the National Assembly, and a new atlas, which would show how the Assembly was redrawing the map of France. Finally, he tried to make peace with his printers.

Paradoxically, Panckoucke found it harder to get the *Méthodique* printed when there were over 200 shops in Paris than when there were 36. If the master printers did not return his copy or shelve it, they demanded 5 to 6 livres more per printed sheet (*feuille d'impression*), that is, 500 to 600 livres more for producing each volume. They explained that they had to pay their workers higher wages. The wage issue needed to be handled with care, for it had brought the production of the *Méthodique* to a halt for a while in late 1785 and 1786, when

49. "Lettre de M. Panckoucke à Messieurs les souscripteurs," *Histoire,* V, 2–3. See also the *Mercure* of Feb. 27, 1790.

50. Avis of April 27, 1789, *Manufactures,* III, xiv.

51. "Mémoire en faveur de M. Panckoucke" in *Mercure,* Dec. 4, 1790, and "Lettre de M. Panckoucke," *Histoire,* V, 2–3.

a dispute over pay in the shop of P.-G. Simon, who was printing the *Dictionnaire des finances,* developed into a general struggle between the compositors, who refused to work on Panckoucke's copy everywhere in Paris, and the masters, who blacklisted and imprisoned Pierre Cadou, the leader of the "cabale."[52] At that time, the masters succeeded in holding down wages. But in 1789 the explosion of political journalism created a scarcity of labor, which made the printers' demands impossible to resist, especially in the thirty or so shops where they had to work at night to put out morning papers. Despite his earlier remarks comparing workers with savages and savages with animals, Panckoucke reacted sympathetically to these demands: "Les ouvriers, il faut en convenir, étaient fort mal payés avant la Révolution: ils ont, avec raison, profité des circonstances pour améliorer leur sort."[53] He agreed to support the increased labor costs but not to allow the master printers to swell their profits and overhead charges (*étoffes*) proportionately. After some hard bargaining, in early 1790, they asked for his final word. He apparently said that he would go as high as 80 livres per printed sheet; but that, it seems, was not high enough, because in 1790 Panckoucke stopped doing business with most of the two dozen different printers who had worked for him before the Revolution. Their names were replaced on the *avis* accompanying the later installments of the *Méthodique* by the names of provincial printers: Frantin of Dijon, Regnault of Lyons, and Couret of Orléans, who was Panckoucke's brother-in-law. Although Panckoucke retained some of his former printers—his old associate Stoupe, the Veuve Herissant, and Laporte, who apparently was a satellite of Panckoucke's—he abandoned his plan to put most of the Parisian printing industry to work on his *Encyclopédie.*[54]

Instead, he took up printing himself. Although he continued to contract work to provincial printers, he set up a shop of

52. Paul Chauvet, *Les ouvriers du livre en France des origines à la révolution de 1789* (Paris, 1959), pp. 193–201.
53. "Lettre de M. Panckoucke," *Histoire,* V, 3.
54. Panckoucke, "Sur l'état actuel de l'imprimerie: Lettre de M. Panckoucke à MM. les libraires et imprimeurs de la capitale" in *Mercure,* March 6, 1790. Panckoucke noted that the *Méthodique* cost him 17 livres per sheet for typesetting and 6 livres per thousand for presswork. He also had to pay half of the labor costs in *étoffes* and a quarter of the labor costs in profit for the master printer. Assuming that the pressrun of the *Méthodique* had by then been reduced to 4,000, its printing costs (excluding paper) would have come to 71 livres 15 sous per *feuille d'impression* or about 7,175 livres per volume.

his own, both for his journals and for the *Méthodique*. The details of this new venture remain obscure, but it was producing books on a large scale by the end of 1790. In June 1790, Panckoucke announced that he had taken measures to prevent further delays in the printing of the *Méthodique,* and at the same time the *Mercure* proclaimed the creation of a "Société typographique nationale," which would solve the problems that had plagued his *Encyclopédie:* "Au milieu des imprimeries que s'élèvent de toutes parts, il n'en est point, à la honte du goût, qu'un bon ouvrage puisse trouver libre, au moins à un prix raisonnable ou ordinaire; il n'en est point qui n'ait abandonné, éloigné, et qui ne refuse encore tout ouvrage de longue haleine, ou qui ne tient pas à un parti et à une circonstance du moment."[55] Panckoucke clearly stood behind the Société typographique nationale, and the Société stood for an attempt to restore the printing practices of the Old Regime. If the revolutionary printers would produce nothing but political ephemera, Panckoucke would manufacture books in the grand old style by himself. Just how he financed, manned, and managed this enterprise is hard to say, but he built it into one of the largest printing establishments in Europe. By 1794 it employed a work force large enough to require two foremen. It had twenty-seven presses, a vast stock of type, including 7,002 pounds of the *petit romain* used in the *Méthodique,* and capital goods worth 58,515 livres. Panckoucke had dropped the grandiloquent title of Société typographique nationale, but since March 1791 he had been issuing *Encyclopédies* under the imprint "chez Panckoucke."[56]

While struggling with the problems of printing his copy, Panckoucke faced even greater difficulties in procuring it, for the Revolution compounded the troubles he had had with his authors since the beginning of the enterprise. In 1787 he signed a new series of contracts that extended the deadline by three to four years for the Encyclopedists who had fallen farther and farther behind in their work after missing the original deadline of July 1785. At the end of 1788, he negotiated some additional agreements binding the writers to produce their copy in

55. *Mercure,* Jan. 12, 1790. See also Panckoucke's Avis of June 14, 1790, in *Manufactures,* III, xxiii.

56. See Robert Darnton, "L'imprimerie de Panckoucke en l'an II," forthcoming in *Revue française d'histoire du livre,* and Panckoucke's Avis of March 21, 1791, in *Manufactures,* III, xxiii. Panckoucke had published a few books under his own imprint as early as 1780 and expanded his shop on a huge scale in 1790.

time for him to finish printing in 1791. This arrangement proved to be more than the twenty authors of the dictionary of medicine could manage. They got their deadline stretched to the end of the 1791 by yet another contract, settled in April 1789.[57] But they had not produced a page by mid-1790, and other authors trailed even farther behind: some had done nothing for nine years, Panckoucke wrote in a printed circular, which he distributed to his staff and subscribers in August 1790. By then he saw the Revolution as the greatest threat to his production schedule; and if he failed to maintain production, he was certain that he would lose subscribers. "Le public d'aujourd'hui n'est point le public de l'Ancien Régime," he warned the authors. Should the installments fail to come out on time, this militant new public would cry swindle, would demand its money back, would drag them all into court. To be sure, many Encyclopedists could defend themselves with a legitimate excuse. They had abandoned his book in order to rush to the aid of the fatherland. But now the revolutionary crisis had passed, Panckoucke argued, and the crisis of the *Méthodique* had worsened. The Encyclopedists should put aside the temptation to follow new careers as journalists and elected officials. They should leave politics to ordinary active citizens; for any educated Frenchman could cope with the affairs of state now that the victories of 1789 were being calmly consolidated under the new constitutional monarchy. But who else could help him raise his immortal edifice to the glory of the arts and sciences? The Encyclopedists should put first things first. If they failed to share his sense of priorities and persisted in their inability to distinguish between things that really mattered and the insubstantial issues of the moment, they had better pay attention to their contracts; Panckoucke could force them to honor their higher responsibilities. He could even ruin their reputations by publishing the texts of the promises they had broken. He hated to take extreme measures, however, and if they made one last effort, they could complete the supreme work of the century by the end of 1792.[58]

57. Avis of May 1787 in *Beaux-Arts*, I, lxxviii; "Lettre de M. Panckoucke, en date de novembre 1788, écrite aux auteurs de l'*Encyclopédie*" in *Mathématiques*, III, xv; and Panckoucke, "Sur le retard que l'*Encyclopédie* a éprouvé de la part de plusieurs auteurs" in *Histoire*, V, 8.

58. Panckoucke printed extracts from the circular of Aug. 5, 1790, in his memoir "Sur le retard que l'*Encyclopédie* a éprouvé" in *Histoire*, V; quotation from p. 9.

Panckoucke implored, cajoled, and threatened in this fashion for another year. He signed new contracts with Thouin, Tessier, Regnier, and Parmentier, who promised to finish the dictionary of agriculture by December 1792; with Guyton de Morveau and Fourcroy, who agreed to complete the dictionary of chemistry by December 1793; and with the twenty medical Encyclopedists, who accepted a new deadline of January 1, 1794. By mid-1791, Panckoucke had negotiated and renegotiated 171 contracts in all, and he hoped to conclude his work in 1794. But he no longer sounded so optimistic about the conclusion of the Revolution, which had just taken a turn for the worse following the king's flight to Varennes. He had treated the Revolution as a short-term crisis and had expected to ward off any damage to the *Méthodique* by a temporary expedient, the increase in the production of the volumes of plates. He also had managed to put out twenty-six volumes of text between January 1789 and July 1791; but they had been ready for printing before the fall of the Bastille, and he had received no fresh copy for twenty-two months. His authors had either been swept up in the revolutionary politics or forced to write for a living by the cancellation of pensions and sinecures. If the *Encyclopédie* were to survive, it would have to continue on a new basis, he concluded; for the Revolution had transformed the old republic of letters and had upset all sorts of relations, including his obligations to the subscribers as well as his authors' commitments to him. "La Révolution, comme nous l'avons déjà dit, a changé toutes les dispositions, tous les actes et traités d'une certaine nature, et ils sont dans le cas d'être modifiés," Panckoucke informed the subscribers in July 1791. "C'est un événement si imprévu que tout homme qui a traité de bonne foi avant cette époque ne peut être tenu à des engagements dont le salut public l'a détourné. Nous devons donc regarder les 22 mois d'événements extraordinaires qui sont arrivés comme nuls, ou à-peu-près nuls, relativement aux travaux littéraires."[59]

Panckoucke made these remarks in an open letter to the subscribers because he needed their support even more than that of the printers and the authors. He explained that he had lost money on every installment published between April 1789

---

59. Ibid., p. 8. See also the remarks in "Lettre de M. Panckoucke," *Histoire*, V, 3–4: "Tous les actes, tous les contrats, toutes les entreprises que cette Révolution a atteints, ne sont-ils pas dans le cas d'être annulés, ou du moins modifiés?"

and July 1791—thirteen in all, with a total deficit of 200,000 livres. There seemed to be no escape from the widening gap between income and expenses because he was also losing subscribers. Some were direct casualties of the Revolution, but most probably decided that Panckoucke could not continue publication or that the *Méthodique* was a luxury to be dispensed with in hard times. Whatever the cause, the earlier desertion rate had increased disastrously after 1789. Panckoucke lost 1,000 subscriptions during the first two years of the Revolution, bringing his total losses to 1,700, or 35 percent of his original customers.[60] He implored the remaining subscribers to stand by him in his hour of need. In the long run, the *Méthodique* was certain to be a great bargain, he assured them. But for the moment it was being attacked on two fronts, production and consumption. The Revolution had brought him nothing but ''des pertes, des sacrifices, des malheurs,'' and he now identified himself increasingly with the original publishers of Diderot's *Encyclopédie,* although their stormy experience under the Old Regime looked mild in comparison with his suffering during the Revolution: ''Cette *Encyclopédie* traversée dès les premiers volumes, plusieurs fois suspendue, où les libraires ont eu, pendant dix ans, leur fortune exposée, et dont deux sont morts de chagrin du procès qu'elle leur a occasionné; cette *Encyclopédie,* dis-je, était pour les éditeurs et les entrepreneurs d'une difficulté infiniment moindre que la nouvelle.'' Nonetheless, if the subscribers remained faithful, he would revise his contracts, readjust his production schedule, and eventually ''terminer ce grand monument qui depuis quelques années fait le tourment de ma vie et que je suis tous les jours au désespoir d'avoir entrepris.''[61]

Although one cannot see through to the reality behind Panckoucke's rhetoric, his situation could not have been hopeless, or he would have abandoned the *Méthodique.* Perhaps he held on to it for personal rather than financial reasons, for he had invested hope and ambition as well as capital in it, and he

60. ''Lettre de M. Panckoucke,'' *Histoire,* V, 2. Panckoucke described his losses in a way which suggested that his subscribers included a significant number of aristocrats. In his ''Lettre,'' p. 4, he explained, ''La Révolution a malheureusement atteint l'*Encyclopédie,* elle m'enlève près de mille souscripteurs, qui par la perte de leur état ou de leur fortune, ou étant absents du royaume, sont dans l'impossibilité de retirer leurs livraisons.'' In his ''Réponse de M. Panckoucke à M. le Baron de . . .,'' *Mathématiques,* III, xvii, he referred to subscribers who were ''distingués . . . qui occupent des places importantes.''
61. Quotations from ''Lettre de M. Panckoucke,'' *Histoire,* V, 3, 7, 5.

wanted to build it into the greatest publication of all time. In any case, the tone of his circulars and notices shifted significantly into the lower key in mid-1791. Gone were the buoyant superlatives of the prerevolutionary period, the gambling spirit of 1789, and the cautious optimism of 1790. Writing in the midst of the Varennes crisis, Panckoucke seemed unsure of what would become of his supreme speculation and what would become of France. Their fates were linked, and their troubles were to increase during the next two years.

In February 1792, when Panckoucke issued his next important notice to the subscribers, the constitutional monarchy that had been patched together after Varennes was falling apart, and France stood on the brink of a disastrous war. The protracted political crises had hurt the economy in general and the book trade in particular, for the publishing industry proved to be peculiarly vulnerable to the economic repercussions of the Revolution. Not only did the new paper currency depreciate, but workers in paper mills and printing shops had to be paid in small-denomination assignats, whose value declined more slowly than that of the larger units. The differential rate of depreciation hit publishers especially hard because the owners of paper mills insisted on being paid in assignats of 5 livres so that they would not lose on the exchange rate when they paid their employees. Ragpickers had tripled their prices (from 90 livres to 240 livres per thousand pounds), and the paper manufacturers also passed that added expense on to the printers, for the proliferation of revolutionary journals had created such a demand for paper that the men who controlled the beginning of the process by which old rags were transformed into printed sheets could ransom the men who worked downstream of them in the production system. Journeymen printers continued to force their wages up. Stitchers and binders almost doubled their charges. And at the receiving end of all these increased costs stood Panckoucke. "Si la Révolution a changé la face de toutes les choses, augmenté d'un tiers le prix de toutes les denrées, fait perdre 50 pour cent aux assignats, établi dans les changes une différence de 25 à 50 pour cent, je ne dois pas, sans doute, être seul victime de tant de changements," he warned the subscribers. He concluded that he would have to increase his own charges from 6 to 9 livres per volume, the provisions of the subscription notwithstanding. Extraordinary circumstances called for extraordinary sacrifices, and nowhere had the Revolution produced more disruption than in the publishing industry.[62]

At this point, Panckoucke may have fallen back on borrowing—at least he said that a Parisian banker called Gastinel, who was handling the finances of the *Méthodique,* had offered to loan him 100,000 livres, and he noted that the speculation had fallen 717,547 livres into the red. He had continued to lose more money with each installment because his subscription losses had increased to 2,000. And he had not even resolved his difficulties with his authors, who found it as hard as ever to cut off their involvement in the Revolution, although they promised once again to finish their manuscripts by 1794 or else to suffer "toutes les peines de droit et d'indemnité, sans qu'ils puissent prétexter le cas de maladies, d'absences, d'affaires, soit publiques, soit particulières, de place dans la nouvelle administration, de fonctions publiques." By this time, the strain of holding the enterprise together had begun to tell. Panckoucke lamented that he had sacrified all his other speculations and had exposed his entire fortune in order to keep the *Méthodique* afloat "dans les circonstances les plus orageuses où se soit jamais trouvé l'Empire français."

Yet he could not shake off the fear of going under. He collapsed into a severe depression, which lasted two weeks and which his doctors diagnosed as a frightful attack of spleen. "Cette cruelle maladie est connue sous le nom de vapeurs; j'ai le malheur d'y être sujet presque toutes les années; les Anglais la connaissent sous le nom de *spleen;* elle provient de deux causes, ou d'une trop grande tension dans la pensée, et de l'inertie de la pensée, ou du défaut d'exercise du corps . . . Le chagrin me dévorait . . . Tous les moments de mon existence étaient douleureux; mon imagination ne voyait les objects que sous les couleurs les plus sombres." Panckoucke may have been trying to evoke sympathy from the subscribers, although he had never before appealed to them in such personal terms. Or he may have been tormented by his vapors in a way that was analogous to the despair that drove financiers to jump off skyscrapers in the 1930s.[63]

In any case, while France went to war and the sans-culottes

---

62. Avis of Feb. 13, 1792, including a circular letter from Panckoucke "A Messieurs les souscripteurs" dated Feb. 11, 1792, *Manufactures,* III, xli.

63. Ibid., quotations from pp. xlii, xlvi.

overthrew the monarchy, Panckoucke nursed his spleen and lapsed into silence. He put out installments in May and July 1792, but he did not issue a statement about his situation until December. By then, he had published only eighty-nine and a half volumes of text and his losses had risen to 900,000 livres, but he still hoped to finish by 1794, although he attached that deadline to a string of ifs—if he could cope with the increased costs (postage, as well as paper, had now doubled in price), if he could overcome a severe paper shortage, and if "les circonstances actuelles" did not deteriorate. He did not mention his health, but he stressed his losses. "C'est qu'il [Panckoucke] est une des grandes victimes de la Révolution, qu'elle lui ôte plus d'un million et le fruit de près de 40 années de pénibles travaux; qu'il a exposé sa fortune entière pour soutenir l'*Encyclopédie*." And he did so in a way to suggest that he deserved well of the Revolution, as if his losses were its gain and he had poured out his capital for reasons of patriotism. "Il a été un des hommes les plus utiles dans la Révolution, en procurant tous les jours de l'occupation à plus de six cents personnes, à cent gens de lettres, soixante graveurs, deux cents ouvriers imprimeurs, et à un plus grand nombre d'ouvriers dans les manufactures de papiers. Les nouveaux malheurs qu'il vient d'éprouver par la suspension des paiements d'une des principales maisons de banque de Paris [Gastinel?] suffiraient seuls pour ôter toute idée de malveillance à son égard."[64] Three weeks later Panckoucke reopened the subscription. Now that France had gone over to republicanism, there was no Garde des Sceaux to absolve him of legal responsibility for this act. He even acknowledged its illegality, but he argued that he had no choice. He could not continue to let the unclaimed installments pile up in his stockrooms while his capital drained away. So he proposed to cancel 200 of the abandoned subscriptions and to sell them again at the subscription price, which now came to 1,474 livres.[65]

Whether that maneuver succeeded and Panckoucke finally stopped the seepage near the 3,000 subscription mark seems doubtful, but it cannot be determined because Panckoucke ceased to provide information about the state of his affairs in his notices of 1793. He published only three small installments in that year, when the Terror took hold of France. In the third,

64. *Mercure*, Dec. 15, 1792.
65. *Mercure*, Jan. 6, 1793.

issued on September 9 at the high point of the ultrarevolutionary Hébertiste movement, he pushed the price of the volumes of text up to 13 livres, more than twice their cost in 1789, "vu le doublement du prix de tous les objets de commerce et des arts."[66] He also eliminated his name from all of his sales material and even from the title pages of the books, which serve as a sort of barometer for the political climate in which he operated.

In the 1780s the title pages had come out bedecked with dedications to ministers and flying the royal imprimateur, "AVEC APPROBATION ET PRIVILÈGE DU ROI." They had carried two addresses, "A PARIS, Chez PANCKOUCKE, Libraire, Hôtel de Thou, rue des Poitevins; A LIEGE, Chez PLOMTEUX, Imprimeur des Etats." By 1791 Plomteux's name had gone, presumably because he had sold his shares back to Panckoucke. After 1791, the transition to the Terror stands out in a comparison of volumes 2 and 3 from the dictionary of philosophy. Neither mentions any privilege. Volume 2, "PAR M. NAIGEON," carries the following address:

> A PARIS,/Chez PANCKOUCKE, Imprimeur-Libraire, hôtel de Thou, rue des/Poitevins [double rule] M. DCC. XCII.

Volume 3, "PAR LE CIT. NAIGEON," appeared:

> A PARIS,/Chez H. Agasse, Imprimeur-Libraire, rue des Poitevins. [double rule] L'AN DEUXIÈME DE LA RÉPUBLIQUE FRANÇAISE/UNE ET INDIVISIBLE.

Panckoucke had abandoned his *Encyclopédie* to his son-in-law.[67]

A contract of January 26, 1794, shows that Panckoucke signed over to Agasse not only the *Méthodique*—its plates, manuscripts, and subscription registers—but also his entire newspaper, bookselling, and printing business. An inventory made six days earlier at his printing shop, 13 rue des Poite-

---

66. Avis of Sept. 9, 1793, in Case Wing Z 45.18, ser. 7, no. 16, Newberry Library. The Avis explained that "le papier, qui, en 1789, ne nous coûtait que 10 livres 10 sols et 11 livres, coûte actuellement 20, 21, et 22 livres, et nous sommes menacés d'une augmentation très prochaine. L'impression et le tirage sont aussi augmentés de plus d'un tiers."

67. In volume VII, p. vi, of the dictionary *Agriculture*, which was published in 1821, Louis-Augustin-Guillaume Bosc, who had kept in contact with the Panckoucke family throughout the Revolution, explained that Panckoucke had turned his business over to Agasse, after being "forcé d'abord de ralentir, ensuite de suspendre l'impression de l'*Encyclopédie méthodique*."

vins, reveals how much he gave up: vast assortments of type, twenty-seven fully equipped presses, and implements of all kinds worth 60,015 livres. More important, Panckoucke surrendered his ambition of producing the supreme book of the century. The century had shifted in its course; Encyclopedism had been swept aside by Jacobinism; and he had finally decided to withdraw from the rush of events to a position in which he could safely tend to his spleen.[68]

At this point the documentation on the *Méthodique* gives out. But Panckoucke's name never again appeared on its title pages. Agasse managed to publish a few more volumes under the Terror, the Thermidoreans, and the Directory, while steering close to the line of the dominant political factions in the *Moniteur*. After Thermidor, Panckoucke stepped once more into the public arena, denouncing Robespierre and campaigning for the creation of a currency strong enough to produce a revival of trade. Although he had been forced to turn in his gold dining service for assignats during the Terror, he had saved a great deal of his fortune. According to the settlement with Agasse in 1794, Panckoucke's assets outweighed his liabilities by 822,000 livres, and in a pamphlet of 1795 he claimed that he still owned real estate—mostly from speculations in nationalized church property—worth almost 300,000 livres. At that time he wrote as a straight Thermidorean: "Le règne de Robespierre a été horrible, celui d'une contre-révolution le serait mille fois davantage."[69] In 1797 he announced his support of the Directory and then began to court Bonaparte, first by publishing an open letter in praise of the Italian campaign, then by proposing to build a toll bridge between the Arsénal and the Jardin des Plantes, which would be decorated with statues of the young hero and the other conquerors of Italy. Although this plan led to nothing except a searing satire by

68. Darnton, "L'imprimerie de Panckoucke en l'an II."

69. Panckoucke, *Mémoire sur les assignats et sur la manière de les considérer dans l'état de baisse actuelle* (Paris, An III, "3ème édition corrigée"), p. 17. Panckoucke said that he had sold a country house near Boulogne for 30,000 livres in specie; that he still owned another house in Boulogne "dont j'ai refusé 200 mille livres en assignats, parce que le mobilier et la bibliothèque non compris, je l'estime 100 mille livres en argent'"; and that he had bought the Abbaye des Prémontrés in Paris for 191,000 livres, of which he still owed 100,000. Ibid., pp. 18, 22, 37. Although it seems unlikely that he lost most of his fortune, his daughter later claimed that he had been ruined by the Revolution, and his ostentatious talk about his wealth could have been an attempt to revive his credit. See also Panckoucke, *Sixième mémoire sur l'assignat* (Paris, 4 frimaire An IV).

Marie-Joseph Chénier, it showed that Panckoucke had revived and was back at his old game of concocting projects and ingratiating himself with the established powers.

He also launched a new journal, *Clef du cabinet des souverains,* and prepared to return to book publishing with an edition of Marmontel's works. But apparently he did not have anything more to do with the *Méthodique,* which remained in the hands of Agasse. Evidently the old ''Atlas de la librairie'' did not feel strong enough to resume his heaviest burden. He died on December 19, 1798, at the age of sixty-two. Having learned to trim and tack under the Old Regime, he had appeared as a Patriot in 1789, a Feuillant under the constitutional monarchy, a Jacobin during the Terror, a Thermidorean after Thermidor, and a Bonapartist during the rise of Napoleon. Although he had not finished the supreme publication to which he had devoted the last twenty years of his life, he had not let it drive him into bankruptcy. But it had ruined his life and his fortune, or so his daughter claimed. She later described it as ''l'entreprise la plus vaste du dix-huitième siècle . . . entreprise au-dessus des forces d'un simple particulier, et qui a coûté à son éditeur sa fortune, sa santé, et sa vie.''[70]

Agasse continued to put out volumes of the *Méthodique* from time to time until 1816, when his name was replaced on the title pages by that of ''Mme. Veuve Agasse.'' Widow Agasse finally stopped the monstrous growth of the book in 1832. A half century earlier, her father had announced that he would finish it in five years and 42 volumes of text. It had expanded to 53, 73, 100, and 128 volumes before he lost control of it, and in the end its text extended to 166½ volumes. Actually the size of a complete set of the *Encyclopédie méthodique,* including all the plates and supplementary material, is difficult to calculate with precision. Catalogue entries show it as a work of 192 volumes in the British Museum, 199 volumes in the Library of Congress, and 200 volumes in the Beinecke Library at Yale.

70. Veuve Agasse, flier entitled *Encyclopédie méthodique ou Bibliothèque universelle de toutes les connaissances humaines,* published sometime in the 1820s, from Case Wing Z 45.18, ser. 7, no. 1, Newberry Library. On Panckoucke's Bonapartist bridge see *Oeuvres de M. J. Chénier* (Paris, 1825), IV, 461–470; and on his plans for publishing Marmontel see S. Lenel, *Un homme de lettres au XVIIIe siècle. Marmontel* (Paris, 1902), p. 544. A well-informed article on Panckoucke in the *Biographie universelle,* ed. J.-F. and L. G. Michaud (Paris, 1811–62), XXXII, 63–64 confirms the remarks of his daughter by noting that he left ''peu de fortune.''

The bibliographical accounts vary according to what is taken to be a volume, what is included in the count, and what set is examined. Although Panckoucke and widow Agasse issued elaborate instructions for assembling and binding the irregular half-volumes into sets, there is no standard version of the book. But it is safe to accept the account of one bibliographer, who described it as composed of ''102 livraisons ou 337 parties, formant 166 volumes et demi de texte et 51 parties renfermant 6439 planches in-quarto.''[71]

An Enlightenment Publisher in a Cultural Revolution

Such was the fate of France's most powerful publisher and his greatest publication during the Revolution. Does it have any underlying meaning? One could argue that it simply illustrates the difficulty of commercial speculation at a time of social upheaval. Production was disrupted, demand declined, and the supporting mechanisms of credit gave way. The Revolution affected the publishing industry with particular intensity because it destroyed the artificial old restraints on the press and created a new demand for political journalism. Panckoucke personified the transition between the old and the new ways of exploiting the printed word, for his *Encyclopédie* belonged to the world of prerevolutionary publishing, while his journals kept pace with the revolutionary press. The *Méthodique* could not have survived without indirect subsidies from the *Moniteur*. Although both were printed in the same shop, they represented incompatible ways of doing business.

This interpretation seems valid enough, but it cannot be confined to questions of economics for it opens onto the broader field of cultural history. The French Revolution was among other things a cultural revolution. Panckoucke had built his publishing empire at the center of the cultural system of the Old Regime. He and the Revolution were bound to come into conflict, no matter how hard he maneuvered and reworked his *combinaisons*. The fundamental principle of culture, as of

71. Johann Georg Theodor Grässe, *Trésor de livres rares et précieux* (Dresden, 1859–69), I, 474. For a detailed and definitive acount of the composition of the work in the 1820s, when Veuve Agasse thought it was complete except for six volumes of text and some plates, see her flier, *Encyclopédie méthodique ou Bibliothèque universelle de toutes les connaissances humaines.*

society, under the Old Regime was privilege—that is, "private law" or the exclusive right to engage in some activity. Far from being restricted to the nobility, privilege ran through all segments of French society, including those where the printed word was a major source of income. The thirty-six privileged printers of Paris maintained a monopoly of their craft in the capital, and they worked in league with the most powerful booksellers, who monopolized the book trade through an exclusive guild. Books themselves carried privileges, or exclusive rights to the reproduction of their texts, which the Parisians administered in such a way as to exclude the members of the provincial guilds. Journalism also was privileged territory, dominated by the Parisian patriciate, because the Parisian publishers of the journals possessed an exclusive right, conveyed by a royal privilege, to cover a certain subject—foreign affairs and the official version of politics in the case of Panckoucke's *Gazette de France* and light literature in the case of his *Mercure*. No journal could be published in France or marketed from outside the kingdom unless its owner received permission from the government, submitted to censorship, and paid an indemnity to the privileged journal whose domain he invaded. The government treated the journals as special concessions, like the *régies* for collecting certain taxes, and assigned pensions on their income. Panckoucke paid more than a hundred pensions worth more than 100,000 livres a year on his journals. They went to the most prominent savants in the country, the same men who wrote for the *Méthodique* and who dominated the academies. And the academies were also exclusive bodies—intellectual guilds, in effect, which conferred status, income, and even, in the case of the Académie française, "immortality." The system operated like a set of interlocking corporations, and the key to it was "protection" or influence-peddling; for the privileges derived from the king and were dispensed through *gens en place*—courtiers and key officials, like the Garde des Sceaux, the Directeur de la librairie, and the Lieutenant-général de police.[72]

72. On Panckoucke's pensions see his essays, "Sur le *Mercure de France* et quelques nouveaux journaux ou papiers-nouvelles" and "Sur les journaux et papiers anglais" in *Mercure*, Oct. 24, 1789, and Jan. 30, 1790, and his short pamphlet, *Lettre à Messieurs les pensionnaires du "Mercure de France"* (Paris, Oct. 15, 1791). Panckoucke claimed that he paid as much in taxes on paper as on pensions; and when he asked the Constituent Assembly for permission to print its proceedings, he stressed that "cent mille écus de redevances qu'il paye au

Panckoucke had become a grand master of this system, and he had based his *Encyclopédie* speculations on a strategy of privilege and protection. Yet he favored liberal reforms, even at the expense of the guild, though not of his own interests, as evidenced by his support of the edicts on the book trade in 1777. His books spread Enlightenment—not radical Rousseauism, to be sure, but the advanced tendencies of science and literature represented by the *Encyclopédie méthodique*. It would be inaccurate to picture him as a reactionary, and it would be anachronistic to assume that progressive elements could not have germinated within the old, closed, corporate culture. But that culture was incompatible with the Revolution, and it had left its mark on the *Méthodique*—in the book's privilege, for example, which must have looked "Gothic" to revolutionary readers:

Louis, par la grâce de Dieu, Roi de France et de Navarre: à nos amés et féaux Conseillers, les Gens tenant nos Cours de Parlement, Maîtres des Requêtes ordinaires de notre Hôtel, Grand Conseil, Prévôt de Paris, Baillis, Sénéchaux, leurs Lieutenants-Civils et autres nos Justiciers qu'il appartiendra: SALUT. Notre aimé le sieur PANCKOUCKE, Libraire à Paris, Nous a fait exposer qu'il désirerait faire imprimer et donner au public un ouvrage intitulé: *Encyclopédie Méthodique;* s'il Nous plaisait lui accorder nos Lettres de Privilège pour ce nécessaires. A CES CAUSES, voulant favorablement traiter l'Exposant, Nous lui avons permis et permettons par ces présentes, de faire imprimer ledit ouvrage . . . Car tel est notre plaisir.[73]

This was the world of royal pleasure and *féaux Conseillers* in which Panckoucke had flourished and which the Revolution obliterated. Panckoucke's success in manipulating it became one of the most damning items in the attacks on him by the revolutionary journalists. When a pamphleteer accused him of currying favor with J. C. P. Lenoir, the former lieutenant général de police of Paris, he replied, "On peut juger si j'ai dû regretter M. de L....., qui, dit-on, me servait si bien. Je n'ai jamais vu qu'une seule fois ce magistrat . . . Je n'en ai jamais obtenu aucune grâce, aucun arrêt de Conseil; et je puis dire avec la plus sincère vérité que depuis que j'existe dans le

gouvernement ou aux auteurs méritent quelques égards.'' *Archives parlementaires de 1787 à 1860*, ed. M. J. Madival, E. Laurent, and E. Clavel (Paris, 1875), VIII, 45, session of May 23, 1789.

73. The privilege, dated June 7, 1780, appears at the beginning of the first volume of the *Recueil de planches*.

commerce, je n'ai sollicité la protection ni l'autorité pour favoriser mes entreprises aux dépens du public.''[74]

Panckoucke's readers could not know that he consistently promoted his speculations by invoking the protection of Lenoir and other top figures in the government, or that the authorities treated him as a *protégé* in their official correspondence.[75] But the quasi-official status of the *Encyclopédie méthodique* stood out on its title pages. The first six volumes of the dictionary of *Arts et métiers mécaniques* proclaimed themselves to be ''Dédiés et présentés à Monsieur Le Noir, Conseiller d'Etat, Lieutenant général de Police.'' (The dedication vanished in the seventh volume, which appeared in 1790.) *Géographie* was dedicated to Vergennes, the foreign minister, until it reached volume 3 (1788), when the dedication went to his successor, the Comte de Montmorin. And Panckoucke's *Tableau* of 1789 suggested that *Géographie* was virtually written from within the Foreign ministry.[76] *Jurisprudence* was dedicated to Miromesnil, the Garde des Sceaux; *Grammaire et littérature* to Le Camus de Néville, the Directeur de la librairie; *Marine* to the Maréchal de Castries, minister of the navy; *Economie politique et diplomatique* to the Baron de Breteuil, minister of the Maison du Roi; and the *Tableau encyclopédique et méthodique des trois règnes de la nature* to Necker, the Directeur général des finances, until 1790, when his name was discreetly dropped. Panckoucke had courted all of the most powerful men in the kingdom; how could one resist the conclusion that he himself was dedicated to the prerevolutionary power structure?

Radical journalists like Desmoulins, Brissot, and Carra con-

74. ''Observations de M. Panckoucke,'' *Mercure*, Nov. 21, 1789, pp. 33–34.

75. In June 1789 Panckoucke petitioned the Garde des Sceaux for the right to import some volumes of the *Méthodique* that were being printed in Liège by Plomteux without passing through the usual inspection at the customs. A memorandum dated June 6, 1789, probably by de Maissemy, the Directeur de la librairie, recommended Panckoucke's request: ''Il me paraît que le sieur Panckoucke fait de trop grandes affaires et tient trop au gouvernement pour vouloir se permettre aucune fraude; la fraude d'ailleurs n'est pas dans le genre de ses spéculations.'' Archives Nationales, V¹553.

76. ''Tableau et apperçu,'' *Mathématiques*, III, 12: ''M. le comte de Vergennes, convaincu de l'importance de la chose, animé d'un zèle éclairé pour les progrès des connaissances utiles, considérant surtout que l'*Encyclopédie* est un ouvrage national, qui demande les secours du gouvernement; ce ministre a daigné nous ouvrir le cabinet des affaires étrangères relativement aux échanges, traités de paix, et stipulations d'Etat à Etat; il nous a muni des documents, notices et renseignements dont nous avons pu avoir besoin dans la confection de cet ouvrage.''

centrated their fire on that very point, and Panckoucke tried to parry their attacks by presenting himself as a champion of progress and reform. "On m'a désigné comme un ennemi de la Révolution actuelle, un partisan du système prohibitif et de la censure," he wrote indignantly in the *Mercure* of November 21, 1789. "J'ai toujours eu l'un et 'autre en horreur. Personne n'a eu plus à en souffir que moi . . . J'ai écrit contre les privilèges exclusifs *éternels,* en défendant les privilèges limités, sans lesquels il ne pourrait exister de propriété ni pour les gens de lettres, ni pour les libraires." Strictly speaking, Panckoucke was right: although he had played his privileges for all they were worth, he had helped to reform the old system of perpetual privileges in books. But the revolutionaries did not settle for reform, while Panckoucke favored halfway measures. He brought to the Revolution the same *esprit de combinaisons* that he had used in his publishing ventures— a preference for maneuvering and devising expedients instead of forcing issues.[77]

The extent to which Panckoucke's penchant for compromise and *combinaisons* betrayed an attachment to the ways of the Old Regime can be appreciated by his attempt to revive the booksellers' guild in 1790. Having freed the press in July 1789, the Revolution had destroyed the basis of the old book trade and had left an administrative vacuum, for it had thrown the censors, police inspectors, and chambres syndicales into instant obsolescence. Associations of booksellers and printers continued to exist but in a legal limbo, unable to defend their monopolies or to enforce their privileges. After the municipal revolution of 1789 transformed the legal authorities in towns throughout France, a great many questions remained unanswered. Publishing no longer belonged to a commercial patriciate, but could anyone print anything? Censorship had disappeared, but did freedom include the right to reprint books that had been covered by privileges? The Direction de la librairie had lost its authority, but would the new authorities permit libel, pornography, blasphemy, and sedition? A whole new industry—one that was vital to the Revolution—

---

77. *Mercure,* Nov. 21, 1789, pp. 81–82. As an example of the journalistic attacks on Panckoucke see *Révolutions de Paris,* Sept. 24, 1791, pp. 587–589; and for Brissot's view of Panckoucke see J.-P. Brissot, *Mémoires,* ed. Claude Perroud (Paris, 1911–12), I, 84–88. Panckoucke described himself as someone characterized by an "esprit de combinaisons" in his notice "Sur le retard que l'*Encyclopédie* a éprouvé de la part de plusieurs auteurs," *Histoire,* V, 11.

had to be organized on a new footing; and so the leading publisher in France felt it necessary to propose a plan to stabilize the situation through a happy combination of the old and the new.[78]

Panckoucke reduced the question concerning the book trade to the problem of tracing "la vraie ligne de démarcation qui sépare la liberté de la licence,"[79] and he proposed to resolve it by restoring the booksellers' guilds on a new basis. The chambres syndicales would give up the right, which the Revolution had effectively destroyed in any case, of inspecting printers' shops and book shipments for prohibited literature. But they would retain their power to police the trade for counterfeit books, for Panckoucke put the protection of property above everything else. Book privileges should be limited to twenty-eight years, as in England, but they should be absolute. The chambres syndicales should work with the municipal governments; and the municipalities, which were then being organized by the Constituent Assembly, should be empowered to purge the book trade of pirates just as they would clean the streets of rioters. Licence was gaining everywhere, and the hand of the executive authority needed to be strengthened.

If Panckoucke seemed eager to revive some of the old police powers, he made it clear that he disapproved of the old guilds' "despotique" character.[80] He believed in free trade: "C'est la concurrence qui fait baisser le prix de toutes les choses." And he thought that no "aristocratique" corporation should be tolerated by a free government.[81] Any man ought to be able to become a master printer if he passed an examination and paid an entrance fee. But the fee would be high enough—perhaps 3,000 or 4,000 livres—to exclude most laborers. In fact, Panckoucke did not want to open up the trade so much as to reduce the "chaos" in it. And after invoking a kind of laissez-faire liberalism in a denunciation of the guilds, he

78. Although the guilds were abolished on the night of Aug. 4, 1789, their abolition was dropped in the final version of the decrees of Aug. 4; it was not formally enacted until March 2, 1791. In fact, however, the press was freed from the control of the guild as well as the state immediately after the storming of the Bastille.

79. Panckoucke, "Sur les chambres syndicales," *Mercure*, Jan. 23, 1790, p. 181. This essay and its sequel, "Sur l'état actuel de l'imprimerie," which appeared in the *Mercure* of March 6, 1790, are the main source for the following discussion.

80. "Sur les chambres syndicales," p. 176.

81. "Sur l'état actuel de l'imprimerie," p. 38.

argued for their rehabilitation as if he were a prophet of the Gothic Revival.

Above all, he stressed the need to win the workers over by paternalistic policies. Working men could not think beyond the gratifications of the present, he explained. They sought quick cash and ruined their health in shops that printed newspapers at night. Worse still, they played off employer against employer, forcing up their wages and neglecting their work. But they could be won back to the old ways of doing things, if the masters cooperated in a plan to domesticate them. First, the masters should grant a standard, temporary increase in wages. Then they should increase the size and stability of the labor pool by training new men in a guild school. For 1,500 livres a year, they could get a priest to teach spelling and rudimentary Latin, and they could easily find a good compositor and a pressman to handle the technical instruction. The school could turn out sixty to eighty workers in six months—and most important, the new men would remain attached to their masters because Panckoucke also proposed that the guild members devote some of their fees and dues to a retirement fund for their employees. A man would never desert a shop if he knew that his master had provided for his old age as well as his education. Moreover, the master would look after the worker's widow and perhaps reserve a job for his son, so loyalty would be built up over several generations and shops would develop into familial "foyers." The masters should also set aside enough of the guild's funds to provide pensions for themselves. And finally, they should undertake a campaign to stamp out piracy, working with the police, academicians, and guild members in other cities; for Panckoucke envisaged a national organization that would unite Parisians and provincials in a common effort to protect literary property. To get his project off the ground, he promised to pay 12,000 livres after the completion of his *Encyclopédie* and 300 livres a year as his own dues.

Although it may have had some connection with his proposal for a "Société typographique nationale" and even with the creation of his own printing business, Panckoucke's plan never amounted to much. But it does reveal the archaic, corporatist elements in the outlook of a progressive publisher; and it also sheds light on Panckoucke's politics, for he associated the need for order in the printing shops with the threat of sedition in general. What horrified him most about the free-booting, high-paying master printers was the material they produced. He recoiled at the thought of them, distilling "dans les ténèbres de la nuit ces poisons avec lesquels on cherche à corrompre l'esprit des peuples et à égarer leur raison."[82] Although he opposed reviving the old censorship, he favored retrospective sanctions against rabble-rousing. Curiously, he argued that France should model its treatment of the press on that of England, where journalists had roused the rabble for years, especially during the Wilkite agitation. But he apparently did not notice the raw and radical strain in British journalism; for he took a quick trip to England in August 1789, and upon his return he mainly expressed his admiration for the firmness of the English authorities, who had just condemned the editor of the *Times* to a fine of £500 and a year in prison for libel. Panckoucke clearly hoped that the Paris Commune would do the same to Desmoulins and Marat. He peppered his plan for restoring the guilds with declamations against "ces écrits incendiaires, calomnieux, ces pamphlets continuellement délateurs, où l'on se permet, sans fondement, d'inculper des personnes en place."[83]

Behind Panckoucke's attachment to guild publishing lay a profound distrust of revolutionary journalism—as could be expected, for he was more vulnerable to attack from the radicals as the man behind the *Mercure* and *Gazette de France* than as the publisher of the *Encyclopédie méthodique*. Soon after the October Days, he published an open letter in the *Mercure* that revealed the extent of his concern. He represented an old style of journalism, he explained. He had built up privileges, paid pensions, and submitted to censorship. The Revolution had destroyed all that, and he realized that he had to adapt to a new order of things. He was willing to compete openly with the new journals that had sprung up everywhere. But they would not settle for free competition. They attacked everything, especially the old journals, with a ferocity that would bring "la destruction de tout ordre, de toute règle." They had tried to steal his writers, to bribe his clerks, to pilfer his subscription lists. They had even offered to provide free

82. Ibid., p. 35.
83. "Sur les chambres syndicales," p. 180. The purpose of Panckoucke's trip is not clear, but it gave rise to a rumor that he was connected with a counter-revolutionary conspiracy, which he denied indignantly in the *Mercure* of Jan. 16, 1790.

copies of their own papers to subscribers who would renounce his. "Il n'y a point d'efforts, de petites ruses, de moyens sourds, qu'on n'ait employés soit directement, soit indirectement pour détruire un établissement où j'ai mis une partie de ma fortune et qui m'a coûté dix années de peines, de soins, et de combinaisons."[84] Furthermore, the dirty tricks of their marketing were less dangerous than the mud-slinging of their politics. The new journalists slandered and declaimed in a way that was leading straight to licence. If France wanted to retain its liberty, it had better emulate England, where there was nothing to fear from "ces pamphlets et ce torrent de feuilles de toutes sortes de formats, qu'on peut imprimer en moins de deux heures et dont tout un faubourg, une ville entière, peuvent être infestés dans un temps très limité."[85]

The English had eliminated such literature, Panckoucke believed, by enforcing libel laws and manipulating taxes, rather than by resorting to censorship. He had discovered a wonderful new species of newspaper during his English trip of 1789. It consisted of huge sheets, covered by up-to-date news and advertisements, utterly different from the small pamphlets called *journaux* in France, which were written like essays and carried little or no advertising. The English newspapers were magnificent enterprises. They employed teams of writers and whole shifts of typographers, who worked to a strict production schedule in specialized printing shops. The papers circulated everywhere, even among the lower classes. Panckoucke was amazed to learn that common laborers took up collections to buy newspapers, which they read and discussed in groups. He had met a small farmer near Bath who was remarkably well informed about events in France. Yet seditious material did not come into the hands of such readers —or so Panckoucke thought—because large-scale journals could not afford to be seditious. They had too much to lose. Their commerce could be ruined if they were condemned for libeling a minister or attacking the established church or the state. Small journals and broadsides might venture heresies, but the government had driven them out of business by levying a tax on paper, a tax on advertisements, and a stamp duty—

84. Panckoucke, "Sur le *Mercure de France* et quelques nouveaux journaux ou papiers-nouvelles," *Mercure*, Oct. 24, 1789, pp. 98–99.
85. Panckoucke, "Sur les journaux et papiers anglais," *Mercure*, Jan. 30, 1790, pp. 233–234.

"un moyen très simple de circonscrire la liberté de la presse dans ces vraies limites," which Panckoucke recommended to the Constituent Assembly.[86]

The English had developed just the combination that appealed to him: big business and moderate politics. In fact, Panckoucke did not understand much about that alien genre of journalism. He may not even have been able to read English. But his report on his trip reveals a great deal about his views on the French press because he congratulated the English for destroying the kind of journalism that he found most threatening in France—the radical, fly-by-night *feuilles volantes.* Such publications had their place, he conceded, but it was at the beginning of the Revolution, when the Old Regime was under attack. The time had come to consolidate the gains of 1789 and to restore order. Journalism should come under the sway of large, respectable enterprises like the *Moniteur,* just as the book trade should be controlled by a wise and wealthy guild—and politics should be limited to the "active" or wealthy citizenry.[87]

Panckoucke did not make his point quite that bluntly, but his pronouncements on the press showed that he supported the wave of reaction that swept over the propertied classes in 1790 and 1791.[88] He published his proposals for reviving the booksellers' guild and taxing radical journals out of existence in January 1790, just when the Constituent Assembly was debating a project of Sieyès for the suppression of libel and sedition in the press. The project never came to anything, but the Assembly took measures against left-wing journals in July 1791; and on August 23, 1791, it passed a general press law that was incorporated into the constitution. The law, proposed by Jacques Thouret (brother of Michel-Augustin Thouret, one of Panckoucke's Encyclopedists), contained some broad provisions against antigovernment propaganda. But it was never effective. The constitutional monarchy never resolved the confusion over how the power of the press was to be channeled within legal limits. And the journalists never stopped taking liberties that looked like licence to Panckoucke. Two of his own journals, the *Mercure* and *Gazette de*

86. Ibid., p. 232.

87. For examples of outspoken comments in this vein see "Sur les chambres syndicales," p. 180 and "Sur le *Mercure de France,*" p. 104.

88. On the character of this reaction see Georges Michon, *Essai sur l'histoire du parti feuillant, Adrien Duport* (Paris, 1924).

France, represented the oldest, most respectable, and most conservative strain of French journalism. They also followed a conservative or ''Feuillant'' line during the early years of the Revolution. So they drew the fire of the radical press, especially during the period after Varennes, when a new struggle for the control of public opinion broke out and Panckoucke found himself cast in the role of a counterrevolutionary.

At first, Panckoucke tried to adopt a neutral position, above the clash of parties and ideologies. The opinions expressed in his papers belonged to his authors, not to him, he insisted: he let them write as they pleased, provided they signed their articles and refrained from libel. Was not that the proper way to respect the freedom of the press? Nonetheless, should all the ''ennemis que la Révolution m'a donnés'' demand that he define his own position, he would satisfy them: ''Propriétaire de journaux, dont les uns passent pour aristocrates, les autres pour démocrates, prétendent-ils me faire un reproche des premiers? Je leur déclare cependant ici par écrit ce que j'ai souvent dit de vive voix: 'Je ne suis ni aristocrate, ni démocrate; je suis, je veux être et mourir citoyen actif de la première monarchie libre et représentative.' ''[89] That may have sounded somewhat less than crystal clear in mid-1791, when the political atmosphere was murkiest. But by the end of the summer the Feuillants had gained control of the situation, and Panckoucke spoke out as though he were one of them.

He even campaigned for a seat in the Legislative Assembly by presenting himself as an unabashed conservative. He had never set foot in a popular society or consorted with the likes of Danton, he declared, and he recognized the need to take a firm hand in government. He would increase the power of the king and restore respect for the clergy and nobility. In fact, he favored the restoration of all the ancient titles and coats of arms. And most of all, he wanted to curb the anarchy of the press; ''La liberté de la presse, telle qu'elle existe, du moins à Paris, est le scandale de l'Europe, la terreur des honnêtes gens, celle d'un peuple égaré, qui semble ne reconnaître ni loi,

89. ''Mémoire pour M. Panckoucke relatif aux journaux dont il est propriétaire,'' *Histoire,* V, 28. Although this memoir was dated July 1, 1791, in the *Méthodique,* Panckoucke had issued it earlier as a pamphlet and had published a still earlier version in the *Mercure* of Dec. 4, 1790.

ni frein, ni autorité.''[90] The Parisian electors rejected this formula for repression, and Panckoucke withdrew from politics. But brief as it was, his political career revealed a consistent pattern in his responses to the Revolution. He wanted to domesticate the working force, to regain control of the book trade, to dominate the press, and to destroy the propaganda of the left. All of these efforts represented so many aspects of a general attempt to contain the Revolution within the limits it had reached by the end of 1789.

At bottom, Panckoucke's polemics with the revolutionary press involved not only conflicts of interest and political principles but also a fundamental disagreement over the function of the printed word. Panckoucke claimed to be proud of his contacts with the philosophes, his success in disseminating their works, and the influence of his *Encyclopédies* on some of the reforms of the National Assembly.[91] But he was appalled at what the printing press turned out after July 1789. He hardly knew what to call it. It did not resemble the journalism that he had discovered in England or that had traditionally flourished in France. Under ''Journaliste,'' the *Encyclopédie méthodique* had reproduced the definition provided by Diderot for the first *Encyclopédie:* ''Auteur qui s'occupe à publier des extraits et des jugements d'ouvrages de littérature, des sciences, et des arts, à mesure qu'ils paraissent.''[92] That species had nearly become extinct after the fall of the Bastille. It had been pushed out of the print shops by a rougher breed of men who wrote daily accounts of political events for papers that contained *news*. Groping for a word to

90. *Lettre de M. Panckoucke à Messieurs le président et électeurs de 1791* (Paris, Sept. 9, 1791), p. 6. See also Panckoucke's *Projet d'une adresse au Roi, tendante à ramener le calme et la paix, à empêcher la guerre, et à rétablir Louis XVI dans l'esprit de la nation* (Paris, Aug. 15, 1791), a work of pure *Feuillantisme*, in which he fulminated against Marat, Fauchet, the Jacobins, the Cordeliers, and ''mille pamphlets, mille journaux odieux, où leurs auteurs se sont permis de vomir les injures les plus atroces contre les princes'' (p. 4). Panckoucke argued that a constitutional monarchy with a strong executive would stamp out such sedition and would also complete the reforms of 1774–1789, as if the happiest conclusion to the Revolution would be a revival of Bourbon reformism.

91. *Lettre de M. Panckoucke à Messieurs le président et électeurs de 1791*, pp. 6, 16–19 and Panckoucke, ''Sur une opinion qui commence à se répandre dans le public, que la Révolution rend inutiles plusieurs dictionnaires de l'*Encyclopédie méthodique*,'' in *Histoire*, V, 21.

92. *Grammaire et littérature* (1784), II, 386.

describe this phenomenon, Panckoucke settled on *papiers-nouvelles,* an anglicism that never stuck but that suggests how strange and unprecedented the explosion of political journalism seemed in 1789.[93] Surrounded by an ocean of these alien newspapers, the *Encyclopédie méthodique* looked as antique as the cathedrals and palaces Panckoucke tried to protect. He still hailed it bravely as "le plus grand monument qui ait jamais été élevé à la gloire des sciences et des arts,"[94] but it stood out like a gigantic anachronism amidst the pamphlets and papers that flooded the literary market place.

How was Panckoucke to defend his monument, when the new journalists attacked it for being "Gothic"? He could not rewrite 100 volumes in order to expunge all vestiges of medievalism, and a text could appear medieval or modern, depending on whether it was written before or after 1789. For example, the last part of the eight-volume dictionary of jurisprudence came out in April 1789, just before the destruction of the old legal system. A few months later its vast collection of articles about seigneurial justice, feudal rights, and the intricacies of "des lois Saliques, Ripuaires, Bourguignonnes, et Lombardes, ainsi que des capitulations des rois de la seconde race" make it look like a museum of antequated usages. Panckoucke tried to make a case for its usefulness by explaining that it could be consulted in disputes arising from the complex legislation against feudalism of August 4–11, 1789.[95] But that argument probably seemed strained to his subscribers, who complained that they did not want eight volumes that had suddenly been made obsolete by decree of the National Assembly. Other dictionaries sometimes sounded downright counterrevolutionary. The dictionary of finance contained plenty of incriminating remarks about the old system of taxation, and the dictionary of history treated the nobility with a respect that seemed reactionary, even when compared with the text of the earlier *Encyclopédie.* Panckoucke's historian, Gabriel-Henri Gaillard of the Académie française, berated his predecessor, the chevalier de Jaucourt, for undermining "une institution politique avantageuse": the prejudice of superiority by birth. "Il cherche à donner du ridicule à celui-là; il l'attaque dans sa source; et

---

93. "Sur le *Mercure de France*," pp. 97–100.
94. "Lettre de M. Panckoucke," *Histoire,* V, 2.
95. Panckoucke, "Sur une opinion," pp. 18–21; quotation from p. 19.

non content d'établir que la nature nous fait tous égaux par la naissance, la mort, et le malheur, il soutient qu'elle a tant contrarié la loi qu'il n'y aurait en effet ni noble, ni routier, si les secrets de la nature étaient dévoilés."[96]

Of course the prerevolutionary *Méthodique* was so enormous that one could find almost any opinion in it—any opinion that could have passed the censorship—but it could not be disassociated from the regime which had given it birth. Nevertheless, Panckoucke tried to erase its birth defects in his sales propaganda of the revolutionary years. Not only did he drop the old dedications, but he tried to construe the whole enterprise as a patriotic venture. In March 1789, for example, he recommended the dictionary of *Police* to the public on the grounds that it had "mérité l'approbation la plus flatteuse de son censeur." In January 1791 he sold it as a manual that would be useful in setting up the revolutionary municipalities.[97] If an author played a prominent part in the Revolution, Panckoucke mentioned it in his *avis*, though he did not capitalize on the ministry of Roland, one of his most important Encyclopedists, probably because he wanted to steer clear of the controversies surrounding the Girondists.[98] He hailed the second volume of Mentelle's dictionary of ancient geography in 1791 for opening up a new view of Roman history, "traitée d'après les principes des droits de l'homme, si bien connus actuellement" in contrast to the work of previous historians, who had written under the influence of "les maximes monstrueuses de notre ancien gouvernment."[99] In 1791 he ostentatiously presented a copy of the *Méthodique* to the National Assembly, along with a gift of 1,000 livres for the National Guard.[100] And by 1793, he sold *Encyclopédies* as if he were campaigning to win the war:

Il sera honorable pour la nation qu'au milieu des troubles qui l'agitent et des combats inévitables qu'elle est obligée de soutenir pour le maintien de la liberté attaquée au dehors avec tant de fureur et au dedans avec tant de perfidie, elle puisse cependant se vanter d'avoir mis à fin le plus grand édifice qu'on ait jamais élevé à la gloire des connaissances

96. *Histoire* (1784), I, i–ii.
97. Avis of March 28, 1789, and Jan. 4, 1791, in *Manufactures*, III, xi, xxxii.
98. See the Avis of Dec. 21, 1789, on P.-L. Lacretelle; of Feb. 8, 1790, on Rabaut de Saint-Etienne; and of Sept. 19, 1791, on Volney, all in *Manufactures*, III.
99. Avis of Nov. 29, 1790, *Manufactures*, III, xxx.
100. *Moniteur*, Aug. 2, 1791.

humaines. Il convenait à la France de garder, même dans des conjonctures si menaçantes, sa suprématie dans les lettres, de faire, au sein des orages, ce que nul autre peuple ne pourrait faire dans le calme de la paix, et de poser d'une main le faîte du temple des sciences, tandis que de l'autre elle combat près du berceau de la liberté.[101]

Whatever the mixture of his metaphors, Panckoucke could sound as fierce as any Jacobin if he needed to. He knew how to bend with the times, to identify his *Encyclopédie* with the cause of the nation, and to make the book seem increasingly radical as the Revolution moved leftward. But his pronouncements of 1790 and 1791 had shown that he really belonged to the right. After he withdrew from politics, he tried to save his greatest speculation and to ride out the Revolution by maneuvering to the left. Although this strategy succeeded, it did not transform the basic character of the *Encyclopédie méthodique*. The book was essentially a product of the Old Regime. Despite Panckoucke's attempts to paint it over in the colors of the republic, it expressed a contradiction that ran throughout French culture in the 1780s: it was progressive but privileged, avant-garde in its science but dedicated to the *gens en place* on its title pages—in short, a *combinaison* of elements that looked incompatible after the rise of Jacobinism.

The Last of the Encyclopedists

That point deserves some emphasis because it has been argued that Encyclopedism led to Jacobinism, not through any philosophical conspiracy such as the one imagined by the abbé Barruel, but by a congruity of outlook.[102] According to

101. *Mercure*, Jan. 6, 1793. Panckoucke's sales talk should not be construed to mean that there was no genuine Jacobinism in the volumes from the revolutionary period. See, for example, the militant republican remarks of Naigeon in *Philosophie* (1792), II, 154.

102. The most challenging general interpretation of science, Encyclopedism, and the French Revolution is Charles C. Gillispie, ''The *Encyclopédie* and the Jacobin Philosophy of Science'' in Marshall Clagett, ed., *Critical Problems in the History of Science* (Madison, Wis., 1959), pp. 255–289. See also L. Pearce Williams, ''The Politics of Science in the French Revolution'' and the critical comments by Henry Guerlac in the same volume, pp. 291–320, as well as L. Pearce Williams, ''Science, Education and the French Revolution,'' *Isis*, XLIV (1953), 311–330. For further elaboration of the Gillispie thesis see Gillispie, ''Science in the French Revolution,'' *Behavioral Science*, IV (1959), 67–73 and *The Edge of Objectivity: An Essay in the History of Scientific Ideas* (Princeton, 1960), chap. 5. For an excellent account, with a full bibliography, of the Academy of Sciences during the Revolution see Roger Hahn, *The Anatomy of a Scientific Institution. The Paris Academy of Sciences, 1666–1803* (Berkeley, 1971).

Encyclopedism, Capitalism, and Revolution

C. C. Gillispie, Diderot's *Encyclopédie* promoted a reaction against the esoteric, mathematical sciences and in favor of a vitalistic or "romantic" version of the biological sciences, and this reaction fed a current of anti-intellectualism that rose with the Jacobins to destroy Lavoisier, the academies, and what became construed as the aristocracy of the mind. The weak point in this interpretation concerns the connection between Encyclopedism and Jacobinism. It is difficult to show how the two "isms" came together, as most of the original Encyclopedists died before 1793 and the Jacobins did not justify their attacks on academicians by citing Diderot's text. But one important point of contact may be sought in the history of the second *Encyclopédie* of the Enlightenment, whose publication stretched from the early 1780s through the entire Revolution and into the nineteenth century.

Although Panckoucke's book differed from Diderot's, it was conceived as a revision and extension of the first *Encyclopédie*, it grew out of a plan that Diderot had originally devised, and it was understood at the time to be an up-dated version of Encyclopedism.[103] As explained above, the *Méthodique* did indeed emphasize biology, but its authors, notably Lamarck, did not treat their subject in the vitalistic or romantic spirit of Diderot and Goethe. The *Méthodique* did not neglect mathematics and physics. And its volumes on chemistry propounded the rigorous, mathematical system that had been developed by Lavoisier as opposed to the vitalism of Venel's articles in the first *Encyclopédie*. Moreover, Panckoucke's chemists were prominent Jacobins: Guyton de Morveau, who served on the Committee of Public Safety, and Fourcroy, who succeeded Marat in the Convention. Panckoucke published his most esoteric dictionaries just when the Revolution reached its hottest point—perhaps because he thought that the Jacobinism of their authors, Monge and Fourcroy, would make them seem legitimate to the revolutionary government. He would hardly have brought out the volumes on physics and chemistry at that time if he expected them to offend the Jacobins, and his most offensive diction-

103. See, for example, the reception given the *Méthodique* in the correspondence of Grimm and La Harpe. *Correspondance littéraire, philosophique et critique par Grimm, Diderot, Raynal, Meister etc.*, ed. Maurice Tourneux (Paris, 1880), XIII, 135 and Jean-François La Harpe, *Correspondance littéraire* (Paris, 1801), III, 301: "C'est une *Encyclopédie* nouvelle, bâtie sur les fondements de l'ancienne."

aries probably were the ones that had won the approval of the censors of Louis XVI, jurisprudence and history in particular. In any case, the Jacobins never accused the *Méthodique* of being undemocratic on the grounds that its science had gone beyond the reach of ordinary readers. Yet its scientific articles were more abstruse than those in the first *Encyclopédie*. In fact, the *Méthodique* expressed a general tendency for knowledge to become specialized and professionalized—to become more esoteric, not less—and instead of fighting that tendency, the Jacobins recruited the specialists to make saltpeter, cannons, and a rational system of weights and measures.

A second attempt to find connections between Encyclopedism and Jacobinism involves group biography.[104] By following the careers of the Encyclopedists who lived long enough to experience the Terror, Frank A. Kafker has tried to discredit the tendency among some historians to portray Diderot's collaborators as radicals who created the ideological basis of Jacobinism. Kafker found that only one of the thirty-eight Encyclopedists alive in 1793 welcomed the Terror, while eight resisted it, and the rest generally withdrew into obscurity, fear, and disgust. Of course it might be unreasonable to expect much congruity between men's reactions to the Revolution and the opinions they had expressed forty years earlier in essays on the arts and sciences. But it is still more difficult to see how the reactions of thirty-eight men could indicate the way all of the original contributors might have felt had they lived into the Revolution. Not only do the thirty-eight represent an insignificant fraction of Diderot's group—and the full size of the group cannot be known, though it may have included 300 persons—but they wrote an insignificant por-

---

104. Frank A. Kafker, ''Les Encyclopédistes et la Terreur,'' *Revue d'histoire moderne et contemporaine*, XIV (1967), 284–295. See also Jacques Proust, *Diderot et l''Encyclopédie''* (Paris, 1967), pp. 38–43 and John Lough, *The Contributors to the ''Encyclopédie''* (London, 1973), especially pp. 51–53. A basic problem in analyzing the original Encyclopedists as a group is the impossibility of knowing how many of them there were. By utilizing Richard Schwab's *Inventory of Diderot's ''Encyclopédie,''* Lough identified 142 contributors, but that is a minimal figure, which could be expanded greatly from the research mentioned below in the Bibliographical Note. Robert Shackleton, for example, has compiled 277 identifications: ''The *Encyclopédie* as an International Phenomenon,'' *Proceedings of the American Philosophical Society*, CXIV (1970), 390. Even so, the authors of about two-fifths of the articles are unknown, and the contributions of the known authors were so uneven as to make statistical statements misleading. One-third of them wrote only one article, in contrast to Jaucourt, who wrote 17,000, or roughly a quarter of the total number of articles.

tion of his *Encyclopédie*. Only four of them contributed extensively, and the really important contributors—the men who wrote the great bulk of the book—all died before the Terror. Like Diderot himself, they belonged to the generation that reached maturity at mid-century. But the second generation of Encyclopedists, the men of the *Méthodique*, came of age with the leaders of the Revolution. Although their work cannot be equated with that of Diderot's collaborators, it can be taken to represent a later stage of Encyclopedism, the stage in which it came into direct contact with Jacobinism. So if anything is to be learned by studying the way a group of intellectuals reacted to the Revolution, the reactions of Panckoucke's Encyclopedists would be most revealing.[105]

Of course one cannot hope to find many revelations about the innermost thoughts of men who lived almost two centuries ago. A few sources like the memoirs of Morellet and Marmontel describe the fears, the fantasies, and even the nightmares of an Encyclopedist trapped in the Terror.[106] But several Encyclopedists continued their work without interruption and without leaving any record of how they felt about the great events of the time. Perhaps they had little interest in what went on outside their studies. Others found their lives invaded by the Revolution, but they responded in contradictory ways. The evidence will not fit into a single picture. After trying to piece it together, one is left with an assortment of unconnected images: Guyton de Morveau mounting in a balloon to observe the enemy's position at the battle of Fleurus; Daubenton dissecting the body of a rhinoceros from the former royal menagerie before a group of deputies to the Convention; Charles, deep in his laboratory in the Tuileries as

105. The following account is based on the information that is tabulated and described in Appendix D. It would be tedious to list all the biographical material consulted, but it is important to acknowledge two biographical dictionaries: *The Dictionary of Scientific Biography*, ed. Charles C. Gillispie (New York, 1970–76), 14 vols. and the more extensive but less reliable *Biographie universelle*, ed. J. F. and L. G. Michaud (Paris, 1811–62), 85 vols. For background on Fourcroy, who was probably the most important Encyclopedist in the reorganization of scientific institutions, see W. A. Smeaton, *Fourcroy, Chemist and Revolutionary* (London, 1962). For the biographies of two other key Encyclopedists see Louis de Launay, *Un grand français. Monge fondateur de l'Ecole Polytechnique* (Paris, 1933) and Keith M. Baker, *Condorcet: From Natural Philosophy to Social Mathematics* (Chicago and London, 1975). For a recent account of doctors in the Revolution see David M. Vess, *Medical Revolution in France* (Gainseville, Fla., 1975).

106. See especially André Morellet, *Mémoires* in *Collection des mémoires relatifs à la Révolution* (Paris, 1820–25), VII, 20–29.

the sans-culottes storm the palace on August 10, 1792 (they spared him, because of his heroic past as a balloonist); Lalande hiding Dupont de Nemours in the Observatory on the same day and, after Thermidor, rising in the Collège de France to denounce "Jacobinical vandalism"; Fourcroy extracting copper for cannons from the bells of former monasteries; Mongez applying his mastery of numismatics to the manufacture of republican coins; Quatremère de Quincy darting in and out of prison and conspiring with monarchists from the time of the Feuillants to the uprising of Vendémiaire; and Vicq d'Azyr, so horrified at the sight of Robespierre during the Festival of the Supreme Being that (according to a rather extravagant article in a later volume of the *Méthodique*) he died from its effects.

Some cases are relatively clear. The contributors to the *Méthodique* included one émigré (Pommereul), one victim of the guillotine (Boucher d'Argis) and three men who barely escaped death after being imprisoned as counterrevolutionary suspects (Desmarets, Ginguené, and Quatremère de Quincy). None of them could be classified as a partisan of the Terror; and the classification of several other Encyclopedists is a matter of record—though the record has aroused a good deal of debate among historians—because they played a prominent part in revolutionary politics. Desmeunier was an influential Patriot in the Constituent Assembly. He then retired from politics but re-emerged, in company with Daubenton, as a Napoleonic senator. In the Legislative Assembly, P.-L. Lacretelle and Quatremère de Quincy rallied to the right-wing Feuillants, while Condorcet, Broussonet, Guyton de Morveau, Garran de Coulon, and Lacuée de Cessac generally backed the Brissotins on the left. As minister of the interior during the early months of the Convention, Roland led the Girondins on the right, while Monge, as minister of the *marine,* generally favored the rise of the Montagnard left. The purge of the Girondins cost Roland and Condorcet their lives (both committed suicide after fleeing from Paris). But their former collaborators on the *Méthodique,* Monge, Fourcroy, and Guyton de Morveau, played a crucial role in organizing the national defense during the Terror. Because they concentrated on technical problems, they are not usually considered hardline Montagnards. But they became deeply involved in Jacobin politics, unlike Garran de Coulon, who withdrew into noncontroversial committee work. For short periods in the spring of 1793 and the winter of 1794–1795, Guyton de Morveau sat on the Committee of Public Safety. And during the Directory, more Encyclopedists took seats near the center of power—Robert, Quatremère de Quincy, Garran de Coulon, and Guyton de Morveau in the Conseil des Cinq-Cents and Fourcroy, Marmontel, and Lacuée de Cessac in the Conseil des Anciens.[107]

Clearly the contributors to the *Encyclopédie* included many political activists and they did not act in any consistent pattern during the Revolution. Not only did they support different factions or parties, but also they almost defy classification because the boundaries between parties shifted and blurred. Some historians even argue that parties did not exist in any coherent form during the Revolution. But that interpretation raises the danger of reducing political differences, which were meaningful for the revolutionaries, to a kind of nominalism.[108] In fact, one can divide the Encyclopedists according to their political sympathies into three general categories: opponents of the Revolution (men who expressed hostility to it from the very beginning), moderates (men who supported the constitutional monarchy and mistrusted the popular revolution), and republicans (men who favored radicalization beyond August 10, 1792, either as Girondins, radical Jacobins, or unaligned individuals). There is virtually no information on ten of the seventy-three principal contributors listed by Panckoucke in 1789. Nine others died before 1793, leaving fifty-four—a reduced but representative sample of the second-generation Encyclopedists.

Fourteen of the fifty-four did not take any clear political stand during the Revolution. Of them, seven apparently continued their careers without interruption, four suffered losses of income or employment, and three lived in obscurity. It

---

107. For a thorough account of the political divisions during the most critical phase of the Revolution see Alison Patrick, *The Men of the First French Republic. Political Alignments in the National Convention of 1792* (Baltimore and London, 1972), which has superseded the revisionist interpretation of M. J. Sydenham, *The Girondins* (London, 1961). A great deal can still be learned about the political involvement of the Encyclopedists from the older work of Aulard and Mathiez and A. Kuscinski, *Les Députés à l'Assemblée Législative de 1791* (Paris, 1900) and *Dictionnaire des Conventionnels* (Paris, 1916–19).

108. See the works of Patrick and Sydenham cited in note 107 and, for further details on the political alignment of the Encyclopedists, the information in Appendix D.

seems likely that half of this group supported the Revolution, at least passively and at least until August 10, 1792; but that is guesswork.

Seven Encyclopedists indicated disapproval of the Revolution at an early stage, but they remained passive in their opposition—except for Pommereul, who emigrated (he had his name struck off the list of émigrés in 1798, however, and became a prefect and baron under Napoleon), and Boucher d'Argis, who denounced the October Days from his position as a magistrate in the Châtelet and was guillotined as a counterrevolutionary on July 23, 1794.

Fifteen Encyclopedists can be considered moderates. At least six of them—Thouret, Lacretelle, Peuchet, Desmeunier, Kéralio, and Quatremère de Quincy—promoted a Feuillant-type of constitutional monarchy. They accepted the destruction of the Old Regime but, like Panckoucke, favored conservative reforms. The others are difficult to classify because they accepted official positions but did not participate actively in the Revolution. Some may have been quite radical—for example, Olivier and Bruguières, whom Roland sent on a scientific expedition to the mideast. But most of them concentrated on their scientific work, with the blessing of various revolutionary governments. These included the biologists of the Muséum d'histoire naturelle: Daubenton, Lamarck, and Thouin. If they did not share the radical views of their republican colleague Fourcroy—and they probably did—they certainly benefited from the Revolution's attempts to encourage botany and agronomy.

Eighteen Encyclopedists were republicans. Of them, five (Monge, Fourcroy, Guyton de Morveau, Doublet, and Chaussier) were associated with the extreme Jacobins; five (Roland, Condorcet, Broussonet, Lacuée de Cessac, and Mongez) favored the Girondins; and the other eight (Garran de Coulon, Ginguené, Mentelle, Robert, Naigeon, Chambon, Bertoli, and de Prony) remained unaligned, although most of them probably sympathized with some shade of Girondism.

This breakdown may give a misleading impression of mathematical precision; but even allowing for indeterminate cases and errors in classification, it suggests two conclusions. First, the Encyclopedists did not act as a group but scattered all over the political spectrum. Second, they did not scatter evenly but tended to cluster in the center-left, that is in a zone bounded by the Feuillants (constitutional monarchists) on the right and the Girondins (moderate republicans) on the left. Few of them supported the Terror but fewer still backed any counterrevolutionary attempt to restore the Old Regime. On the whole, they were more radical than one might expect.

Perhaps the Encyclopedists' involvement in the Revolution can be studied more fruitfully by examining institutions rather than political factions and by concentrating on the period after Thermidor rather than on the first five years of the revolutionary decade. The Jacobin Encyclopedists had adapted chemistry to weaponry, developed military medicine, and applied mathematics to the national defense. Their success reinforced an attempt, from 1793 through the Empire, to reorganize knowledge for service to the state. Of course the Encyclopedists had served the state before 1789, as academicians, censors, professors, and administrators; and they might have been expected to defend their posts against attack from the revolutionaries. But most of them helped to tear down the old intellectual institutions and to erect new ones in their place. Lamarck, Daubenton, Thouin, and Fourcroy helped transform the old Jardin du Roi into the Muséum d'histoire naturelle. Vicq d'Azyr and Antoine Louis began to reshape the medical profession from the Comité de salubrité of the Constituent Assembly, and Fourcroy and Thouret continued their work in 1794–1795 by organizing the Ecoles de santé. Five of the original twelve professors in the Parisian Ecole, later renamed the Ecole de médecine de Paris, had written on medicine for the *Méthodique*. The first professors of the Ecole normale also included a strong contingent of Encyclopedists: Monge, Thouin, and Mentelle. Monge was the driving spirit behind the creation of the Ecole polytechnique, where he was joined by four of his collaborators on the *Méthodique:* Fourcroy, Guyton de Morveau, Chaussier, and de Prony. Monge, Fourcroy, and Ginguené also played a decisive part on the Comité d'instruction publique of the Convention, which reorganized the system of higher education in France by creating not only the grandes écoles of Paris, but also the series of écoles centrales in the departments, where four other Encyclopedists—Desmarets, Mentelle, Grivel, and Bonnaterre—took up professorships. And finally, almost all of the Encyclopedists who had belonged to the royal academies took seats in the Institut, which was created as a "living encyclopedia" in 1795.[109]

The careers of the Encyclopedists show an extraordinary pattern of continuity, across regimes and over political divisions, from the old institutions to the new. The lines lead directly from the Jardin du Roi to the Muséum d'histoire naturelle, from the Société royale de médecine to the Ecoles de santé, from the technical schools of the monarchy to the grandes écoles of the republic, and from the royal academies to the republican Institut. Thirty Encyclopedists—half the number alive in 1796—joined the Institut, and twenty-nine had belonged to the academies of Paris. The intelligentsia that had dominated French culture during the early 1780s re-emerged, stronger than ever, in the later 1790s. But the staying-power of the intellectual elite does not illustrate that overused French proverb "Plus ça change, plus c'est la même chose," because the new institutions differed significantly from the old. Although their membership remained remarkably consistent, they eliminated all vestiges of privilege, corporatism, and aristocracy. The Revolution swept away the ancient ceremonies and hierarchical distinctions, the honorary academicians and the genteel amateurs. It perpetuated an old elite but under new terms: the openness of careers to talent and the dominance of professionalism. Those conditions had existed before 1789 but in muted form, mixed up with masses for Saint Louis, panegyrics to Louis XIV, and patronage from the Gentlemen of the King's Bedchamber. In the educational system of the republic, the Encyclopedists appeared as experts: each man had a field, and each field had its place within a modernized curriculum. Similarly, the Encyclopedists of the Institut sat in "classes," grouped according to their expertise. The Revolution had not eliminated intellectual elitism but had cast it in a new form, wiping out privilege and advancing professionalization.

109. Hahn, *The Anatomy of a Scientific Institution*, p. 297. In his original proposal for an Institut National of Sept. 10, 1791, Talleyrand seemed to echo Panckoucke's sales talk for the *Méthodique:* "L'institut présenterait une sorte d'encyclopédie toujours étudiante et toujours enseignante; et Paris verrait dans ses murs le monument le plus complet et le plus magnifique qui jamais ait été élevé aux sciences." *Archives parlementaires*, XXX, 465. For more information about the participation of the Encyclopedists in the intellectual institutions of the republic see Appendix D.

Although this transformation occurred abruptly and violently in the years 1793–1796, its origins went back to the Old Regime—and in part to the *Encyclopédie méthodique*. Panckoucke did not believe that his book represented an advance over Diderot's because it spoke out louder against *l'infâme* and in favor of social equality—in fact, it was more cautious than its predecessor on religious and political questions. But he thought that it expressed a more progressive view of knowledge; for he had divided the topography of learning into fields and had assigned an expert to each of them, with instructions to produce the most advanced work that was possible within the professional boundaries. When it came to marketing his product, Panckoucke had put together *combinaisons* of protection and privilege, a strategy that looked reactionary after 1789. If viewed from the perspective of the sociology of knowledge, however, his venture can be seen as advanced: he organized the material of the *Encyclopédie* in the same way that the Encyclopedists were organized in the Institut—according to strictly professional standards.

Personally, Panckoucke remained a conservative, more so than most of his authors and despite his radical posturing after 1792. But his *Encyclopédie* cannot be identified with any explicit ideology, either Jacobinical or counterrevolutionary, and his authors scattered into different political camps. It was not a common political faith that united the Encyclopedists and gave cohesion to their work but rather an underlying tide that swept all learning in the direction of professionalism. Encyclopedism as an "ism" remained complex and contradictory. But as a phase of intellectual development in the late eighteenth century, it expressed a tendency for knowledge to concentrate among experts and for experts to be drawn into the service of the state—a tendency that had gathered force under Louis XVI, that became crucial for the *salut public* in 1793–1794, and that did not disappear from history after the French Revolution.

X

ʏʏʏʏʏʏʏʏʏʏʏ

CONCLUSION

Whether or not the *Encyclopédie* was, as its publishers believed, the greatest undertaking in the history of publishing, it developed into one of the great enterprises of the eighteenth century. The first folio editions were multimillion-livre speculations; the quarto and octavo editions surpassed them by far in size; and the *Encyclopédie méthodique* grew to such colossal proportions that it made its gigantic predecessors look small. The sheer scale of *Encyclopédie* publishing suggests the importance of Encyclopedism, for as its friends and enemies agreed, the book stood for something even larger than itself, a movement, an "ism." It had come to embody the Enlightenment. By studying how the *Encyclopédie* emerged from the projects of its publishers, one can watch the Enlightenment materialize, passing from a stage of abstract speculation by authors and entrepreneurs to one of concrete acquisition by a vast public of interested readers. The papers of the STN reveal every phase of this process. They provide all the information necessary to follow the production and diffusion of the quarto *Encyclopédie,* the largest by far in the eighteenth century, and they show how the quarto was linked to all the other editions, forming a continuous effort to bring *Encyclopédies* to an ever-widening public from 1750 to 1800.

The Production and Diffusion of Enlightenment

The manufacture of the quarto has been discussed in some detail because it sheds light on the way most books were pro-

duced in the era of the hand press. At that time, the raw
material of literature had far more importance than it does
in modern publishing. Not only did paper account for as much
as 75 percent of production costs but also its quality had a
great influence on the decisions of consumers. Book buying
in the eighteenth century differed considerably from what it is
today because the men of the Old Regime devoted a great deal
of attention to the physical aspect of books. They cared about
the material of the page as well as the message printed on it.
And so publishing history should take account of the life
cycle of paper, a complex story, which began with ragpickers
begging old linens at the back doors of bourgeois houses and
ended with the same rags returning through the front doors,
transformed into pages of the *Encyclopédie*. Thanks to the
records of the STN, one can follow the flow of paper from individual mills, through the printing shop, and into copies of
the quarto in libraries today. One can trace watermarks to
particular mills with particular styles of manufacturing and
pick out bits of thread, which had once lined gentlemen's un-
derwear and ladies' petticoats. One can even identify the fingerprints on the quarto's pages. And by searching through
some forgotten pathways of working-class history, one can
link them with the lives of the men who produced the book—
hard lives, lived on the road, between printing house and printing house; for the master printers ran through batches of
workers in the same way that they consumed paper: they ordered *assortiments* of men for particular jobs and discarded
them when the work was done.

While "le bourgeois," as the men called their master, fixed
his sights on profits and losses, and while the foreman struggled to maintain some order in the work-flow of the shops,
the journeymen pursued their own purposes. They did not
force their pace or struggle to become "bourgeois" themselves, because they knew that the only mobility available to
them was geographical. If they needed to make it easier to
pull the bar of the press, they overinked the formes. If they
wanted to relieve the monotony of filling and emptying their
composing sticks, they took a break in the bistro. And if they
got tired of the local wine or angry at the foreman, they struck
out for another shop farther down the road. To know that
Desgranges manufactured a particular sheet of paper soon
after his mill had been flooded, that Champy set the type for

it while his family feared he was dying of consumption, that young Kindelem ran it off at the press after running off himself with the shop girl of his previous employer—to see beyond the book into the lives of the men who made it is to sense the vastness of the human experience which the *Encyclopédie* embodied. A whole world had to be set into motion to bring the book into being. Ragpickers, chestnut gatherers, financiers, and philosophers all played a part in the making of a work whose corporeal existence corresponded to its intellectual message. As a physical object and as a vehicle of ideas, the *Encyclopédie* synthesized a thousand arts and sciences; it represented the Enlightenment, body and soul.

The role of copy editors and advertisers in this process deserves special study, because it shows how the text was treated and how it was presented to the public in the course of its reproduction and diffusion. Far from evincing any concern for the integrity of the original version, the abbé Laserre cut it and shaped it to suit his own purposes and those of Duplain. Thus the "mass" editions of the *Encyclopédie*—the quarto and the octavo, which followed the quarto word for word— have a flavor of their own: they taste of Laserre's Lyons in addition to the basic blend that Diderot had concocted. Of course Diderot's original publisher also had adulterated the text. Publishers in general had a cavalier attitude toward the written word in the eighteenth century. Books that now look like classics were thrown together casually and reshaped from edition to edition or even in the course of one printing, as when Duplain deflated the size of his volumes to 800 pages by cutting the text and then, once his cuts had been discovered, expanded them to 1,000 pages apiece.

Similarly, publishers said what they pleased in their advertising. The publicity for the *Encyclopédie,* from the first prospectus of 1751 to the last flier about the *Méthodique* around 1830, is a series of half truths, falsehoods, broken promises, and fake announcements about phony editions. The publishers lied so often and so casually that one wonders whether they ever considered honesty as a policy. It probably never occurred to them that they should feel responsible for keeping the public accurately informed. Indeed, the French state encouraged them to use false information in their prospectuses and title pages, so that it could turn a blind eye to books that the clergy and parlements wanted it to confiscate.

Conclusion

The official notion of *permissions tacites* as a rubric covering *librairie venant de l'étranger* was a legal fiction that everyone knew to be a falsehood, just as everyone "in-the-know" disbelieved Voltaire's protestations that he had not written his pamphlets. Eighteenth-century publishing was no gentlemanly game, played according to some kind of honor code; as a Swiss bookseller put it, it was "brigandage."[1] The publishers lived in a different world from that of their modern counterparts, and they operated from different premises. Lacking the protection of adequate copyrights, surrounded by pirates, hounded by spies, and threatened by traitors, they could not afford to be truthful. So they told the public whatever they thought would sell books. Their behavior will not gladden the heart of anyone who objects to dishonest labeling and manipulative publicity today because it shows that false advertising has a long history. There was no golden age of honesty before the advent of Madison Avenue, at least not in the elegant eighteenth century.[2]

If read as historical documents, slanted advertisements can be more revealing than straight *avis* because they show how sellers thought their products would appeal to the public. In the propaganda for the quarto, the publishers emphasized that their customers would get a compendium of modern knowledge and a synthesis of modern philosophy, all in one. In this respect, they carried out the strategy of Diderot and d'Alembert, who wanted to promote *philosophie* by identifying it with knowledge. To ask whether the *Encyclopédie* was a reference work or a manifesto of Enlightenment is to pose a false problem, for it was meant to combine those characteristics, and it was presented as a combination of them, by its promoters as well as its authors. Insofar as one can know anything about the response of the readers, it seems that they, too, saw the

1. Serini of Basel to STN, Nov. 29, 1777.
2. The character of the *Encyclopédie* advertisements can be appreciated from their context. One in the *Gazette de Leyde* of June 27, 1777, appeared just above an Avis by a sieur Pastel, who announced that "il a trouvé un remède des plus assurés et des plus efficaces, dont il est seul possesseur, pour guérir non seulement les maladies vénériennes les plus invétérées et abandonnées par le mercure; mais il est encore spécifique dans les maladies chroniques, qui ont été de tout temps incurables; telles sont les humeurs froides, le scorbut, le lait répandu des femmes et les dartres de toute nature, qu'il guérit, non d'une manière douteuse, mais radicale." Similarly a notice for bear grease, a sure-fire cure for baldness, appeared on the back of an advertisement for the quarto in the *Morning Herald* of London, which Jean-Baptiste d'Arnal clipped and sent to the STN in a letter of April 19, 1782.

Encyclopédie in this fashion. They wanted *philosophie* as much as information, and they did not treat the *Encyclopédie* as modern encyclopedias are treated—that is, as a neutral compilation about everything from A to Z. The contemporary understanding of the book should be taken seriously because it shows the extent to which the *Encyclopédie* was identified with the Enlightenment in the eighteenth century. The publishers predicated their sales campaign on that identification. They expected the public to buy the book for the reasons they cited in their advertising: a quarto on the shelf would proclaim its owner's standing as a man of knowledge and a philosophe. *Philosophie* had become fashionable by 1777; the commercialization of intellectual vogues had become quite advanced; and the advance had occurred along lines laid out by Diderot and d'Alembert. In short, the Enlightenment seemed to be penetrating rather far into French society, but how far?

Book consumption can serve as only a crude indicator of tastes and values among the reading public, and it may appear impertinent to talk about "consuming" books in the first place. But the purchase of a book is a significant act, when considered culturally as well as economically. It provides some indication of the spread of ideas beyond the intellectual milieu within which intellectual history is usually circumscribed. And as there has never been a study of the sales of any eighteenth-century book, a sales analysis of the most important work of the Enlightenment ought to be worthwhile.

The price of the *Encyclopédie* set a limit to its diffusion, for the book remained beyond the purchasing power of peasants and artisans, even though some of them might have consulted it in *cabinets littéraires*. But as the *Encyclopédie* progressed from edition to edition, its format decreased in size, it contained fewer plates, its paper declined in quality, and its price went down. And as the publishing consortia succeeded one another, they cast their nets more and more widely, reaching out with each new edition to remoter sections of the reading public. By the time they launched the quarto, they proclaimed that the *Encyclopédie* had ceased to be a luxury item and had come within the range of ordinary readers. But who were the *Encyclopédie* readers of eighteenth-century France, and what part of the kingdom and the social order did they inhabit?

To identify them, it is necessary to compile statistics from the subscription list of the quarto, which covers about three-fifths of the *Encyclopédies* that existed in France and nearly one-third of those everywhere in the world before 1789. The map of subscriptions (see Figure 5) shows that the quartos spread throughout the country, but they sold better in some places than in others—better in towns than in villages, better along commercial arteries than in the hinterlands, better in the valleys of the Rhône and the Garonne than in those of the Loire and the Meuse, and best of all in the great provincial capitals: Lyons, Montpellier, Toulouse, Bordeaux, Rennes, Caen, Nancy, Dijon, and Besançon. The only areas that the quarto did not reach were Brittany beyond Rennes, the Landes below Bordeaux, and the rural region of the southwest encircled by the Loire, the Cher, and the Dordogne. It is perilous to argue from geographical to social distribution, but some of the puzzling points on the map can be clarified by the correspondence of the booksellers. After weighing both sorts of evidence, it seems clear that the demand for the *Encyclopédie* came primarily from ancient cities that had acquired rich endowments of ecclesiastical and educational institutions during the late Middle Ages or that rose with the Bourbon monarchy to become administrative and cultural centers—seats of parlements, academies, and intendancies. The quarto did not sell well in the cities of the future, where the stirrings of industrialization could already be felt.

The two extremes in the market for the *Encyclopédie* are represented by Besançon, an old-fashioned provincial capital of about 28,000 inhabitants, which absorbed 338 quartos, and Lille, a burgeoning industrial center of 61,000, which absorbed only 28.[3] If the STN's correspondents are to be believed, the explanation for this disparity is simple: manufacturers and merchants had no interest in literature. Actually the subscribers of the quarto did include a few merchants, certainly in Marseilles and probably in Lyons and Bordeaux, although their poor showing in the north and the northeast suggests that they may have belonged to the commercial oligarchies of the older trading cities rather than to any emerging industrial society. In the case of the Franche-Comté,

3. The master subscription list attributes 338 quartos to Besançon, but only 137 of them appear in the list of individual subscribers published by Lépagnez. Other Comtois subscribers on Lépagnez's list accounted for another 116 quartos. The remainder of the 338 cannot be traced.

253 subscribers or 65 percent of the total can be identified. Only 15 of them were merchants. The vast majority came from the traditional elite: men of the robe, led by the councilors of the *parlement*, and men of the sword, led by the officers in the garrison of Besançon. Royal officials subscribed so heavily that the book seems to have penetrated the entire administration of the province. In the small towns, it appealed to an intelligentsia of lawyers, administrators, and even curés. In Besançon, it went into the libraries of *parlementaires*, civil and military officers, lawyers, doctors, and priests. Half the Bisontin subscribers came from the first two estates, although the eventual readership of the book probably extended down to the lower middle classes, thanks to borrowing and Lépagnez's *cabinet littéraire*. In general, however, the *Encyclopédie* did not seep into the base of society: it circulated through the middle sectors and saturated those at the top.

This view of a top-heavy diffusion process corresponds to the strategy of Enlightenment formulated by Voltaire and d'Alembert—an Enlightenment from above, which would filter down through the superstructure from the salons and academies into the world of small-town notables and country gentlemen—but not farther.[4] Thus the *Encyclopédie* began as a luxury limited primarily to the elite of the court and capital. But after it assumed a more modest form and acquired a price that suited middle-class budgets, it spread through the *bourgeoisie d'Ancien Régime*, a bourgeoisie that lived off *rentes*, offices, and services rather than industry and commerce. The modern capitalist bourgeoisie also could have afforded the later *Encyclopédie*, and a few enlightened merchants did buy it—but so few that they seem trivial in comparison with the *privilégiés* and professional men, who bought most of the copies. Voltaire's prescription for the Enlightenment therefore appears to be rather close to what actually happened—closer than the interpretation of some of the most eminent historians in France today, who usually rivet the Enlightenment to the industrializing bourgeoisie and treat the *Encyclopédie* as the expression of class consciousness.[5]

4. This strategy emerges clearly in the correspondence between the two philosophes, which has been studied thoroughly by John N. Pappas, *Voltaire and d'Alembert*, Indiana University Humanities Series, no. 50 (Bloomington, 1962).

5. See, for example, Ernest Labrousse, *Histoire économique et sociale de la France* (Paris, 1970), II, 716–725 and Albert Soboul, *Encyclopédie ou Diction-*

Conclusion

Nothing could have been more cutthroat and capitalistic than the *Encyclopédie* as an enterprise, but the audience of the *Encyclopédie* did not consists of capitalists. The readers of the book came from the sectors of society that were to crumble quickest in 1789, from the world of the parlements and bailliages, from the Bourbon bureaucracy and the army and the church. It may seem paradoxical that a progressive ideology should have infiltrated the most archaic and eroded segments of the social structure, but the Revolution began with a paradox—with collapse at the top before upheaval from below. And although some of the *Encyclopédie* subscribers may have been devastated by the Revolution, most of them probably gained by it, at least in the long run, for it ultimately came under the control of lawyers and notables who directed it in their own interests and who continued to dominate France for the next hundred years, if not longer.

The diffusion of the *Encyclopédie* also has relevance for another basic issue in the interpretation of French history: the debate between those who see the Enlightenment as a broad movement that modified public opinion on a large scale and those who view it as a relatively superficial phenomenon limited to a small circle of intellectuals. The first thesis goes back to polemicists such as the abbé Barruel, who attributed the Revolution to a conspiracy of philosophes and free masons; but it was incorporated in serious historical writing by Tocqueville and subsequently elaborated by literary historians such as Paul Hazard, Gustave Lanson, and to a certain extent Daniel Mornet. The second thesis has recently gained the upper hand, owing to the influence of the social and cultural history developed by the so-called *"Annales* school." Taking their cue from Lucien Febvre, who stood literary history on its head with a paradoxical idea, *le livre retarde*,[6] the *Annales* historians studied production statistically; they

---

naire raisonné des sciences, des arts, et des métiers. Textes choisis (Paris, 1952), pp. 7–24 and Soboul, *Précis d'histoire de la Révolution française* (Paris, 1962), 52–59.

6. The phrase was formulated by Alphonse Dupront in ''Livre et culture dans la société française du 18e siècle: réflexion sur une enquête'' in François Furet and others, *Livre et société* (Paris and The Hague, 1965–70), I, 232, 219; but the idea goes back to a thesis developed by Febvre and Henri-Jean Martin in *L'apparition du livre* (Paris, 1958). On inertia and cultural retardation see Furet's Avertissement; Furet, ''La 'librairie' du royaume de France au 18e siècle''; and Julien Brancolini and Marie-Thérèse Bouyssy, ''La vie provinciale du livre à la fin de l'Ancien Régime,'' all in *Livre et société*.

found that "inertia" stifled "innovation" in the general literary culture of the Old Regime. Despite a declining interest in theology and an increase of publications in science and belles-lettres, most Frenchmen continued to read the classics and the religious books that their fathers and grandfathers had read: the Enlightenment did not upset the long-term, deep-running currents of traditional culture.

This disagreement about the overall impact of the book as a force in history cannot be resolved by a study of one work. But the publishing history of the *Encyclopédie* may shift the grounds of the debate by revealing the richness of an untapped source. Unlike the printed texts consulted by the literary historians and the state records analyzed by the *Annalistes,* the papers of publishers bring the researcher into direct contact with the world of books as it existed in the eighteenth century. Of course the papers of the STN have their own bias: they lean toward modernity and away from traditional literature, and it will not be possible to get a full view of *l'histoire du livre* in the Old Regime until they are studied systematically with other sources, including the papers of the Chambre syndicale and the Direction de la librairie in Paris. But a preliminary conclusion may be permitted at this point. After reading the 50,000 letters exchanged by the STN with booksellers everywhere in Europe, one comes away with the conviction that Voltaire and Rousseau did speak to an enormous public after all and that the history of the *Encyclopédie,* when studied by the methods of the *Annales,* leads to Tocquevillian conclusions. The story of how the *Encyclopédie* became a best seller demonstrates the appeal of the Enlightenment on a massive scale, among the upper and middle ranges of French society, if not the "masses" who made the Revolution in 1789.

But that story cannot be confined to France. The STN boasted that it had sold its *Encyclopédie* everywhere between "les deux bouts de l'Europe,"[7] and a few sets even reached some remote areas of Africa and America. Actually, most of the *Encyclopédies* sold outside of France came from the other editions, but by following the references to them in the correspondence about the quarto, one can get a general idea of their relative importance in different parts of the Continent. The *Encyclopédie d'Yverdon* went primarily to the Low

7. STN to Perregaux, Jan. 11, 1778.

Countries and to western Germany; the Italian editions remained mostly in Italy, although a few copies turned up as far afield as London and Copenhagen; and the octavo sold well everywhere—so well, in fact, that the cheap little *Encyclopédie de poche* probably was the form in which most readers made the acquaintance of Diderot's text beyond the French borders. Although little is known about the diffusion of the first two folio editions, they probably went into a great many libraries of the courts and country houses scattered throughout Europe. As the publishers themselves observed, the stately folios and the diminutive octavos represented two extremes in the social diffusion of the *Encyclopédie*.

Unfortunately, however, the booksellers did not discuss their clientele thoroughly enough for one to develop a clear picture of it, except in a few cases. Whenever the booksellers of central and eastern Europe referred to their customers, they named aristocrats. Evidently the sales of the *Encyclopédie* beyond the Elbe and along the Danube corresponded to a Frenchified cultural cosmopolitanism that remained restricted to a tiny elite. But the book penetrated more deeply into the social order of the western countries. In parts of Italy and The Netherlands it probably reached small-town lawyers and officials, as it did in France. Those were the areas where the density of *Encyclopédies* was thickest and where the booksellers also reported a heavy demand for the works of Rousseau, Voltaire, and even d'Holbach. The demand existed on the Iberian Peninsula as well, but church and state combined to stifle it. Contrary to what one might expect, the French book trade suffered more in the Spain of Charles III and the Portugal of Pombal than in any other part of Europe, including Italy, where the Index looked like a best-seller list.[8]

Although it is impossible to provide statistics for all of

8. The German market is the most difficult to characterize. Although the booksellers' letters suggest that a fairly heavy demand for *Encyclopédies* existed, at least in the Rhineland, German historians argue that pietism remained too strong for the *Encyclopédie* to have had much appeal, especially among the northern middle classes. See F. Schalk, "Le rayonnement de l'*Encyclopédie* en Allemagne," *Cahiers de l'Association internationale des Etudes françaises*, no. 2 (May 1952), 85–91 and Rolf Engelsing, *Der Bürger als Leser. Lesergeschichte in Deutschland 1500–1800* (Stuttgart, 1974), pp. 121–136. Salvatore Bongi, in "L'*Enciclopedia* in Lucca," *Archivio storico italiano*, 3d ser., XVIII (1873), 90, claims that the two Italian editions sold well in Italy but only among the elite: "Principi, patrizi, prelati e frati fecero a gara per accogliere festosamente il filosofico." But he cites no evidence, and his statement does not apply to the cheaper quarto and octavo editions, which sold well south of the Alps.

Europe comparable to those for France, one can get an intimate view of the realities of the book trade by studying the reports of eighteenth-century booksellers. The booksellers operated in the thick of the market place, far closer to the play of supply and demand than any historian can hope to get. Of course that very closeness may have blinded them to certain factors, such as their own tastes and values. But if one reads enough of their letters, the idiosyncrasies cancel themselves out and a general picture emerges.

The Enlightenment appears as a movement that radiated out of Paris to the cultural avant-garde in choice spots throughout Europe. By 1770 it had begun to pass through successive waves of popularization, reaching into every corner of the French provinces and stirring up interest among groups located everywhere in the middle sectors of French society. As it passed beyond France, it concentrated in certain areas where the cultural current ran thickest, like the Low Countries and the Rhineland. Then it thinned out, spreading across the north European plains to the Scandinavian fiords and the Russian steppes until finally it reached remote outposts like Lex's bookshop in Warsaw and Rüdiger's in Moscow. Thanks to the booksellers' letters, one can follow the books down the Rhine to the great warehouses of Amsterdam and Ostend and from there to the shops of Bertrand in Lisbon, White in Dublin, and Philibert in Copenhagen. One can pursue them up the Baltic to Muller in Saint Petersburg, at least until November, when the ice closed in, cutting off the flow of literature for half a year—except for a few volumes hauled across the snow from Leipzig by sled. Moving into more temperate zones, one can accompany the shipments up the Elbe and the Moldau, from Virchaux in Hamburg to Gerle in Prague. One can watch them cross the Alps to Reycends in Turin and descend the Rhône to Mossy and Gravier in Marseilles and Genoa. And finally one can trace them down the Danube to Weingand in Pest, where Paris seemed centuries away in contrast to the immediacy of the Ottoman Empire and the unremitting warfare on the eastern front of western culture. The booksellers realized that they were participating in a vast process by which ideas coursed through commercial arteries and trickled and seeped into the furthest reaches of the Continent. They knew they were agents of Enlightenment, not because they felt committed to the diffusion of

Lumières, but because they made a business of it. As the STN wrote to Bruere of Homburg,

Jamais entreprise de ce genre et de cette force n'a eu plus de succès et n'a été menée avec autant de célérité. En moins de 2½ ans et après avoir renouvelé par deux fois la souscription, nous avons imprimé 8,000 exemplaires de cette *Encyclopédie,* dont il ne nous reste qu'un petit nombre à placer. Le public semblait attendre avec une sorte d'impatience qu'il pût être servi à cet égard par des bibliopoles moins rapaces que les premiers éditeurs. Nous nous sommes piqués, nos associés et nous, de le satisfaire à cet égard, et vous jugerez, Monsieur, que si les lumières philosophiques manquent dans ce meilleur des mondes, ce ne sera pas certainement notre faute.[9]

Enlightenment Publishing and the Spirit of Capitalism

Publishing was a brutal business in the eighteenth century. After ten years of hard knocks on the literary market place, the Neuchâtelois decided that "ce métier-ci donne plus de bile que d'autres."[10] Their bile rose because of the need to cope with the tricks of their trade. As the history of the *Encyclopédie* shows, the most prominent publishers of the Enlightenment operated by bribery and extortion, by falsifying accounts and stealing subscription lists, by spying on each other and manipulating Machiavellian alliances that gave full play to treachery and intrigue. The struggle to sell *Encyclopédies* in the 1770s and 1780s had a baroque flavor, which may have derived from the "booty capitalism" of the sixteenth and seventeenth centuries. But there was also a modern element in the commercial history of the *Encyclopédie*—more modern, in fact, than the style of business described in its articles on trade. To open the *Encyclopédie* at COMMERCE, NÉGOCE, or LETTRE DE CHANGE is to enter the archaic world of Savary's *Parfait négociant.* But to read the correspondence of the *Encyclopédie* publishers is to find oneself in the *Comédie humaine.*

Capitalism, whether of the booty or Balzacian variety, is built on the principle of linking supply and demand. So the scramble to supply *Encyclopédies* suggests that the demand for the book had spread throughout the length and breadth of France. It was the richness of the market that touched off the

9. STN to J. G. Bruere of Homburg, Aug. 19, 1779.
10. Bosset to STN, May 12, 1780.

clawing and scratching among the publishers. And the ferocity of their combat confirms the impressions one gets from the statistics of their sales: the reading public was hungry for Encyclopedism.

In the rush to exploit the market for *Encyclopédies,* the publishers found it impossible to keep to a straight course. They had to change directions according to rapid shifts in circumstances: hence the crooked line of evolution that led from the first edition to the second folio, the Neuchâtel reprint, the Suard *refonte,* the three quartos, the two octavos, the Liégeois *Encyclopédie par ordre des matières,* and the *Encyclopédie méthodique.* Three of those encyclopedias never got beyond the planning stage. Threatened by competing editions or attracted by more promising speculations, the publishers consigned them to the limbo of unexecuted projects, where they kept company with d'Alembert's unwritten history of the *Encyclopédie,* Raynal's unfinished treatise on Protestantism, Voltaire's unpublished tract on the ministry of Turgot, Morellet's incomplete *Dictionnaire du commerce,* and many other works of the Enlightenment-that-might-have-been. The Enlightenment that actually broke into print represented only a part of the output envisaged by its entrepreneurs. Publishers had to pick up projects and drop them, while juggling as much as they could manage. Their profession called for quick decision-making and slight of hand, for a mistake could bring down an entire business. Even Panckoucke, who carried an inventory worth 1,400,000 livres, came close to collapse for a while in 1777; and smaller men went under frequently, especially during the economic crisis on the eve of the Revolution.[11]

High risks for high stakes: such was the premise of Enlightenment publishing. Having accepted it, the publishers of

11. On Panckoucke's fortune and his near bankruptcy see Panckoucke to STN, June 16, 1777, and Nov. 22, 1779. The government granted him a temporary suspension of payments by an *arrêt de surséance* of April 4, 1777. He did not make use of it, but he went through a period of panic described in his *Lettre de M. Panckoucke à Messieurs le président et électeurs de 1791* (Paris, Sept. 9, 1791), p. 11: "Je venais d'éprouver une faillite de trois cents quarante mille livres. Je me crus perdu; le trouble me saisit, mes amis, M. de Buffon, sollicitèrent eux-mêmes cet arrêt. Je l'obtins; mais bientôt, revenu de ma première terreur, je n'en ai fait aucun usage." The papers of the STN and the bankruptcy records in the Archives de Paris contain a great deal of information on the economic crisis in publishing during the 1780s.

the *Encyclopédie* gambled according to two different strategies. They either attempted to sweep the market with a sensational new product, or they tried to sell the old text in cheaper versions and larger quantities. Panckoucke favored the former plan. He treated the quarto as a detour in the pursuit of his grand design for a remodeled *Encyclopédie,* while Duplain recognized that the quarto could become an unprecedented best seller and subordinated everything to the exploitation of it. The relative failure of Panckoucke's policy and the success of Duplain's suggest the same conclusion: the demand for Enlightenment literature had moved outside the narrow circle of Diderot's original clientele to a more popular public.

By studying the strategy of publishing as an enterprise, it is possible to penetrate the mentality of publishers as entrepreneurs. But to make sense of the evidence, it is necessary to lay aside preconceptions about "economic man" and to observe the men at work. The documentation is rich enough for one to follow them, step by step, as they planned policies and reached decisions. One can reconstruct their calculations by using the same memoranda and notes that they used—scraps of paper covered with scribbling: "aperçu de l'entreprise," "ce qui doit nous revenir," "calcul des bénéfices." And in reworking the arithmetic, one can get a sense of the almost mathematical rationality with which they tried to maximize profits. They had no doubts about the power of their own profit motives. Panckoucke described Duplain as "un homme avide . . . et qui aime l'argent avec fureur." Duplain told Panckoucke that the Neuchâtelois had an insatiable appetite for money: "Vos Suisses sont des gens affamés." And the Swiss, who saw nothing but greed in Panckoucke, recognized that they all were driven by the same passion for "l'argent, qui est le grand mobile de tout."[12] If the *Encyclopédie* publishers acted like other businessmen of their age, it would seem that eighteenth-century entrepreneurs sought their fortune with an uninhibited avarice that is hard to imagine today—just as it is difficult to appreciate the lust for plunder of the Norsemen, the primitive delight in gems among the Merovingians, or

12. Panckoucke to STN, March 7, 1779; Duplain to Panckoucke, Dec. 27, 1779, from a copy in Panckoucke to STN, Jan. 2, 1780; and Ostervald and Bosset to STN, April 10, 1780.

other expressions of extinct mentalities.[13] It is not that modern capitalism has made greed outmoded, but the pre-modern capitalists thought and felt in a way that now seems foreign.

Of course the publishers were moved by something more than utilitarian calculus and a raw love of riches. Jean-Pierre Bérenger of the Société typographique de Lausanne wanted to squeeze the last sou out of the octavo *Encyclopédie* in order to withdraw into a small chalet, with an orchard and flowers and fields, where he could watch his grandchildren gambol. Duplain had grander dreams—a luxurious marriage, a château, nobility. He shifted his wealth from commerce and industry to land and status, speculating in the long run for his own social promotion. And once he had left commerce, he never looked back, not even after the death of his wife. ''M. Duplain a absolument quitté toutes affaires,''[14] his successor, Amable Le Roy, informed the STN in 1784. ''Il me charge de vous répondre qu'il est devenu tout à fait étranger à la librairie, à laquelle il renonce pour toujours.'' Duplain's case was not unique. Le Roy himself tried to imitate it;[15] and it serves as a sort of parable about the slow pace of economic expansion in France. Duplain, the perfect bourgeois capitalist turned out to be a pseudonobleman. The modern robber baron operated within an archaic value system, as other men have done in other eras; for those who seem most ahead of their time in some respects can appear backward in others, and *mentalités* can include incongruous mixtures of advanced and antiquated attitudes.[16] Panckoucke, it is true, speculated in the spirit of nineteenth-century empire building. But in his sympathies, he remained attached to the Old Regime—the enlightened Old Regime, which had promoted his investment in Enlightenment. When viewed from the perspective of the

13. Marc Bloch, *La Société féodale. La formation des liens de dépendance* (Paris, 1949), p. 34 and Georges Duby and Robert Mandrou, *Histoire de la civilisation française* (Paris, 1958), I, 17.

14. Le Roy to STN, Jan. 29, 1784. This is the last reference to Duplain in the STN papers. Research in Lyons has failed to turn up any trace of him after 1784.

15. Ibid. ''Je viens de contracter un mariage, arrête le cours de mes voyages . . . Mon intention, d'après le mariage que je viens de faire, étant de céder mon commerce à mon frère, il ne convient ni à lui ni à moi de diriger notre industrie sur d'autres articles que ceux de notre fonds.''

16. See Jacques Le Goff, ''Les mentalités, une histoire ambiguë,'' *Faire de l'histoire* (Paris, 1974), III, 79.

Encyclopédie, therefore, the entrepreneurial spirit looks varied and complex. It combined progressive and anachronistic elements in ways that illustrate the uneven march of capitalism.

The *Encyclopédie* and the State

It took daring to publish the first edition of the *Encyclopédie*. The publishers of the second edition also had to brave some persecution by the state. But the men behind the three quarto "editions" faced dangers that were merely commercial—attacks from pirates without and traitors within. From the reign of Louis XV to the reign of Louis XVI official policy changed, and French authorities tended increasingly to treat the *Encyclopédie* as an economic rather than an ideological phenomenon.

It would have been impossible, in any case, for Duplain to manufacture and market his books clandestinely. He ran too big a business. Unlike the previous publishers, who had printed the second edition in Geneva, he produced most of the quarto volumes in Lyons, using dozens of presses, several warehouses, and whole armies of clerks, printers, and shippers. He made sure that the local authorities—De Flesselles the intendant and La Tourette the book inspector—received complimentary copies; and they cooperated with him while their superiors worked hand in glove with Panckoucke. The Directeur de la librairie and the "Atlas de la librairie" virtually collaborated. Panckoucke issued instructions, in Néville's name, to La Tourette, who expedited the quarto shipments instead of confiscating them. In fact, middlemen considered the shipments safe enough to be used as camouflage for prohibited books the officials were trying to seize. And while the customs agents, police, and book inspectors waved the quartos on, they confiscated rival editions—until Panckoucke directed them to open the French market to the octavo. Far from arousing any opposition among the French authorities, the quarto was advertised, shipped, and sold everywhere in the country with their active support.[17]

17. The quarto publishers capitalized on this support in their sales campaign. See STN to Barthes of Versailles: "L'on peut, sans courir aucun risque, s'occuper hautement de cet objet, puisque nous avons permission même exclusivement d'introduire et de débiter notre édition dans le royaume."

Only twice did the quarto publishers express any doubts about their immunity from persecution in France. In the summer of 1778, Duplain warned his partners that the local clergy might denounce him: "Nos magasins sont pleins, de manière qu'à la moindre délation au clergé nous serions pris comme des rats dans une souricière."[18] But he emphasized that danger in order to force the associates to grant him a generous "insurance" clause in the contract for the third edition, so his statements should be taken as examples of his bargaining technique rather than as indications of a genuine threat—especially as the clergy of Lyons remained quiet throughout his entire noisy operation.

The Parlement of Paris was a more formidable danger, however, not merely because of its frequent attempts to suppress the works of the Enlightenment but also because of its support for Panckoucke's enemies in the Parisian booksellers' guild during the agitation over the reforms of 1777. In August 1779, Panckoucke informed the STN that the parlement might try to crush the quarto: "L'affaire des libraires concernant les arrêts [that is, the edicts of August 30, 1777, reforming the book trade] est au Parlement. On y a dénoncé notre *Encyclopédie* in-quarto. Je viens d'écrire à Duplain pour faire mettre tout en sûreté et tout suspendre. La librairie de Paris est enragée."[19] The STN actually welcomed this bad news. It shot off a letter to Duplain, suggesting that he transfer all the printing to the safety of its workshop; and it exulted that "si notre édition venait à être proscrite, il y a apparence qu'elle se vendrait mieux."[20] Duplain, however, dismissed the threat as a "fausse alarme" and kept his lucrative operation going in Lyons.[21] He believed that the parlement meant to attack the reform edicts, not the quarto. But Panckoucke had received his information from a very high source: "On m'a assuré que la dénonciation avait été faite," he repeated in a letter to the STN. "Le magistrat [le Camus de Néville] lui-

---

18. Duplain to STN, July 24, 1778. On the "insurance" clause see Chapter III.

19. Panckoucke to STN, Aug. 18, 1779.

20. STN to Bosset, Aug. 28, 1779.

21. D'Arnal to STN, Aug. 29, 1779, and Duplain to STN, Sept. 2, 1779: "Nous faisons peu de cas d'une dénonciation au Parlement, qui d'ailleurs présente peu d'objets de crainte, vu qu'il entre en féries jusqu'au mois de décembre."

même l'a cru, et par bonté il a bien voulu m'en donner avis. J'ai même eu la précaution de faire mettre ici tout à part, dans un magasin séparé.''[22] Although nothing ever came of this scare, it showed the limits to Panckoucke's "protections." He had complete support in Versailles. But the government officials preferred to pull strings behind the scenes rather than to clash openly with independent bodies like the parlement and the clergy. The *Encyclopédie* still had enemies, but their influence had declined since the 1750s and had ceased to exist in the government, where Panckoucke found his strongest allies.

There may have been an element of "enlightened despotism" in this alliance. Néville, Vergennes, Lenoir, and the other top administrators of Louis XVI sympathized with the rational, reformist principles that the *Encyclopédie* had come to embody. But Panckoucke made sure that they favored his enterprises by lobbying and influence-peddling. A codicil to the Traité de Dijon authorized him to clear a path for the quarto in France by distributing 240 livres before the appearance of each volume, and an anonymous pamphlet accused Duplain of buying Néville for 40,000 livres. Although there is no solid evidence that the publishers bribed officials in the way that they bribed competitors, they opened up the corridors of power by distributing free *Encyclopédies* and money —a common practice, which Panckoucke urged the STN to adopt.[23] Actually, Panckoucke did not need to bribe his way into the king's council because his importance as a press baron sufficed to give him entry to Versailles. Booksellers considered him "le favori de tous les ministres";[24] and ministers worked with him as if he were "un fonctionnaire ayant aussi un portefeuille," especially during the crisis connected with

22. Panckoucke to STN, Sept. 10, 1779.

23. The attack on Duplain is in a libelous but well-informed *Lettre d'un libraire de Lyon à un libraire de Paris* (1779), p. 1: "Je vous ai mandé dans le temps, et toute la librairie de Lyon en est informée, que Duplain a donné 40,000 livres [to Néville] pour avoir la permission d'imprimer l'*Encyclopédie*." The STN papers contain a great many remarks about gifts to officials. When the STN asked for help in getting the release of some confiscated copies of its pirated edition of the *Description des arts et métiers,* for instance, Panckoucke advised it to soften up Néville's first secretary: "Il est bon enfant. Envoyez-lui quelques douceurs. Faites-lui connaître l'intention où vous êtes d'augmenter sa bibliothèque, et si l'appât réussit, vous obtiendrez plus facilement par cette voie que par toute autre ce que vous désirez." Panckoucke to STN, Aug. 18, 1779.

24. Le Roy to STN, Dec. 17, 1783.

the reforms of 1777.[25] As a reformer, Panckoucke helped trim the unlimited privileges of the publishers, though as a publisher he refused to give an inch in the defense of his privilege for the *Encyclopédie*. Politics and lobbying in the Old Regime often involved such contradictions and complexities. But whatever the ambiguity of Panckoucke's position, it permitted him to direct his speculations on the *Encyclopédie* from the epicenter of legitimate power in France.

That point deserves some emphasis because the Enlightenment and the French state have often been interpreted as enemies; and the official condemnation of the *Encyclopédie*, notwithstanding Malesherbes's success in saving it, has often been cited as the supreme example of their enemity. That interpretation fails to take account of a change in the tone of French administration during the last fifteen years of the Old Regime. The persecution of the *Encyclopédie* in the 1750s turned into protection in the 1770s. Malesherbes's successor, Le Camus de Néville, actively promoted the quarto editions. And Panckoucke, the successor to the publishers of the first edition, based his speculation on the support of the government. From his first skirmishes with the pirates of Lyons, Geneva, and Avignon to his final settlements with the consortia of Liège, Lausanne, and Bern, Panckoucke relied on a strategy of official protection and privilege to defend his market against interlopers. Although Barret and Grabit forced him to pay ransom, Panckoucke's victory in the quarto–octavo war demonstrated the effectiveness of his policy. The octavo publishers sued for peace and sacrificed 24,000 livres for the right to sell their book in France. As the STN remarked, Panckoucke held the keys to the kingdom.[26]

In throwing their weight behind Panckoucke, the officials of Louis XVI treated the *Encyclopédie* as a commodity. They let the book lapse into a state of ideological neutrality as far as the law was concerned and admitted it into the game of interest-playing and power-broking that characterized high finance and high politics under the Old Regime. Panckoucke, who played that game better than anyone in the publishing business, succeeded in wheeling the vast machinery of the state around to a position from which it could be employed to

25. D. J. Garat, *Mémoires historiques sur la vie de M. Suard, sur ses écrits et sur le XVIIIe siècle* (Paris, 1820), II, 274.
26. STN to Panckoucke, Dec. 7, 1777.

defend his interests. The very force that had once been used to crush the *Encyclopédie* had become crucial in its diffusion.

What had been a pronounced tendency in the quarto venture became a predominant characteristic in the *Méthodique,* which shed the last vestiges of clandestineness and appeared fully decked out in a royal privilege. Panckoucke's ultimate *Encyclopédie* was printed openly in almost every shop in Paris, advertised and sold as a quasi-official publication, and written by men who not only owed their careers to the state but also included a heavy proportion of royal censors. The authors of the *Méthodique* may have taken a kind of Erastian view of learning: they put their knowledge in the service of the monarchy, and with a few exceptions they were willing to do the same for the republic. For some of them, this position provided a way of getting on with their work after 1789. For others, it meant a genuine commitment to revolutionary ideals. But in all cases, it went along with a trend toward professionalization that was sweeping over the old republic of letters. The *privilégiés* became professionals, the savants turned into civil servants. If the experience of the second generation of Encyclopedists does not demonstrate any direct link between Encyclopedism and Jacobinism, it shows an important strain of continuity in the intellectual elite from the Old Regime to the Revolution.

The Cultural Revolution

The publishing history of the *Encyclopédie,* as it evolved from edition to edition and conquered book markets everywhere, demonstrates that the supreme work of the Enlightenment was a best seller, but it does not provide easy answers to questions about the ideological origins of the French Revolution. The *Encyclopédie* was so vast and varied that one cannot know how it affected its readers, and one certainly cannot assume that hours of staring at the well-rounded letters on its starched and crinkly pages would infect anyone with Jacobinism. Nonetheless, the book did represent something coherent, an "ism," to the reading public of the eighteenth century. It showed that knowledge was ordered, not random; that the ordering principle was reason working on sense data, not revelation speaking through tradition; and that rational standards, when applied to contemporary institutions, would expose absurdity and iniquity everywhere. This message permeated the book, even the technical articles, for the details about grinding pins and constructing water wheels took on larger significance if seen in the light of the Preliminary Discourse and certain key articles that emphasized the need to order the everyday world according to rational principles derived from experience itself. The officials of Louis XV perceived this message clearly when they condemned the *Encyclopédie,* and the deputies of the Constituent Assembly expressed it when they redesigned the country's political, administrative, legal, and ecclesiastical systems. The diffusion of the *Encyclopédie* among a general public of lawyers, officials, and local notables—the sort of men who led the Revolution—indicates the extent to which the value system of the Old Regime was being undermined by an incompatible ideology.

Unfortunately, however, talk about "value systems" and "isms" tends to float off into speculation concerning ineffable climates of opinion, and so much was in the air in 1789 that one cannot trace connections between the revolutionary explosion and the sales of the *Encyclopédie.* In fact, as a general proposition, one can never argue from sales patterns of books to behavior patterns of human beings. It might not be safe to venture beyond the general proposition that the widespread diffusion of *Encyclopédies* symptomized a widespread disposition to question the ideological basis of the Old Regime and, in some cases, to accept radical change. Of course the ministers of Louis XVI, like the deputies to the Constituent Assembly, promoted some radical reforms, and they also helped to promote Panckoucke's *Encyclopédies.* Encyclopedist ideas seem to have been accepted by an important segment of the prerevolutionary elite and to have reinforced a strain of enlightened statecraft that ran from the reign of Louis XVI right through the Revolution and into the Empire. The re-emergence of the Encyclopedists in the official intelligentsia of the Directory also illustrates a strain of continuity from the 1780s to the 1790s. And their fragmentation during the early phases of the Revolution points up the impossibility of identifying Encyclopedism with any revolutionary party. Far from leading directly to Jacobinism, it did not pose a real threat to the state. In fact, if other forces had not destroyed the Old Regime, Encyclopedism might have been assimilated in France, and the kingdom might have ridden out the Enlightenment, just as other societies have survived sea changes in their value systems.

One way to bring these general questions about the relation between Encyclopedism and revolution down to earth, where they can be examined more closely if not resolved, is to study the history of the *Encyclopédie méthodique*. The second *Encyclopédie* was an extension of the first, and it marks the point where the "ism" and the Revolution came into contact. By 1789 the fate of the book seemed to be bound up with the fate of the regime, for the *Méthodique* had taken on a semiofficial character. It was a product of protection and privilege, published by the grace of the king, censored by royal officials, dedicated to the king's principal ministers, written in large part by royal academicians, and printed by twenty-five of the thirty-six printers who held a royally granted monopoly on book production in Paris. To follow Panckoucke and his *Encyclopédie* into the Revolution is to watch a cultural system being overthrown.

Although Panckoucke welcomed the Revolution at first, he realized soon after the fall of the Bastille that it would damage his supreme speculation. By 1790, it had upset his arrangements with his authors, printers, and subscribers. And by 1794 it had driven him to the brink of bankruptcy, to a state of nervous exhaustion, and perhaps even into hiding, for he turned over his affairs to his son-in-law and withdrew into semiseclusion until the end of the Terror. Of course his personal experience cannot be equated with the fate of Encyclopedism, but it exemplifies an underlying process, a basic change in the role of the printed word, which had a profound effect on the *Encyclopédie* and on the Revolution as well.

From the invention of movable type to the fall of the Bastille, the government had adopted different methods to control the press. It had hanged printers and booksellers, imprisoned them, made them officers of the University of Paris, and in the end organized them into a guild, with an exclusive right to produce books and a responsibility to police the book trade. In the late seventeenth century, the government limited printing in Paris to thirty-six masterships and gave the Parisian booksellers control over most book privileges. This typically Colbertist stroke of legislation created a privileged patriciate that monopolized legal book production throughout the eighteenth century, cooperating with royal censors and book inspectors and waging war against provincials and pirates like Duplain and the STN. Privilege in printing collapsed with the Bastille in 1789, and at the same time an insatiable appetite for news grew up in the reading public. About 350 newspapers burst into print in Paris during the first two years of the Revolution, crowding the genteel journals of the Old Regime off the market place. To produce them, and an equally great avalanche of pamphlets, about 200 printing shops sprang up.[27] They revolutionized the printing industry in Paris, and their influence reverberated everywhere—in the rag trade of Burgundy, throughout the paper mills of Auvergne, and all around the typographers' *tours de France*. The workers forced up wages, while their employers scrambled to profit from the increased demand for printed matter. Costs rose by 30–40 percent in two years, according to Panckoucke, and the republic of letters, like the labor market, dissolved in chaos. Many distinguished authors suddenly lost their pensions, their sinecures, and their influence—unless they, too, rushed into revolutionary journalism or politics. A rough crowd of interlopers took over the press, tossed censorship and privilege to the winds, and gave the public what it wanted—not books but political pamphlets and newspapers.

Because of their attraction to politics, historians have never paid much attention to the revolution in publishing. It cannot be explored adequately here, but its general character stands out from the story of Panckoucke's attempts to steer his *Encyclopédie* through a series of crises from 1782 to 1794. His basic problem, as he described it in 1791, came from the fact that he had planned to produce his book in a system that had suddenly been destroyed. He reacted to this predicament in three ways: he tried to maneuver around his immediate difficulties, to take the lead in reorganizing the publishing industry, and to reinforce the efforts of conservatives to contain the Revolution within the limits of the constitutional monarchy.

His maneuvering took the form of *combinaisons*—expedi-

27. ''Lettre de M. Panckoucke à Messieurs les souscripteurs'' in *Histoire*, V, 3: 'On m'a assuré, et on m'a offert de m'en donner la liste, qu'il y a actuellement dans Paris plus de deux cents imprimeries. Il n'y en avait que trente-six privilégiées avant la Révolution . . . Presque toutes sont occupées de journaux.''

ents to keep up the flow of copy, maintain the output of the presses, and stop the drain of subscriptions. Every time the speculation began to fall apart, Panckoucke patched it up in a new way. He kept it together, but he had to subsidize it from his journals, and it probably cost him about a million livres, as he claimed. In the end, the Revolution destroyed his publishing empire, even if it did not drive him into bankruptcy or to the guillotine. Panckoucke's proposals to reorganize the book trade in 1790 did not go beyond his position in 1777: he wanted to open up the publishing industry to provincial dealers and to restrict the duration of book privileges, while reinforcing their effectiveness. This stand had seemed liberal during the reform agitation of the 1770s, and it had created bad feelings between Panckoucke and the patriarchs of the booksellers' guild. By 1790, it looked conservative: the Revolution had destroyed all restraints on the book trade, and Panckoucke's main concern was to restore order in it—essentially by restoring the guild. He argued that a reorganized guild would keep up standards, protect literary property, domesticate the labor force, and help suppress *licence*. He did not go so far as to demand the return of exclusive masterships and censorship. But he thought that the French could wipe out radical propaganda in the English manner—by prohibitive taxes on radical news sheets and severe laws against libel and sedition. With the marginal, radical element eliminated, the trade would be dominated by books like the *Méthodique* and journals like the *Moniteur*.

Panckoucke associated the "insurrections dans l'imprimerie" with sedition in general.[28] His revolutionary politics were fought out in print. He had dominated the press of the Old Regime, and the dominant politicians of the Revolution— Mirabeau, Brissot, Robespierre, Desmoulins, Hébert, and many others—were journalists. To them, Panckoucke represented a new species of aristocrat, a press baron. To him they stood for licence. The conflict came out in the open during the crisis of 1791, when Panckoucke replied to the attacks on him by equating disorder in the press with disorder in the streets—a Feuillant formula for repression. In campaigning for the Legislative Assembly, he made it clear that he thought the Revolution had gone far enough by the end of 1789: liberty

28. "Mémoire en faveur de M. Panckoucke relatif aux journaux dont il est propriétaire" in *Mercure*, Dec. 4, 1790, p. 9.

had been won, licence must be destroyed, property protected, and the direction of affairs placed in the hands of men of substance—the respectable element in publishing and the "active"elite in politics. After 1792, Panckoucke disguised his true colors, tacked to the left, ducked into obscurity, bobbed up on the right, and veered off into Bonapartism. He was a trimmer, but he had a sense of direction. With all its zigs and zags, his career illustrates a line of evolution in politics that led from enlightened reform to the rule of the rich, from the France of Malesherbes to the France of Balzac.

Panckoucke represents the transition between those two worlds and shows how uneven and painful it could be. Although he seemed to incarnate the entrepreneurial spirit under the Old Regime, he did not greet the Revolution as an opportunity for unlimited freedom of trade on the literary market place. Instead, he tried to channel the flow of books through chambres syndicales. He remained corporate and conservative in his thinking, while the Revolution destroyed the corporate structure of culture—the guilds, academies, and privileged corps of all kinds that had dominated painting, music, architecture, drama, science, and literature. Panckoucke had learned to live with privilege, to turn it to his own advantage, to peddle influence and manipulate protections in a world that collapsed in 1789. He tried to re-establish order by reviving old forms, like the booksellers' guild, but the revolutionaries would not accept compromises and *combinaisons*. They eradicated all kinds of privilege—not only in academies and printing shops but in the army and the church, in politics and law, and down to the way the humblest peasant disposed of his crops.

Panckoucke's *Encyclopédie* and his Encyclopedists illustrate the same process of cultural transformation. The *Méthodique* mirrored the social organization of knowledge under the Old Regime. Through academies, guilds, censors, and protectors, the official culture of the kingdom stamped its influence directly on the book. At the same time, the Encyclopedists produced the most advanced survey of the sciences that could be conceived at that time. After 1789, the official forms seemed incompatible with the Encyclopedic content. Panckoucke stripped the book of its ministerial dedications and its royal imprimatur and sold it as an expression of national superiority in matters of intellect. In its new tricolor wrapping, the *Encyclopédie* corresponded to the new shape that the Revolution was imposing on the world of learning and to the new pattern in the careers of the Encyclopedists as well. For after scattering during the early phase of the Revolution, the Encyclopedists regrouped in the Grandes Ecoles and the Institut, where they were organized according to their special fields, just as Panckoucke had assigned them to his dictionaries. They came together as professionals in the service of the nation, not as epigoni in a privileged corps. The group was transformed, though the individuals remained the same.

Similarly, men were reshaping institutions according to new principles everywhere in France and reordering their mental world while changing the institutions—that was the general process exemplified in the history of the *Encyclopédie*. One can trace lines of continuity from Diderot's editorship to Panckoucke's, from Enlightenment publishing to revolutionary journalism, from royal academies to the Institut National, and perhaps even from Encyclopedism to Jacobinism, but the ruptures are equally significant because they show how a cultural system shattered. The Revolution destroyed privilege, the fundamental principle of the Old Regime, and then it built a new order around the principles of liberty and equality. Those abstractions may sound empty today, but they were full of meaning for the revolutionary generation of Frenchmen. The history of the *Encyclopédie* shows how they became expressed in print, disseminated in the social order, embodied in institutions, and incorporated in a new vision of the world.

YYYYYYYYYYYYY
APPENDICES
BIBLIOGRAPHICAL NOTE
INDEX
ΛΛΛΛΛΛΛΛΛΛΛΛΛ

A

ᵞᵞᵞᵞᵞᵞᵞᵞᵞᵞᵞ

CONTRACTS OF THE ENCYCLOPÉDIE PUBLISHERS, 1776–1780

These documents come from two dossiers, "Encyclopédie" (ms. 1233) and "Procès STN contre Duplain" (ms. 1220), in the STN papers. They have been arranged in chronological order, and in the few cases where no title appears on the manuscript a title has been added in English.

ᵞᵞᵞᵞᵞᵞᵞᵞᵞᵞᵞ

I. Traité de Société entre M. C. J. Panckoucke, libraire à Paris, & la Société Typographique de Neuchâtel en Suisse du 3 juillet 1776

L'an mille sept cent soixante et seize, le troisième jour du mois de juillet, les soussignés Messieurs Frédéric Ostervald, ancien banneret de la ville de Neuchâtel, Bosset DeLuze, membre du Grand Conseil de ladite ville, et Jean Elie Bertrand, professeur en belles lettres, d'une part, et Monsieur Charles Joseph Panckoucke, libraire à Paris, actuellement à Neuchâtel, d'autre part, propriétaire des droits et cuivres du *Dictionnaire encyclopédique,* y compris les discours servant d'explication, par l'acquisition qu'il en a faite conjointement avec les sieurs Dessaint et Chauchat, des sieurs Le Breton, David et Briasson, par acte du seize décembre, mille sept cent soixante-huit, est devenu seul propriétaire desdits droits et cuivres par la cession que lui ont faite lesdits sieurs Dessaint et Chauchat, par acte sous seing privé du premier juillet, mille sept cent soixante-neuf, et vingt-quatrième mai, mille sept cent soixante et dix, laquelle acquisition a été confirmée par un privilège du Roi de France de douze années en date du vingt mai, mille sept cent soixante et seize, sous le titre de *Recueil de Planches sur les Sciences, Arts et Métiers,* sont convenus de ce qui suit.

Ledit sieur Panckoucke a associé, comme de fait il associe, lesdits sieurs Ostervald, Bosset DeLuze, et Bertrand, ce acceptant, pour moitié dans la totalité des cuivres, droits, et privilège du *Dictionnaire encyclopédique,* tant pour le présent que pour l'avenir, ainsi que dans une nouvelle édition à deux mille cent cinquante dudit *Dictionnaire* orné d'un frontispice et des portraits de Messieurs Diderot et d'Alembert. La présente vente et cession d'intérêt de moitié est faite pour le prix et somme de cent huit mille livres, argent de France, payable en seize billets de six mille sept cent cinquante livres, même valeur, échéant au premier avril, premier juillet, premier octobre, premier décembre, mille sept [cent] soixante dix-sept, dix-huit, dix-neuf et quatre-vingt, que ledit sieur Panckoucke reconnaît avoir reçus.

Le sieur Panckoucke déclare que les trois premiers volumes de cette nouvelle édition sont actuellement imprimés dans ses magasins, ainsi que le premier volume des planches composé de cent quarante-six cuivres, et que la dépense de ces deux objets est de soixante-dix mille huit cent livres, dont le sieur Panckoucke a fait les avances et dont la moitié monte à trente-cinq mille quatre cent livres, que lesdits sieurs acquéreurs lui ont remis en six de leurs billets, le

premier	au	premier août	1777	de	⧺ 6000.—
second	au	1er novembre	1777		⧺ 6000.—
troisième au		1er février	1778		⧺ 6000.—
quatrième au		1er mai	1778		⧺ 6000.—
cinquième au		1er août	1778		⧺ 6000.—
sixième	au	1er novembre	1778		⧺ 5400.—
				ensemble	⧺ 35400.—

Et au moyen des deux paiements ci-dessus, lesdits sieurs acquéreurs n'auront rien à payer pour l'impression, papiers des trois premiers volumes de discours, le tome premier de planches, ainsi que pour le frontispice et les deux portraits de Messieurs Diderot et d'Alembert, qui sont à la tête du premier volume.

Et au sujet de la présente société, lesdits sieurs soussignés ont arrêté entre eux les articles suivants :

1. Lesdits sieurs acquéreurs ne pourront résigner leurs intérêts sans le consentement par écrit dudit sieur Panckoucke, et, réciproquement, ledit sieur Panckoucke s'engage de rester intéressé dans l'entreprise au moins pour un tiers.

2. Pour simplifier les redditions de compte, les associés sont convenus de fixer un prix invariable à tous les objets de dépense comme suit. Le prix de gravure de chaque nouvelle planche, qui en contient deux et quelquefois trois des anciennes, à soixante livres, dessins et retouches compris ; le prix de la retouche des planches anciennes qu'on croira pouvoir conserver à vingt-cinq livres ; le prix du tirage de chaque millier de planches à quinze livres ; le prix du papier des planches à quinze livres la rame ; le prix de l'impression à deux mille cent cinquante de chaque feuille du discours concernant les planches à trente-huit livres ; le prix de chaque feuille du discours à commencer par le tome quatrième à trente-quatre livres ; et le papier des volumes de discours à dix livres.

3. Il sera alloué en dépenses communes aux trois associés de Neuchâtel la somme de quinze cent livres pendant six années, à commencer du premier janvier prochain, pour frais de commis, port de lettres, magasinage, etc.

4. Les frais de prospectus, de lettres circulaires, avis de gazettes formeront un objet de dépenses communes.

5. Monsieur Panckoucke se charge de faire les avances, de diriger, soigner, et payer l'impression de tous les volumes de planches, dont il lui sera tenu compte au prix fixé article 2 ; il ne pourra répéter pour cet objet aucun faux frais quelconque. Il s'oblige expressément de prendre les arrangements convenables pour que ces volumes paraissent au temps fixé par le *Prospectus,* que les sieurs associés ont signé et qui demeure annexé au présent acte.

6. Messieurs Ostervald, Bosset, et Bertrand se chargent de leurs côtés de faire les avances nécessaires pour l'impression des quatorze volumes de discours à deux mille cent cinquante, de les faire corriger, assembler, collationner, emmagasiner, de payer les frais de commis, teneur de livres et autres, dont

il leur sera tenu compte, ainsi qu'il est dit article 3 et 4. A cet effet, ils se procureront les caractères nécessaires pour que les volumes paraissent exactement au temps fixé par le *Prospectus*. Ils s'obligent expressément à employer les mêmes caractères de Monsieur Fournier le jeune qui ont servi aux trois premiers volumes, d'en avoir toujours une quantité suffisante pour remplacer à mesure les caractères qui s'useront et de se servir du papier de France grand bâtard fin conforme à celui du *Prospectus* signé aujourd'hui à double par les parties, lequel servira d'échantillon.

7. La vente de cette nouvelle édition se fera particulièrement à Neuchâtel, les prospectus, avis porteront le nom de Neuchâtel, Monsieur Panckoucke s'obligeant de faire passer en détail auxdits acquéreurs la liste de tous les correspondants et celle des banquiers, notaires et principaux négociants de l'Europe.

8. Les associés rendront compte réciproquement des ventes et souscriptions, soit aux libraires, soit aux particuliers, sur le pied qu'elles auront été faites et suivant qu'il constera par leur correspondance.

9. L'expédition des trois premiers volumes de discours et du premier des planches se fera de Paris, l'expédition des autres volumes se fera de Paris et Neuchâtel. On fera parvenir à Paris un certain nombre de volumes de discours, et à Neuchâtel un certain nombre de volumes de planches pour faciliter les expéditions de part et d'autre.

10. La Société tiendra compte des avances qui seront faites sur le pied de six pour cent, et afin de mettre dans cette entreprise le plus grand ordre, chacun des associés fournira son compte de débours tous les mois, à compter du premier juin prochain, ainsi que l'état de ses recettes, tant en argent qu'en billets, sur lesquels se fera le remboursement des avances, à mesure qu'il y aura des fonds et au prorata desdites avances. Dans le cas que les fonds qui rentreraient ne se trouveraient pas suffisants pour rembourser Messieurs Ostervald, Bosset et Bertrand, qui sont dans le cas des plus grandes avances, de l'excédant de leurs avances sur celles de Monsieur Panckoucke, celui-ci après chaque livraison sera tenu de leur remettre la moitié de cet excédant en lettres sur Paris ou Lyon à 2 usances, bien entendu que cette clause n'aura lieu que trois mois après avoir fait la seconde livraison, et l'on n'envisagera

comme avances, tant pour les planches que pour le discours, que les impressions faites.

11. Les associés fixent le prix de cette nouvelle édition à six cent livres pour libraires et à sept cent vingt livres pour le particulier, et aucun des associés n'en pourra vendre au-dessous de ce prix sous quelque prétexte que ce soit. On donnera en outre aux libraires le onzième exemplaire gratis et aux particuliers le sixième. La distribution des paiements pour libraires se fera à la manière suivante, chaque volume de discours à seize livres, chaque volume de planches à trente livres, excepté le onzième, qui sera du prix de vingt-huit livres. Le prix pour le particulier sera conforme au *Prospectus*.

12. Les livraisons se feront aussi conformément au *Prospectus*. Le crédit aux libraires sera de six mois pour tout nombre au-dessous de trois. On étendra le crédit pour les nombres plus considérables. Monsieur Panckoucke s'en rapportera sur ce point à la prudence de ses coassociés. Les paiements se feront par des traites tirées de Neuchâtel en faisant chaque expédition. On fera payer les particuliers comptant le plus qu'il sera possible.

13. On fera tous les trois mois un inventaire général des magasins pour en faire le rencontre, et après la dernière livraison les exemplaires invendus resteront en magasin pour être vendus pour le compte de la Société, à moins que d'un consentement unanime elle n'en dispose autrement.

14. Il sera libre à Messieurs Ostervald, Bosset, et Bertrand de retirer leurs billets sous l'escompte de six pour cent.

15. On ne publiera aucun avis ni prospectus relativement à cette entreprise avant le premier janvier prochain, et on gardera à cet égard le plus profond secret jusqu'à la publication dudit *Prospectus*.

16. On passera en dépenses communes une somme de six cent livres, argent de France, qui sera payée à la chambre de charité de cette ville pour la distribuer aux pauvres.

17. Monsieur Panckoucke fournira un exemplaire de *L'Encyclopédie* de Paris, ancienne édition, pour servir de copie à l'impression.

18. Dans le cas où les associés de Neuchâtel trouveraient convenable à leur intérêt de réimprimer les Suppléments qu'on vient de publier à Paris pour servir de suite à cette édition,

Monsieur Panckoucke s'engage de leur fournir les cuivres de ces Suppléments retouchés, moyennant le prix de vingt-cinq livres par retouche. Cette réimpression des Suppléments ne pourra commencer que dans deux ans depuis la date du présent acte.

19. Dans le cas où l'on jugerait convenable de publier dans quelques années une nouvelle édition corrigée dudit *Dictionnaire encyclopédique,* dans laquelle on fondrait tous les Suppléments, cette nouvelle édition se ferait de compte à demi par tous les soussignés.

20. En cas de mort de l'un de nous pendant la durée de cette entreprise, il est convenu que nos héritiers et ayants cause seront astreints aux mêmes devoirs, et entreront dans tous les droits résultants de la présente convention.

21. S'il survenait quelques difficultés entre les associés relativement à cette affaire, il ne sera permis à aucun de nous sous quelque prétexte que ce soit de se traduire devant aucun tribunal; mais nous nous obligeons réciproquement de soumettre nos difficultés à deux amis communs, lesquels arbitres pourront en cas de différence d'avis se choisir eux-mêmes un surarbitre, et au jugement desquels nous nous obligeons de nous soumettre définitivement et sans appel.

22. La Société donnera gratis les exemplaires qui seront nécessaires pour l'établissement et la circulation de l'ouvrage.

23. Les associés sont convenus de travailler de concert pour obtenir du Roi qu'il favorise l'impression de cet ouvrage à Neuchâtel.

24. Enfin, comme l'usage général de cette ville est de passer ''sous seing privé'' tous les actes et traités de société de la nature de celui-ci, nous convenons de signer le présent acte double pour valoir comme s'il était fait par main de notaire, avec réserve expresse de le faire rédiger devant notaire au cas que cela convienne à l'une ou l'autre des parties et sur sa première réquisition. Ainsi fait et passé à double de bonne foi à Neuchâtel le troisième juillet 1776.

Frédéric Ostervald   C. Panckoucke
Bosset DeLuze
Jean Elie Bertrand

Appendix A

II. Paris, 14 août 1776, Accord fait avec
M. Suard au sujet de la rédaction

Les soussignés, Monsieur Suard de l'Académie Française et
Monsieur Charles Joseph Panckoucke, libraire, rue des Poitevins, d'autre part, sont convenus de ce qui suit.

1. M. Suard est convenu et consent de se charger de la révision
et rédaction d'une nouvelle édition du *Dictionnaire encyclopédique,* avec les Suppléments, aussi bien que les articles des
encyclopédies étrangères qui seront jugés dignes d'être con-
servés. Ledit M. Suard fera ce travail conjointement avec
M. d'Alembert et M. le Marquis de Condorcet, secrétaire
perpétuel de l'Académie des Sciences de Paris, que l'on annoncera avoir présidé avec lui à l'édition. M. Suard s'oblige de
fondre dans cette nouvelle édition une partie des matériaux
qu'il avait rassemblés pour le dictionnaire particulier qu'il se
proposait de faire sur la langue française, d'après l'approbation des savants nommés ci-dessus.

2. Le travail de M. Suard et de ses coassociés embrassera
particulièrement les volumes de discours. Il aura l'attention
de faire rapporter les planches aux discours et le discours aux
planches, en examinant soigneusement chaque figure, chaque
planche, et chaque lettre de renvoi pour corriger toutes les
fautes qui se trouvent sur cet objet dans l'ancienne *Encyclopédie.* Outre les coassociés ci-dessus, M. Suard s'associera
plusieurs hommes de lettres d'un mérite reconnu et distingué,
chacun dans leur partie, comme Messieurs de Saint Lambert,
Thomas, l'abbé Morellet, l'abbé d'Arnaud, Marmontel, de la
Harpe, Petit, Louis, etc., etc.

3. La copie de chaque volume sera payée comptant à M. Suard
la somme de cinq mille livres, savoir mille livres par mois, à
commencer du premier moment où il aura livré la copie d'une
portion du premier volume, et ainsi de suite de mois en mois.
Le surplus de ce qui lui sera dû sera payé à la fin de chaque

année, en trois paiements égaux, à six, douze et dix-huit mois. Plus il lui sera payé une somme de douze cent livres par an, pendant cinq ans, pour un copiste chargé de transcrire non seulement une partie de son travail, mais encore le travail des différents collaborateurs, à qui on donnera cette facilité pour accélérer la besogne. M. Suard ne pourra exiger aucun paiement pour les volumes de planches.

4. Sur ce paiement de mille livres par mois, M. Suard payera ses associés et les différentes personnes qu'il aura employées, M. Panckoucke ne voulant traiter qu'avec lui seul pour les détails et les frais de l'entreprise. M. Suard s'oblige sur les sommes que doit lui payer M. Panckoucke d'employer au moins celle de quarante mille livres pour les paiements des différentes personnes qu'il employera et il en justifiera par quittances.

5. Il sera délivré à M. Suard un exemplaire de l'ouvrage pour lui et un pour chacun des collaborateurs qui auront part à la somme de quarante mille livres stipulée ci-dessus.

6. M. Suard s'engage très expressément de ne point laisser languir la copie et d'en fournir au moins trois volumes chaque année. Il s'oblige que les tomes un et deux de discours seront complets et entre les mains de l'imprimeur au premier mai, mille sept cent soixante et dix-sept, mais comme la réussite de cette entreprise dépend de l'exécution exacte de ses engagements, il est expressément réservé qu'au cas que M. Suard ne livrât point aux imprimeurs la copie de l'ouvrage dans le temps fixé ci-dessus, en sorte que l'impression se trouvât suspendue, il sera tenu de payer aux imprimeurs le chômage sur le pied de cinq cent livres par semaine.

7. M. Panckoucke s'oblige de fournir à M. Suard pour son travail deux exemplaires de *L'Encyclopédie* avec les Suppléments et un exemplaire de *L'Encyclopédie d'Yverdon*. Ainsi fait et passé de bonne foi à double à Paris, le quatorze août, mille sept cent soixante et seize.

Approuvé l'écriture   Suard   Panckoucke

Appendix A

III. Paris, 31 août 1776,
Addition au traité avec M. Panckoucke

Addition à l'acte fait et passé à Neuchâtel le troisième juillet 1776, entre Messieurs Fréd. Ostervald, Bosset DeLuze, et Jean Elie Bertrand et Charles Panckoucke.

1. Messieurs Fréd. Ostervald, Bosset DeLuze, Jean Elie Bertrand, ayant pris lecture d'un mémoire signé de Messieurs d'Alembert, le Marquis de Condorcet, Suard, en date du 27e juillet 1776 concernant la refonte de *L'Encyclopédie,* sont d'avis, ainsi que ces messieurs, de suivre le plan de cette refonte et d'abandonner le projet qu'ils avaient de réimprimer *L'Encyclopédie* telle qu'elle était.

2. Messieurs Ostervald, Bosset DeLuze et Bertrand, comme intéressés pour la majeure partie dans cette entreprise, chargent Monsieur Panckoucke de traiter avec Monsieur Suard, conformément au projet qu'ils lui envoyent aujourd'hui pour le travail et le prix de cette refonte.

3. Monsieur Panckoucke s'oblige de leur envoyer à Neuchâtel les discours préliminaires, préfaces, tables et autres articles des trois premiers volumes actuellement imprimés, qu'on consent de sacrifier, lesquels discours, préfaces, tables ne pouvant servir pour la nouvelle édition corrigée, et le surplus desdits trois volumes sera vendu à la rame, pour le compte de la Société, le frontispice et les deux portraits seront joints au premier volume de planches.

4. Messieurs Ostervald, Bosset DeLuze et Bertrand, sur les représentations que leur a faites Monsieur Panckoucke, qu'il était nécessaire pour éviter toute concurrence d'intéresser dans cette entreprise quelques libraires, tant de Hollande que d'autres pays, ont consenti que Monsieur Panckoucke ne gardât que trois douzièmes dans ladite entreprise, et ils s'obligent de lui rétrocéder un des six douzièmes de leur intérêt, aux conditions convenues dans leur lettre du premier

août et dans celle de Monsieur Panckoucke du premier du même mois de cette année.

5. Les associés de Neuchâtel seront chargés de l'impression des trois volumes qu'on consent de réimprimer au prix et conditions prises pour les autres, notamment article 6 du traité fait à Neuchâtel le 3e juillet de cette année.

6. L'expédition de ces trois volumes se fera de Neuchâtel, en sorte qu'il n'y aura que les volumes de planches qui s'expédieront de Paris.

7. Les trois associés de Neuchâtel de concert avec Monsieur Panckoucke, tant en son nom qu'en celui des autres intéressés à qui il pourrait céder des parts dans cette entreprise, fixent le prix de cette nouvelle édition corrigée et refondue avec le Supplément à 24 livres le volume de discours in-folio, et 36 livres le volume de planches pour les particuliers, et pour les libraires à 20 livres le volume de discours et trente livres le volume de planches.

8. Monsieur Panckoucke, ayant actuellement environ trois cent exemplaires de l'édition de Genève à placer, cette nouvelle édition corrigée ne pourra pas être annoncée avant le premier juillet de l'année prochaine, temps auquel paraîtront les deux premiers volumes de discours et le premier des planches, de sorte que toutes les livraisons se feront conformément au prospectus, avec cette seule différence que ce qui devait être mis en vente au premier janvier 1777 paraîtra au premier juillet et ainsi de suite.

9. Les clauses 17 et 18 du traité fait à Neuchâtel le 3e juillet de cette année sont annulées comme inutiles.

10. Toutes les autres clauses dudit acte auxquelles on n'a point dérogé par le présent seront exécutées dans leur entier.

11. Monsieur Suard s'oblige de fournir la copie de chaque volume dans le plus grand ordre, prêt à être mis sous presse, de manière qu'il n'y ait rien à changer ni pour le style ni pour le fond des choses. Cependant, comme il pourrait se glisser quelques erreurs, transpositions, inadvertances, etc., dont un homme de lettres seul pourrait s'apercevoir, Monsieur le Professeur Bertrand, un des associés à cette entreprise, consent de se charger de la révision de toutes les feuilles avant, et après, l'impression, et pour ce travail il lui sera payé une somme de vingt mille livres, savoir quatre mille livres au

premier janvier 1778, et ainsi de suite d'année en année, jusqu'à l'entier paiement.

Ainsi fait à double à Neuchâtel, ce 31e août 1776
  Frédéric Ostervald Bosset DeLuze Jean Elie Bertrand
  C. Panckoucke

ʏʏʏʏʏʏʏʏʏʏʏ

IV. 3 janvier 1777:
Seconde Addition à l'Acte du 3 juillet 1776

Seconde addition à l'acte fait et passé à double à Neuchâtel le 3e juillet 1776 entre Messieurs Frédéric Ostervald, Bosset DeLuze et Jean Elie Bertrand, lesquels composent la Société Typographique de Neuchâtel, d'une part, et Monsieur Ch. Panckoucke, d'autre part.

1. Les soussignés, ayant connaissance que Monsieur Duplain, libraire à Lyon, avait répandu sous le nom du Sieur Pellet, libraire à Genève, et depuis sous celui de Nouffer et Compagnie, libraires de la même ville, le prospectus d'une prétendue nouvelle édition refondue de *L'Encyclopédie* à deux colonnes et en 32 volumes in-quarto, dont trois de planches, nous sommes convenus de réduire à mille exemplaires les deux mille que nous étions convenus de tirer de cet ouvrage tel que nous l'avions projeté, c'est-à-dire, corrigé et refondu, avec les Suppléments et les planches réduites, à teneur de la première addition à notre traité, laquelle a été signée le 31e août dernier.

2. Le prix de l'impression pour ces mille exemplaires sera fixé à 26 livres au lieu de 34 livres, et l'impression du discours concernant les planches sera fixée à 30 livres au lieu de 38 livres.

3. A côté de cette édition de mille exemplaires in-folio, nous convenons d'en donner une du même caractère in-quarto, à une ou deux colonnes, savoir 36 ou 40 volumes de discours, chacun de 120 feuilles, avec 3 ou 4 volumes de planches, et d'en porter le nombre à trois mille cent cinquante exemplaires.

4. Toutes les clauses de l'acte concernant l'édition in-folio auront lieu pour l'in-quarto, et ils fixent le prix de chaque

feuille à 42 livres au lieu de 34 livres et l'impression du discours des planches à 46 livres au lieu de 38 livres.

5. Le prix de chaque volume in-quarto sera de 12 livres pour
les particuliers et 10 livres pour les libraires, le sixième et le
onzième gratis comme pour l'in-folio.

6. Monsieur Panckoucke fera faire avec la plus grande diligence à Paris les premiers volumes de planches qui doivent
servir aux deux éditions, in-folio et in-quarto, et la gravure,
l'impression et le papier de ces planches lui seront payés
comme il est dit à l'article deux de l'acte de société. Fait à
double à Neuchâtel, ce troisième janvier mille sept cent
soixante dix-sept.

C. Panckoucke
    La Société Typographique de Neuchâtel en Suisse

ΥΥΥΥΥΥΥΥΥΥΥΥ

V. Neuchâtel, 3 janvier 1777, Supplément à la seconde addition faite à l'acte du 3e juillet 1776

Supplément à la seconde addition à l'acte fait et passé à double
à Neuchâtel, le 3e juillet 1776, entre Messieurs Frédéric Ostervald, Bosset DeLuze et Jean Elie Bertrand, lesquels composent la Société Typographique de Neuchâtel, d'une part, et
Monsieur Panckoucke, d'autre part, lequel supplément contient des articles particuliers à la Société Typographique et à
Monsieur Panckoucke.

1. Monsieur Panckoucke a rétrocédé à Messieurs de la Société Typographique de Neuchâtel le douzième qu'ils lui
avaient cédé par la clause 4 de la première addition, de sorte
que ces Messieurs restent intéressés pour la moitié dans
l'entreprise.

2. Les billets mentionnés dans l'acte de société du 3e juillet
1776 ont été annulés et convertis en 36 billets de ₶3066–13–4
payables à l'ordre de Monsieur Panckoucke au domicile de
Monsieur Batilliot l'aîné, banquier à Paris, les premiers de
chaque mois des années 1778 et 1779 et 1780. Lesdits 36 billets

forment ensemble la somme de ╫110,400, dont Monsieur Panckoucke s'est contenté. Fait à double à Neuchâtel, le 3e janvier 1777.

    C. Panckoucke
        La Société Typographique de Neuchâtel en Suisse

ʏʏʏʏʏʏʏʏʏʏʏ

VI. Traité de Dijon

14 janvier 1777

Les soussignés Joseph Duplain et Compagnie, libraire de Lyon, d'une part, et Charles Joseph Panckoucke, libraire de Paris et propriétaire des droits, cuivres du *Dictionnaire encyclopédique* et du privilège du recueil de planches sur les sciences, arts et métiers, ont fait entre eux le traité de société suivant.

1. Messieurs Duplain et Cie., ayant eu le projet d'une édition de l'*Encyclopédie* in-quarto à trente-deux volumes, dont le prospectus s'est répandu dans toutes les provinces, ont associé le sieur Panckoucke, ce acceptant, dans ledit projet pour moitié et dans toutes les souscriptions qu'ils ont reçues, et réciproquement le sr. Panckoucke leur a transmis tous les droits qu'il peut transmettre pour l'exécution de ladite entreprise.

2. Ladite édition de l'*Encyclopédie* en 32 volumes in-quarto, caractère de philosophie, comprendra les Suppléments qu'on vient de publier à Paris, sans addition ni correction; on ne fera qu'ajouter dans l'order alphabétique tous les articles du Supplément.

3. Ladite édition ne devant être composée que de trois volumes de planches, M. Panckoucke fournira gratis à la société tous les cuivres de son édition in-folio qui pourront servir à celle-ci, particulièrement les cuivres des sciences mathématiques comme la géométrie, le physique, la mécanique, etc. La retouche de ces planches lui sera payée ainsi qu'il sera dit plus bas.

4. Cette édition in-4° sera tirée à quatre mille, l'impression s'en fera à Genève et dans différentes villes de la Suisse à la convenance de M. Duplain, et dans le cas où on en imprimerait quelques volumes à Lyon, ils seront expédiés à Genève, afin que tout soit mis dans un magasin commun.

5. La vente s'en fera particulièrement sous le nom du sieur Pellet, libraire à Genève.

6. On mettra actuellement sous presse et l'impression sera poussée avec la plus grande célérité, de manière qu'on puisse publier au moins huit volumes in-quarto, de cent dix à cent vingt feuilles, chaque année.

7. La première livraison de quatre volumes ne sera mise en vente qu'au premier juillet, la deuxième de pareil nombre de volumes au mois de décembre et ainsi de six mois en six mois.

8. Pour simplifier les redditions de compte et mettre dans cette association la plus grande clarté, on est convenu de fixer un prix invariablement à tout. Le prix de l'impression de la feuille de quatre mille deux cent cinquante est fixé à cinquante-quatre livres, savoir, le premier mil à trente livres et le tirage de chaque mil à huit livres ; le prix de papier de dix-huit à vingt livres à neuf livres comptant la rame ; le prix de la retouche des planches anciennes que doit fournir M. Panckoucke à vingt-cinq livres et entretenu jusqu'ici quatre mille de tirage ; le prix de chaque nouvelle planche à soixante livres ; le prix du tirage de chaque planche pour mille à quinze livres ; le prix du papier des planches à quatorze livres ; le prix de l'impression des explications des planches à cinquante-quatre livres la feuille. Dans le cas d'un plus grand tirage, l'augmentation sera proportionnelle.

9. Il sera payé par chaque volume au rédacteur une somme de six cents livres.

10. M. Duplain et Cie. étant chargé de tous les détails de l'impression des volumes de discours, M. Panckoucke le sera de celui de l'impression, retouche, tirage des volumes de planches, dont l'expédition se fera de Paris.

11. Le prix du magasin, séchage, assemblage, commis teneur de livres, frais de voyage, commis de M. Duplain est fixé à deux mille livres par an pendant quatre ans. Les frais de prospectus, ports de lettres, etc. et autres seront alloués séparément.

12. Les associés fixent le prix de cette édition à dix livres aux particuliers pour les volumes de discours, dix-huit livres pour les volumes de planches, à sept livres dix sols pour les libraires, quinze livres dix sols pour les planches et les treizièmes exemplaires gratis. C'est à ces divers prix qu'ils se rendront compte.

13. Les associés se fourniront tous les mois le détail des souscriptions placées.

14. On arrêtera tous les six mois le compte en recette et dépense, et dans le cas où la recette excédera la dépense, comme on a lieu de l'espérer, le partage du bénéfice se fera entre les associés.

15. M. Duplain étant actuellement en recette pour les souscriptions qu'il a reçues, se charge de faire les avances de cette édition, et pour sûreté de la moitié des avances que devrait faire M. Panckoucke,. ce dernier s'oblige à l'instant que l'impression sera commencée de fournir à M. Duplain une obligation de vingt mille livres, payable à un an, laquelle restera entre leurs mains, et dans le cas où la dépense excéderait les premières recettes, M. Panckoucke tiendra compte des intérêts à raison de cinq pour cent par an, et au bout de l'année ce billet sera renouvelé.

16. Cette entreprise se faisant aux risques, pertes et fortunes des associés, dans le cas où elle essuyerait des pertes, elles leur seraient communes.

17. M. Panckoucke ayant actuellement une société formée pour une nouvelle édition de l'*Encyclopédie,* laquelle doit être refondue et revue en entier par ses anciens auteurs et les gens de lettres les plus considérables de la capitale, promet de ne publier aucun prospectus à ce sujet d'ici à deux ans à compter de la date du présent acte, sous peine de dépens, dommages et intérêt.

18. Il sera libre à MM. Duplain et Cie. de s'intéresser pour trois douzièmes dans cette entreprise du Sr. Panckoucke, aux charges, clauses et conditions des personnes qui y sont déjà intéressées, lorsque l'entreprise commencera.

19. M. Panckoucke se charge de l'entrée à Paris pour les exemplaires qu'on pourra y placer et les associés se donnent réciproquement leur parole d'honneur pour le secret du présent acte. Fait double sous leurs signatures privées, avec promesse d'en passer acte par devant notaire, à la première

réquisition de l'un d'eux, à Dijon, quatorze janvier mil sept cent soixante et dix-sept

ont signé   C. Panckoucke et Joseph Duplain et Cie.

Rétrocession du 3 février 1777

Je soussigné déclare que l'intérêt de moitié que M. Duplain m'a cédé dans la susdite édition in-quarto appartient à la société que j'ai formée pour une nouvelle édition de l'*Encyclopédie,* dont il est fait mention article 17, de sorte que MM. de la Société Typographique de Neuchâtel, qui sont intéressés pour moitié dans l'édition projétée de l'*Encyclopédie* seront intéressés pour un quart dans l'édition du sr. Duplain et n'auront aucun fonds à faire pour ledit quart.

Paris, 3 février 1777   C. Panckoucke

Addition du 23 janvier 1777

Addition au traité sous signature privée passée à Dijon le quatorze janvier, présent mois, entre le sr. Panckoucke et le sr. Joseph Duplain et Cie.

Attendu les difficultés qu'éprouve l'exécution dudit acte, le ministère le regardant comme contraire aux intérêts des gens de lettres, M. Thomas Leroi, associé aux Srs. Duplain et Cie., étant de retour à Paris pour lever les difficultés élevées au sujet du susdit acte, a chargé le sr. Panckoucke de faire toutes les démarches convenables pour surmonter les obstacles qui se rencontrent en cette occasion, et à cet effet il l'autorise à offrir à qui il appartiendra une somme de cent pistoles par chaque volume de discours, à l'effet d'obtenir les facilités nécessaires pour l'entrée de cette édition en France, à condition que ledit paiement de cent pistoles n'aura lieu qu'à fur et mesure que les volumes paraîtront, et le paiement formera un objet de dépenses communes. Fait à Paris ce 23 janvier 1777

a signé   Thomas Leroi

Appendix A

ɣɣɣɣɣɣɣɣɣɣɣɣ

VII. Paris, 28e mars 1777, Accession de La Société Typographique au Traité de Dijon entre MM. Panckoucke, libraire à Paris, & Joseph Duplain, libraire à Lyon

Nous soussignés le Banneret Ostervald, Bosset DeLuze, et le Professeur Bertrand, formant La Société Typographique de Neuchâtel en Suisse, nous nous engageons envers M. Panckoucke de souscrire à l'acte passé entre lui et Messieurs Joseph Duplain et Compagnie, libraire à Lyon, en date du quatorze janvier, mil sept cent soixante et dix-sept, relativement à une édition de l'*Encyclopédie* en 32 volumes in-quarto, et nous ratifions tous les articles dudit acte aux conditions suivantes.

1. que M. Duplain et Cie. nous donneront à imprimer à Neuchâtel trois volumes de ladite édition, ainsi qu'il le promet par sa lettre du 16 mars courant, lesquels volumes seront payés aux prix stipulés dans l'acte et à l'instant que nous les fournirons assemblés et collationnés, nous obligeant expressément de les imprimer avec un caractère mat de philosophie, sur beau papier d'Auvergne, et enfin conformément au modèle que nous fera passer M. Duplain, promettant en outre de veiller à la correction des épreuves et les fournir six mois après que M. Duplain nous en aura remis la copie. Nous nous soumettons encore relativement à ces trois volumes d'y ajouter les Suppléments, si ce travail convient à M. Duplain et s'il n'a point pris des engagements à ce sujet. Dans le cas où nous serions chargés de cette refonte, pour ces trois volumes ou plus, ils nous seront payés chacun six cent livres, conformément audit acte du 14 janvier 1777.

2. que Messieurs Duplain et Compagnie s'engageront par écrit, ainsi qu'ils le promettent par leur lettre du 20e mars 1777, de ne point contrefaire soit directement, soit indirectement notre nouvelle édition refondue, dans le cas où ils ne

prendraient pas d'intérêt dans notre dite édition, conformément à la clause 18 de l'acte de Dijon.

3. Nous confirmons par le présent acte tous les articles mentionnés dans nos traités antérieurs, savoir l'acte du 3e juillet 1776, la première addition du 31 août 1776, et la 2e du 3 janvier 1777.

Fait double à Paris ce 28 mars, mil sept cent soixante dix-sept

La Société Typographique de Neuchâtel en Suisse

C. Panckoucke

ΥΥΥΥΥΥΥΥΥΥΥΥ

VIII. Paris, 28e mars 1777, Troisième addition à l'acte du 3e juillet 1776

Troisième addition à l'acte fait et passé à double à Neuchâtel le trois juillet, mille sept cent soixante-seize, entre Messieurs Frédéric Ostervald, Bosset DeLuze, et Jean Elie Bertrand, d'une part, et Charles Joseph Panckoucke, d'autre part.

1. Messieurs de la Société Typographique de Neuchâtel nommés ci-dessus ont rétrocédé à Monsieur Panckoucke le douzième mentionné dans l'article 7 de la seconde addition datée du 3e janvier 1777, en sorte qu'ils ne sont plus intéressés dans l'acte du 3e juillet 1776 que pour cinq douzièmes, et ils n'ont consenti à cette rétrocession d'un douzième qu'à condition qu'il serait cédé à Monsieur Suard, éditeur principal de l'entreprise.

2. Par cette rétrocession d'un douzième, les trente-six billets de 3066^{l}13^{s}4^d chacun, mentionnés dans ladite deuxième addition du 3e janvier 1777, ont été annulés et convertis en quarante-huit billets à ordre, savoir

douze de mille livres chacun en 1778		╫12000.—
douze de deux mille livres chacun en 1779		╫24000.—
douze de deux mille livres chacun en 1780		╫24000.—
douze de deux mille six cent soixante-six livres en 1781		╫29326.—
un de deux mille six cent soixante-quatorze livres en 1781		╫ 2674.—
		╫92000.—

Nous disons la somme de quatre-vingt-douze mille livres faisant le prix de cinq douzièmes.

3. Nous confirmons par le présent acte tous les articles mentionnés dans nos traités antérieurs, savoir l'acte du 3e juillet 1776, la première addition du 31e août 1776, et la seconde du 3e janvier 1777. Fait à double à Paris ce 28e mars 1777.

La Société Typographique de Neuchâtel en Suisse

C. Panckoucke

ʏʏʏʏʏʏʏʏʏʏ

IX. Copie de l'engagement de M. Duplain en date du 28 mai 1777

Nous soussignés Joseph Duplain et Cie. nous obligeons de donner à imprimer à la Société Typographique de Neuchâtel trois volumes de l'*Encyclopédie* in-quarto, aux conditions du traité passé à Dijon, comme aussi de ne point imprimer ni directement ni indirectement l'édition que M. Panckoucke prépare de la même *Encyclopédie* avec augmentations et corrections, à laquelle nous avons une part, si, dans le temps, elle nous convient.

Lyon, 28 mai 1777   sig. Jos. Duplain et Cie.

Je soussigné déclare le présent conforme à l'original.

C. Panckoucke

ʏʏʏʏʏʏʏʏʏʏ

X. Copie du Traité pour la *Table analytique* entre M. Duplain et M. Panckoucke [September 29, 1777]

Nous soussignés Jos. Duplain et Compagnie libraires à Lyon d'une part et Ch. Panckoucke libraire à Paris d'autre part sont convenus de ce qui suit :

Le sieur Panckoucke, propriétaire d'une Table analytique et raisonnée des matières de l'*Encyclopédie* et des *Supplé-*

ments, laquelle doit former un volume in-folio en deux parties, s'oblige d'envoyer à MM. Duplain et Compagnie à mesure de l'impression de ladite Table les feuilles qui sortiront de dessous presse. Lesdites feuilles seront remises à M. l'abbé Laserre, qui se charge d'y faire le travail et les corrections nécessaires pour en composer une table en deux volumes in-quarto petit caractère, laquelle servira de table à l'édition de l'*Encyclopédie* in-quarto sous le nom de Pellet dans laquelle les sieurs J. Duplain et Panckoucke se sont intéressés. Les sieurs Jos. Duplain et Compagnie et Ch. Panckoucke s'associent de moitié dans l'entreprise de ladite table in-quarto, qui se fera à frais communs et la vente par M. Duplain seul, qui l'annoncera au public par souscription lorsque les derniers volumes de l'*Encyclopédie* in-quarto seront prêts à paraître.

On déterminera alors le prix de ces deux volumes de tables ainsi que celui de l'impression et du papier. M. Duplain ne pourra compter aucuns frais d'écritures, de commis et de gestion particuliers.

Il est convenu qu'on donnera à M. de Laserre pour son travail deux mille quatre cents livres pour les deux volumes, et il est entendu que cette table in-quarto ne pourra avoir plus de deux volumes.

M. Panckoucke fournira le cuivre qui [contient] l'explication détaillée du système des connaissances humaines, et ce cuivre sera retouché et tiré aux frais des associés.

Les soussignés engagent réciproquement leur parole d'honneur de garder le secret sur l'exécution de cette table et sur les conventions particulières de l'acte.

Lyon 29 septembre 1777. Signé Ch. P. J. Dup. & C.
Je certifie le présent conforme à l'original
Lyon 13 février 1779 Panckoucke

ᲧᲧᲧᲧᲧᲧᲧᲧᲧᲧᲧ

XI. Acte du 30 septembre 1777

Je soussigné C. Panckoucke, libraire de Paris, en conséquence de la déclaration du sr. Joseph Duplain et Compagnie qu'ils ont placé quatre mille quatre cent sept souscriptions de l'*Encyclopédie* in-4° sous le nom de Pellet, et, après m'en être assuré par moi-même d'après le registre des souscriptions, j'ai consenti comme en effet je consens qu'ils tirent encore un nombre de trois rames dix mains par feuille, et qu'il leur sera tenu compte des prix suivants, savoir

L'impression par feuille pour les volumes à recomposer et tirer au nombre de trois rames dix mains à trente-trois livres.

Pour l'augmentation de tirage sur chaque feuille qu'on ne recompose pas et sur lesquelles on tirera trois rames dix mains, après le nombre premier de huit rames seize mains, ladite augmentation à dix-sept livres dix sols.

Et pour chaque rame de papier employée à cette partie seulement de trois rames dix mains, neuf livres dix sols au lieu de neuf livres cinq sols portées dans une addition du premier acte passé à Dijon, du quinze mai 1777.

Je soussigné C. Panckoucke déclare encore que j'approuve l'augmentation que lesdits sieurs Joseph Duplain et Compagnie ont promis à Monsieur l'abbé de Laserre de deux cent cinquante livres par volume de plus, vu l'augmentation du nombre de feuilles de chaque volume, qui sont aujourd'hui de 110 à 115 feuilles et qui ne devaient être d'abord que de 90 à 95 feuilles, vu aussi la promesse et l'engagement réitéré pris par M. de Laserre de fournir la totalité de la copie dans toute l'année 1779 au plus tard, et considérant que par cette raison il est obligé de payer à ses frais de nouveaux aides. Ladite augmentation de deux cent cinquante livres ne pourra être augmentée sous aucun prétexte quelconque.

Lyon, ce 30 septembre 1777.   signé C. Panckoucke

Je soussigné reconnais que Monsieur Panckoucke m'a remis le présent consentement. Lyon, ce 30 septembre 1777   signé Joseph Duplain

Je déclare la présente copie conforme à l'original, qui est dans mes mains.

Lyon, 8 octobre 1777   C. Panckoucke

ᵧᵧᵧᵧᵧᵧᵧᵧᵧᵧᵧ

XII. Copie du traité conclu avec Benard graveur pour les planches de l'*Encyclopédie* in-4° Paris, 28 décembre 1777

Les soussignés Monsieur Panckoucke, libraire à Paris, d'une part, et Monsieur Benard, graveur, rue St. Thomas, porte St. Jacques, d'autre part, sont convenus de ce qui suit.

M. Benard se charge de réduire des figures etc. de 560 in-folio, première édition, pour les graver et faire graver avec la plus parfaite exactitude possible dans les figures et lettres, en 280 planches doubles de l'in-4°, qui composeront trois volumes de ce dernier format, lesquels sont particulièrement destinés pour une édition in-4° publiée à Genève sous le nom de Pellet, en 32 volumes, ces trois volumes de planches compris.

Il reconnaît et s'engage à examiner les planches de chaque volume après un premier tirage de quinze cent pour faire reprendre les parties de gravure qui faibliront, afin que le deuxième tirage de quinze cent soit parfaitement semblable au premier.

Après le tirage de 3000 épreuves de chaque cuivre en deux parties, et dont le sr. Benard répond, il fera planer tous les cuivres de chacun des trois volumes pour en faire une retouche entière de figures et de lettres absolument conformes à celles des trois premiers mille d'impression de toutes ces planches, lesquelles seront tirées une seconde fois au nombre de 3000 de chaque cuivre, dont ledit sr. Benard reconnaît de reprendre la gravure de ces cuivres qui ne pourront pas être tirés au deuxième nombre de trois mille dont il répond.

Pour lesdites réductions de dessins, de gravure de ces mêmes dessins diminués au moins de moitié du nombre de ceux des anciennes planches in-folio, de l'entretien de cette gravure pour deux tirages de 1500 chaque cuivre, et aussi pour le planage et l'entière retouche des figures et lettres de chaque planche après le premier tirage de 3000, à l'effet d'en faire un deuxième tirage de 3000 : M. Panckoucke s'engage de payer comptant au sr. Benard, à chaque livraison de vingt planches terminées, la somme de soixante-six livres chaque planche double de l'in-4°, les doubles comptés pour deux ; et la somme de vingt-cinq livres par planche à chaque livraison de trente planches replanées et retouchées après le premier tirage de 3000. Il s'engage aussi de lui payer comptant, en recevant la livraison de vingt planches, une somme de 9 livres pour chacune des quarante-sept planches du tome premier, qu'il a remises précédement, lesquelles ne lui ont été payées que 57 livres chaque planche double de l'in-quarto.

M. Panckoucke se charge également de remettre au sr. Benard deux exemplaires de cette nouvelle édition en 32 volumes, le premier lui étant dû de droit comme rédacteur et graveur des planches de cette édition, et qu'il lui délivrera à fur et mesure que les volumes paraîtront ; et le second comme indemnité de la condition sans équivalent dont ledit sr. Benard est chargé par le présent traité de répondre d'un tirage de 3000 de chacun des cuivres retouchés composant les trois volumes de planches de cette édition in-4°. Fait double, à Paris, le 28 decembre 1777

signé   Panckoucke
Benard

J'approuve le présent acte tant en mon nom qu'en celui de M. Duplain et Cie.

Paris, le 28 décembre 1777   signé   Merlino de Giverdy

Le présent acte annule le précédent, lequel était signé de MM. Leroy et de Giverdy.

ʏʏʏʏʏʏʏʏʏʏʏʏ

XIII. Agreement to Buy Off Barret and Grabit, June 24, 1778

Nous soussignés Joseph Duplain et Cie., libraires à Lyon d'une part, et Joseph Sulpice Grabit et Jean Marie Barret, libraires à la même ville d'autre part, sommes convenus de ce qui suit.

571

1. que nous Grabit et Barret convenons et promettons de cesser et discontinuer l'édition de L'*Encyclopédie* que nous avions commencée.

2. de remettre au sieur Duplain et Cie. six feuilles déjà faites de ladite édition au nombre de quatre rames et huit mains.

3. de n'entrer, favoriser, ni nous intéresser dans aucune édition qui pourrait être entreprise de cet ouvrage sous peine de vingt mille écus de dédommagement, et moi, Barret, en ma qualité d'imprimeur, de ne pouvoir en imprimer sous la même peine.

4. en conséquence du désistement ci-dessus de la part desdits srs. Barret et Grabit et pour les indemniser, tant des frais et avances qu'ils ont déjà faits que du bénéfice qu'ils auraient pu faire, soit en continuant leur édition, soit en participant à celle du sr. Duplain, dans laquelle il leur offrait de prendre part, ledit sr. Duplain s'engage à compter auxdits srs. Grabit et Barret la somme de vingt-sept mille livres, dont trois mille en ce courant paiement de Pâques et vingt-quatre mille livres en paiement de Pâques, mil sept cent soixante dix-neuf.

5. attendu que les srs. Duplain et Cie. ne donnent l'indemnité ci-dessus que pour éviter la concurrence qui naîtrait indispensablement de la part de l'édition des srs. Grabit et Barret, et afin de mieux assurer le succès de la leur, dans le cas où la première édition du sr. Duplain tirée à douze rames six mains vînt à ne pas être placée entièrement et que le sr. Duplain pût en représenter cinq cents exemplaires invendus, ce cas arrivant par l'effet de quelque critique faite dudit ouvrage ou autre cause imprévue avant le terme de l'échéance des promesses du sr. Duplain, les srs. Grabit et Barret se désistent des dernières vingt-quatre mille livres, dont ils déchargent le sr. Duplain et Cie., se contentant du premier paiement de trois mille livres.

6. le cas arrivant où le sr. Duplain pourrait représenter les cinq cents exemplaires invendus de sa première édition mentionnée ci-dessus, les srs. Barret et Grabit auront la liberté d'acheter, s'ils le jugent à propos, lesdits cinq cents exemplaires à raison de la moitié du prix de la souscription pour le libraire, et moyennant ledit achat de la part desdits srs. Grabit et Barret, le susdit paiement de vingt-quatre mille livres s'effectuera et sera exécuté par le sr. Duplain.

7. les srs. Grabit et Barret reconnaissent avoir reçu des srs. Duplain et Cie. en conséquence des présentes conventions leur promesse en date de ce jour à l'ordre des srs. Grabit et Barret de la somme de trois mille livres, payable en ce courant paiement de Pâques, et pour les vingt-quatre mille livres restantes, ils remettront leurs promesses de ladite somme payable au paiement de Pâques, mil sept cent soixante dix-neuf, à l'ordre des srs. Grabit et Barret entre les mains de M. Montoimat, notaire de cette ville, pour y être déposée jusqu'à l'époque dudit paiement. Ainsi convenu et fait double à Lyon le vingt-quatre juin, mil sept cent soixante dix-huit.

Conforme à l'original, Paris, 14 octobre 1778.

C. Panckoucke

ΥΥΥΥΥΥΥΥΥΥΥΥ

XIV. Copie du Traité signé à Paris le 10 octobre 1778 entre MM. Panckoucke et Duplain touchant la 3ème édition

Nous soussignés Charles Joseph Panckoucke, libraire de Paris, d'une part et Merlino de Ghiverdy, traitant tant en son nom qu'en celui des srs. Joseph Duplain et Compagnie, desquels il est fondé de procuration en date du 31 octobre 1777 signée Lagnier notaire et son confrère, d'autre part, sont convenus de ce qui suit.

1. La première édition de l'*Encyclopédie* 4° tirée à 4000, dont il est fait mention dans l'acte de Dijon du 14 février 1777, étant épuisée ; la 2e, dont il est fait mention dans un acte du 30 septembre 1777, tirée à trois rames, dix mains, étant pareillement épuisée ; les soussignés sont convenus de faire une 3e édition à quatre rames, seize mains, à laquelle MM. Duplain et Compagnie promettent de donner tous leurs soins ; et afin que cette troisième édition soit parfaitement complète, lesdits srs. Duplain et Compagnie feront choix d'un réviseur auquel il sera accordé mille livres en commençant, mille livres au milieu de l'ouvrage, et mille livres à la fin : en tout trois mille livres.

2. Le prix de l'impression de cette 3e édition sera comme les précédentes, de trente livres pour le premier mil, et huit livres du mil suivant : et comme l'on tire à quatre rames, seize mains, le prix de l'impression de chaque feuille à ce nombre est fixé à quarante-quatre livres ; et le prix de chaque rame de papier à dix livres, à cause du prix de l'augmentation sur le papier.

3. L'impression de cette troisième édition se fera à Lyon et autres villes de France ; dans le cas que les srs. Duplain et Compagnie jugeraient convenable d'en faire faire quelques volumes à Genève et en Suisse, la Société payera les frais de port de Genève à Lyon sur les quittances et lettres de voiture que produiront les srs. Duplain et Compagnie. Les frais de prospectus, ports de lettres et autres seront alloués séparément.

4. Quoique par l'acte de Dijon, art. 11 on ait fixé le prix de magasin, de séchage, assemblage, commis teneur de livres, frais de voyage, commis de Mr. Duplain, à deux mille livres par an, pendant quatre ans ; cependant comme l'on tire aujourd'hui toutes ces éditions à huit mille, il est convenu que l'article des dépenses ci-dessus serait porté à seize mille livres en quelque temps que l'édition finisse.

5. M. Duplain et Compagnie voulant d'après l'acte de Dijon, du 14 janvier 1777, art. 4, porter en dépense commune le prix du transport à Genève des volumes imprimés en France de l'*Encyclopédie* pour les trois éditions, il est convenu qu'on tiendra compte auxdits srs. Duplain et Compagnie du prix du transport des volumes de l'*Encyclopédie* imprimés à Lyon, et autres villes de France, comme si réellement ils eussent été transportés à Genève ; et ce prix se fixera d'après le taux le plus bas de cette espèce de transport. Au moyen de ce consentement, MM. Duplain et Cie. seront seuls et personellement responsables de tous les événements quelconques et frais que pourraient occasionner la saisie des exemplaires de ces éditions, soit à Lyon, soit dans ses magasins.

Le sr. Panckoucke n'a consenti à la présente clause qu'autant que la Société Typographique de Neuchâtel y donnera son approbation ; et dans le cas de refus de sa part, la clause 4e et la présente clause seront nulles, et MM. Duplain et Cie. resteront dans toutes leurs prétentions, tant par rapport à ces frais de transport que relativement au prix de magasin, de séchage, assemblage, commis teneur de livres, frais de voyage, commis de MM. Duplain et Cie. qui ont été alloués par l'acte de Dijon, art. 11e à 2000 livres par an pendant quatre ans pour quatre mille, lesquels frais MM. Duplain et Cie. prétendent former pour eux un objet de dépense annuelle infiniment plus considérable.

Dans le cas de contestation, soit à cet égard, soit pour tout autre objet relatif à cette entreprise, il est expressément convenu, et les associés s'y obligent sur leur parole d'honneur, de remettre la décision de toutes les difficultés quelconques à l'arbitrage de trois ou de cinq avocats de Lyon au choix des parties intéressées et de s'y soumettre sans appel.

6. M. Panckoucke donne son consentement à l'acte passé entre les srs. Duplain et Cie. d'une part et Joseph Sulpice Grabit et Jean-Marie Barret d'autre part, en date du 24 juin 1778.

7. M. Duplain ayant fait quelques dépenses pour des presses, elles seront un objet de dépense commune, comme aussi les voyages des ouvriers que les circonstances ont obligé de faire venir de Paris.

8. Monsieur Panckoucke consent que l'on donne gratis les volumes qui excéderont le nombre de trente-six, suivant que l'on s'y est engagé vis-à-vis des souscripteurs.

Fait double entre nous à Paris ce dix octobre mille sept cent soixante dix-huit. signé C. Panckoucke
Merlino De Ghyverdy

ᵞᵞᵞᵞᵞᵞᵞᵞᵞᵞᵞᵞ

XV. Acte passé à Lyon le 10 février 1779

Je soussigné Charles Panckoucke libraire à Paris et Joseph Duplain et Compagnie libraires à Lyon avons fait un relevé des registres tenus pour l'*Encyclopédie* en présence de MM. de la Société typographique de Neuchâtel et avons reconnu qu'il avait placé le nombre entier des deux premières éditions de l'*Encyclopédie* tirées au nombre de douze rames six mains dont ledit sieur Joseph Duplain et Compagnie s'obligent de nous tenir compte à la forme des traités sur le pied de sept livres dix sols le volume de discours en feuilles et de quinze livres dix sols le volume de planches aussi en feuilles, déduction faite des treizièmes exemplaires que nous donnons gratis; et nous avons reconnu que la totalité des dépenses pour ces deux premières éditions et même pour la troisième, qu'on tire à quatre rames quinze mains, étant à peu près faite, M. Duplain et Compagnie sont encore en avances.

Les soussignés sont convenus relativement à cette troisième édition de la mettre en société, à l'exception de mille exemplaires dont le sieur Ch. Panckoucke prend cinq cents pour son compte particulier et le sieur Joseph Duplain et Compagnie cinq cents aussi pour son compte particulier. Fait à double à Lyon le 10 février 1779.

Signé Ch. Panckoucke
J. Duplain & Ce.
la Société typographique de Neuchâtel

Je soussigné déclare la présente copie conforme à l'original, qui est resté dans mes mains. Lyon 13 février 1779.

C. Panckoucke

ᵧᵧᵧᵧᵧᵧᵧᵧᵧᵧᵧᵧᵧ

XVI. The Panckoucke–STN Agreement of February 13, 1779

Le sieur Panckoucke ayant pris par l'acte ci-dessus la quantité de 500 exemplaires de la 3me. édition de l'*Encyclopédie,* ce pour le compte de sa compagnie, Messieurs de la Société typographique de Neuchâtel intéressés pour 5/12, je dis cinq douzièmes, dans lesdits cinq cents exemplaires, sont convenus, comme en effet ils conviennent, de prendre pour leur compte à leurs risques, périls et fortune la quantité de deux cent huit exemplaires entiers formant les cinq douzièmes dudit nombre de cinq cents. Au moyen de ce partage le sieur Panckoucke et la Société typographique s'obligent réciproquement de maintenir rigoureusement le prix de cet ouvrage ainsi qu'il est établi dans le prospectus pour les libraires.

De plus le sieur Panckoucke désirant obliger Messieurs de la Société typographique a consenti comme en effet il consent intéresser ladite Société typographique ce acceptant pour cinq vingt-quatrièmes dans la Table des matières de l'*Encyclopédie* in-quarto, promettant ladite Société de s'en rapporter aux arrangements que ledit sieur Panckoucke prendra avec M. Joseph Duplain et Compagnie relativement à l'impression et le débit de cette Table, conformément à l'acte passé à Lyon le 29e septembre 1777, qu'ils approuvent et dont copie leur a été remise; fait à double à Lyon ce treizième février mille sept cent soixante dix-neuf.

Panckoucke   La Société typographique de Neuchâtel en Suisse

P.S. Au moyen du partage des exemplaires ci-dessus et de la cession que fait Monsieur Panckoucke d'un intérêt dans la Table des matières de l'*Encyclopédie,* Messieurs de la Société typographique de Neuchâtel prennent de nouveau l'engagement solidaire et un seul pour le tout d'acquitter leurs billets à ordre montant à quatre-vingt douze mille livres mentionnés dans un acte passé à Paris le vingt-huit mars mille sept cent soixante dix-sept; fait à double à Lyon ce treizième février mille sept cent soixante dix-neuf.

Panckoucke
                 La Société typographique de Neuchâtel en Suisse

ⲧⲧⲧⲧⲧⲧⲧⲧⲧⲧⲧⲧ

XVII. The Perrin Agreement, August 3 and 13, 1779

Nous soussignés Charles Panckoucke libraire de Paris et Duplain de Ste Albine sommes convenus ce ce qui suit, savoir:
Moi Joseph Duplain après avoir rendu compte de vive voix à M. Panckoucke de l'état de la vente du *Dictionnaire de l'Encyclopédie,* je l'ai engagé pour terminer cette opération d'accepter l'offre qu'on me fait de prendre environ trois cent quatre-vingt exemplaires qui restent de cet ouvrage à Lyon et environ deux cents à Paris qui appartiennent à M. Panckoucke au prix de quatre livres chaque volume de discours payable et de huit livres chaque volume de planches, le tout en blanc, ce qui fait monter chaque exemplaire à cent cinquante-six livres, sous condition que se réserve l'acheteur qu'on n'annoncera aucune édition augmentée de ce livre jusqu'

au mois d'août 1780, sous peine de tout dépens, dommages et intérêts; et moi Charles Panckoucke croyant qu'il convient à la compagnie de terminer de cette manière cette affaire, afin de recevoir le produit net de l'objet à la fin de cette année en espèces et billets qu'il aura donné, autorise Monsieur Duplain à accepter l'offre qu'on lui fait de prendre ce qui reste de l'*Encyclopédie* au prix de 156 livres chaque exemplaire payable en valeurs à ma satisfaction à échoir dans le courant de février mille sept cent quatre-vingt, me réservant seulement d'obtenir la ratification de la Société typographique de Neuchâtel, mes associés; et si je ne donne pas opposition de leur part aux présentes dans quinze jours à compter du cinq août 1779, c'est-à-dire le 20 de ce mois, la présente autorisation aura son plein et entier effet pour le sr. Duplain, et la vente qu'il fera des exemplaires restants sera bien et dûment faite pour le compte de la compagnie intéressée à l'*Encyclopédie*.

Fait double à Paris ce 3e août 1779.

Signé   Duplain de Ste. Albine
Panckoucke

Je soussigné Joseph Duplain reconnais avoir vendu et livré à M. Perrin commissionnaire à Lyon le nombre de quatre cent vingt-deux exemplaires de l'*Encyclopédie* complets, qui sont à Lyon appartenant à la Compagnie intéressée à ladite *Encyclopédie* et en outre le nombre de cent soixante exemplaires dudit livre complets, qui appartiennent au seul M. Panckoucke, lequel M. Panckoucke tient à sa disposition à Paris. Je reconnais de plus avoir reçu dudit M. Perrin la somme de soixante cinq mille huit cent trente-deux livres et sa promesse en février prochain pour les quatre cent vingt-deux exemplaires de l'*Encyclopédie,* qui sont à Lyon, que je lui ai livrés, à l'exception des trois volumes de planches et du tome 36, que je lui livrerai avant l'expiration de 3 mois. Quant aux cent soixante exemplaires qui sont à Paris, Mr. Panckoucke les lui livrera avant, à lui ou ses ayant cause, contre des valeurs à la satisfaction dudit sieur Panckoucke payables dans Paris et à échoir en février prochain pour la somme de vingt-quatre mille soixante livres à raison de 156 livres chaque exemplaire; fait à double à Lyon 13e août 1779.   Signé J. Duplain

Signé Perrin

Je certifie ce que dessus conforme à l'original que j'ai entre mains.   Signé Joseph Duplain

Appendix A

XVIII. The Formal Settlement of February 12, 1780

Nous soussignés Joseph Duplain d'une part et Charles Panckoucke, libraire de Paris, de l'autre, sommes convenus de ce qui suit, savoir :

1. Moi Charles Panckoucke, après avoir pris communication des registres du sieur Duplain tenus relativement à l'*Encyclopédie* du sieur Pellet dans laquelle j'avais un intérêt de moitié, après avoir reconnu que l'état des ventes que le sieur Duplain m'avait donné était très juste et conforme à l'exacte vérité, après m'être édifié sur la vente des quatre cent vingt-deux exemplaires, que de mon ordre Monsieur Duplain avait vendus au sr. Perrin et avoir parfaitement reconnu qu'ils existaient et que si M. Duplain nous a fait cession du bénéfice qu'il avait droit d'exiger sur cette affaire qui lui appartenait, c'est un effet de la générosité de ses procédés, en conséquence pour témoigner au sr. Duplain ma satisfaction, je lui ai cédé par les présentes tous mes droits et prétentions, tous ceux de mes divers associés sur la Table de ladite *Encyclopédie,* qui s'est publiée et annoncée en six volumes in-quarto sous le nom de M. Le Roy, dans laquelle je m'étais réservé un intérêt de moitié, que je lui cède, me réservant tous mes droits sur la Table in-folio, deux volumes.

2. Je lui maintiens en tant que de besoin son intérêt dans l'*Encyclopédie méthodique* à la forme de l'acte passé à ce sujet.

3. Je lui abandonne irrévocablement, tant en mon nom qu'à celui de mes associés, tous mes droits et prétentions sur l'entreprise de l'*Encyclopédie* du sieur Pellet, sur les bénéfices qui en résulteront, ainsi que tous les exemplaires invendus qui restent chez le sr. Duplain et tous les défets qui sont provenus de ladite édition, n'entendant comprendre dans cette cession les cuivres, qui me resteront ainsi que le privilège.

Et moi Joseph Duplain, en conséquence des cessions que me fait M. Panckoucke consens

1. de lui donner pour prix de ces divers objets une somme de cent septante six mille livres, en mes promesses à son ordre que je lui ai à l'instant remis, savoir
un tiers en décembre 1780
un tiers en octobre 1781
et le solde en août 1782

2. de fournir pendant six mois à dater de ce jour à M. Panckoucke et ses associés tous les défets que lui ou eux demanderont pour compléter leurs exemplaires.

3. de régler le compte des planches de M. Panckoucke demain la journée.

4. Je consens de me charger de la liquidation de ladite entreprise à mes perils et risques, sans avoir rien à répéter contre M. Panckoucke ni ses divers associés, sous quelque prétexte que ce soit, comprenant dans ladite liquidation toutes les difficultés qu'on pouvait me faire de quelque nature qu'elles puissent être, qui pourraient être élevées par les souscripteurs, et toutes les non valeurs résultantes des payements suspendus ou retardés, de l'événement desquels il restera seul responsable.

5. de lui abandonner la convention et les billets de la Société typographique de Lausanne que j'endosserai à ses périls et risques, moyennant que le sieur Panckoucke remplira les charges et conventions passées avec ladite Société typographique de Lausanne. Fait double et promis d'exécuter le présent traité suivant sa forme et teneur à Lyon ce 12 février 1780.

Signé Joseph Duplain et C. Panckoucke
Je déclare la présente copie conforme à l'original. Lyon ce 13e février 1780. C. Panckoucke

ᵧᵧᵧᵧᵧᵧᵧᵧᵧᵧᵧ

XIX. Documents Concerning the Informal Settlement

1. Relevé des registres des souscriptions donné par MM. Regnault et Rosset de la part de M. Duplain le 6e février 1780 [by Duplain]

Des 2 premières éditions		. . . 6009 pour 5570		
à déduire donné gratis		22	22	
		5987 pour 5548		
Ci	5987 pour 5548 à 294 l.			. . . 1,631,112 l.
3me. édition	580 pr.	526 à 294 l.		. . . 154,644
vendu par le				
traité fait à Paris	422 pr	422 à 156 l.		65,832
à M. Panckoucke	500 pr	500		
à J. Duplain	500 pr	500		
	7989	7496		
donné gratis	22	22		
déduit ci-dessus	8011	7518		. . . 1,851,588 l.

2. Produit net de l'entreprise tel qu'il doit être réellement [by Bosset]

Chaque volume de 124 feuilles, les 36 volumes 4464 feuilles, l'impression 115 l., le papier 161–10. Chaque feuille coûte de papier et d'impression 276–10 :

les 4464 feuilles		1,232,060 l.
volume d'explications		17,999
planches		119,423
l'abbé Laserre & au réviseur		33,600
Barret		27,000
Port à Genève	20,000 ⎫	
faux frais	30,000 ⎬	86,000
change & escompte	36,000 ⎭	
Dépense		1,516,082 l.

L'édition a produit   8411 exemplaires
          à déduire :
défets   400 ⎤
exemplaires partagés   1000 ⎬   1822
vendus à Perrin   422 ⎦

reste   6589
gratis   529
                                     6060

Recette
lesquels 6060 à 294 l.   . . . 1,781,640 1.
          422 à 156 l.   65,832

                                                           1,847,472

à revenir sur Perrin   48,828 ⎤
200 exemplaires en magasin   30,000 ⎬   98,828
vente de défets, Jossinet   20,000 ⎦   1,946,300
à déduire la dépense   1,516,082

                                                             430,218

Il faut ajouter ici les Tables que l'on
offre de reprendre au même prix   50,000

                                                             480,218 l.

          3. Tableau de ce qui devrait nous revenir de l'entreprise
               [by Bosset]
Par compte fait il doit y avoir à partager la somme
de   . . . 480,000
à quoi l'on peut encore ajouter le reste de la 3e
édition, qui est sûrement vendu ; c'est-à-dire
la ½ des 400 que nous comptons être en magasin
ou en défets :
fait   200 exemplaires dont déduit les 13èmes
fait   15 exemplaires

reste 185 exemplaires à 294   . . . 54,390

(NB surtout si nous lui abandonnons les défets)   534,390

D'où il résulte qu'en comptant les 34,390 l. tant pour l'escompte et
l'agio prétendu de M. Merlino, quoique c'est peut-être nous qui avons
des intérêts à répéter, il y aurait 500,000 l. à partager. Mais comme il

y a peut-être 100,000 l. à rentrer et qui rentreront sans doute d'ici au premier juillet, nous disons qu'il doit nous payer :

200,000 l. pour ce payement de rois

50,000 pour celui de Pâques échéant le 3e juillet.

Voilà, ce nous semble, les premières propositions d'accommodement que nous pouvons lui faire et dont notre avis n'est point de nous départir à moins qu'en nous donnant l'inventaire général que nous demandons il nous mette dans le cas de nous relâcher ou de lui abandonner la Table analytique.

4. Premier Mémoire de M. Plomteux

M. Duplain porte la totalité des 3 éditions de l'*Encyclopédie* non compris ceux à provenir des défets à 8011 exemplaires. Par le relevé des registres des souscriptions il conste qu'on avait placé en février 1779, y compris les exemplaires gratis   7373 exemplaires

Dans la liste des souscriptions qui vient
d'être produite par le sr. Duplain on
trouve non porté dans le relevé de
février 1779   137
                                              ―――――
                                              7510

Il ne restait donc réellement d'invendus à cette
époque de février 1779 que   501
                                              ―――――
                                              8011 exemplaires

Le sieur Duplain en cachant à sa compagnie cette situation avantageuse, en l'intimidant au contraire sur la cessation des demandes, trouvait le moyen de l'engager au partage de 1000 exemplaires. Il a non seulement vendu pour son compte particulier les 500 exemplaires de sa part; mais par une nouvelle perfidie il a trouvé le moyen d'escroquer sous le nom supposé de Perrin 422 exemplaires, qui restaient invendus selon lui. Il les a obtenus en exagérant le discrédit de leur édition au prix de 156 livres, tandis qu'ils étaient déjà placés ainsi que ces 500 au prix de 294 livres en souscriptions reçues pour et du nom de la Société.

Monsieur Panckoucke n'ayant nulle part à ces ventes, privé de son bénéfice, perdant moitié sur les 422 exemplaires du supposé Perrin, exposé à devoir donner au rabais les 500 exemplaires restants de l'édition dont on a eu l'art de le charger seul, et frustré par toutes ces ruses de plus du quart de sa propriété, n'en est pas moins chargé par son avide associé de tous les frais des trois éditions portées à force d'artifice à des sommes énormes, qui assurent au sieur Duplain plus de 100,000 écus de bénéfice, tandis qu'il ne reste à Monsieur Panckoucke, victime de sa bonne foi et de sa confiance, que les tristes regrets d'avoir sacrifié trois volumes déjà imprimés, plus de 9000 livres payés pour copies, ses cuivres, ses privilèges, et son travail de plus de douze années, qui aurait rendu ces titres respectables à tout autre que le sieur Duplain, travail qui assurait le succès d'une entreprise dont les trois quarts du bénéfice ne peuvent rassasier son insatiable cupidité.

Telle est la situation de M. Panckoucke, qui, réduit à la dure nécessité de livrer son associé infidèle à la vindicte publique, préfère de lui faire des sacrifices qui puissent le sauver de l'opprobre.

5. Ostervald and Bosset in Lyons to Mme. Bertrand in Neuchâtel, February 13, 1780

. . . Nous nous empressons, Madame, de vous communiquer la fin de notre combat avec Duplain, qui heureusement est terminé sans sang répandu.

Après avoir épuisé à nous quatre toutes nos réponses, appuyés de d'Arnal comme troupe auxiliaire; après avoir été obligés de dresser nous mêmes nos comptes sur les chiffons de Duplain et le dépouillement de nos actes; après nous être convaincus que la vente de Perrin était une vente supposée et une escroquerie de 48,000 livres qu'il voulait soustraire à la société et que le registre des souscriptions reçues qu'il venait de nous remettre comparé avec celui que nous relevâmes furtivement et sans qu'il s'en doutât il y a un an était faux; après que l'écrivain [Bosset] a été obligé de faire hier une descente chez lui à la tête des gens de justice, savoir commissaire, huissier, et procureur pour demander que ses livres soient compulsés juridiquement de même que ses magasins, ce qui a opéré l'aveu de l'escroquerie de ces 48,000 livres; l'après-midi ayant mis de ses propres parents et amis à ses trousses pour lui faire comprendre que nous allions le perde de réputation tant ici qu'à Paris, nous sommes parvenus à lui faire donner par le canal de M. Panckoucke la somme de 200,000 livres en ses billets [placés] dans 10, 15 et 20 mois, ce qui par le tableau que nous avions de l'entreprise et des accessoires énormes dont elle était chargée nous fait tous regarder comme un bonheur d'avoir terminé ainsi. S'étant toujours obstiné à ne vouloir donner que 128 [mille] livres, alléguant une perte considérable à essuyer sur la queue de cette affaire et des procès à soutenir contre quantité de souscripteurs etc. etc., il a fallu pour l'amener à cette somme lui abandonner le profit à faire sur la Table analytique, qui a été évalué à 25 mille livres, dont il avait déjà traité avec Le Roy, de façon que vous voyez, Madame, que bien malgré nous, il faut renoncer à cette entreprise. Il a fallu de plus y comprendre encore 24 mille livres que MM. de Lausanne payeront pour l'entrée de leur 8°, mais qui regardent seul M. Panckoucke, soit la Société de Liège, en faveur de qui nous nous sommes dépouillés volontairement de nos droits sur les privilèges de l'*Encyclopédie,* comme vous le savez, en sorte que M. Duplain donne 176 mille livres et que c'est M. Panckoucke par les 24 mille livres qui complète les 200 mille livres. Voici, Madame, de tout cela ce qui en résulte pour nous : nos 5/12 de ces 200 mille livres font   . . . 83,666 livres, qui avec ce que nous avons précédemment reçu de M. Panckoucke et notre part des 3 volumes vendus à la rame, déduction faite de ce que l'on avait payé à M. Suard, feraient à peu près le montant des 90,000 livres que nous avons payés à M. Panckoucke pour entrer dans ses privilèges en nos billets, dont la moitié environ est encore à payer dans le courant de cette année et la suivante, au moyen de quoi qu'en attendant le succès de l'*Encyclopédie méthodique* à laquelle M. Panckoucke travaille à force à Paris et laquelle est sous presse, nous avons de profit réel le montant de nos 208 exemplaires, qui évalués à 250 livres ferait environ 50,000, ce qui est en effet bien modique, après avoir été presque convaincu que nous pourrions doubler cette somme, et qui l'aurait été certainement sans les fautes impardonnables qu'il y a eu dans la gestion de cette affaire et dans les traités . . .

B

ʏʏʏʏʏʏʏʏʏʏʏ

SUBSCRIPTIONS TO THE
QUARTO *ENCYCLOPÉDIE*

The following list covers all but one of the 8,011 subscriptions the quarto publishers took as the basis for settling their accounts in Lyons in February 1780. Except for a few cases, it gives the names of retailers, not individual subscribers, for the publishers marketed almost all of the three quarto editions through bookdealers, who sold subscriptions to their local customers at the retail price and then bought the *Encyclopédies* wholesale through Duplain, the STN, or Panckoucke.

The list was transcribed from the STN's copy of Duplain's secret subscription register (ms. 1220 in the Bibliothèque de la Ville de Neuchâtel), which shows how he falsified the accounts. It was then checked against other information in the Duplain dossiers, correspondence, and account books of the STN. Finally, entries were grouped by city so that one can see the number of subscriptions sold by each retailer.

Although some of the retailing was done by persons other than booksellers, part I of the list may serve as a general guide to the most active dealers in the French book trade around 1777. (It should be compared with the *Manuel de l'auteur et du libraire,* Paris, 1777, which also lists booksellers by city, though not very accurately.) Part II covers subscriptions sold outside France. Part III concerns subscriptions sold directly to certain individuals by the publishers, presumably at the retail price. And part IV identifies the 25 copies given away by Duplain and Panckoucke, either as compensation for their collaborators or as offerings to their protectors.

As the publishers kept the wholesaling in their own hands,

the booksellers confined the retailing to their own clientele. Their letters confirm that they sold the quartos in their local markets instead of trading them among themselves or exporting them, as they did when they handled books wholesale. The list therefore gives an accurate picture of the quarto's geographical distribution—except in one place : Paris.

The Parisian sales probably came to something less than the 575 subscriptions mentioned on the list because Panckoucke most likely did not market all of his quartos in the capital. When he and the STN divided up the 500 last copies of the third edition, they acquired a stock that they could have sold to private individuals at the retail price or to other booksellers at the wholesale price. The 208 quartos that the STN received as its share of the 500 do not pose a problem; they have been traced through the STN's account books and entered in the list below (85 went to France and 123 to other European countries). But it is impossible to know what became of the 292 copies kept by Panckoucke. According to a note by Bosset at the end of the STN's list, Panckoucke was also to receive 52 quartos that Duplain had debited to the STN and that the STN had refused to accept for reasons explained in Chapter VII. Finally, Panckoucke sold 20 copies of the first two editions, making his total 364. Although his letters explain that he tried to market most of his quartos in Paris, he probably sold some to dealers located elsewhere. The list, therefore, gives a somewhat exaggerated impression of quarto consumption in Paris. Taken as a whole, however, it shows the distribution of about 60 percent of the *Encyclopédies* that existed in France and about a third of those in existence everywhere in the world before the French Revolution.

I. *French subscriptions*

Abbeville (26)		Angers (109)	
Pintiau	26	Parizot	108
		Merlet	1
Aire (8)		Argentan (3)	
Dubois	8	Lefrançois	3
Aix (6)		Arras (26)	
Pacard	6	Topino	26
Alençon (34)		Auch (65)	
Jouhann	34	Lacaza	65
Amiens (59)		Aurillac (13)	
Mastin	59	Armand	13

Autun (39)
Habert 39

Auxerre (10)
Fournier 10

Auxonne (1)
Laziers 1

Avignon (55)
Aubanel 6
Guichard 40
Niel 9

Bar-le-Duc (13)
Robert 13

Bayonne (16)
Trebose 14
Fauvel Duhard 2

Beaune (26)
Bernard 26

Beauvais (8)
Gaudet 8

Bergerac (13)
Bargeas 13

Bergues (1)
Vandelvegh 1

Besançon (338)
Lépagnez cadet 338

Billom (2)
De Bompar 1
Tiffalier 1

Bordeaux (356)
Bergeret 58
Chappuis frères 88
Gauvry 6
Labottière frères 165
Philipot 39

Boulogne-sur-Mer (34)
Battu 34

Bourg-en-Bresse (91)
Robert & Gautier 91

Bourg-Saint-Andéol (4)
Bandit 2
Guirmeau 2

Bourges (20)
Debeury 20

Brest (20)
Malas 20

Caen (221)
Le Roy 208
Manoury 13

Cambrai (57)
Berthoud 57

Carpentras (2)
De Châteaubon 2

Castelnaudary (27)
Borelly 27

Castres (28)
Le Potier 28

Chalon-sur-Saône (67)
De Livany 67

Châlons-sur-Marne (1)
Pavier 1

Champagne (2)
Le Marquis 2

Chartres (77)
Jouanne 77

Châtillon (39)
Cornillac Lamberc 39

Clermont (13)
Del Croz 13

Colmar (2)
Neukirck 2

Dijon (152)
Benoit 4
Capel 131
Frontin 17

Dole (52)
Chaboz 52

Douai (14)
Tesse 14

Embrun (3)
Moyse 3

Evreux (65)
Ancelle 65

Falaise (45)
    Bouquet — 45

Ganges (1)
    Pomaret — 1

Grenoble (80)
    Cuchet — 13
    Veuve Giroud — 66
    Ducros de Châteaubon — 1

Guéret (19)
    Piot — 19

Joinville (1)
    De Gaulle — 1

La Fère (15)
    Lunhit — 8
    Gentillon — 7

La Flèche (39)
———

Langres (26)
    Rouyer — 26

Laon (17)
    Melleville — 17

La Rochelle (59)
    Chabosseau — 13
    Pavie — 43
    Ranson — 3

Le Havre (52)
    Batry — 52

Le Mans (40)
    Robert la Commune — 14
    Monnoyer — 26

Le Puy (39)
    Boisserand — 39

Lille (28)
    Jacquiez — 28

Limoges (3)
    Marc Dubois — 3

Lisieux (27)
    De Launay — 23
    Mistral — 4

Loudun (1)
    Malherbe — 1

Lunéville (1)
    Richard — 1

Laigle (3)
    Glacon — 3

Lyon (1079)
    Arles l'aîné — 25
    Barret — 15
    Bousquet — 64
    Cizeron — 9
    Descherny — 1
    Faucheux — 2
    Grabit — 14
    Jacquenod — 65
    Perrin — 62
    Perisse — 77
    Rosset — 94
    Vatar — 13
    Cellier — 26
    Audambron de
      Salacy & Jossinet — 585
    Esparron — 1
    Chaulu — 1
    Pascal — 40
    Guiget (abbé de
      la Croix-Rousse) — 1
    Charlet — 1
    Gay — 2
    Barraud — 1

Macon (17)
    Garsin — 17

Mantes (8)
    Testard — 8

Marmande (1)
    Baillas de Lombarède — 1

Marseille (228)
    Lallemand — 13
    Caldesaigues — 52
    Mossy — 52
    Roullet — 28
    Allié — 1
    Sube & Laporte — 78
    Ricard — 4

Meaux (30)
    Charles — 26
    Prudhomme — 4

Melun (1)		Noyon (26)	
Prévost	1	Despalle	26
Metz (22)		Orléans (52)	
Gerlache	13	Couret de Villeneuve	39
Marchal	9	Letourmy	13
Millau (8)		Paris (575)	
Montbroussoux	6	Andriette & Ferino	17
Cousin de Mauvoisin	2	Razuret	20
Montargis (26)		E. Rose	50
Gille	26	Sepolino	13
Montauban (105)		Bougy	14
Beaumont	26	Boscary	1
Crosilhes	78	De la Rive	1
Cazamex	1	Esprit	9
		Le Comte d'Orsey	1
Montbrisson (6)		Monnier	42
Baumont	6	Panckoucke	364
Montpellier (169)		Chavissiez	1
Cézary	26	De la Motte	1
Rigaud Pons		Bacher	1
et Cie	143	Garçon	3
Morlaix (1)		Le Comte	1
Nicole	1	Boisgibault	2
		Huguenin	1
Mortagne (22)		Perregaux	1
Le Peguchet	22	Quandet de Lachenal	1
Moulins (52)		Batilliot	30
Enaut	52	Perigueux (36)	
Nancy (121)		Dubreuil	36
Babin	91	Peronne (15)	
Henry	2	Laisney	15
Bonthoux	1	Perpignan (52)	
Mathieu	26	Goully	52
Le Gros	1	Poitiers (65)	
Nantes (38)		Chevrier	65
Veuve Brun	4	Reims (24)	
Despilly	8	Petit	21
Vatar	26	Prevoteau	3
Nîmes (212)		Rennes (218)	
Buchet	104	Blouet	84
Gaude père et fils	105	Robiquet	91
Verot	3	Remelein	43
Niort (58)		Rethel (40)	
Elies	58	Migny	40

Riom (46)
De Gouette	11
Coste	35

Roanne (26)
Boisserand	26

Rochefort (27)
Romme	27

Roquemaure (7)
Giraudy	7

Rouen (125)
Le Boucher	108
Ab. Lucas	16
Bourgeois	1

Saint-Chamond (2)
Dugas	2

Saint-Dizier (3)
Fournier	3

Saint-Etienne (13)
Bernard	13

Saint-Flour (24)
Veuve Sardine & Fils	24

Saint-Lô (7)
Fabulet	7

Saint-Omer (5)
Huguet	5

Saint-Quentin (16)
Hautoy	13
Desnoyer	1
Rigaut	1
Harlé	1

Saintes (26)
De Lys	26

Saumur (1)
De Goury	1

Sedan (2)
Bechet de Balan	2

Sète (13)
Michel	13

Soissons (52)
Varoquier	52

Strasbourg (16)
Maynaurd	2
Turkheim	13
Gay	1

Tarbes (52)
Bourdin	52

Thiers (39)
Bernard	39

Toul (1)
Carez	1

Toulon (22)
Surre	21
Serjeans	1

Toulouse (451)
Gaston	130
Dalles & Vihac	6
Monavit	69
Sacarau	233
Robert	13

Tours (65)
Billaut fils	65

Troyes (53)
Sainton	52
André	1

Tulle (4)
Chirac	4

Valence (65)
Aurel	39
Muguet	26

Valenciennes (17)
Giart	13
Henry	4

Verdun (13)
Mondon	13

Versailles (5)
Le Comte de Neuilly	1
De Barthès	1
Deloubignac	1
Le Gommelin	1
Diesbach	1

Vichy (2)
Giraud (curé)	1
Rouganne	1

Villefranche (37)
Vedeilhé	37

II. *Non-French subscriptions*

Amsterdam (5)
Vlanc	1
Du Sanchoy	1
Rey	1
Changuion	2

Basel (1)
Tourneisen	1

Brussels (46)
Ricour	26
Le May l'aîné	20

Copenhagen (4)
Philibert	3
Schlegel	1

Dublin (13)
Luke White	13

Frankfurt am Main (2)
Holweg & Laue	2

Geneva (284)
Nouffer & Bassompierre	177
Pellet	41
Scala	13
Teron l'aîné	52
Bruncke	1

Genoa (32)
Gravier	32

Hamburg (2)
Virchaux	2

Haarlem (1)
Bosck	1

The Hague (13)
Bosset	13

Homburg (3)
Bruere	3

Lausanne (3)
Grasset	1
Pott	2

Leyden (3)
Murey	1
Luzac & Vandamm	2

Liège (52)
De Mazeau	26
Plomteux	26

Lisbon (1)
Bertrand	1

London (13)
Durand	13

Madrid (3)
Sancha	3

Maestricht (1)
Dufour & Roux	1

Mannheim (27)
Fontaine	26
La Nouvelle Librairie	1

Mantua (5)
Rome (professor)	5

Milan (1)
Carli	1

Moscow (4)
Rudiger	4

Munich (4)
Fritz	2
Daun (Count)	1
De La Luzerne	1

Naples (16)
Société typographique	16

Neuchâtel (39)
Société typographique	36
Convert	1
Saint-Robert	1
Fornanchoz	1

Nyon (2)
Michel	2

Pest (1)
Weingand & Köpf	1

Prague (5)
Gerle	5

Saint Petersburg (8)
  Weitbrecht 5
  Muller 3
Soleure (1)
  de Berville 1
Turin (53)
  Guibert & Orgeas 13
  Reycends frères 39
  Laurent 1
Utrecht (1)
  Spruyt 1
Venice (1)
  Zenaut 1
Warsaw (31)
  Gröll 18
  Lex 13
Worms (2)
  Wiechenhagen 2
Ypres (8)
  De Clerq 8

III. *Sales to individuals*

Ollier 6
De Meyremand 1
Le Comte de la Tour du Pin 1
Fritz 1
Le Curé de Cherier 1
Veuve Rousset 1
Dusers Chevalier de
  Saint-Louis 1
Genevois Dusoizon 1
Le Morhier 1
De Moutille 1
De Fontenay 1
De Mouty 1

Jouty 1
Cathelin 1
De la Loge 1
Duchaffaut 1
De la Chèze 1
De Fisson 1
Du Châtelet 1
Damberieu 1
Vasselier 4
Berage 8

IV. *Free copies*

Panckoucke 4
De Flesselles intendant 1
La Tourette 1
Moyroud 1
Plingues 1
Prost de Royer 1
Miège 1
Laserre 1
Le Roy 1
Académie de Lyon 1
Mahieu 1
Morel 2
Merlino 1
Couterez 1
Revol 1
Champeaux 3
De Gumin 2
Duplain P. Berage 1

V. *Totals*

French subscriptions 7,257
Non-French subscriptions 691
Sales to individuals 37
Free copies 25
8,010

C

ΥΥΥΥΥΥΥΥΥΥΥΥ

INCIDENCE OF SUBSCRIPTIONS
IN MAJOR FRENCH CITIES

The following table covers the thirty-seven cities in which
at least 50 subscriptions were sold and the thirty-six cities
with a population of at least 20,000. In order to illustrate the
relation between sales and size among the cities, they have
been ranked, both by number of subscriptions sold and by
population. Dashes indicate that a city had under 20,000 inhabitants or under 50 subscriptions, and an x in the last three
columns indicates that a city was the seat of a parlement, an
academy, or an intendancy. The academies of Bourg, Valence,
and Orléans were not founded until 1784, so x is in parentheses
after their names in the academy column. Avignon appears on
the table, although it was a papal territory, and Perpignan
appears with a parlement, although its court was formally a
Conseil supérieur. The population figures come from the
census of 1806, which is the best source of information about
the relative importance of all French cities and towns, even
when projected back to 1780: see René Le Mée, "Population
agglomérée, population éparse au début du XIXe siècle," in
Annales de démographie historique (1971), pp. 455–510.

	Number of Subscriptions	Population (1806)	Rank by Subscriptions	Rank by Population	Parlement	Academy	Capital of Généralité
ı	1,079	111,840	1	2		x	x
s	575	580,609	2	1	x	x	x
ouse	451	51,689	3	8	x	x	x
leaux	356	92,966	4	4	x	x	x
nçon	338	28,721	5	21	x	x	x
seille	228	99,169	6	3		x	
ı	221	36,231	7	14		x	x
ıes	218	29,225	8	19	x		x
es	212	41,195	9	11		x	
tpellier	169	33,264	10	15		x	x
n	152	22,026	11	29	x	x	x
en	125	86,672	12	5	x	x	x
cy	121	30,532	13	18	x	x	x
ers	109	29,187	14	20		x	
tauban	105	23,973	15	25		x	x
rg	91	7,417	16	—		(x)	
ıoble	80	22,129	17	28	x	x	x
rtres	77	13,809	18	—			
on-sur-ıône	67	11,204	19	—			
ı	65	8,918	20	—			x
ux	65	9,511	20	—			
ıers	65	21,465	20	33			x
ıs	65	21,703	20	32			x
ıce	65	8,212	20	—		(x)	
ens	59	39,853	21	12		x	x
ochelle	59	18,346	22	—		x	x
t	58	15,066	23	—			
brai	57	15,608	24	—			
non	55	23,789	25	26			
es	53	27,196	26	27			
	52	8,462	27	—			
lavre	52	19,482	27	—			
lins	52	14,101	27	—			
ans	52	42,651	27	10		(x)	x
ignan	52	12,499	27	—	x		x
ons	52	8,126	27	—		x	x
es	52	7,934	27	—			

Appendix C

Large Cities with Few Subscriptions, Other Than Those Above.

City	Number of Subscriptions	Population (1806)	Rank by Subscriptions	Rank by Population	Parlement	Academy	Ca ta Gé ra
Nantes	38	77,226	—	6			
Lille	28	61,467	—	7			
Reims	24	31,779	—	16			
Metz	22	39,133	—	13	x	x	
Toulon	22	28,170	—	22			
Brest	20	22,130	—	27			
Clermont-Ferrand	13	30,982	—	17			
Aix	6	21,960	—	30	x		
Saint-Omer	5	20,362	—	35			
Versailles	5	26,974	—	24			
Limoges	3	21,757	—	31			
Strasbourg	16	51,464	—	9			
Lorient	0	20,553	—	34			
Arles	0	20,151	—	36		x	

D

CONTRIBUTORS TO THE
ENCYCLOPÉDIE MÉTHODIQUE

The following table represents the seventy-three main contributors to the *Encyclopédie méthodique* as Panckoucke listed them in 1789 in the dictionary of *Mathématiques*, III, xxviii. In order to give some idea of the structure of the *Méthodique* as a whole, their names have been placed under the titles of the dictionaries they wrote. Thus the Encyclopedists who contributed to more than one dictionary appear more than once, and some dictionaries, such as *Médecine,* which had a large number of authors, take up a disproportionately large amount of space. However, those dictionaries were also the largest in size: *Médecine,* with eighteen authors in 1789, eventually ran to thirteen volumes. So the contributions of Panckoucke's authors were roughly comparable, unlike those of Diderot's, which varied from one to many hundreds of articles.

By 1789, Panckoucke had finished the framework of the *Méthodique* and his authors had written about half of its text, but it was not completed until 1832. The table, therefore, represents the work as it existed in its maturity, at a key point in its development, but it does not show how the *Méthodique* evolved throughout the half century of its existence. The authors listed in 1789 did write the bulk of the book, however, and they can be taken to represent the second generation of Encyclopedists, the generation that experienced the French Revolution.

Their experience varied enormously, but their reactions to the Revolution can be classified roughly into three categories. The Antirevolutionaries expressed hostility to the Revolution from the beginning, though few of them opposed it openly. The Moderates supported the constitutional monarchy and even, in some cases, the republic. But none of them participated actively in politics beyond August 10, 1792, although they often continued their work in some kind of official position and a few became involved in public life under the Directory. The Republicans did not believe the Revolution had gone too far in 1792, but they drew the line at different stages thereafter. The more radical of them have been identified with a *J* for Jacobin, the more moderate with a *G* for Girondin, and the rest, whose political affiliation is unclear, have been identified by an *X*.

It should be emphasized that political identifications of this sort involve a good deal of guesswork and disagreement among historians. The table is intended to give a general idea of how the Encyclopedists responded to the Revolution, not to be a precise scorecard of party alignments. It omits all indications concerning the politics of the fourteen Encyclopedists whose reactions to the Revolution cannot be determined: Andry, Goulin, Hallé, Huzard, Verdier, Grivel, Clairbois, Lacombe, Lévesque, Framery, Charles, Jeanroi, Duhamel, and Gaillard. About half of these seem to have continued their careers without interruption, so they probably were not hostile to the new order. Nine Encyclopedists died before 1793: Watelet, Colombier, Fougeroux de Bondaroy, Morvilliers, Beauzée, Bergier, Louis, Baudeau, and Delacroix. Very little information can be provided for ten other contributors: Caille, Chamseru, Dehorne, La Porte, Mauduyt, Saillant, de Surgy, Blondeau, Jabro, and Le Rasle. Most of them were prominent professional men (six were doctors), but they did not achieve enough fame to make more than a fleeting appearance in printed sources. The analysis of the Encyclopedists' response to the Revolution is therefore limited to fifty-four cases: fourteen whose reactions cannot be determined, seven who opposed the Revolution from the beginning, fifteen who tended to favor a constitutional monarchy, and eighteen who supported the Republic—five as Girondins, five as more radical Jacobins, and eight in some other fashion.

The phrases following the author's name are Panckoucke's and show how he identified the members of his team. The main source for tracing institutional affiliations was the *Almanachs royaux* and *nationaux* from the 1780s and 1790s.

Dictionary and contributor	Académie française	Académie des sciences	Other Parisian academy	Société royale de médecine	Société royale d'agriculture	Censeur royal	Institut	Ecole de Santé	Other revolutionary Ecole	Anti-revolutionary	Moderate	Republican
Mathématiques												
Abbé C. Bossut, de l'Académie des sciences, 1730–1814.		x										
M.-J.-A.-N. Caritat, marquis de Condorcet, de l'Académie des sciences, 1743–1794.	x	x	x			x	x			x		G
J.-A.-C. Charles, de l'Académie des sciences, 1746–1823.		x					x					
J.-J. Lefrançois de Lalande, de l'Académie des sciences, 1732–1807.		x				x	x			x		
Physique												
G. Monge, de l'Académie des sciences, 1746–1818.		x					x		x			J
Médecine												
F. Vicq d'Azyr, de l'Académie des sciences, 1748–1794.	x	x		x	x						x	
C.-L.-F. Andry, docteur en médecine, 1741–1829.				x								

Dictionary and contributor	Académie française	Académie des sciences	Other Parisian academy	Société royale de médecine	Société royale d'agriculture	Censeur royal	Institut	Ecole de Santé	Other revolutionary Ecole	Anti-revolutionary	Moderate	Republican
C. A. Caille, docteur en médecine, ?–1831.												
N. Chambon de Montaux, docteur en médecine, 1748–1826.				x								x
J.-F.-J. Roussille de Chamseru, docteur en médecine, 1750?–1823.				x								
J. Colombier, docteur en médecine, 1736–1789.				x								
A.-F. de Fourcroy, docteur en médecine, 1755–1809.		x		x	x	x	x	x				J
J. De Horne, docteur en médecine.				x	x	x	x	x	x			
F. Doublet, docteur en médecine, 1751–1795.				x				x				J
J. Goulin, docteur en médecine, 1728–1799.												
J.-N. Hallé, docteur en médecine, 1754–1822.				x			x	x				

J.-B. Huzard, docteur en médecine, 1755–1839.

D. Jeanroi, docteur en médecine, 1730–1816.

J. J. De Laporte, docteur en médecine.

Mauduyt de la Varenne, docteur en médecine.

C. J. Saillant, docteur en médecine, 1747–1814.

M.-A. Thouret, docteur en médecine, 1749–1810.

J. Verdier, docteur en médecine, 1735–1820.

Anatomie

Vicq d'Azyr : see above.

Chirurgie

A. Louis, secrétaire perpétuel de l'Académie de chirurgie, 1723–1792.

Chimie, métallurgie, pharmacie

L.-B. Guyton de Morveau, avocat-général honoraire, 1737–1816.

J.-P.-F.-G. Duhamel, de l'Académie des sciences, 1730–1816.

(continued)

Dictionary and contributor	Académie française	Académie des sciences	Other Parisian academy	Société royale de médecine	Société royale d'agriculture	Censeur royal	Institut	Ecole de Santé	Other revolutionary Ecole	Anti-revolutionary	Moderate	Republican
F. Chaussier, professeur de chimie, 1746–1828.			x				x	x	x			J
Agriculture												
Abbé A.-H. Tessier, de l'Académie des sciences, 1741–1837.		x		x	x	x	x	x	x		x	
A. Thouin, jardinier en chef du Jardin du roi, de l'Académie des sciences, 1747–1823.		x				x	x		x		x	
Bois et Forêts												
A.-D. Fougeroux de Bondaroy, de l'Académie des sciences, 1732–1789.		x			x							
Histoire naturelle												
L.-J.-M. Daubenton, de l'Académie des sciences, 1716–1800.		x		x	x		x		x		x	
Abbé P.-J. Bonnaterre, 1752–1804.									x	x	x	

G

Les animaux quadrupèdes				
Daubenton : see above.				
Mauduyt : see above.				
P.-M.-A. Broussonet, de l'Académie des sciences, 1761–1807.		x	x	
Les poissons				
Daubenton : see above.				
Les insectes				
Mauduyt : see above.				
G.-A. Olivier, docteur en médecine, 1756–1814.			x	x
Les vers, coquillages, zoophytes				
J.-G. Bruguières, docteur en médecine, 1750–1799.			x	x
Botanique				
J.-B.-P.-A. de Lamarck, de l'Académie des sciences, 1744–1829.			x	x
Minéraux				
Daubenton : see above.				
Géographie physique				
N. Desmarets, de l'Académie des sciences, 1725–1815.	x		x	x

(continued)

Dictionary and contributor	Académie française	Académie des sciences	Other Parisian academy	Société royale de médecine	Société royale d'agriculture	Censeur royal	Institut	Ecole de Santé	Other revolutionary Ecole	Anti-revolutionary	Moderate	Republican
Géographie et histoire anciennes												
E. Mentelle, géographe de Monseigneur Comte d'Artois, 1730–1815.						x	x		x		x	
Géographie moderne												
F. Robert, géographe ordinaire du roi, 1737–1819.												x
N. Masson de Morvilliers, secrétaire du government de Normandie, 1740?–1789.												
Antiquités												
Abbé A. Mongez, garde des antiques de Sainte Geneviève, 1747–1835.			x				x				G	
Histoire												
G.-H. Gaillard, de l'Académie française, 1726–1806.	x		x			x	x					

Théologie
N.-S. Bergier, confesseur de Monsieur, 1718–1790.

Philosophie
J.-A. Naigeon, 1738–1810.

Métaphysique, logique
P.-L. Lacretelle, avocat, 1751–1824.

Morale
Lacretelle : see above.

Education
Lacretelle : see above.

Grammaire, littérature
J.-F. Marmontel, de l'Académie française, 1723–1799.
N. Beauzée, de l'Académie française, 1717–1789.

Jurisprudence
A.-J. Boucher d'Argis, avocat, 1750–1794.
Le Rasle, avocat.
J.-V. Delacroix, avocat, 1743–1792.
J.-P. Garran de Coulon, avocat, 1748–1816.

(*continued*)

Dictionary and contributor	Académie française	Académie des sciences	Other Parisian academy	Société royale de médecine	Société royale d'agriculture	Censeur royal	Institut	Ecole de Santé	Other Ecole	Anti-revolutionary	Moderate	Republican
P.-P.-N. Henrion de Pansey, avocat, 1742–1829.										x		
Abbé A.-R.-C. Bertolio, avocat, 1750?–1812.												x
Police et municipalité												
J. Peuchet, avocat, 1758–1830.											x	
Finances												
De Surgy, premier commis des Finances.						x						
Economie politique et diplomatique												
J.-N. Desmeunier, censeur royal, 1751–1814.						x						
G. Grivel, avocat, 1735–1810.									x		x	
Commerce												
Abbé N. Baudeau, 1730–1792?												
Grivel: see above.												

	1	2	3	4	5	6
Marine						
H.-S. Vial du Clairbois, ingénieur-constructeur, 1733–1816.						
Blondeau, professeur de marine.						
Art militaire						
L.-F.-G. de Kéralio, de l'Académie des Inscriptions, 1731–1793.	x					
J.-G. de Lacuée, comte de Cessac, capitaine d'infanterie, 1752–1841.		x	x			G
Jabro, lieutenant-colonel des grenadiers royaux.						
Artillerie						
F.-R.-J. de Pommereul, capitaine d'artillerie, 1745–1823.				x		
Ponts et chaussées						
G.-C.-F.-M. Riche de Prony, inspecteur des ponts et chaussées, 1755–1839.		x	x			x
Arts académiques, manège, escrime, danse, natation, no author listed.						
Vénerie, chasse, pêches						
J. Lacombe, avocat, 1724–1801.					x	

(continued)

Dictionary and contributor	Académie française	Académie des sciences	Other Parisian academy	Société royale de médecine	Société royale d'agriculture	Censeur royal	Institut	Ecole de Santé	Other revolutionary Ecole	Anti-revolutionary	Moderate	Republican
Beaux-arts												
C.-H. Watelet, de l'Académie française, 1718–1786.	x											
P.-C. Lévesque, de l'Académie des beaux-arts de Saint-Pétersbourg, 1736–1812.			x									
Musique												
J.-B.-A. Suard, de l'Académie française, 1734–1817.	x					x				x		
N.-E. Framery, intendant de la musique de Monseigneur Comte d'Artois, 1745–1810.							x					
P.-L. Ginguené, 1748–1815.							x					x
Architecture												
A.-C. Quatremère de Quincy, 1755–1849.							x				x	

Arts et métiers
J.-M. Roland de la Platière, inspecteur
des manufactures, 1732–1793.

												G
Totals	7	15	7	18	8	12	30	5	12	7	15	18

BIBLIOGRAPHICAL NOTE

This book is based primarily on the papers of the Société typographique de Neuchâtel in the Bibliothèque de la ville de Neuchâtel, Switzerland. Although it reveals a great deal about publishing and the book trade in the Old Regime, the view from Neuchâtel can be distorted and therefore has been supplemented by research in the following archives:

PARIS: Bibliothèque nationale, Archives de la Chambre syndicale des libraires et imprimeurs de Paris, 21863-4 (Panckoucke and the *Encyclopédie méthodique*), 21933-4 (confiscations of prohibited books), 21958, 21966-7, 22001 (book privileges). Collection Anisson-Duperron, 22073 (Duplain and Veuve Dessaint), 22086 (the *Méthodique*), 22100 (the confiscation of volumes 1-3 of Panckoucke's folio *Encyclopédie*).

Archives nationales, V^1549, 553, V^61145 (the Direction de la librairie and the *Méthodique*).

Archives de Paris (formerly the Archives du Département de la Seine), 5AZ 2009, 8AZ 278 (Panckoucke correspondence).

Bibliothèque historique de la ville de Paris, ms. 770-1, 776, 779, 815 (Panckoucke correspondence).

GENEVA: Bibliothèque publique et universitaire, ms. suppl. 148 (Panckoucke papers).

Archives d'Etat, Commerce F61-3 (Gosse papers).

AMSTERDAM: Bibliotheek van de vereeniging ter bevordering van de belangen des boekhandels, dossier Marc Michel Rey (the *Supplément* and the Geneva Folio).

Universiteits-Bibliotheek, Schenking Diederichs (Panckoucke correspondence).

611

Bibliographical Note

OXFORD: Bodleian Library, ms. French c.31, d.31, Don. d. 135 (Panckoucke correspondence).

CHICAGO, ILLINOIS: Newberry Library, Case Wing ms. Z 311.P188 (Fougeroux de Bondaroy and Panckoucke), Z 45. 18, ser. 7 (circulars and catalogues of booksellers).

LAWRENCE, KANSAS: Kenneth Spencer Research Library, ms. 99 (contracts for the *Méthodique*).

As the literature on the Encyclopedists and the *Encyclopédie* has grown beyond the stack capacity of many libraries today, it cannot be summarized here. For bibliographical orientation and a good survey of the subject see Jacques Proust, *Diderot et l'Encyclopédie* (Paris, 1967) and Arthur Wilson, *Diderot* (New York, 1972).

Despite the proliferation of this scholarship, the publishing history of the *Encyclopédie* has remained obscure, owing to a lack of sources. For basic information about the origins and evolution of the first edition see Franco Venturi, *Le Origini dell'Enciclopedia* (Florence, 1946); Douglas H. Gordon and Norman L. Torrey, *The Censoring of Diderot's Encyclopédie and the Re-established Text* (New York, 1947); and R. N. Schwab, "Inventory of Diderot's *Encyclopédie*," *Studies on Voltaire and the Eighteenth Century,* LXXX (1971). Some fragments of the papers of the original publishers were published by Louis-Philippe May, "Histoire et sources de l'*Encyclopédie* d'après le registre de délibérations et de comptes des éditeurs et un mémoire inédit," *Revue de synthèse,* XV (1938), 1–109. But the documentation is too thin to support any firm economic interpretation such as those advanced by Ralph H. Bowen in "The *Encyclopédie* as a Business Venture," *From the Ancien Régime to the Popular Front: Essays in the History of Modern France in Honor of Shepard B. Clough,* ed. Charles K. Warner (New York and London, 1969), pp. 1–22 and Norman L. Torrey, "L'*Encyclopédie* de Diderot, une grande aventure dans le domaine de l'édition," *Revue d'histoire littéraire de la France,* LI (1951), 306–317. What little can be concluded about the business history and readership of the first edition from a thorough sifting of this material is best presented in John Lough, "Luneau de Boisjermain v. the Publishers of the *Encyclopédie*," *Studies on Voltaire and the Eighteenth Century,* XXIII (1963), 115–173. See also Frank A. Kafker, "The Fortunes and Misfortunes of a Leading

Bibliographical Note

French Bookseller-Printer: André-François Le Breton, Chief Publisher of the *Encyclopédie,*" *Studies in Eighteenth-Century Culture,* V (1976), 371–385.

The publishing history of the subsequent editions began to be perceptible after George B. Watts discovered some key letters and notarial archives in Geneva. Watts, "Forgotten Folio Editions of the *Encyclopédie,*" *French Review,* XXVII (1953–54), 22–29, 243–244; "The Swiss Editions of the *Encyclopédie,*" *Harvard Library Bulletin,* IX (1955), 213–235; "The Genevan Folio Reprinting of the *Encyclopédie,*" *Proceedings of the American Philosophical Society,* CV (1961), 361–367; and "The *Supplément* and the *Table analytique et raisonnée* of the *Encyclopédie,*" *French Review,* XXVIII (1954–55), 4–19. By studying some papers of the Société typographique de Bouillon and Marc Michel Rey, Fernand Clément and Raymond F. Birn unraveled the story behind the publication of the *Supplément* and its relation to the Geneva folio *Encyclopédie.* Clément, "Pierre Rousseau et l'édition des *Suppléments* de l'*Encyclopédie,*" *Revue des sciences humaines de la faculté des lettres de l'Université de Lille,* LXXXVI (1957), 133–143; Birn, "Pierre Rousseau and the philosophes of Bouillon," *Studies on Voltaire and the Eighteenth Century,* XXIX (1964). John Lough brought this material together in a series of well-documented articles, collected in *Essays on the Encyclopédie of Diderot and d'Alembert* (London, 1968) and *The Encyclopédie in Eighteenth-Century England and Other Studies* (Newcastle upon Tyne, 1970). On the quarto and octavo editions see Robert Darnton, "The *Encyclopédie* Wars of Prerevolutionary France," *American Historical Review,* LXXVIII (1973), 1331–1352, a preliminary sketch for the present volume. The problems of the "missing" quarto and octavo editions and the scrap editions are pursued further in Darnton, "True and False Editions of the *Encyclopédie,* a Bibliographical Imbroglio," forthcoming in the proceedings of the Colloque international sur l'histoire de l'imprimerie et du livre à Genève.

The publishers and protectors of the Italian editions have been studied in Salvatore Bongi, "L'*Enciclopedia* in Lucca," *Archivio storico italiano,* 3d ser., XVIII (1873), 64–90; Ettore Levi-Malvano, "Les éditions toscanes de l'*Encyclopédie,*" *Revue de littérature comparée,* III (1923), 213–256; and Adriana Lay, *Un editore illuminista: Giuseppe Aubert nel carteg-*

gio con Beccaria e Verri (Turin, 1973). Unless new sources are unearthed, however, it seems unlikely that much can be known about the production and diffusion of the Italian *Encyclopédies*. Information about the *Encyclopédie d'Yverdon* is somewhat richer, thanks to the research of J. P. Perret, *Les imprimeries d'Yverdon au XVIIe et au XVIIIe siècle* (Lausanne, 1945) and E. Maccabez, *F. B. de Félice (1723–1789) et son Encyclopédie (Yverdon, 1770–1780)* (Basle, 1903).

Only a few of the papers of the *Encyclopédie méthodique* have turned up in widely scattered sites such as Amsterdam, Oxford, Paris, Chicago, and Lawrence, Kansas. But the text of the *Méthodique* contains enough material, mainly reprinted *avis* to subscribers, for one to piece together its publishing history. In this case, too, Watts opened the way for further research with an article, "The *Encyclopédie méthodique*," *Publications of the Modern Language Association of America*, LXXIII (1958), 348–366 and several essays on Panckoucke, culminating in a biography, "Charles Joseph Panckoucke, 'l'Atlas de la librairie française,' " *Studies on Voltaire and the Eighteenth Century*, LXVIII (Geneva, 1969). On Panckoucke see also David I. Kulstein, "The Ideas of Charles-Joseph Panckoucke, Publisher of the *Moniteur Universel*, on the French Revolution," *French Historical Studies*, IV (1966), 304–319 and Suzanne Tucoo-Chala, "La diffusion des lumières dans la seconde moitié du XVIIIe siècle: Ch.-J. Panckoucke, un libraire éclairé (1760–1799)," *Dix-huitième siècle* (1974), pp. 115–128. Mme. Tucoo-Chala's *Charles-Joseph Panckoucke & la librairie française 1736–1789* (Pau and Paris, 1977) contains a full account of Panckoucke's career. Unfortunately, it appeared after the completion of this study, and its discussion of the *Encyclopédie* ventures is inaccurate.

Although there are no data on the diffusion of the *Encyclopédie* aside from those presented above in Chapter VI, many scholars have attempted to sketch the spread of Encyclopedism by drawing on literary sources. Several of these studies were published in a special issue of *Cahiers de l'Association internationale des études françaises*, no. 2 (May 1952). See especially the articles by Jean Fabre on Poland, by Jean Sarrailh on Spain, and by Gilbert Chinard on America and also two subsequent and more thorough works, Roland Mortier, *Diderot en Allemagne (1750–1850)* (Paris, 1954) and

Charly Guyot, *Le rayonnement de l'Encyclopédie en Suisse française* (Neuchâtel, 1955).

Sociological analyses of the Encyclopedists as a group tend to become snarled in unrepresentative data and faulty classification schemes. The most modest and useful study is John Lough, *The Contributors to the Encyclopédie* (London, 1973). For contrasting views of the character and number of Diderot's collaborators see Proust, *Diderot et l'Encyclopédie*, chap. 1 and appendix I; Robert Shackleton, "The *Encyclopédie* and Freemasonry," *The Age of Enlightenment: Studies Presented to Theodore Besterman* (London, 1967), 223–237 and *The Encyclopédie and the Clerks* (Oxford, 1970); Frank A. Kafker, "A List of Contributors to Diderot's Encyclopedia," *French Historical Studies,* III (1963), 106–122 and "Les Encyclopédistes et la Terreur," *Revue d'histoire moderne et contemporaine,* XIV (1967), 284–295; Louis-Philippe May, "Note sur les origines maçonniques de l'*Encyclopédie* suivie de la liste des Encyclopédistes," *Revue de synthèse,* XVII (1939), 181–190; and Takeo Kubawara, Syunsuke Turumi, and Kiniti Higuti, *Les collaborateurs de l'Encyclopédie, les conditions de leur organisation* (Kyoto, 1951).

French research in *histoire du livre* derives in part from the work of Daniel Mornet, especially his statistical study of eighteenth-century libraries, "Les enseignements des bibliothèques privées (1750–1780)," *Revue d'histoire littéraire de la France,* XVII (1917), 449–496. Mornet found enough copies of the *Encyclopédie* and of related works like Bayle's *Dictionaire* to conclude (p. 455), "Le XVIIIe siècle fut très certainement, par une tendance profonde, un siècle encyclopédique." Since then, however, a great many statistical studies—of *inventaires après décès,* library catalogues, requests for royal *privilèges* and *permissions tacites,* and articles in eighteenth-century periodicals—have emphasized the archaic, unenlightened element in eighteenth-century literary culture. The most important example of this tendency is François Furet and others, *Livre et société dans la France du XVIIIe siècle* (Paris and The Hague, 1965 and 1970), 2 vols. For a survey of the literature in this new historical subgenre by two of its best practitioners see Roger Chartier and Daniel Roche, "Le livre. Un changement de perspective," *Faire de l'histoire* (Paris, 1974), III, 115–136, and for further details see the special issue of the *Revue française d'histoire du livre,* new ser., no. 16 (July–Sept. 1977). The quantitative strain in French writing is taken to an extreme by Robert Estivals, *La statistique bibliographique de la France sous la monarchie au XVIIIe siècle* (Paris and The Hague, 1965). But there are more conventional studies, most of them regional in character; for example Madeleine Ventre, *L'imprimerie et la librairie en Languedoc au dernier siècle de l'ancien régime 1700–1789* (Paris and The Hague, 1958); Jean Quéniart, *L'imprimerie et la librairie à Rouen au XVIIIe siècle* (Paris, 1969); and René Moulinas, *L'imprimerie, la librairie et la presse à Avignon au XVIIIe siècle* (Grenoble, 1974). For the background to the eighteenth-century book trade, the essential work is Henri-Jean Martin, *Livre, pouvoirs et société à Paris au XVIIe siècle (1598–1701)* (Geneva, 1969). David T. Pottinger, *The French Book Trade in the Ancien Regime* (Cambridge, Mass., 1958) contains only a superficial account, based on printed sources.

For a concise general survey of the rich literature on book production and analytical bibliography see Philip Gaskell, *A New Introduction to Bibliography* (Oxford, 1972), which can be consulted for further reading in the works of Sir Walter Greg, Fredson Bowers, R. B. McKerrow, Graham Pollard, and others. The most important studies used in the preparation of this volume were D. F. McKenzie, "Printers of the Mind," *Studies in Bibliography,* XXII (1969), 1–75; McKenzie, *The Cambridge University Press, 1696–1712* (Cambridge, Eng., 1966), 2 vols.; Leon Voet, *The Golden Compasses* (Amsterdam, 1969 and 1972), 2 vols.; and Raymond de Roover, "The Business Organization of the Plantin Press in the Setting of Sixteenth-Century Antwerp," *De gulden passer,* XXIV (1956) 104–120. Still more important were printing manuals from the eighteenth century: A.-F. Momoro, *Traité élémentaire de l'imprimerie ou le manuel de l'imprimeur* (Paris, 1793); S. Boulard, *Le manuel de l'imprimeur* (Paris, 1791); numerous articles in the *Encyclopédie,* which have been reprinted by Giles Barber as *Book Making in Diderot's Encyclopédie* (Westmead, Farnborough, Eng., 1973); and Nicolas Contat (dit Le Brun), *Anecdotes typographiques d'un garçon imprimeur,* ed. Giles Barber (forthcoming, Oxford Bibliographical Society, 1979), the autobiography of an eighteenth-century printing shop foreman. A good deal can be learned about the lives of journeyman printers from Paul Chauvet, *Les ouvriers du livre en France des origines à la révolution de 1789* (Paris, 1959). But for a thorough account, based on sources emanating directly from the printing shop, one must await the publication of Jacques Rychner's thesis on the Société typographique de Neuchâtel. That thesis should be as important for bibliographical studies as D. F. McKenzie's work on the Cambridge University Press. For an enticing preview of it see Rychner, "A l'ombre des Lumières: coup d'oeil sur la main d'oeuvre de quelques imprimeries du XVIIIe siècle," *Studies on Voltaire and the Eighteenth Century*, CLV (1976), 1925–1955.

INDEX

Index

Index

Index

Index

Index

Thouin, André, 423, 429, 431, 488, 516, 517
Thouret, Michel-Augustin, 516, 517
Toulouse, 131–133
Tournes, Samuel de, 18, 21–23, 45
Toussaint, François-Vincent, 10
Turgot, Anne-Robert-Jacques, 10, 27, 66, 432

Venel, Gabriel-François, 511
Verdier, Jean, 433
Vergennes, Charles Gravier, comte de, 72, 156, 499
Vernange, Louis, 181–182, 207–208

Vial du Clairbois, Honoré-Sébastien, 432
Vicq d'Azyr, Félix, 431, 432, 436, 439–440, 468, 476; career during the Revolution, 514, 517
Voltaire, François-Marie Arouet, 10, 13, 17; his *Questions sur l'Encyclopédie*, 20, 62, 417–418; his manuscripts, 403–409

Watelet, Claude-Henri, 433, 435–436, 446, 467
Watts, George B., 6

Yverdon, 19